ENZYMES:

Biochemistry, Biotechnology and Clinical Chemistry

Second Edition

"Talking of education, people have now a-days" (said he) "got a strange opinion that every thing should be taught by lectures. Now, I cannot see that lectures can do so much good as reading the books from which the lectures are taken. I know nothing that can be best taught by lectures, except where experiments are to be shewn. You may teach chymestry by lectures — You might teach making of shoes by lectures!"

James Boswell: *Life of Samuel Johnson, 1766*

ABOUT THE AUTHORS

Trevor Palmer was born in South Yorkshire and graduated from Cambridge University in 1966 with an honours degree in biochemistry, being influenced by (amongst others) Peter Sykes in organic chemistry and Malcolm Dixon in enzymology. He then worked as a clinical biochemist at the Queen Elizabeth Hospital for Children, linked to the Institute of Child Health, University of London, obtaining a PhD for research into inherited disorders. From this emerged the two main interests of his subsequent career, enzymology and evolution, the latter stimulating a further interest in the long-term effects of natural catastrophes. He moved to Nottingham Trent University (then Trent Polytechnic) in 1974, initially as a lecturer in biochemistry, before becoming Head of Department of Life Sciences (1987), Dean of the Faculty of Science and Mathematics (1992), Senior Dean of the University (1998) and Pro Vice-Chancellor for Academic Development (2002), returning to predominantly academic activity as Emeritus Professor in 2006. His books include *Understanding Enzymes* (1981), *Principles of Enzymology for Technological Applications* (1993), *Controversy – Catastrophism and Evolution* (1999) and *Perilous Planet Earth* (2003). His wife, Jan, teaches psychology and sociology (and is currently a part-time PhD student at Leicester University). Their son, James, is carrying out postdoctoral studies as a Leverhulme Fellow at Nottingham University and their daughter, Caroline, is researching for a PhD at Sheffield University.

Philip L. Bonner went to school in Coventry, West Midlands, before graduating from the University of Sussex in 1978 with an honours degree in biochemistry. He then worked as a research assistant at Glaxo plc on Merseyside before leaving to take up a Research Assistant/Demonstrator post at Trent Polytechnic, where he obtained a PhD for research concerning enzymes associated with seed germination. Several postdoctoral appointments followed, at Bristol, Lancaster and Central Lancashire Universities, working on a variety of topics including relaxin, aspartate kinase and phospholipase C, before he was appointed as Senior Lecturer at Nottingham Trent University in 1991. There, he has maintained his research interests in enzymology and analytical biochemistry, working on the role of transglutaminase in plant/animal tissue and methods to isolate and characterise post-translationally-modified MHC peptides. His first single-author book, on protein purification, was published in 2007. His wife, Liz, is a manager of an occupational therapist team in Nottingham and their daughter, Francesca, is at junior school.

ENZYMES:
Biochemistry, Biotechnology and Clinical Chemistry
Second Edition

Trevor Palmer, BA, PhD, CBiol, FIBiol, FIBMS, FHEA
Emeritus Professor in Life Sciences
Nottingham Trent University

Philip L. Bonner, BSc, PhD
Senior Lecturer in Biochemistry
Nottingham Trent University

Oxford Cambridge Philadelphia New Delhi

For:
Caroline, Francesca, James, Jan and Liz

Published by Woodhead Publishing Limited,
80 High Street, Sawston, Cambridge CB22 3HJ, UK
www.woodheadpublishing.com

Woodhead Publishing, 1518 Walnut Street, Suite 1100, Philadelphia,
PA 19102-3406, USA

Woodhead Publishing India Private Limited, G-2, Vardaan House, 7/28 Ansari Road,
Daryaganj, New Delhi – 110002, India
www.woodheadpublishingindia.com

First edition published by Horwood Publishing Limited, 2001
Second edition published by Horwood Publishing Limited, 2007
Reprinted by Woodhead Publishing Limited, 2011

British Library Cataloguing in Publication Data
A catalogue record for this book is available from the British Library

ISBN 978-1-904275-27-5

Printed by Lightning Source

Table of Contents

Part 2 : Kinetic and chemical mechanisms of enzyme-catalysed reactions

6 An introduction to bioenergetics, catalysis and kinetics

7 Kinetics of single-substrate enzyme-catalysed reactions

Authors' Preface

This book was written, as all textbooks should be, with the requirements of the student firmly in mind. It is intended to provide an informative introduction to enzymology, and to give a balanced, reasonably-detailed, account of all the various theoretical and applied aspects of the subject which are likely to be included in an honours degree course. Furthermore, some of the later chapters may serve as a bridge to more advanced texts for students wishing to proceed further in this area of biochemistry.

Although the book is intended mainly for students taking first degree courses which have a substantial biochemistry component, large portions may be of value to students on comparable courses in biological sciences, biomedical sciences or forensic sciences, and even to ones enrolled on, in one direction, foundation programmes, or, in the other, MSc or other advanced courses who are approaching the subject of enzymology for the first time (or the first time in many years).

No previous knowledge of biochemistry, and little of chemistry, is assumed. Most scientific terms are defined and placed in context when they are first introduced. Enzymology inevitably involves a certain amount of elementary mathematics, and some of the equations which are derived may appear somewhat complicated at first sight; however, once the initial biochemical assumptions have been understood, the derivations usually follow on the basis of simple logic, without involving any difficult mathematical manipulations. Numerical and other problems (with answers) are included, to test and reinforce the student's grasp of certain points. These problems generally use hypothetical data, although the results are often based on findings reported in the biochemical literature.

If the size of the book is to be kept reasonable, some things of value have to be left out. The chief aim of this particular book is to help the student understand the concepts involved in enzymology, and the historical context in which they were worked out. It is not a reference book for practising enzymologists, so no comprehensive tables of data or long, finely-detailed accounts are included. Instead, an attempt has been made to give a perspective of each topic, and examples are quoted where appropriate. Credit has been given wherever possible to those responsible for the development of the subject, but many names deserving of mention have been excluded for reasons of space.

Individual scientific papers have not, in general, been referred to, but at the end of each chapter is a list of relevant books and articles, to provide context and an up-to date viewpoint, from which references to the original papers may usually be obtained.

As with any book at this level, certain topics have been presented in a simplified (possibly even over-simplified) form. However, a considerable effort has been made to avoid giving a distorted account of any topic. It is hoped that this book can provide a foundation for those wishing to pursue more advances studies, and that nothing learned from it will have to be 'unlearned' later. There are good reasons for thinking that this is a realistic hope.

For this second edition of *Enzymes*, we have revised and updated the first edition, reducing coverage of techniques whose use is declining to make room for discussion of topics of greater current and future interest, e.g. expanded bed chromatography, affinity precipitation, immobilized metal affinity chromatography, hydroxyapatite chromatography, hydrophobic charge induction chromatography, lectin affinity chromatography, covalent chromatography, membrane technology, capillary electrophoresis, absorbance fluorescence and lumimetric methods, high-throughput screening methods, 6-His tag and fusion protein technology, mass spectrometry and the use of protein arrays. A completely new section has been added on the use of enzymatic analysis in forensic science, and the final chapter of the first edition has been split into two to allow greater discussion of the rapidly-expanding areas of genomics, bioinformatics and proteomics. Elsewhere, coverage of protein structure, synthesis and function and mechanisms of enzyme activity has been revised to take into account recent developments (e.g. concerning the mechanism of action of lysozyme).

Acknowledgements. In the preparation of this new edition, we are grateful for the help of many people, including Lesley Atherton, Mark Crowley, Nick Howard, Elaine James, Caroline Palmer, Jan Palmer and Karen Roberts. However, any errors of fact or interpretation which may have crept into the book are entirely our own responsibility, and we would be grateful if we could be informed about them. Finally, we would like to acknowledge the helpful cooperation of the staff of Horwood Publishing and, in particular, to express our gratitude to, and admiration for, the distinguished scientific publisher, the late Ellis Horwood, without whom this book would never have come into being.

Trevor Palmer and Philip Bonner, 2007

Part 1

Structure and Function of Enzymes

1

An Introduction to Enzymes

1.1 WHAT ARE ENZYMES?

Enzymes are biological catalysts. They increase the rate of chemical reactions taking place within living cells without themselves suffering any overall change. The reactants of enzyme-catalysed reactions are termed **substrates**. Each enzyme is quite specific in character, acting on a particular substrate or substrates to produce a particular product or products.

All enzymes are proteins. However, without the presence of a non-protein component called a **cofactor**, many enzyme proteins lack catalytic activity. When this is the case, the inactive protein component of an enzyme is termed the **apoenzyme**, and the active enzyme, including cofactor, the **holoenzyme**. The cofactor may be an organic molecule, when it is known as a **coenzyme**, or it may be a metal ion. Some enzymes bind cofactors more tightly than others. When a cofactor is bound so tightly that it is difficult to remove without damaging the enzyme, it is sometimes called a **prosthetic group**.

To summarize diagrammatically:

ENZYME < INACTIVE PROTEIN + COFACTOR < COENZYME / METAL ION
ENZYME < ACTIVE PROTEIN

As we shall see later, both the protein and cofactor components may be directly involved in the catalytic processes taking place.

1.2 A BRIEF HISTORY OF ENZYMES

Until the nineteenth century, it was considered that processes such as the souring of milk and the fermentation of sugar to alcohol could only take place through the action of a living organism. In 1833, the active agent breaking down the sugar was partially isolated and given the name **diastase** (now known as **amylase**).

A little later, a substance which digested dietary protein was extracted from gastric juice and called **pepsin.** These and other active preparations were given the general name **ferments.** Justus von Liebig recognized that these ferments could be non-living materials obtained from living cells, but Louis Pasteur and others still maintained that ferments must contain living material.

While this dispute continued, the term ferment was gradually replaced by the name **enzyme.** This was first proposed by Wilhelm Kühne in 1878, and comes from the Greek, *enzumé (ένζυμη)*, meaning 'in yeast'. Appropriately, it was in yeast that a factor was discovered which settled the argument in favour of the inanimate theory of catalysis: brothers Eduard and Hans Büchner showed, in 1897, that sugar fermentation could take place when a yeast cell extract was added even though no living cells were present.

In 1926, James Sumner crystallized urease from jack-bean extracts and, in the next few years, many other enzymes were purified and crystallized. Once pure enzymes were available, their structure and properties could be determined, and the findings form the material for most of this book.

Today, enzymes still form a major subject for academic research. They are investigated in hospitals as an aid to diagnosis and, because of their specificity of action, are of great value as analytical reagents. Enzymes are still widely used in industry, continuing and extending many processes which have been used since the dawn of history.

1.3 THE NAMING AND CLASSIFICATION OF ENZYMES

1.3.1 Why classify enzymes?

There is a long tradition of giving enzymes names ending in '-ase'. The only major exceptions to this are the **proteolytic enzymes**, i.e. ones involved in the breakdown of proteins, whose names usually end with '-in', e.g. **trypsin.**

The names of enzymes usually indicate the substrate involved. Thus, **lactase** catalyses the hydrolysis of the disaccharide **lactose** to its component monosaccharides, **glucose** and **galactose**:

$$\underset{\text{lactose}}{C_{12}H_{22}O_{11}} + H_2O \rightleftharpoons \underset{\text{glucose}}{C_6H_{12}O_6} + \underset{\text{galactose}}{C_6H_{12}O_6} \quad (1.1)$$

The name lactase is a contraction of the clumsy, but more precise, lactosase. The former is used because it sounds better but it introduces a possible trap for the unwary because it could easily suggest an enzyme acting on the substrate lactate. There is nothing in the name of this enzyme or many others to indicate the type of reaction being catalysed. **Fumarase,** for example, by analogy with lactase might be supposed to catalyse a hydrolytic reaction, but, in fact, it *hydrates* **fumarate** to form **malate**:

$$\underset{\text{fumarate}}{^{-}O_2C.CH{=}CH.CO_2^-} + H_2O \rightleftharpoons \underset{\text{malate}}{^{-}O_2C.CHOH.CH_2CO_2^-} \quad (1.2)$$

The names of other enzymes, e.g. **transcarboxylase,** indicate the nature of the reaction without specifying the substrates (which in the case of transcarboxylase are methylmalonyl-CoA and pyruvate). Some names, such as **catalase**, indicate neither the substrate nor the reaction (catalase mediates the decomposition of hydrogen peroxide).

Needless to say, whenever a new enzyme has been characterized, great care has usually been taken not to give it exactly the same name as an enzyme catalysing a different reaction. Also, the names of many enzymes make clear the substrate and the nature of the reaction being catalysed. For example, there is little ambiguity about the reaction catalysed by **malate dehydrogenase**. This enzyme mediates the removal of hydrogen from malate to produce oxaloacetate:

$$\underset{\text{malate}}{{}^{-}O_2C.\underset{\displaystyle OH}{\underset{|}{C}}H.CH_2.CO_2^-} + NAD^+ \rightleftharpoons \underset{\text{oxaloacetate}}{{}^{-}O_2C.\underset{\displaystyle O}{\underset{\|}{C}}.CH_2.CO_2^-} + NADH + H^+ \quad (1.3)$$

However, malate dehydrogenase, like many other enzymes, has been known by more than one name.

So, because of the lack of consistency in the nomenclature, it became apparent as the list of known enzymes rapidly grew that there was a need for a systematic way of naming and classifying enzymes. A commission was appointed by the International Union of Biochemistry (later re-named the International Union of Biochemistry and Molecular Biology, IUBMB), and its report, published in 1964, forms the basis of the currently accepted system. Revised editions of the report were published in 1972, 1978, 1984 and 1992. An electronic version is now maintained by the IUBMB on an accessible web-site, and this is updated on a regular basis.

1.3.2 The Enzyme Commission's system of classification

The Enzyme Commission divided enzymes into six main classes, on the basis of the total reaction catalysed. Each enzyme was assigned a code number, consisting of four elements, separated by dots. The first digit shows to which of the main classes the enzyme belongs, as follows:

First digit	*Enzyme class*	*Type of reaction catalysed*
1	Oxidoreductases	Oxidation/Reduction reactions
2	Transferases	Transfer of an atom or group between two molecules (excluding reactions in other classes)
3	Hydrolases	Hydrolysis reactions
4	Lyases	Removal of a group from substrate (not by hydrolysis)
5	Isomerases	Isomerization reactions
6	Ligases	The synthetic joining of two molecules, coupled with the breakdown of the pyrophosphate bond in a nucleoside triphosphate

The second and third digits in the code further describe the kind of reaction being catalysed. There is no general rule, because the meanings of these digits are defined separately for each of the main classes. Some examples are given later in this chapter. Note that, for convenience, and in line with normal practice, some structures are written in a slightly simplified form in the lists provided. So, for example, in the case of the acyl group, which is transferred in reactions catalysed by E.C. 2.3 enzymes, it should be understood that the structure written –COR represents:

$$-\underset{\underset{\displaystyle O}{\|}}{C}-R$$

Enzymes catalysing very similar but non-identical reactions, e.g. the hydrolysis of different carboxylic acid esters, will have the same first three digits in their code. The fourth digit distinguishes between them by defining the actual substrate, e.g. the actual carboxylic acid ester being hydrolysed.

However, it should be noted that **isoenzymes,** that is to say, different enzymes catalysing identical reactions, will have the same four figure classification. There are, for example, five different isoenzymes of **lactate dehydrogenase** within the human body and these will have an identical code. The classification, therefore, provides only the basis for a unique identification of an enzyme. The particular isoenzyme and its source still have to be specified.

It should also be noted that all reactions catalysed by enzymes are reversible to some degree and the classification which would be given to the enzyme for the catalysis of the forward reaction would not be the same as that for the reverse reaction. The classification used is that of the most important direction from the biochemical point of view, or according to some convention defined by the Commission. For example, for oxidation/reduction involving the interconversion of NADH and NAD^+ (see section 11.5.2) the classification is usually based on the direction where NAD^+ is the electron acceptor rather than that where NADH is the electron donor.

Some problems are given at the end of this chapter to help the student become familiar with this system of classification.

1.3.3 The Enzyme Commission's recommendations on nomenclature

The Commission assigned to each enzyme a systematic name in addition to its existing trivial name. This systematic name includes the name of the substrate or substrates in full and a word ending in '-ase' indicating the nature of the process catalysed. This word is either one of the six main classes of enzymes or a subdivision of one of them. When a reaction involves two types of overall change, e.g. oxidation and decarboxylation, the second function is indicated in brackets, e.g. oxidoreductase (decarboxylating). Examples are given below.

The systematic name and the Enzyme Commission (E.C.) classification number unambiguously describe the reaction catalysed by an enzyme and should always be included in a report of an investigation of an enzyme, together with the source of enzyme, e.g. rat liver mitochondria.

However, these names are likely to be long and unwieldy. Trivial names may, therefore, be used in a communication, once they have been introduced and defined in terms of the systematic name and E.C. number. Trivial names are also inevitably used in everyday situations in the laboratory. The Enzyme Commission made recommendations as to which trivial names were acceptable, altering those which were considered vague or misleading. Thus, 'fumarase', mentioned above, was considered unsatisfactory and was replaced by 'fumarate hydratase'.

1.3.4 The six main classes of enzymes

Main Class 1: Oxidoreductases

These enzymes catalyse the transfer of H atoms, O atoms or electrons from one substrate to another. The second digit in the code number of oxidoreductases indicates the donor of the reducing equivalents (hydrogen or electrons) involved in the reaction. For example:

Second digit	*Hydrogen or electron donor*
1	alcohol (>CHOH)
2	aldehyde or ketone (>C=O)
3	–CH.CH–
4	primary amine (>$CHNH_2$ or >$CHNH_3^+$)
5	secondary amine (>CHNH–)
6	NADH or NADPH (only when some other redox catalyst is the acceptor)

The third digit refers to the hydrogen or electron acceptor, as follows:

Third digit	*Hydrogen or electron acceptor*
1	NAD^+ or $NADP^+$
2	Fe^{3+} (e.g. cytochrome)
3	O_2
99	An otherwise unclassified acceptor

Trivial names of oxidoreductases include oxidases (transfer of H to O_2) and dehydrogenases (transfer of H to an acceptor other than O_2). These often indicate the identity of the donor and/or acceptor. Here are some examples:

(S)-lactate: NAD^+ oxidoreductase (E.C. 1.1.1.27), trivial name lactate dehydrogenase, catalyses the reaction:

$$\underset{\text{(S)-lactate}}{CH_3.\underset{\substack{|\\ OH}}{C}H.CO_2^-} + NAD^+ \rightleftharpoons \underset{\text{pyruvate}}{CH_3.\underset{\substack{\|\\ O}}{C}.CO_2^-} + NADH + H^+ \qquad (1.4)$$

Note that it is the alcohol group of lactate, rather than the carboxyl group, which is involved in the reaction and this is indicated in the classification.

Isocitrate: NAD^+ oxidoreductase (decarboxylating) (E.C. 1.1.1.41), trivial name isocitrate dehydrogenase, catalyses:

$$\begin{array}{c} {}^-O_2C.CH_2.\underset{\displaystyle\underset{\displaystyle {}^-O_2C}{|}}{C}H.\underset{\displaystyle\underset{\displaystyle OH}{|}}{C}H.CO_2^- + NAD^+ \\ \text{isocitrate} \end{array} \rightleftharpoons \begin{array}{c} {}^-O_2C.CH_2.CH.\underset{\displaystyle\underset{\displaystyle O}{\|}}{C}.CO_2^- \\ \text{2-oxoglutarate} \end{array} + NADH + H^+ + CO_2 \qquad (1.5)$$

D-amino acid: oxygen oxidoreductase (deaminating) (E.C. 1.4.3.3), trivial name D-amino acid oxidase, catalyses:

$$\begin{array}{c} R.\underset{\displaystyle\underset{\displaystyle {}^+NH_3}{|}}{C}H.CO_2^- \\ \text{D-amino acid} \end{array} + H_2O + O_2 \rightleftharpoons \begin{array}{c} R.\underset{\displaystyle\underset{\displaystyle O}{\|}}{C}.CO_2^- \\ \text{oxo acid} \end{array} + {}^+NH_4 + H_2O_2 \qquad (1.6)$$

Note that this enzyme is less specific than most and will act on any D-amino acid.

Main Class 2: *Transferases*
These catalyse reactions of the type:

$$AX + B \rightleftharpoons BX + A$$

but specifically exclude oxidoreductase and hydrolase reactions. In general, the Enzyme Commission recommends that the names of transferases should end 'X-transferase', where X is the group transferred, although a name ending 'trans-X-ase' is an acceptable alternative. The second digit in the classification describes the type of group transferred. For example:

Second digit	*Group transferred*
1	1-carbon group
2	aldehyde or ketone group (>C=O)
3	acyl group (–COR)
4	glycosyl (carbohydrate group)
7	phosphate group

In general, the third digit further describes the group transferred. Thus:
E.C. 2.1.1 enzyme are methyltransferases (transfer $-CH_3$), whereas
E.C. 2.1.2 enzymes are hydroxymethyltransferases (transfer $-CH_2OH$) and
E.C. 2.1.3 enzymes are carboxyl transferases (transfer –COOH)
or carbamoyl transferases (transfer $-CONH_2$).

Similarly,
E.C. 2.4.1 enzymes are hexosyltransferases (transfer hexose units), and
E.C. 2.4.2 enzymes are pentosyltransferases (transfer pentose units).

The exception to this general rule for transferases is where there is transfer of phosphate groups: these cannot be described further, so there is opportunity to indicate the acceptor.

E.C. 2.7.1 enzymes are phosphotransferases with an alcohol group as acceptor,
E.C. 2.7.2 enzymes are phosphotransferases with a carboxyl group as acceptor,
E.C. 2.7.3 enzymes are phosphotransferases with a nitrogenous group as acceptor.

Phosphotransferases usually have a trivial name ending in '-kinase'. Some examples of transferases are:

(S)-2-methyl-3-oxopropanoyl-CoA: pyruvate carboxyltransferase (E.C. 2.1.3.1) (trivial name: methylmalonyl-CoA carboxyltransferase, formerly transcarboxylase) which catalyses the transfer of a carboxyl group from methylmalonyl-CoA to pyruvate:

$$\underset{\text{methylmalonyl-CoA}}{CH_3.\underset{|}{CH}(CO_2^-).COSCoA} + \underset{\text{pyruvate}}{CH_3CO.CO_2^-} \rightleftharpoons \underset{\text{propionyl-CoA}}{CH_3.CH_2.COSCoA} + \underset{\text{oxaloacetate}}{\underset{|}{CH_2}(CO_2^-).CO.CO_2^-} \quad (1.7)$$

ATP: D-hexose-6-phosphotransferase (E.C. 2.7.1.1) (trivial name: hexokinase) which catalyses:

$$\underset{\text{D-hexose}}{C_5H_9O_5.CH_2OH} + ATP \rightleftharpoons \underset{\text{D-hexose-6-phosphate}}{C_5H_9O_5.CH_2OPO_3^{2-}} + ADP \quad (1.8)$$

This enzyme will transfer phosphate to a variety of D-hexoses.

Main Class 3: Hydrolases

These enzymes catalyse hydrolytic reactions of the form:

$$A\text{–}X + H_2O \rightleftharpoons X\text{–}OH + HA$$

They are classified according to the type of bond hydrolysed. For example:

Second digit	*Bond hydrolysed*
1	ester
2	glycosidic (linking carbohydrate units)
4	peptide (–CONH–) (see chapter 2)
5	C–N bonds other than peptides

The third digit further describes the type of bond hydrolysed. Thus:
E.C. 3.1.1 enzymes are carboxylic ester (–COO–) hydrolases,
E.C. 3.1.2 enzymes are thiol ester (–COS–) hydrolases,
E.C. 3.1.3 enzymes are phosphoric monoester ($-O-PO_3^{2-}$) hydrolases,
E.C. 3.1.4 enzymes are phosphoric diester ($-O-PO_2^- -O-$) hydrolases.

For example, orthophosphoric monoester phosphohydrolase (E.C. 3.1.3.1) (alkaline phosphatase) catalyses:

$$\underset{\text{organic phosphate}}{R-O-PO_3^{2-}} + H_2O \rightleftharpoons R-OH + \underset{\text{inorganic phosphate}}{HO-PO_3^{2-}} \quad (1.9)$$

Alkaline phosphatases are relatively non-specific, and act on a variety of substrates at alkaline pH.

The trivial names of hydrolases are recommended to be the only ones to consist simply of the name of the substrate plus '-ase'.

Main Class 4: Lyases

These enzymes catalyse the non-hydrolytic removal of groups from substrates, often leaving double bonds.

The second digit in the classification indicates the bond broken, for example:

Second digit	*Bond broken*
1	C–C
2	C–O
3	C–N
4	C–S

The third digit refers to the type of group removed. Thus, for the C–C lyases:

Third digit	*Group removed*
1	carboxyl group (i.e. CO_2)
2	aldehyde group (–CH=O)
3	ketoacid group ($-CO.CO_2^-$)

For example, L-histidine carboxy-lyase (E.C. 4.1.1.22) (trivial name: histidine decarboxylase, catalyses:

$$\underset{\text{histidine}}{C_3N_2H_3.CH_2\underset{\substack{|\\ CO_2^-}}{C}H.NH_3^+} \rightleftharpoons \underset{\text{histamine}}{C_3N_2H_3.CH_2.CH_2.NH_3^+} + CO_2 \quad (1.10)$$

(Note the importance of the hyphen and the extra 'y' in the systematic name, because carboxy-lyase and carboxylase do not mean the same thing: carboxylase simply refers to the involvement of CO_2 in a reaction without being specific.)

Also classified as lyases are enzymes catalysing reactions whose biochemically important direction is the reverse of the above, i.e. addition across double bonds. These may have the trivial name **synthase** or, if water is added across the double bond, **hydratase,** as discussed earlier in the example of fumarate hydratase (fumarase), the systematic name of this particular enzyme being (S)-malate hydro-lyase (E.C. 4.2.1.2).

Main Class 5: Isomerases
Enzymes catalysing isomerization processes are classified according to the type of reaction involved. For example:

Second digit	*Type of reaction*
1	Racemization or epimerization (inversion at an asymmetric carbon atom)
2	cis-trans isomerization
3	intramolecular oxidoreductases
4	intramolecular transfer reaction

The third digit describes the type of molecule undergoing isomerization. Thus, for racemases and epimerases:

Third digit	*Substrate*
1	amino acids
2	hydroxy acids
3	carbohydrates

An example is alanine racemase (E.C. 5.1.1.1) which catalyses:

$$\text{L-alanine} \rightleftharpoons \text{D-alanine} \qquad (1.11)$$

Main Class 6: Ligases
These enzymes catalyse the synthesis of new bonds, coupled to the breakdown of ATP or other nucleotide triphosphates. The reactions are of the form:

$$X + Y + ATP \rightleftharpoons X\text{–}Y + ADP + P_i$$

or

$$X + Y + ATP \rightleftharpoons X\text{–}Y + AMP + (PP)_i$$

The second digit in the code indicates the type of bond synthesized. For example:

Second digit	*Bond synthesized*
1	C–O
2	C–S
3	C–N
4	C–C

The third digit further describes the bond being formed. Thus

E.C. 6.3.1 enzymes are acid-ammonia ligases (amide, $-CONH_2$, synthases) and
E.C. 6.3.2 enzymes are acid-amino acid ligases (peptide, –CONH–, synthases).
Prior to 1984, such enzymes could also be known as synthetases.

An example is L-glutamate: ammonia ligase (E.C. 6.3.1.2), trivial name: glutamate-ammonia ligase, formerly glutamate synthetase, which catalyses:

$$\underset{\text{L-glutamate}}{\underset{^{-}\text{O}\qquad\quad ^{+}\text{NH}_3}{\text{O=C.CH}_2\text{CH}_2\text{.CH.CO}_2^-}} + \text{ATP} + \text{NH}_3 \rightleftharpoons \underset{\text{L-glutamine}}{\underset{\text{NH}_2\qquad\quad ^{+}\text{NH}_3}{\text{O=C.CH}_2\text{CH}_2\text{.CH.CO}_2^-}} + \text{ADP} + \text{P}_i \qquad (1.12)$$

SUMMARY OF CHAPTER 1

Enzymes are proteins which catalyse, in a highly specific way, chemical reactions taking place within the living cell. Often a further, non-protein, component called a cofactor is required before an enzyme has catalytic activity.

Enzymes have been used for many centuries, although their true nature has only become known relatively recently, and they are still of great importance in scientific research, clinical diagnosis and industry.

Because of the lack of consistency and occasional lack of clarity in the names of enzymes, an Enzyme Commission appointed by the International Union of Biochemistry (now the International Union of Biochemistry and Molecular Biology) has given all known enzymes a systematic name and a four-figure classification. These, together with the source of the enzyme concerned, should be quoted in any report.

FURTHER READING

Enzyme Nomenclature Recommendations (1992) *of the Nomenclature Committee of the International Union of Biochemistry and Molecular Biology,* Academic Press (note that the updated electronic version can be accessed at the following address: http://www.chem.qmul.ac.uk/iubmb/enzyme).

Nomenclature Database, on the World Wide Web pages of the Swiss Institute of Bioinformatics (SIB). See http:www.expasy.ch/enzyme.

Purich, D. L. and Allison, R. D. (2003), *The Enzyme Reference: A Comprehensive Guidebook to Enzyme Nomenclature, Reactions and Methods*, Academic Press.

PROBLEMS

1.1 Give the systematic names and the first three digits in the E.C. classifications of the enzymes catalysing the following reactions:

(Note: it should be possible to deduce the classification and make a reasonable attempt at the systematic name from the information given in Chapter 1.)

(a) $$\underset{\text{acyl choline}}{\text{R.}\underset{\text{O}}{\underset{\|}{\text{C}}}\text{O.CH}_2\text{CH}_2\text{.}\overset{+}{\text{N}}(\text{CH}_3)_3} + \text{H}_2\text{O} \rightleftharpoons \underset{\text{acid anion}}{\text{R.}\underset{\text{O}}{\underset{\|}{\text{C}}}\text{O}^-} + \underset{\text{choline}}{\text{HOCH}_2\text{CH}_2\text{.}\overset{+}{\text{N}}(\text{CH}_3)_3}$$

(b) $H_2N.C(=O).OPO_3^{2-} + H_3\overset{+}{N}(CH_2)_3.CH(^{+}NH_3).CO_2^- \rightleftharpoons H_2N.C(=O).NH(CH_2)_3.CH(^{+}NH_3).CO_2^- + P_i$

carbamoyl phosphate, L-ornithine, citrulline

(c) $ATP + H_3\overset{+}{N}.CH(CH_3)CO_2^- + H_3\overset{+}{N}.CH(CH_3)CO_2^- \rightleftharpoons H_3\overset{+}{N}.CH(CH_3).C(=O).N(H).CH(CH_3)CO_2^- + ADP + P_i$

D-alanine, D-alanine, D-alanyl-alanine

(d) $CH_2OH.C(=O).CH_2OH + NADH + H^+ \rightleftharpoons CH_2OH.CHOH.CH_2OH + NAD^+$

dihydroxyacetone, glycerol

(e) $CH_2OPO_3^{2-}.C(=O).CHOH.CHOH.CHOH.CH_2OPO_3^{2-} \rightleftharpoons CH_2OPO_3^{2-}.C(=O).CH_2OH + CHO.CHOH.CH_2OPO_3^{2-}$

D-frucrose-1,6-bisphosphate, dihydroxyacetone-phosphate, D-glyceraldehyde-3-phosphate

(f) NADH + 2 ferricytochrome $b_5 \rightleftharpoons NAD^+$ + 2 ferrocytochrome b_5

(g) UDP-galactose $\rightleftharpoons$ UDP-glucose

(glucose and galactose are aldohexoses differing in configuration at C4)

1.2 Give the E.C. classification of the enzyme catalysing the following reactions: (this question has been designed to encourage the student to become familiar with the Enzyme Commission's recommendations and can only be answered satisfactorily by reference to their report or to a detailed account of it.)

(a) $CH_3.\underset{\underset{O}{\|}}{C}.CO_2^- + ATP + CO_2 + H_2O \rightleftharpoons {}^-O_2C.CH_2.\underset{\underset{O}{\|}}{C}.CO_2^- + ADP + P_i$

pyruvate — oxaloacetate

(b) $H_2S + 3NADP^+ + 3H_2O \rightleftharpoons$ sulphite + 3NADPH

(c) ATP + AMP $\rightleftharpoons$ ADP + ADP

(d) $HC{=}O$ / $CHOH$ / $CH_2OPO_3^{2-}$ $\rightleftharpoons$ CH_2OH / $C{=}O$ / $CH_2OPO_3^{2-}$

D-glyceraldehyde-3-P — dihydroxyacetone-P

(e) Hydrolysis of dipeptides in cytosol, e.g.

$H_3\overset{+}{N}.CH_2.\underset{\underset{O}{\|}}{C}.\overset{H}{N}.CH_2.CO_2^- + H_2O \rightleftharpoons H_3\overset{+}{N}.CH_2.CO_2^- + H_3\overset{+}{N}.CH_2.CO_2^-$

glycylglycine — glycine — glycine

(f) Endohydrolysis of α-1,4 glucosidic links in polysaccharides containing 3 or more α-1,4 linked D-glucose units.

(g)

$H_3\overset{+}{N}.\underset{|\,CO_2^-}{CH}.(CH_2)_3.NH.\underset{\|\,{}^+NH_2}{C}.NH.\underset{|\,CH_2CO_2^-}{\overset{CO_2^-\,|}{CH}} + H_2O \rightleftharpoons H_3\overset{+}{N}.\underset{|\,CO_2^-}{CH}.(CH_2)_3.NH.\underset{\|\,{}^+NH_2}{C}.NH + CHCO_2^- {=} CHCO_2^-$

N-(L-arginino)succinate — L-arginine — fumarate

2

The Structure of Proteins

2.1 INTRODUCTION

Since all enzymes are proteins, a knowledge of protein structure is clearly a prerequisite to any understanding of enzymes.

Proteins are **macromolecules** (i.e. large molecules) with molecular weights of at least several thousand. They are found in abundance in living organisms, making up more than half the dry weight of cells. Two distinct types are known: **fibrous** and **globular** proteins.

Fibrous proteins are insoluble in water and are physically tough, which enables them to play a structural role. Examples include α-keratin (a component of hair, nails and feathers) and collagen (the main fibrous element of skin, bone and tendon). In contrast, **globular proteins** are generally soluble in water and may be crystallized from solution. They have a functional role in living organisms. All enzymes are globular proteins.

Unlike polysaccharides and lipids, which may be hoarded by cells solely as a store of fuel, each protein in a cell has some precise purpose which is related to its shape and structure. Nevertheless, should the need arise, proteins may be broken down, either to provide energy or to supply raw materials for the synthesis of other macromolecules.

All proteins consist of **amino acid** units, joined in series. The sequence of amino acids in a protein is specific, being determined by the structure of the genetic material of the cell (see section 3.1), and this gives each protein unique properties. Some proteins are composed entirely of these amino acid building blocks and are termed **simple proteins**. Others, called **conjugated proteins**, contain extra material, which is firmly bound to one or more of the amino acid units. Conjugated proteins are classified according to the nature of the additional component. Thus, a nucleoprotein contains a nucleic acid; a lipoprotein a lipid; a glycoprotein an oligosaccharide; a haemoprotein an iron protoporphyrin; a flavoprotein a flavin nucleotide; and a metalloprotein a metal. Enzymes may be either simple or conjugated proteins.

2.2 AMINO ACIDS, THE BUILDING BLOCKS OF PROTEINS

2.2.1 Structure and classification of amino acids

Amino acids, by definition, are organic compounds which contain within the same molecule an amino group (–NH_2 or >NH) and a carboxyl group (–COOH). Thus they have properties of both bases and acids.

The amino group in all but one of the twenty amino acids commonly found in proteins is a primary one (–NH_2), the exception being proline, which contains a secondary amino group (>NH). The carbon atoms of organic molecules containing a carboxyl group may be identified with Greek letters as follows:

γ-carbon atom α-carbon atom

$$-C-C-C-C(CO_2H) \leftarrow \text{carboxyl group}$$

δ-carbon atom β-carbon atom

All the amino acids commonly found in proteins are **α-amino acids**, since the amino group is on the α-carbon atom. The general formula is:

$$H_2N-\overset{H}{\underset{R}{C}}-\underset{O}{\overset{\|}{C}}-OH \text{, or, more conveniently, } H_2N.CHR.CO_2H$$

The symbol R represents the rest of the molecule, often called the **side chain.** The amino and carboxyl groups attached to the α-carbon atom are termed the α-amino and α-carboxyl groups, to distinguish them from similar groups which may be present as part of the side chain. Proline, whose α-amino group forms part of an imino ring, is actually an **imino** acid, and has a formula slightly different from the general one given above (see Fig. 2.1).

The α-amino acids may have polar or non-polar side chains. A **polar** molecule or group has a degree of ionic character and is **hydrophilic,** i.e. it is quite soluble in water because its structure may be stabilized by hydrogen bonding in aqueous solution. Polar groups may be acidic, basic or neutral. A **non-polar** molecule or group is entirely covalent in character and is **hydrophobic,** i.e. it is relatively insoluble in aqueous solvents but more soluble in organic solvents such as diethyl ether. The side chains of the amino acids commonly found in proteins, classified according to their polar or non-polar characteristics, are shown in Fig. 2.1.

It will be seen that the side chain of **histidine** contains an **imidazole** ring, while that of **tryptophan** includes a double-ringed structure called an **indole**. One of these rings is an aromatic benzene ring, so tryptophan, in common with phenylalanine and tyrosine, may be called an **aromatic** amino acid. In **tyrosine**, the aromatic ring is linked to -OH to form a **phenolic** group. **Glutamic acid** and **aspartic acid** contain a **carboxyl** group in their side chains, which is converted to an **amide** group in **glutamine** and **asparagine.**

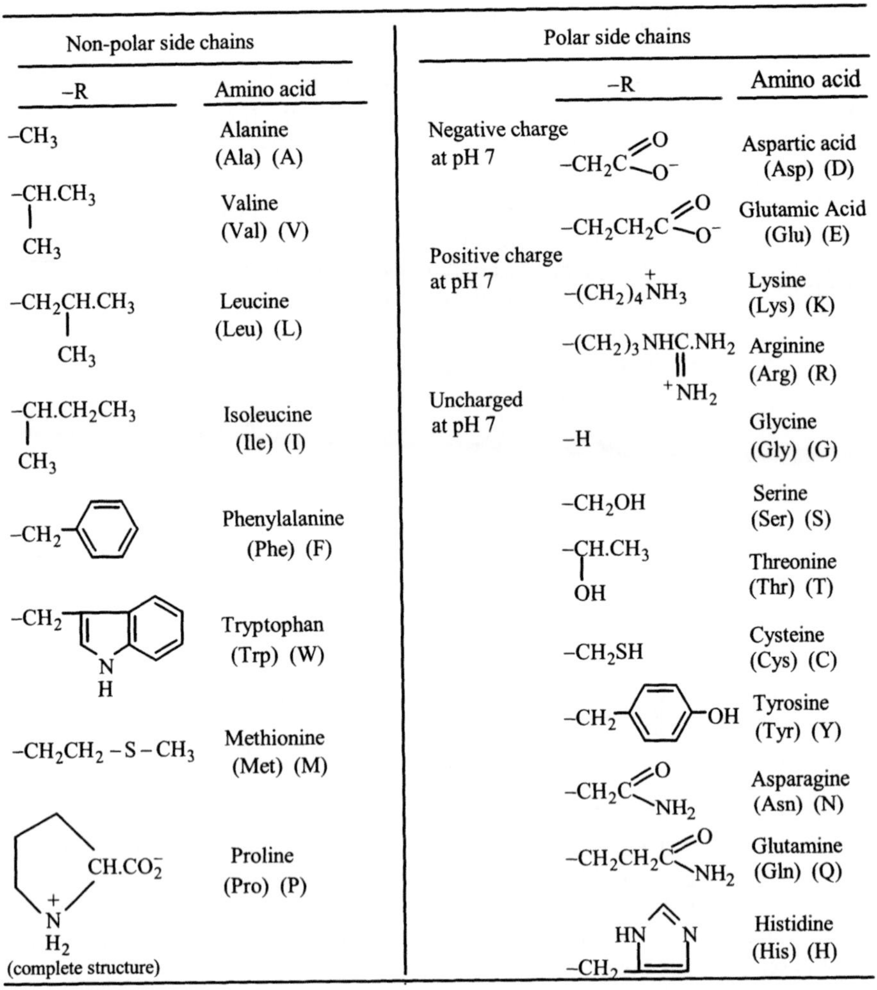

Non-polar side chains

–R	Amino acid
$-CH_3$	Alanine (Ala) (A)
$-CH.CH_3$ $\vert$ CH_3	Valine (Val) (V)
$-CH_2CH.CH_3$ $\vert$ CH_3	Leucine (Leu) (L)
$-CH.CH_2CH_3$ $\vert$ CH_3	Isoleucine (Ile) (I)
$-CH_2-$[benzene ring]	Phenylalanine (Phe) (F)
$-CH_2-$[indole ring, N–H]	Tryptophan (Trp) (W)
$-CH_2CH_2-S-CH_3$	Methionine (Met) (M)
[ring] $CH.CO_2^-$, $\overset{+}{N}H_2$ (complete structure)	Proline (Pro) (P)

Polar side chains

	–R	Amino acid
Negative charge at pH 7	$-CH_2C(=O)O^-$	Aspartic acid (Asp) (D)
	$-CH_2CH_2C(=O)O^-$	Glutamic Acid (Glu) (E)
Positive charge at pH 7	$-(CH_2)_4\overset{+}{N}H_3$	Lysine (Lys) (K)
	$-(CH_2)_3NHC.NH_2$ $\Vert$ $^+NH_2$	Arginine (Arg) (R)
Uncharged at pH 7	$-H$	Glycine (Gly) (G)
	$-CH_2OH$	Serine (Ser) (S)
	$-CH.CH_3$ $\vert$ OH	Threonine (Thr) (T)
	$-CH_2SH$	Cysteine (Cys) (C)
	$-CH_2-$[benzene ring]$-OH$	Tyrosine (Tyr) (Y)
	$-CH_2C(=O)NH_2$	Asparagine (Asn) (N)
	$-CH_2CH_2C(=O)NH_2$	Glutamine (Gln) (Q)
	$-CH_2-$[imidazole ring, HN, N]	Histidine (His) (H)

Fig. 2.1 – The side chains of the twenty amino acids commonly found in proteins. Note that several polar side chains contain ionizable groups, the degree of ionization being pH-dependent (see section 2.3.2); only the form which predominates at pH 7 is shown here. Included in the figure are the standard three-letter and one-letter symbols for each amino acid.

The side chains of **lysine** and **arginine** contain **amino** groups, which in the case of **arginine** forms part of a **guanidine** structure. The R groups of **valine, leucine** and **isoleucine** have a **branched-chain aliphatic hydrocarbon** structure while **proline,** as mentioned previously, is an **imino acid. Methionine** and **cysteine** contain **sulphur,** which in the case of **cysteine** is present as part of a **sulphydryl** (-SH) group. Cysteine is readily oxidized to form the dimeric compound, **cystine,** the two component cysteine units being linked by a **disulphide bridge.**

$$
\begin{array}{c}
H_2N.CH.CO_2H \\
| \\
CH_2 \\
| \\
S \\
| \\
S \\
| \\
CH_2 \\
| \\
H_2N.CH.CO_2H
\end{array}
\qquad \text{cystine}
$$

Thus, amino acids with a considerable variety of side chain characteristics are found in proteins. As we shall see later, this explains the range of properties shown by these macromolecules.

2.2.2 Stereochemistry of amino acids

Each carbon atom in a molecule can form four single covalent bonds with other atoms. These are often represented at right angles to each other on a single plane, as in section 2.2.1. However, it must be realized that this is done entirely for convenience, since a page of a book is two-dimensional and thus lends itself to a two-dimensional representation of structure. In fact the four bonds are evenly distributed in three-dimensional space, which means they point to the four corners of a regular tetrahedron, each bond forming an angle of 109° with each of the other bonds.

If we consider the bonds involving the α-carbon of an amino acid, we see that two different spatial arrangements, or **stereoisomeric forms,** are possible: the structure depicted in Fig. 2.2a cannot be superimposed on that in Fig. 2.2b by rotation of the molecules. The α-carbon atom is covalently linked to four different atoms or groups, so it is **asymmetric**: no plane drawn through this carbon atom can divide the molecule into two parts in such a way that each half is the exact mirror image of the other. As a consequence of this, two mirror image forms of the complete molecule can exist. Such forms are termed **optical isomers,** since one will usually rotate the plane of polarized light passing through it to the right, and the other to the left.

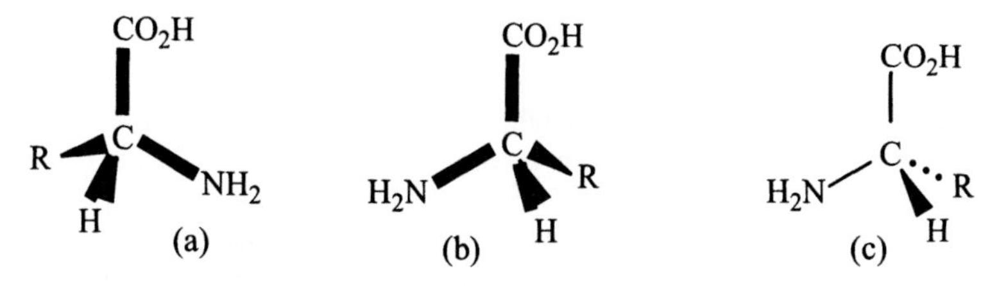

Fig. 2.2 – Three-dimensional arrangements about the α-carbon atom for (a) a D-amino acid and (b) an L-amino acid. A bond coming out of the plane of the page towards the reader is indicated by a thickening of the line; one going away from the reader is represented by a narrowing of the line. Note that, by convention, a bond going away from the reader may also be indicated by a dotted line, as in the L-amino acid structure given in (c).

The molecule shown in Fig. 2.2a is *defined* as a **D-amino acid** and the one in Fig. 2.2b as an **L-amino acid.** This says nothing about how each isomer will affect the plane of polarized light, a property which has to be determined by experiment. Thus L-alanine is found to rotate polarized light to the right, but L-leucine rotates it to the left. All the common amino acids, with the exception of glycine, exist as optical isomers. Glycine does not have an asymmetric carbon atom since in this case there are two hydrogen atoms attached to the α-carbon (R = H). Threonine and isoleucine possess two asymmetric carbon atoms, but this extra complication need not concern us here.

If amino acids are synthesized by an uncatalysed chemical process, a **racemic mixture** (one containing equal amounts of L- and D-isomers) is produced, the isomeric forms being almost indistinguishable from a chemical point of view. However, proteins are built almost exclusively of L-amino acids, and most naturally occurring amino acids are in this same isomeric form. The explanation is that protein biosynthesis (see section 3.1) and most other metabolic processes are mediated by enzymes which are specific for a particular isomeric form of the substrate (see section 4.1). This is essential for ensuring the high degree of three-dimensional organization which is found in structures within cells. It is presumably evolutionary chance which has determined that life as we know it is based on L- rather than D-amino acids.

2.3 THE BASIS OF PROTEIN STRUCTURE

2.3.1 Levels of protein structure

Four separate levels of protein structure can be determined: these are the primary, secondary, tertiary and quaternary structures.

The **primary structure** is the sequence of amino acids making up the protein: a **peptide bond** connects the α-carboxyl group of each amino acid to the α-amino group of the next in the chain

$$H_2N.CHR'.CO_2H + H_2N.CHR''.CO_2H \rightarrow H_2N.CHR'.\underbrace{\overset{\displaystyle O}{\overset{\|}{C}}-\underset{\displaystyle H}{\underset{|}{N}}}_{\text{peptide bond}}.CHR''.CO_2H + H_2O \qquad (2.1)$$

Since a molecule of water is lost when two free amino acid molecules undergo this condensation reaction, only their **residues** are linked. A molecule consisting of two amino acid residues joined by a peptide bond is called a **dipeptide.** Several residues linked in this way form an **oligopeptide,** while a chain of many amino acid residues linked together is termed a **polypeptide.** The **covalent backbone** of such a structure consists of α-carbon atoms linked by peptide bonds, the R groups sticking out from the chain. Each peptide chain has one free amino end (the **N-terminus,** which is regarded as belonging to the first amino acid in a peptide chain of any length) and one free carboxyl end (the **C-terminus**). All the other α-amino and α-carboxyl groups present are involved in peptide bonds. For example:

Proteins may contain one or more polypeptide chains, each one having a specific primary structure. Although a two-dimensional representation of a polypeptide chain can give the impression that the backbone is linear, it should be understood that this is not so. The even distribution in three-dimensional space of the single covalent bonds about the carbon and nitrogen atoms in the backbone means that no two bonds emerging from the same atom will be diametrically opposite each other (see section 2.2.2). Molecules may rotate freely about single covalent bonds, so an unlimited number of arrangements of a polypeptide chain in space are possible. However, some of these will be more stable than others, so are more likely to exist. **Secondary structure** refers to regular, repeating patterns formed by the backbone of at least part of a polypeptide chain and stabilized by hydrogen bonding.

Certain amino acids cannot be accommodated in these regular arrangements, so the secondary structure is disrupted wherever they occur. Again the possibility of free rotation about a bond at each point of disruption suggests that a great number of different structures could result, but in fact each polypeptide chain is found to have a single, characteristic, three-dimensional structure. This is termed the **tertiary structure** and, once formed, it may be stabilized by bonding between amino acids which find themselves in close proximity. It should be noted that amino acids which are widely separated in the primary structure may be close together in space, because of the twists of the polypeptide chain.

Several identical or non-identical polypeptide chains may then be linked together to form the actual protein. The complete three-dimensional structure, including the interactions between the component polypeptide chains, is termed the **quaternary structure.**

2.3.2 Bonds involved in the maintenance of protein structure

A **single covalent bond** is formed by the sharing of a pair of electrons between two atoms, each atom contributing one electron to the pair. By means of such a sharing arrangement, involving one or more covalent bonds, an atom can achieve an arrangement of electrons round its nucleus identical to that of an inert gas and thus become chemically stable. Two atoms of given identity, when linked by a single covalent bond, are located a characteristic distance apart, this distance being known as the **bond length**. If two atoms share *two* pairs of electrons between them, then a **double covalent bond** is formed. In this case the bond length is less than for the equivalent single bond. Molecules can rotate about single covalent bonds but not about double covalent bonds, which are more rigid.

The primary structure of a protein consists of amino acid residues linked by covalent peptide bonds. Covalent disulphide bridges (–S–S–), linking cysteine residues, are often involved in the maintenance of tertiary structure. In a very few instances, disulphide bridges may also link the separate polypeptide components of a protein (see sections 3.1.4 and 5.1.2).

An alternative way by which an atom might achieve stability by obtaining the same electron structure as an inert gas is for it to gain or lose a number of electrons, i.e. to form an **ion**. Ions which are formed by loss of electrons from an atom will have a net positive charge and are called **cations** while those formed by the addition of electrons will have a negative charge and are termed **anions**. The magnitude of the charge will depend on the number of electrons transferred.

An **electrostatic interaction** occurs between each pair of ions in the same medium. The force (F) between two ions A and B in dilute solution is given by **Coulomb's law**:

$$F = Z_A Z_B e^2 / Dr^2 \tag{2.2}$$

where Z_A is the number of unit charges carried by ion A, Z_B the number carried by ion B, e is one unit of electronic charge, r is the distance between the two ions and D is the dielectric constant of the medium. Ions with like charge repel each other while those of opposite charge attract.

Thus the tertiary and quaternary structures of proteins could involve electrostatic linkages between amino acids with side chains of opposite charge (e.g. between lysine and glutamic acid). In fact, in an aqueous environment it is energetically more favourable for a charged group to form linkages with surrounding water molecules rather than with another charged group in the protein. However such linkages do occur in hydrophobic regions of proteins (see below) and so could play an important role in the stabilization of the three-dimensional structure.

It often happens that covalent bond formation does not lead to an *equal* sharing of a pair of electrons between two atoms. The electrons may be associated with one of the components more than the other, producing a slight separation of charge between the atoms, called a **dipole effect.** The atom having the greater association with the shared pair of electrons will have a partial negative charge and can thus form weak electrostatic linkages, as can the other atom, which will have a partial positive charge.

The most important example of this phenomenon is the **hydrogen bond.** The oxygen atoms of –OH or –C=O groups have a slight negative charge, while the hydrogen atoms of >NH or –OH groups have a slight positive charge. Hence a weak electrostatic linkage can be formed between the oxygen atom in one group and the hydrogen atom in another, e.g. –C=O····H–O–. The bond energy involved is small, but sufficient to add stability to a structure.

All the water molecules present in an aqueous medium link by means of hydrogen bonding to produce a huge three-dimensional network. Hydrogen bonds can be formed between groups in polypeptides and the surrounding water molecules, as well as between different components within polypeptide chains. Such bonds can help to stabilize the secondary, tertiary and quaternary structures of proteins.

Although the arrangement of electrons about an atom may, on average, be symmetrical, the constant fluctuations in electron distribution mean that the arrangement is likely to be asymmetrical at any given instant. Hence a dipole exists, however momentarily, and this induces a corresponding effect in all neighbouring atoms, causing them to attract each other. This is true of all atoms, even those of inert gases.

However, when two atoms come into *very* close proximity, the repulsion between their respective clouds of surrounding electrons is greater than the induced attraction. There is an optimal distance between two non-bonding atoms, known as the **van der Waals contact distance,** when the forces of attraction and repulsion are equal. These forces are known as **London dispersion forces** and the weak linkages resulting from dipole effects are sometimes termed **van der Waals bonds.** These play an important part in governing which three-dimensional structure is taken up by a protein.

Non-polar, or hydrophobic bonds also have a considerable influence on protein structure. These bonds are not formed as a result of any direct interaction between atoms and may be best considered from the point of view of the complete protein/solvent mixture. The network of hydrogen bonds linking water molecules to each other confers great stability, so the most stable structure for a protein in aqueous solution will be that which gives the greatest possibility of hydrogen bonding between the protein molecule and the surrounding water molecules. Non-polar side chains of amino acids cannot form hydrogen bonds, so the contact between these and the water molecules must be minimized. Two such side chains in close proximity will tend to come even closer, forcing out all water molecules from between them, so that a single non-polar region is formed from the two originally present. Many non-polar side chains may be incorporated into a single non-polar zone, creating a hydrophobic **micro-environment** that is quite different from the micro-environments in other parts of the protein molecule.

In the case of metalloenzymes a further type of bond, the **co-ordinate bond,** needs to be mentioned. Like the covalent bond, this involves the sharing of a pair of electrons between two atoms, but in this case both electrons come originally from the same atom. A metal atom can accept pairs of electrons in this way from donor groups, or **ligands,** until it has the required number of electrons at a particular level. (A ligand is simply something which binds, the word having the same Latin root as the name of the group of enzymes called ligases.) The electrons which may be lost by a metal atom to form an ion are at a different level from those involved in co-ordinate bond formation, so the two processes are quite distinct.

The bonds involved in the maintenance of protein structure will be discussed further, in the light of experimental evidence, in section 2.5.2. In general, the three-dimensional structure taken up by a protein will be that which is energetically most favourable, taking into account all possible interactions involving the types of bond discussed in the present section.

2.4 THE DETERMINATION OF PRIMARY STRUCTURE

2.4.1 The isolation of each polypeptide chain

The first step in the determination of protein structure is to find out how many different types of polypeptide chain are present in the intact protein. Since each polypeptide chain has an N-terminus and a C-terminus, this should be the same as finding out how many different N-terminal or C-terminal amino acids are present.

In view of the possibility that two otherwise dissimilar polypeptide chains might have, for example, the same amino acid at the N-terminus, it is usually best to determine the number of different N-terminal amino acids *and* the number of different C-terminal amino acids. If these are not the same, the larger of the two numbers would be taken to indicate the number of different polypeptide chains present. Another reason for doing this is the possibility that a terminal amino acid might be buried within the protein molecule and thus not be accessible to the reagents used.

The identity of **N-terminal amino acids** can be determined by the use of **dansyl chloride** or of 1-fluoro-2-4-dinitrobenzene (known as **Sanger's reagent**, after Fred Sanger, who introduced it in 1945). Both of these reagents form addition compounds with free amino groups. In a polypeptide chain, the only free α-amino group belongs to the N-terminal amino acid. Hence, after treatment of a protein with one of these reagents, and subsequent complete hydrolysis to the constituent amino acids (e.g. with 6 M HCl at 105°C for 24 h), the only α-N-substituted amino acids present will be those originally at an N-terminus.

In the **dansyl chloride** procedure, the reaction sequence is:

(2.3)

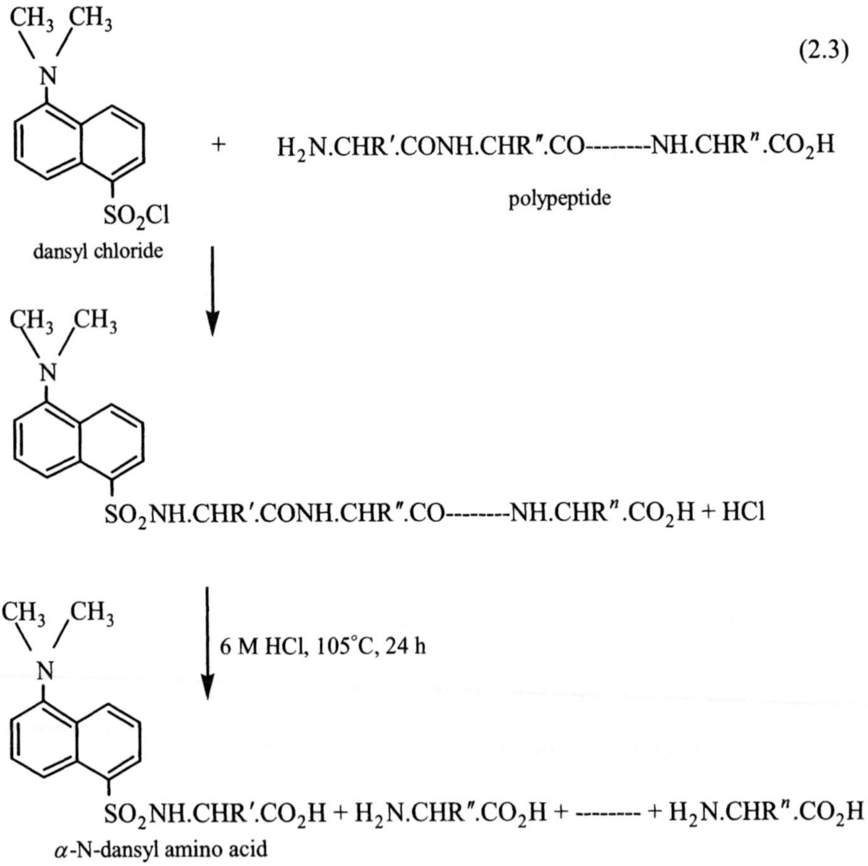

CH_3 CH_3
N
SO_2Cl
dansyl chloride

\+ $H_2N.CHR'.CONH.CHR''.CO$--------$NH.CHR^n.CO_2H$

polypeptide

CH_3 CH_3
N
$SO_2NH.CHR'.CONH.CHR''.CO$--------$NH.CHR^n.CO_2H + HCl$

6 M HCl, 105°C, 24 h

CH_3 CH_3
N
$SO_2NH.CHR'.CO_2H + H_2N.CHR''.CO_2H +$ -------- $+ H_2N.CHR^n.CO_2H$

α-N-dansyl amino acid

The structures of **Sanger's reagent** and the corresponding α-N-substituted amino acid are:

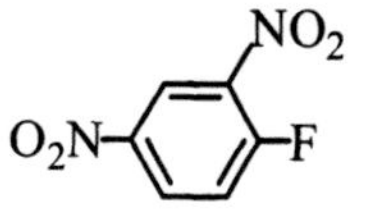

$$O_2N-C_6H_3(NO_2)-F \qquad O_2N-C_6H_3(NO_2)-NH.CHR.CO_2H$$

1-fluoro-2,4-dinitrobenzene (FDNB) 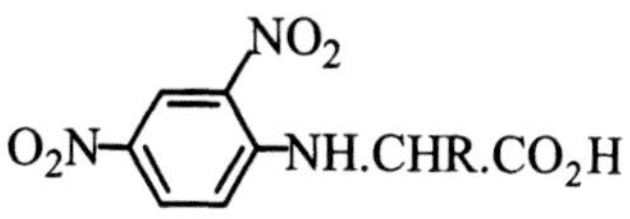 *α*-N-2,4-dinitrophenyl amino acid (*α*-N-DNP amino acid)

The α-N-DNP amino acids can be identified by paper or thin layer chromatography, or by high performance liquid chromatography (HPLC), using α-N-DNP amino acids of known identity as markers. These compounds are yellow, so there is no need to use staining reagents. Amino groups present in the side chains of amino acids (e.g. of lysine) will also react with Sanger's reagent, but the products can be distinguished chromatographically from α-N-DNP amino acids, so no confusion will result. The dansyl chloride technique is similar but has an added advantage in that dansyl amino acids are highly fluorescent and hence can be identified in very small quantities by HPLC.

The **C-terminal amino acids** may, in corresponding fashion, be identified by treating the protein with a reagent which attacks free carboxyl groups, e.g. with the reducing agent **sodium borohydride.** This requires preliminary esterification of the carboxyl groups and protection of the free amino groups by acetylation, neither of these processes being shown in the following simplified scheme:

$$H_2N.CHR'.CONH.CHR''.CO\text{--------}NH.CHR''.CO_2H \qquad (2.4)$$

$$\downarrow NaBH_4$$

$$H_2N.CHR'.CONH.CHR''.CO\text{--------}NH.CHR''.CH_2OH$$

$$\downarrow 6\ M\ HCl,\ 105^{\circ}C,\ 24\ h$$

$$H_2N.CHR'.CO_2H + H_2N.CHR''.CO_2H + \text{--------} + H_2N.CHR''.CH_2OH$$

α-amino alcohol

Only C-terminal amino acids will be converted to α-amino alcohols. Side chain carboxyl groups (e.g. in glutamic and aspartic acids) will also be reduced, with the subsequent production of amino alcohols, but these will not be α-amino alcohols, so can be distinguished by chromatography.

Alternatively, the polypeptide chain may be treated with anhydrous **hydrazine** (NH_2NH_2) at high temperatures for several hours in the presence of a catalyst. As peptide bonds are broken, each freed carboxyl group reacts to form a hydrazide (–CONH.NH_2). However, the C-terminal group, being free from the start, does not react in this way, so can be identified from amongst the products by chromatography.

By means of these techniques, the N-terminal and C-terminal amino acids in a protein can be identified, enabling the number of different polypeptide chains present to be deduced. The linkages between the various polypeptide chains may then be broken in a variety of ways. For example, **disulphide bridges** may be cleaved by treatment with **performic acid,** each cystine unit being oxidized to two cysteic acid units without the breakage of any peptide bonds.

$$\underset{\text{cystine residue}}{\text{CH.CH}_2-\text{S}-\text{S}-\text{CH}_2\text{.CH}} \xrightarrow{\text{performic acid}} \underset{\substack{\text{cysteic acid}\\ \text{residue}}}{\text{CH.CH}_2\text{.SO}_3\text{H}} + \underset{\substack{\text{cysteic acid}\\ \text{residue}}}{\text{HO}_3\text{S.CH}_2\text{.CH}} \quad (2.5)$$

Performic acid also oxidizes methionine and tryptophan residues, if these are present, so an alternative approach is to cleave the disulphide bridge by reduction, e.g. with 2-mercaptoethanol (2-ME), and then alkylate the sulphydryl groups produced to prevent them reforming –S–S– bonds. **Non-covalent bonds** may be broken at **extremes of pH,** at **high salt concentrations,** or in presence of reagents such as **urea or guanidine hydrochloride.**

Each of the polypeptide chains known to be present can then be separated from the others by chromatographic or electrophoretic techniques, as described in section 16.2.2. Hence a pure specimen of each polypeptide chain maybe obtained.

A great many enzymes consist of a number of identical polypeptide chains linked only by non-covalent bonds (see section 5.2). This is indicated by the finding of only a single N-terminal and a single C-terminal amino acid, by the failure to separate any polypeptide chains from any others after the breaking of non-covalent linkages, *and* by demonstrating that a several-fold decrease in molecular weight occurs when these linkages are broken (see section 16.3).

2.4.2 Determination of the amino acid composition of each polypeptide chain

The **molecular weight** (i.e. **relative molecular mass,** M_r, defined as the ratio of the mass of the molecule to 1/12th that of a ^{12}C atom) of the polypeptide should first be determined, using such techniques as size-exclusion chromatography (SEC) or ultracentrifugation (see section 16.3). The polypeptide is then completely hydrolysed to its component amino acids and the concentration of each determined.

A common procedure for the **quantitative determination of amino acids** is **ion exchange chromatography (IEX)**. The sample is applied to a cation exchange column, e.g. sulphonated ($-SO_2O^-$) polystyrene, and buffers of increasing pH and salt concentration are pumped through. In general, amino acids with acidic or neutral polar side chains are eluted from the column before those with basic or non-polar side chains, largely according to their relative attraction for the charges on the ion exchange resin, conditions being chosen so that each amino acid is eluted separately. For a long time, only glass columns were used, but developments in HPLC technology led to the utilization of stainless steel columns. The column eluate is mixed with a reagent such as **ninhydrin,** which reacts with most amino acids to give a blue-purple colour, the absorbance being measured at 570 nm:

$$\text{ninhydrin} + H_2N.CHR.CO_2H \longrightarrow \text{hydrindantin} + NH_3 + CO_2 + RCHO \qquad (2.6)$$

$$\text{ninhydrin} + NH_3 + \text{hydrindantin} \longrightarrow \text{Ruheman's purple}$$

The imino acid, proline, reacts differently, giving a yellow colour, the absorbance usually being measured at 440 nm. Hence, if the colour produced is monitored continuously at both 570 nm and 440 nm, and if the instrument has been pre-calibrated with amino acid standards, it is possible to determine the concentration of each amino acid present. As an alternative to ninhydrin, more sensitive fluorimetric reagents, such as *o*-phthalaldehyde (OPA), may be employed.

Reversed-phase HPLC, where the stationary phase is a C_8 or C_{18} hydrocarbon, may be used to analyse pre-derivatized (e.g. DNP or OPA) amino acids in less than 30 minutes, with mixtures of aqueous and organic solvents as the moving phase.

Before the results of amino acid analysis are interpreted, the type of hydrolysis used needs to be taken into consideration. Acid hydrolysis (6 M HCl, 105°C, 24 h) leads to complete breakdown of the polypeptide chain to amino acids, but some tryptophan is lost in the process. Also, the amides glutamine and asparagine are hydrolysed to the corresponding acids, liberating ammonia. Measurement of the ammonia produced gives the total amount of amide initially present.

Hydrolysis of the amides also occurs if alkali is used instead of acid. Such alkaline hydrolysis, e.g. with 5 M NaOH, again leads to total breakdown of the polypeptide. Under these conditions there is complete recovery of tryptophan, but only partial recovery of several other amino acids, including cysteine, cystine, serine and threonine.

Thus the results of both acid and alkaline hydrolysis are required to determine the total concentration of all amino acids present, and even then only the combined amide content can be assessed. With this proviso, these results, and those of the molecular weight determinations, can be used to give a reasonable estimate of the number of molecules of each amino acid in the polypeptide.

2.4.3 Determination of the amino acid sequence of each polypeptide chain

The **N-terminal amino acid** of a polypeptide may be identified by the use of **Sanger's reagent or dansyl chloride,** as described in section 2.4.1. However, an even more valuable reagent is that introduced in 1950 by Pehr Edman, since it allows extra information to be obtained.

Edman's reagent (phenylisothiocyanate) forms an addition compound with a free amino group under alkaline conditions, so will attach itself to the α-amino group of the N-terminal amino acid of a polypeptide chain, exactly like the other reagents mentioned. Its special property is that if the conditions are then made mildly acidic, the addition compound formed between the reagent and the N-terminal amino acid will become detached from the polypeptide chain without any other peptide bond being broken.

(2.7)

$$C_6H_5\text{-N=C=S} + H_2N.CHR'.CONH.CHR''.CO\text{--------}NH.CHR^{n}.CO_2H$$

phenylisothiocyanate polypeptide

$$\downarrow OH^-$$

$$C_6H_5\text{-NH.C(=S).NH.CHR'.CONH.CHR''.CO--------NH.CHR}^{n}.CO_2H$$

$$\downarrow H^+ \text{(in organic solvent)}$$

$$C_6H_5\text{-N} \langle \text{C(=O)—CHR'—NH—C(=S)} \rangle + H_2N.CHR''.CO\text{--------}NH.CHR^{n}.CO_2H$$

phenylthiohydantoin
(PTH) amino acid

The PTH amino acid, and thus the original N-terminus, can be identified by chromatography, while the rest of the polypeptide chain remains intact to be investigated further. The procedure can be repeated to reveal the new N-terminal amino acid (that with side chain R″), and so on. Protein sequencers involving separation of PTH amino acids by reversed-phase HPLC became commercially available during the 1980s (an example being the Beckman System 890M).

As an alternative to carrying out the full Edman procedure, phenylisothiocyanate may be used just to remove the N-terminal amino acid, with the new N-terminus thus created being identified by the more sensitive dansyl method.

The C-terminal amino acid may be identified by the use of a reducing agent or hydrazine (section 2.4.1), but an alternative approach employs carboxypeptidase enzymes, which cleave peptide bonds sequentially, starting from the C-terminus. The order in which particular free amino acids appear in solution after treatment gives the sequence from this end. Thus, in theory, it is possible to start at either the N-terminus or the C-terminus of a polypeptide chain and determine the entire amino acid sequence. However, in practice, the size and complexity of such chains increase the problems of analysis. For this reason, long polypeptide chains are usually split into smaller, more manageable, units by the use of specific reagents before sequence analysis is attempted. These reagents include cyanogen bromide and a variety of proteolytic enzymes (i.e. enzymes which cleave peptide bonds). Consider the following polypeptide chain:

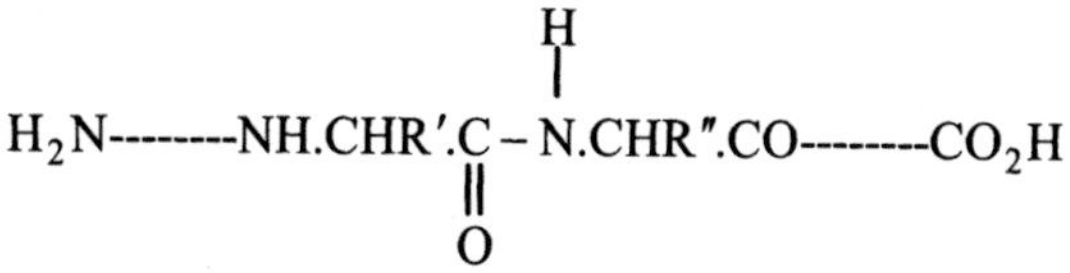

$$\begin{array}{c} \qquad\qquad\qquad\qquad\qquad\quad \mathrm{H} \\ \qquad\qquad\qquad\qquad\qquad\quad | \\ \mathrm{H_2N\text{--------}NH.CHR'.\underset{\underset{O}{\|}}{C}-N.CHR''.CO\text{--------}CO_2H} \end{array}$$

The peptide bond shown, which represents any peptide bond in the molecule, may be broken by a specific reagent if R′ or R″ has a specific identity. For example, it is hydrolysed by the enzyme **trypsin,** at pH 7-9, where R′ is a **lysine** or **arginine** side chain, or by the enzyme **chymotrypsin,** again at pH 7-9, where R′ is a **phenylalanine, tryptophan** or **tyrosine** side chain. Other proteolytic enzymes have a less clearly defined specificity (section 5.1.3). **Cyanogen bromide (CNBr)** cleaves the bond where R′ is the side chain of methionine, by the following reaction:

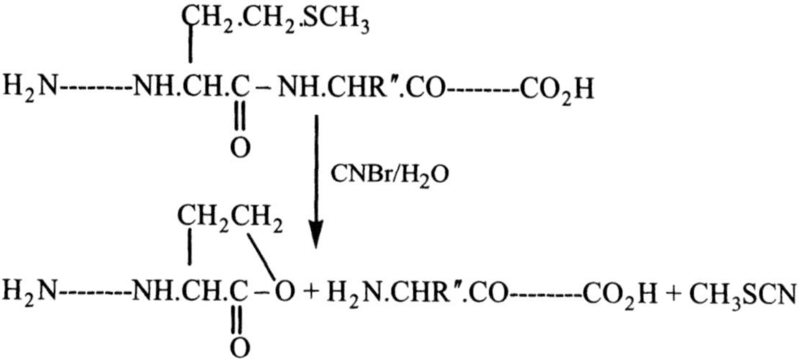

$$\begin{array}{c} \mathrm{H_2N\text{--------}NH.\overset{\overset{CH_2.CH_2.SCH_3}{|}}{CH}.\underset{\underset{O}{\|}}{C}-NH.CHR''.CO\text{--------}CO_2H} \\ \Big\downarrow \ \mathrm{CNBr/H_2O} \\ \mathrm{H_2N\text{--------}NH.\overset{\overset{CH_2CH_2}{|\quad\diagdown}}{CH}.\underset{\underset{O}{\|}}{C}-O + H_2N.CHR''.CO\text{--------}CO_2H + CH_3SCN} \end{array} \tag{2.8}$$

Thus, action by anyone of these reagents on a polypeptide chain produces a number of peptide fragments, which may be separated from each other by chromatography or electrophoresis techniques. Then, since each fragment is relatively small, it is usually possible to determine its complete amino acid sequence by use of the methods discussed above.

An alternative technique, first applied to the investigation of peptides during the 1980s and now widely used, is **mass spectrometry (MS)** (see section 21.2.1). An ion source is used to generate peptide ions in the gas-phase, and these are then separated on the basis of their mass-to-charge ratio. As conditions are chosen so that most of the peptide ions will have a charge = 1, the separation is in fact largely on the basis of mass.

Peptides consisting of up to around 25 amino acid residues may be sequenced directly by means of **tandem mass spectrometry (MS-MS)** (see Fig. 21.4). One mass spectrometer is used to select a particular peptide ion from a mixture and direct it into a collision-cell containing chemically-inert atoms such as helium, which results in limited cleavage of peptide bonds. The molecular masses of the products are then determined by the use of a second mass spectrometer. With the help of computers to analyse the results, the sequence of a peptide can be determined within a few minutes. This is much quicker than can be achieved using the Edman degradation technique. However, the amino acids leucine and isoleucine have the same molecular mass, so cannot be distinguished by mass spectrometry.

After a family of peptide fragments have been sequenced, by whatever means, the next task is to find the order in which they join together to form the complete polypeptide. Partial hydrolysis is again performed on a sample of intact polypeptide, this time using a different specific reagent. The resulting peptide fragments are separated as before and the amino acid sequence of each is determined. **Overlapping sequences** between the first and second sets of peptide fragments enable the complete primary structure to be deduced, as in the following example.

A peptide consisting of 11 amino acid residues is known to have alanine at its N-terminus. Hydrolysis by trypsin gives the following three peptide fragments (each depicted conventionally with the N-terminus to the left):

Pro-Trp-Gly-Arg   Arg-Glu-Phe-Asp-Lys   Thr-Ser

Hydrolysis by chymotrypsin also gives three fragments:

Gly-Arg-Thr-Ser   Ala-Glu-Phe   Asp-Lys-Pro-Trp

The complete sequence may be deduced by starting with the known N-terminus and looking for overlaps between the fragments, as follows:

```
Ala-Glu-Phe                                  chymotrypsin fragment
Ala-Glu-Phe-Asp-Lys                          trypsin fragment
            Asp-Lys-Pro-Trp                  chymotrypsin fragment
                    Pro-Trp-Gly-Arg          trypsin fragment
                            Gly-Arg-Thr-Ser  chymotrypsin fragment
                                    Thr-Ser  trypsin fragment
-------------------------------------------  ---------------------
Ala-Glu-Phe-Asp-Lys-Pro-Trp-Gly-Arg-Thr-Ser  complete sequence
-------------------------------------------  ---------------------
```

The amides, glutamine and asparagine, are easily hydrolysed to their respective acids and may thus be wrongly identified in the determination of amino acid sequence. The result may be checked by subjecting intact polypeptide to partial hydrolysis by proteolytic enzyme under the mildest possible conditions, in order to minimize hydrolysis of any amide present, and then quickly investigating the peptide fragments for the presence of amide.

This may be done by electrophoresis, or by performing complete hydrolysis on a fragment and determining the ammonia produced. For example, the fragment Ala-Gln-Phe would not leave the origin on electrophoresis at neutral pH, whereas Ala-Glu-Phe would move towards the anode. The first of these fragments would yield one molecule of ammonia on complete hydrolysis, but no ammonia would be produced by hydrolysis of the other.

Once the complete primary structure of a polypeptide has been determined, the amino acid residues may be numbered, always starting with the N-terminus as residue number 1 and working sequentially from this.

2.4.4 Determination of the positions of disulphide bridges

Since the amino acid sequence of a polypeptide chain is most conveniently determined if the peptide bonds of the backbone are the only covalent bonds present which link amino acid residues, the procedures discussed above usually commence with the splitting of all inter- and intra-chain disulphide bridges (see section 2.4.1). Therefore, they can give no indication as to which particular cysteine units are linked by each disulphide bridge. In order to elucidate this, it is necessary to start again with a sample of intact protein.

Partial hydrolysis is performed to break some of the peptide bonds without disturbing any of the disulphide bridges. However, under certain hydrolysis conditions, disulphide bridges may be cleaved and reformed, not necessarily reconnecting the original partners. This is minimized by using 5 M H_2SO_4 for partial hydrolysis or by using the enzyme pepsin at pH 2: pepsin breaks the peptide bonds between a wide range of amino acids, but particularly those between two hydrophobic residues.

The peptide fragments formed may be separated from each other as before, and each is then treated with performic acid. Those peptides containing a disulphide bridge will break into two smaller fragments as the –S–S– bond is oxidized, enabling the fragments to be separated and the sequence of each determined. Thus the sequence of amino acids around each end of a particular disulphide bridge is made known and, by reference to the previously worked out primary structures, the position of the bridges in the intact protein can be deduced.

2.4.5 Some results of experimental investigation of primary structure

One of the first to use the general approach described above was Fred Sanger, who in 1953 elucidated the complete primary structure of **insulin,** an achievement which was rewarded with the Nobel Prize. Insulin was found to consist of two polypeptide chains: an A chain, of 21 amino acid residues, and a B chain, of 30 residues. The chains are linked by two disulphide bridges and there is also an intra-chain disulphide bridge in the A chain (see section 3.1.4).

The first enzyme to have its complete amino acid sequence determined was bovine pancreatic **ribonuclease A**, the result of work by Derek Smyth, William Stein, Stanford Moore (1963) and others. This enzyme consists of a single polypeptide chain of 124 amino acid residues; with four intra-chain disulphide bridges being present.

In general, work on a variety of **globular proteins** has revealed that these usually incorporate all 20 amino acids without there being any recurring features in the primary structure. The amino acid sequence is absolutely specific, so that an error of synthesis resulting in one amino acid residue being out of place can affect the functioning of the protein. Let us consider the example of **haemoglobin,** an iron-containing protein which occurs in erythrocytes (red blood cells) and acts as a carrier for oxygen and carbon dioxide. Haemoglobin consists of four polypeptide chains: two α-chains (each of 141 amino acid residues) and two β-chains (each of 146 residues). Well over one hundred different abnormal structures, resulting from genetic mutations, have been described in humans. Some of these mutations are harmless, the abnormal haemoglobin molecule functioning as well as the normal one, but some affect the patient very severely indeed. In **sickle-cell anaemia,** where the erythrocytes are sickle-shaped and are broken down more easily than normal, the abnormality is at position 6 (from the N-terminus) of the β-chain of haemoglobin, where **valine** is present instead of **glutamate.** Normal haemoglobin molecules obtained from different species also show some differences in primary structure, these being fewer the more closely related the species. Haemoglobin has been investigated in more detail than most proteins because it is readily available in large amounts, but the same general conclusions have been found with other globular proteins.

Fibrous proteins, on the other hand, often contain only three or four different amino acid residues, and recurring sequences of amino acids are frequently found. For example, **fibroin,** the protein in silk, consists of only glycine, serine and alanine residues, the glycine residues occurring alternately throughout the molecule. Hence it has not yet proved possible to specify the complete primary structure for a fibrous protein.

2.4.6 Indirect determination of primary structure

Increasingly, primary structure is not being determined by the direct procedures outlined above, but deduced instead from a knowledge of the structure of the genetic material containing the information for the synthesis of the protein in question. At this stage of the book, we have not considered enough relevant factors to be able to discuss this further, but details are given in section 3.1.6.

2.5 THE DETERMINATION OF PROTEIN STRUCTURE BY X-RAY CRYSTALLOGRAPHY

2.5.1 The principles of X-ray crystallography

Crystals, including those of globular proteins, consist of repetitions of a basic structural component called a **unit cell,** which may be a single molecule or a symmetrical arrangement of several molecules. Thus each atom in a crystal must lie in a specific position with regard to all other atoms in the crystal, enabling the structure to be determined by X-ray diffraction analysis. This consists of directing a beam of X-rays of a single wavelength at a crystal and studying the characteristics of the emerging rays.

Most rays pass straight through the crystal without being affected, but those which come into contact with an atom in the crystal are scattered by the clouds of electrons surrounding it. More precisely, these electrons act as secondary sources of X-rays, which then radiate out from the atom in all directions. The intensity of the X-rays leaving an atom of high electron density, such as a heavy metal, is much greater than for those leaving an atom of low electron density, such as hydrogen. Thus, areas of high electron density can be said to scatter X-rays more strongly than areas of low electron density, but it should be realized that in each case the radiation emerges from the scattering centre with spherical symmetry. X-rays, like other forms of electromagnetic radiation, are best regarded as **waves** of characteristic **length** and **amplitude** (Fig. 2.3). The **intensity** of a ray is proportional to the square of its amplitude. If two rays of identical wavelength are directed along a common path so that they are exactly **in phase,** i.e. the crests and troughs of the waves correspond exactly, then they will combine to give a ray of the same wavelength and phase, but greater amplitude (Fig. 2.3a). The amplitude, and hence intensity, obtained under these conditions will be the maximum that can be obtained by combination of these two rays. If the two rays are one quarter of a cycle out of phase (Fig. 2.3b), the intensity of the combined ray will be about one quarter of the maximum possible value, and the phase of the combined ray will be a combination of the phases of its component rays. If the two rays are exactly half a cycle out of phase (Fig 2.3c), the waves will cancel each other out, and the intensity of the combined ray will be zero.

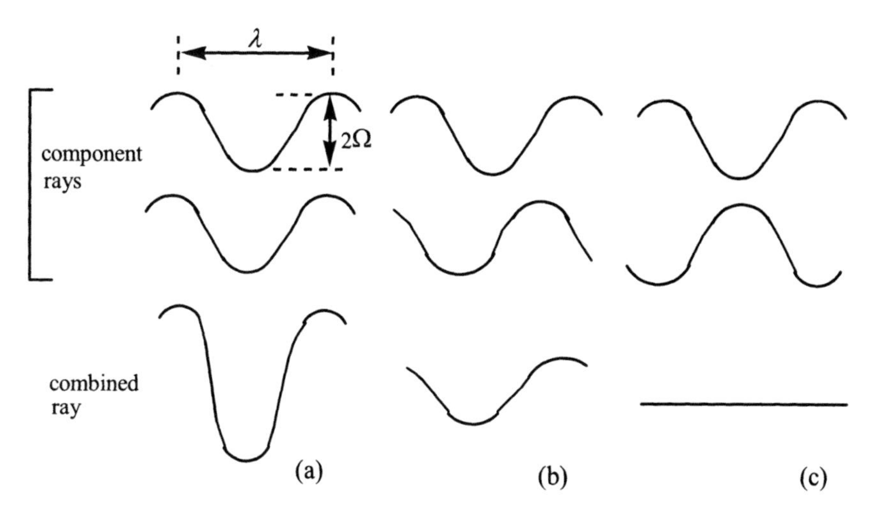

Fig. 2.3 – The combination of rays of identical wavelength (λ) and amplitude (Ω) when directed along a common path, where component rays are (a) exactly in phase, (b) one quarter of a cycle out of phase, and (c) half a cycle out of phase.

The scattered X-rays emerging from a crystal can combine in this way: rays emerging at certain angles to the incident ray will combine to give rays of maximum intensity, while those emerging at other angles will cancel each other out. The results can be observed by placing a photographic plate behind the crystal to register the impact of emerging rays.

In general, development of the plate will show a spot at the centre, due to the undeflected X-rays, which of course will all be in phase; this is surrounded by a pattern of other spots, corresponding to the angles where emerging rays combine to give intensity maxima. The overall effect is known as a **diffraction pattern.**

Before discussing X-ray crystallography in more detail, let us consider why we cannot observe the atoms in a molecule by the use of optical or electron microscopy. Vision consists of two processes: beams of light (another form of electromagnetic radiation) which strike an object are scattered by the atoms exactly as discussed above; and these scattered rays are brought back together (**focused**) by the lens in the eye to produce an image of the object on the retina. A magnified image may be produced by the use of further lenses (**optical microscopy**), enabling features to be clearly distinguished (**resolved)** which are too close to be seen separately by the unaided eye. The limit of resolution in microscopy depends on the wavelength of the type of electromagnetic radiation used and the focusing properties of the instrument. With optical microscopy, the limit of resolution is about half the wavelength of the light used. Hence individual atoms, which are separated in a molecule by distances in the order of 1–2 Å (1 Å = 0.1 nm), cannot be resolved by an optical microscope, since the wavelength of visible light is in excess of 4000 Å. **Electron microscopes** give much greater resolving power than optical microscopes, but despite the very low wavelengths of electron beams, individual atoms still cannot be visualized because of the generally poor performance of the electromagnetic lenses used in electron microscopy.

X-rays similarly have wavelengths much smaller than those of light rays. In fact they are of the same order of magnitude as inter-atomic distances. However, no procedure has yet been devised for focusing X-rays, so no image can be produced. Nevertheless, the detailed structure of a crystal scattering X-rays can be deduced from the diffraction patterns obtained.

Each unit cell in a crystal may contain many atoms in a complex arrangement, but let us for the moment consider it simply as a region of high electron density which can act as a scattering centre for X-rays. Thus the crystal consists of a regular arrangement of major scattering centres, each corresponding to a unit cell, as shown in Fig. 2.4.

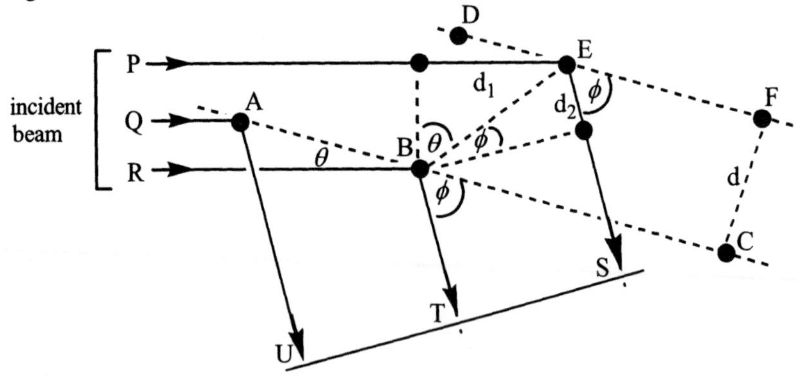

Fig. 2.4 – An incident beam of X-rays striking a crystal at an angle θ to planes of scattering centres ABC and DEF. Rays are scattered with spherical symmetry, only those emerging at an angle ϕ to the planes of scattering centres being shown (see text for discussion).

First of all let us look at the plane of scattering centres containing A, B and C, which is inclined at an angle θ to the incident beam of X-rays. Some rays will be scattered by the electron-dense regions in the plane, while most will pass straight through. Each scattered ray has the same wavelength and phase as the incident ray, so in respect of these properties can be regarded as simply being deflected. Rays will be scattered in all directions, so some will emerge at an angle ϕ to the plane. Those leaving A and B at the same angle to the incident beam will reach a given point having travelled exactly the same distance through space (so that distance QAU = distance RBT) only when $\phi = \theta$. This is known as the **reflection condition** since the phenomenon of reflection at a planar surface is also characterized by these angles being equal. Therefore, all rays **reflected** by the scattering centres on the plane ABC, i.e. all those emerging at an angle such that $\phi = \theta$, will be exactly in phase and will combine to give a ray of maximum intensity.

Now let us consider a second plane of scattering centres, containing D, E and F, which is parallel to the plane ABC and separated from it by a distance d. Rays leaving B and E in the same direction can never reach a given point having travelled the same distance through space, since distance PES must be greater than distance RBT. However, if the difference in distance ($d_1 + d_2 = d \sin \theta + d \sin \phi$) is exactly a whole number of wavelengths, the emerging waves will still be exactly in phase. Hence, waves which emerge at the same angle to the incident beam from scattering centres on different but parallel planes will combine to give a ray of maximum intensity if $n\lambda = d \sin \theta + d \sin \phi$, when n is a whole number, λ is the wavelength of the rays, and the other terms are as defined in Fig. 2.4.

So, to summarize, all the scattering centres in a single plane will combine to give a diffracted ray of maximum intensity at the angle where the reflection condition is met, while centres in different planes will combine to give rays of maximum intensity at angles where $n\lambda = d \sin \theta + d \sin \phi$. If these two conditions are put together, rays emerging from all the scattering centres on any number of parallel planes a distance d apart will combine to give an intensity maximum where $n\lambda = 2d \sin \theta$. This was first stated by William and Lawrence Bragg (father and son) in 1913 and is known as the **Bragg condition.**

Thus it can be seen that regular repeating units are essential for the establishment of clear diffraction patterns, since patterns from different scattering centres may reinforce each other under these conditions. Crystals are rotated in a beam of X-rays, allowing time in each position for the investigation of the diffraction pattern, until the pattern obtained indicates that planes of scattering centres are inclined at a suitable angle to the incident beam for reinforcement to take place. A clear diffraction pattern may be obtained without rotating the specimen if this is not a single crystal but is composed of separate regions of repeating units set at random angles to each other, and thus to the incident beam. This is often the case where the specimen is a powder of fine crystals or a natural fibre. In general, the greater the repeating distances within a specimen, the closer are the intensity maxima on the photographic plate.

If clear diffraction patterns can be obtained with a crystal in three different orientations, then the dimensions of the unit cell and the arrangement of unit cells within the crystal can be deduced from those maxima nearest the centre, which correspond to the largest repeat distances. Turning our attention to the structure of the molecule or molecules making up the unit cell, we must immediately realize that not all the atoms in a complex molecule can lie on the same plane. Hence, although we have hitherto considered a unit cell to be a single scattering centre lying on a specific plane, in fact it may consist of a large number of component scattering centres (atoms), some of which lie at various distances in front of the plane and some at various distances behind the plane. This inevitably affects the diffraction patterns obtained.

If the intensity (and hence amplitude) and the phase is known for each X-ray causing a spot in the diffraction pattern, then a three-dimensional contour map showing the distribution of electron density within the unit cell can be drawn up using a mathematical procedure called **Fourier synthesis.** From this, the structure of the molecule can be deduced. The intensity of an X-ray can be determined from a photograph of a diffraction pattern or measured directly using a Geiger-Müller counter. However, there is no direct way of determining the phase. This has been called the **phase problem.**

The phase problem can be overcome in either of two ways. A **model** may be built of a possible molecular structure and the theoretical diffraction patterns this would give are compared to those actually obtained. This may be useful in explaining certain repeating features, but otherwise the number of possible structures of a complex molecule is too immense to enable this method to be very successful if employed alone. The alternative method is that of **isomorphous replacement,** introduced by Max Perutz in 1954. A heavy metal atom, such as mercury or uranium, is attached to a specific site of each molecule in the crystal without altering the three-dimensional structure of the molecule. The heavy metal atoms, being regions of very high electron density, will cause appreciable changes to the amplitude and phase of the rays producing a diffraction pattern. If the intensities of the spots are compared to those of the spots in the original diffraction pattern, it is possible to deduce the location of the substituted atoms within the unit cells. The contribution of rays from the heavy metal atoms to each spot in the diffraction pattern may then be calculated, in terms of both phase and amplitude. This enables two possible solutions of the phase problem to be obtained for each spot, one where the phase of the original ray is in advance of that of the ray from the substituted atom, and one where it is an equal distance behind. Which of the alternative solutions is correct may be determined by substituting with a heavy metal atom in a different place, or possibly several different places. Thus it is possible to deduce the complete structure of a molecule.

Low resolution analysis (to about 5 Å), showing the main features of the molecular structure but not the fine detail, may be performed using only the spots near the centre of the diffraction pattern. For **high resolution analysis,** showing the complete structure of the molecule, all the spots must be used. Apart from the extra labour involved, which usually necessitates the use of computers, high resolution analysis is hindered by the fact that the outermost spots in a diffraction pattern are of lower intensity than those nearer the centre.

Isomorphous replacement almost inevitably introduces some changes, however slight, in the three-dimensional structure and this becomes more significant the higher the resolution attempted. Also, hydrogen atoms are extremely weak scatterers of X-rays, so are very difficult to pinpoint by these techniques. It has often been found advantageous to use the **model-building** and **isomorphous replacement** techniques to complement each other. Low resolution studies indicate the general shape of the molecule and, from this information, models can be built to help elucidate the fine structure. X-rays are commonly obtained by accelerating **electrons** released from an **incandescent tungsten filament** against a **copper** target. This produces rays of approximate wavelength 1.5 Å.

With regard to more recent developments, X-rays selected from the electromagnetic radiation emitted by highly expensive devices called **synchrotrons or electron storage rings** are of higher intensity than those from a conventional source. This is advantageous for the structural determination of proteins of high molecular weight and of those proteins whose crystalline structure is unstable over the relatively long periods required for exposure to weak radiation. Another possibility is the use of **neutron beams,** which are scattered by atomic nuclei rather than electrons. Thus they may be employed for high resolution analysis, since they are scattered strongly by hydrogen atoms. Also, they cause very little radiation damage to macromolecules, enabling irradiation to be carried out for far longer periods than is possible with X-rays.

2.5.2 Some results of X-ray crystallography

In 1939, Linus Pauling, Robert Corey and their co-workers began a systematic investigation of the three-dimensional structures of amino acids, dipeptides and other molecules to provide data with a view to the eventual elucidation of protein structure. X-ray diffraction analysis soon showed that the C–N bond length on a peptide bond was shorter than would be expected for a single covalent bond. Therefore, some degree of double bond character must be present, the actual structure being between the two extremes shown below:

$$-\overset{\overset{\displaystyle O}{\|}}{C}-\underset{\underset{\displaystyle H}{|}}{N}- \longleftrightarrow -\overset{\overset{\displaystyle O^-}{|}}{C}=\underset{\underset{\displaystyle H}{|}}{\overset{+}{N}}- \qquad (2.9)$$

A consequence of this partial double bond character is that rotation about the bond is restricted and all of the atoms involved lie in the same plane. Two isomeric arrangements are possible: the *trans* form (as shown above), with the oxygen and hydrogen atoms diametrically opposed, and the *cis* form, with these atoms adjacent. In fact, only the *trans* isomer is found, a more detailed representation of this being given in Fig. 2.5.

The most significant factor accounting for the stability of the *trans* form is the spacing between the α-carbon atom and the oxygen atom (2.8 Å), which is only marginally less than the van der Waals contact distance between these atoms (3.4 Å), so repulsion is slight (see section 2.3.2).

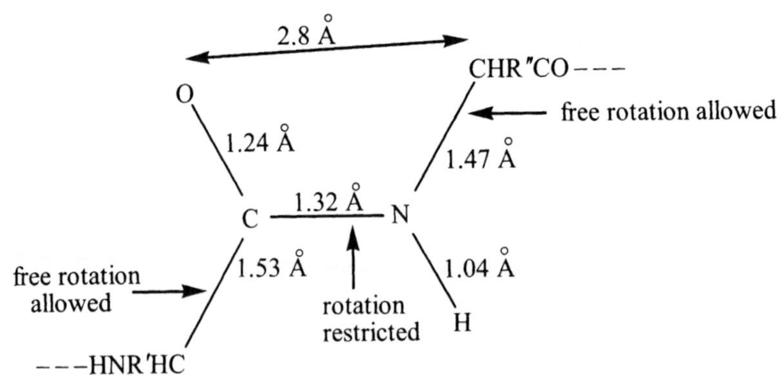

Fig. 2.5 – The dimensions of the peptide bond.

In the unstable *cis* form, the two α-carbon atoms would be adjacent to each other, separated by a distance (2.8 Å) much less than the van der Waals contact distance between two carbon atoms (4.0 Å), so repulsive forces would be great.

Pauling and Corey and their colleagues also noted that, in crystals, there is a high degree of hydrogen bonding between oxygen atoms in one peptide bond and nitrogen atoms in another. The distance between such atoms is often about 2.9 Å, much less than the van der Waals contact distance between non-bonded oxygen and nitrogen atoms, thus indicating the presence of the hydrogen bond. Furthermore, the N–H···O linkage is usually approximately linear.

On the basis of these findings they suggested various theoretical types of secondary structure which might be found in proteins. In particular, both right- and left-handed **α-helices** seemed consistent with the available data, but the arrangement of London dispersion forces was more favourable in the former structure.

William Astbury had already used X-ray scattering to demonstrate regular features in the structures of several fibrous proteins, and in some cases these were found to be consistent with the postulated right handed α-helix. The main features of the X-ray diffraction pattern of mammalian **α-keratin** (Fig. 2.6a) showed a periodicity of 5.2 Å along the axis of the fibres, which was close to the theoretical distance (5.4 Å) between each turn of the α-helix, and a periodicity of 9.7 Å at right angles to this, possibly the distance between adjacent helices.

An α-helix contains approximately 3.6 amino acid residues per turn. This results in each peptide oxygen and nitrogen atom being in a suitable position to form hydrogen bonds with the corresponding atoms in the next turn of the helix (Fig. 2.6b); these hydrogen bonds are all approximately parallel to the axis of the helix.

Other types of secondary structure have also been demonstrated. If α-keratin is stretched under humid conditions it is converted to β-keratin, which has characteristic periodicities of 3.3 Å along the axis of the fibre and 4.7 Å and 9.7 Å perpendicular to this, evidence of a more extended form of secondary structure called a **β-pleated sheet**. In this structure, hydrogen bonds can again be formed between oxygen and nitrogen atoms in different peptide bonds, but in this instance they are perpendicular to the axis of the fibre.

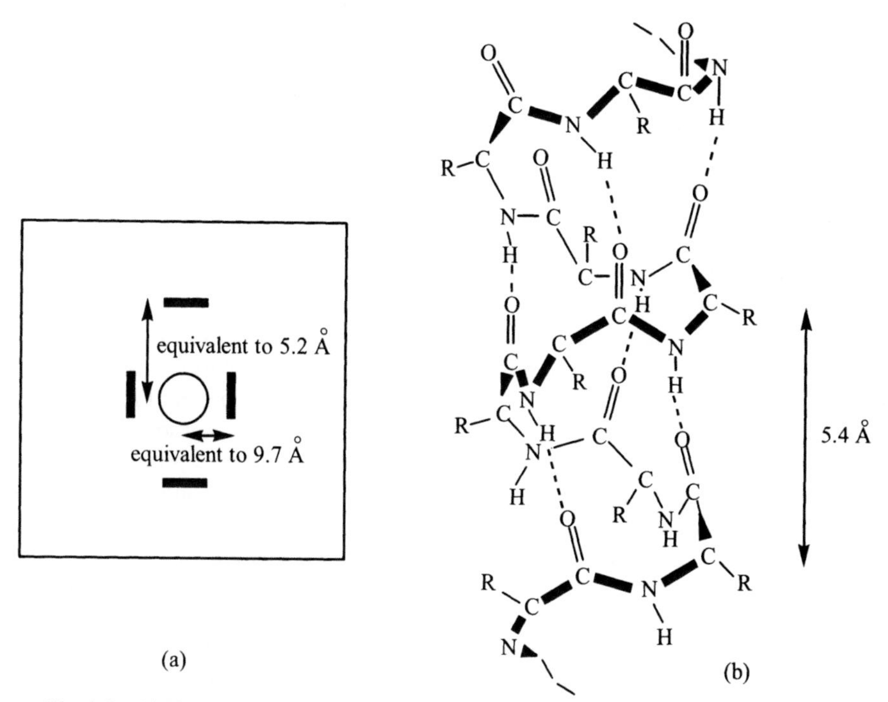

Fig. 2.6 – (a) The main features of the X-ray diffraction pattern of α-keratin, indicating a helical structure, and (b) a polypeptide chain in the form of a right-handed α-helix (N.B. the α-hydrogen atom has been omitted from the diagram to minimize congestion). The periodicity along the axis of α-keratin is almost but not quite equal to that expected in a right-handed α-helix structure

The β-pleated sheet structure is unstable in β-keratin formed by stretching mammalian α-keratin, but not in naturally-occurring bird or reptile β-keratin, or in other fibrous proteins such as silk fibroin. Keratin is said to form **parallel β-pleated sheets,** since the N-termini of adjacent polypeptide chains lie in the same direction, whereas silk fibroin forms **anti-parallel β-pleated sheets,** the N-termini of adjacent chains being in opposite directions (Fig. 2.7).

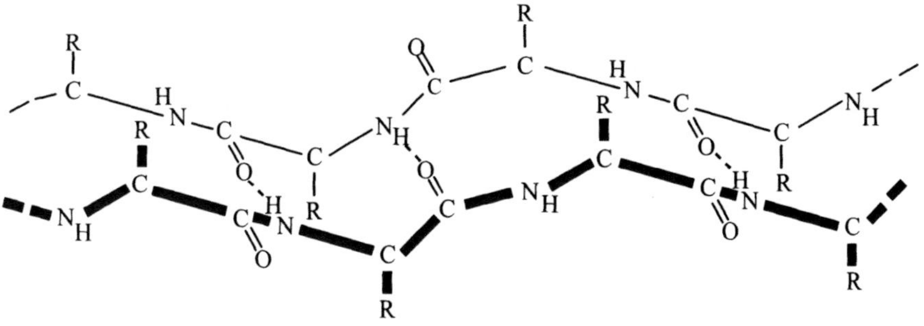

Fig. 2.7 – A section of anti-parallel β-pleated sheet (N.B. as with Fig. 2.6, the α-hydrogen atom has been omitted from the diagram).

Anti-parallel β-pleated sheets may also be formed by the doubling-back of a single polypeptide chain. Sections of polypeptide chains which provide the framework for β-pleated sheets are often termed **β-strands**.

The stability of these α-helix and β-pleated sheet structures depends on the nature of the amino acid side chains present, large or charged ones tending to be disruptive. Hence fibrous proteins, which usually consist only of amino acids with small and unchanged side chains, have well-developed secondary structures. Proline, because of the restricted rotation resulting from its ring structure, is another amino acid which cannot form part of α-helix or β-pleated sheet, but it is incorporated into the unique triple-helix structure of the fibrous protein **collagen.**

In globular proteins, the disruption of an ordered secondary structure by certain amino acid residues results in the formation of loops, giving rise to features which may be regarded as constituting a **super-secondary structure**. In particular, characteristic folding patterns called **motifs** are formed by grouping the side chains of adjacent α-helices or β-strands closely together (Fig. 2.8). These folding motifs have been found to be common in the structures of many different proteins.

For example, a calcium-binding helix-turn-helix motif (two sections of α-helices separated by a loop region) was identified by Robert Kretsinger in 1973, involving the 5th and 6th helices (termed helices E and F) of parvalbumin, and other calcium-binding proteins (e.g. troponin C and calmodulin) with an appropriate amino acid content also show this characteristic **EF hand** motif. Other examples of common protein-folding motifs include the **β-hairpin** (formed by two adjacent antiparallel β-strands) and the **Greek key** (formed in areas of four adjacent antiparallel β-strands).

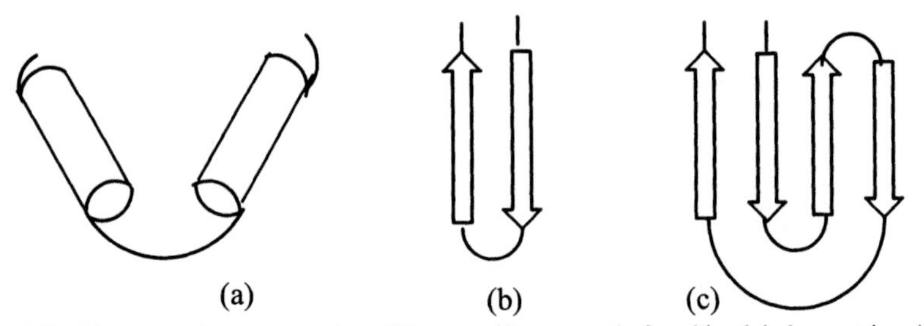

Fig. 2.8 – Diagrammatic representation of three motifs commonly found in globular proteins: (a) an EF hand; (b) a β-hairpin; and (c) a Greek key. Helices are represented by cylinders and β-strands by arrows pointing towards the C-terminus of the polypeptide chain.

Several adjacent motifs within a protein structure may fold to form a close association referred to as a globular **domain**, representing a region that has characteristic shape and function. The same domain may be present in different proteins (see Fig. 2.9), and domains have been demonstrated to be involved in the cellular location of proteins, protein-protein and protein-nucleic acid interactions, sites of catalytic activity and other important functions (see section 3.1.5).

The first globular protein to have its complete three-dimensional structure elucidated by X-ray crystallography was sperm whale **myoglobin** (in 1960). This close relative of haemoglobin is a single polypeptide chain of 153 amino acid residues. Despite its small size, over 10 000 diffraction spots had to be accurately analysed to give a resolution to 2 Å.

Domain	*Interaction and Role*	*Examples*
Pleckstrin binding (PH) (120 residues)	interacts with phosphoinositides, helping to recruit proteins to the membrane surface	Phospholipase C β,γ,δ β-Spectrin Btk, βARK
Src homology 2 (SH2) (100 residues)	interacts with phosphorylated tyrosine in signal transduction	Src, Abl PI3 kinase Phospholipase C γ
Src homology 3 (SH3) (60 residues)	interacts with proline-rich peptides in proteins, in the assembly of protein complexes	Src, CDC25 CDC24, PI3 kinase Phospholipase C γ
Nucleotide binding (NBD) (160 residues)	interacts with nucleotides, in the binding of nucleotides to proteins	ABC-transporters G-proteins
Hydrophobic (8-15 residues)	interacts with phospholipid bilayer	G-protein-coupled receptors glycophorin insulin receptors transmembrane proteins

Fig. 2.9 – Some common domains found in proteins, their length in terms of amino acid residues, their cellular role and examples of proteins which contain them.

Shortly afterwards, the structural analysis of **haemoglobin** itself was completed. Each of the four component polypeptide chains was found to have a tertiary structure almost identical to that of myoglobin. John Kendrew, for his work on myoglobin structure, and Max Perutz, for his studies on haemoglobin, were rewarded with the Nobel Prize in 1962. The structure of **lysozyme,** an enzyme from egg white consisting of 129 amino acid residues in a single chain, was given by Colin (C. C. F.) Blake, David Phillips and colleagues in 1965.

These and other studies on globular proteins have shown that a limited degree of secondary structure is usually present. An all-α domain, noted in myoglobin and haemoglobin, in which a series of α-helices, linked by small loops of polypeptide chain, fold back upon themselves to form a globular shape, has been termed the **globin fold**. Lysozyme, similarly, has about 25% of its amino acids in α-helical zones, but also has some small sections of β-pleated sheet (Fig. 2.10). Some other globular proteins have more extensive β-pleated sheet regions (see Chapter 11).

The amino acids at the positions where the secondary structure is disrupted are, as expected, those with large side chains, e.g. proline and leucine, or those with charged side chains where two or more with like charge are close together. The molecules are very compact, with space for very few water molecules within the interior. Most of the amino acids with non-polar side chains are found within the interior of the molecule, where they are unlikely to come into contact with water, while those with polar side chains are usually exposed to the solvent.

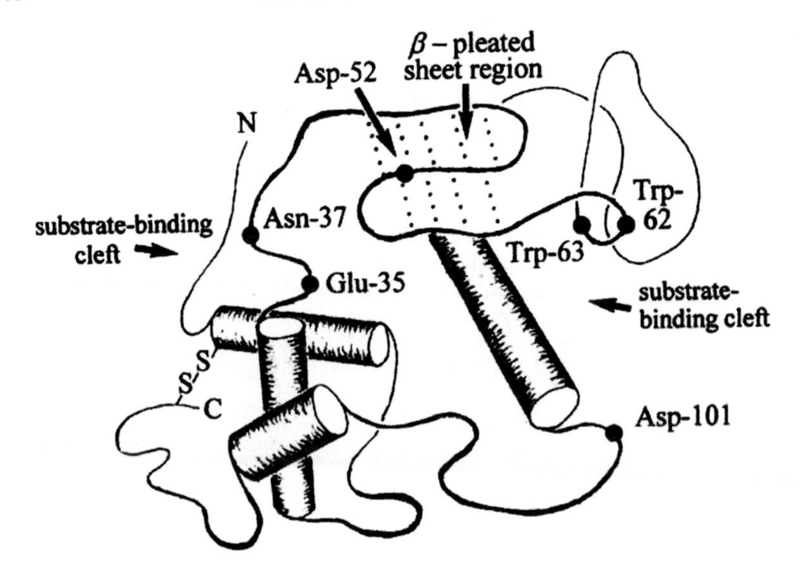

Fig. 2.10 – A simplified representation of the three-dimensional structure of egg-white lysozyme, as revealed by the X-ray diffraction studies of Phillips and colleagues (1965). Only the backbone of the polypeptide chain is shown, and α-helical regions are represented by cylinders. The amino acid side chains would fill most of the available space within a molecule, but a clearly-defined cleft for the binding of the substrate is apparent, running from one side of the molecule to the other. The positions of the N- and C-termini, and of certain important amino acid residues, are indicated.

2.6 THE INVESTIGATION OF PROTEIN STRUCTURE IN SOLUTION

X-ray diffraction analysis is not suitable for the investigation of proteins in solution, since the molecules are not fixed in a regular arrangement. However other techniques may be used to give information as to the structures of proteins in solution, in particular as to the degree of secondary structure present.

There will be differences in both the infrared and ultraviolet spectra between a polypeptide chain in an **α-helix** conformation and one existing as a **random coil** (i.e. one without regular, repeating three-dimensional features). These are due to the presence or absence of hydrogen bonding between atoms in different peptide bonds. Also, since a right handed α-helix is an asymmetric structure, there will be differences in optical rotation between a polypeptide in such a conformation and one consisting of the same amino acid residues in a random coil.

Historically, optical rotation investigations were of importance in helping to demonstrate that polypeptide chains could exist as α-helices in solution. Such a structure is most readily formed if all amino acid side chains present are small and uncharged, as is the case with polyalanine, a synthetic polypeptide consisting entirely of L-alanine residues. If all the side chains are large, as with polyisoleucine, no α-helix is formed. In the case of synthetic polypeptides with ionizable side chains, e.g. polyglutamic acid, the structure in solution varies with pH. At acid pH, the glutamic acid side chains are uncharged (see section 3.2.2) and an α-helix is formed.

However at alkaline pH the side chains all have a negative charge; these repel each other and the α-helix is disrupted, as shown by optical rotation measurements. The reverse effect is found with polylysine, whose side chains are uncharged at alkaline pH but have a positive charge at acid pH. Today, direct measurements of optical rotation are rarely carried out, as similar information can be obtained more easily by the use of **circular dichroism spectroscopy**, which measures the differential absorption of right and left circularly polarized light over a range of wavelengths.

Spectrophotometry (see section 18.1.3) may give useful information about protein structure since peptide bonds, aromatic and imidazole side chains and disulphides all give absorbance bands in the ultraviolet range which may vary according to the conformation of the protein and the micro-environment of the absorbing group. **Spectrofluorimetry** (see section 18.1.4) may also be of value, for example, in investigations of the fluorescence of tryptophan side chains. The **electron spin resonance (ESR)** technique, which detects unpaired electrons, is useful for the investigation of metal ions in enzymes. However, the technique which currently gives the best structural information about proteins in solution is **nuclear magnetic resonance (NMR)** spectrometry.

The NMR technique detects atoms with an odd number of protons in their nuclei, so giving them a magnetic moment. If a kilogauss magnetic field is applied, such nuclei will precess around it with a frequency depending on the magnetic moment (μ) and the magnitude of the applied field (H_0). If a radio frequency (around 100 MHz) field is then generated so that its magnetic vector rotates perpendicular to the kilogauss field, and the conditions are adjusted, resonance will occur when the frequency of oscillation of the field corresponds with the precession of a nuclear dipole, and this can enable the nuclei concerned to move to higher energy levels. For the single proton in the hydrogen nucleus, for example, there are two possible orientations: one, aligned with the kilogauss field, has an energy level given by $-\mu H_0$ whilst the other, aligned against the field, has an energy level of $+\mu H_0$. To move from the orientation of lower energy to the other, therefore, requires an energy input of $2\mu H_0$ and the resonance frequency (v) is given by $hv = 2\mu H_0$, where h is Planck's constant. This resonance is detected as the absorption of energy ($= 2\mu H_0$) by the proton from the radiofrequency field. In general, the **intensity** of a resonance absorption line is directly proportional to the number of nuclei in an identical environment. However, **chemical shifts** in resonance frequency for identical nuclei in different electronic environments can be detected, as can the splitting of resonance peaks into multiplets of fine structure because of the interaction of neighbouring nuclear spins. Switching off the radiofrequency field and studying the characteristics of the change back (termed **relaxation)** to the original distribution of nuclei at different energy levels can also give information about interactions between neighbouring nuclei. Amongst the nuclei which can be investigated by NMR techniques are those of ^{1}H, ^{13}C and ^{15}N. Of course, before isotopes such as ^{13}C and ^{15}N, which do not occur naturally, can be used in the investigation of protein structure, they must first be incorporated into the protein, for example by utilizing bacteria to synthesize it in a medium rich in the isotope.

All data obtained by the type of investigation outlined in this section are consistent with the assumption that three-dimensional structures found in protein crystals may also occur in solution (see section 10.2).

SUMMARY OF CHAPTER 2

Proteins consist of L-amino acid residues linked by peptide bonds. The sequence of amino acids in each polypeptide chain constitutes the primary structure of the protein. This can be determined by a systematic use of chemical and enzymic procedures, often now supplemented by mass spectrometry. Regular, repeating three-dimensional features constitute the secondary structure. This is largely uninterrupted in fibrous, structural proteins but disrupted at many points in globular, functional proteins, including enzymes. The overall three-dimensional structure of each polypeptide chain is termed the tertiary structure. Proteins may consist of one or more polypeptide chains, the complete structure being called the quaternary structure.

The three-dimensional structure of proteins in fibres and crystals may be determined by X-ray diffraction analysis. The structures found in crystals may also occur in solution, as shown by NMR spectrometry and other techniques.

FURTHER READING

Berg, J. M., Tymoczko, J. L. and Stryer, L. (2006), *Biochemistry*, 6th edn., Freeman (Chapters 2, 3).

Biemann, K. (1992), Mass spectrometry of peptides and proteins, *Annual Review of Biochemistry,* **61**, 977-1010.

Branden, C. and Tooze, J. (1998), *Introduction to Protein Structure,* 2nd edn., Garland.

Chadarevian, S. de (1999), Protein sequencing and the making of molecular genetics, *Trends in Biochemical Sciences,* **24**, 203-207.

Clore, G. M. and Gronenborn, A. M. (eds.) (1993), *NMR of Proteins,* Macmillan.

Clore, G. M. and Gronenborn, A. M. (1998), Determining the structures of large proteins and protein complexes by NMR, *Trends in Biochemical Sciences,* **16**, 22-34.

Creighton, E. (ed.) (1997), *Protein Structure -A Practical Approach,* 2nd edn., IRL.

Drenth, J. (1999), *Principles of Protein X-Ray Crystallography,* Springer.

Fersht, A. (1999), *Structure and Mechanism in Protein Science,* Freeman (Chapt. 1).

Johnson L. N. and Petsko, G. A. (1999), David Phillips and the origin of structural enzymology, *Trends in Biochemical Sciences,* **24**, 287-289.

Methods in Enzymology, **47** (1977), **48, 49** (1978), **61** (1979), **91** (1983), **130, 131** (1986): Enzyme structure; **114, 115** (1985): Diffraction methods; **176, 177** (1989), **239** (1994), **338, 339** (2001), **394** (2005): Nuclear magnetic resonance; **276, 277** (1997), **368, 374** (2003): Macromolecular crystallography; **402** (2005), Biological mass spectrometry; Academic Press.

Nelson, D. L. and Cox, M. M. (2004), *Lehninger Principles of Biochemistry,* 4th edn., Worth (Chapters 3, 4).

Price, N. C. and Stevens, L. (1999), *Fundamentals of* Enzymology, 3rd edn., Oxford University Press (Chapter 3).

Rattle, H. (1995), *An NMR Primer for Life Sciences*, Partnership Press.

Roberts, G. K. C. (1993), *NMR of Macromolecules - A Practical Approach,* IRL.
Tugarinov, V., Hwang, P. M. and Kay, L. E. (2004), Nuclear magnetic resonance spectroscopy of high-molecular-weight proteins, *Annual Review of Biochemistry*, **73**, 107-146.
Voet, D. and Voet, J. G. (2004), *Biochemistry,* 3rd edn., Wiley (Chapters 4, 7, 8).
Wilson, K. and Walker, J. (2000), *Principles and Techniques of Practical Biochemistry,* 5th edn., Cambridge University Press (Chapter 6).

PROBLEMS

2.1 A peptide consisting of 18 amino acid residues, with glycine at the N-terminus, gave the following four peptide fragments on digestion with trypsin:

Ser-Phe-Val-Leu-Lys; Ala-Phe-Ser-Lys; Gly-Ala-Thr-Arg; and Ile-Trp-Glu-Thr-Ser.

Hydrolysis of the intact peptide with chymotrypsin gave the following four fragments:

Ser-Lys-Ile-Trp; Gly-Ala-Thr-Arg-Ser-Phe; Glu-Thr-Ser; and Val-Leu-Lys-Ala-Phe.

On fresh hydrolysis of the peptide with chymotrypsin, the fragment initially identified as Glu-Thr-Ser remained at the origin on electrophoresis at neutral pH.
Deduce the primary structure of the peptide.

2.2 A series of investigations were carried out on two peptide chains originally linked by one or more disulphide bridges. The N-termini of the peptides were found to be valine and proline.
The disulphide bridges were cleaved by performic acid, two cysteic acid (Cys.SO_3H) residues being formed from each cystine residue initially present. Subsequent hydrolysis with trypsin gave the following five fragments:

Val-Ser-Lys   Pro-Arg   Glu-Trp-Cys.SO_3H-Gly
Cys.SO_3H-Leu-Tyr-Cys.SO_3H-Arg   Gly-Cys.SO_3H-Phe-Ile-Ala

Hydrolysis of the original peptides with pepsin, following treatment with performic acid, also gave five fragments:

Val-Ser-Lys-Cys.SO_3H-Leu   Ile-Ala   Trp-Cys.SO_3H-Gly
Pro-Arg-Gly-Cys.SO_3H-Phe   Tyr-Cys.SO_3H-Arg-Glu

A sample of the original specimen was hydrolysed with pepsin without first treating it with performic acid. The fragments produced were then separated from each other. One of these fragments, on subsequent treatment with performic acid, split further to give the following:

Val-Ser-Lys-Cys.SO_3H-Leu   Trp-Cys.SO_3H-Gly

What can be deduced from these results?

3

The Biosynthesis and Properties of Proteins

3.1 THE BIOSYNTHESIS OF PROTEINS

3.1.1 The central dogma of molecular genetics

As we have seen in section 2.4.5, the amino acid sequence of each globular protein in a given species is absolutely specific. This implies that the biosynthesis of proteins is under genetic control and hence connected with **nucleic acid** structure and function.

Nucleic acids are macromolecules composed of sequences of **mononucleotides**. Each mononucleotide consists of a **nitrogenous base** (either a two-ringed **purine** or a single-ringed **pyrimidine**), a **pentose sugar** and a **phosphate** group. The sugar-phosphate component forms the recurring unit in the backbone of a polynucleotide (or nucleic acid) chain:

```
 base                   base                   base
  |                      |                      |
–pentose – phosphate – pentose – phosphate – pentose – phosphate –
```

In a molecule of **ribonucleic acid (RNA)**, the pentose is always **ribose**; in **deoxyribonucleic acid (DNA)** it is **2′-deoxyribose**. (The number given to an atom in the pentose unit of a nucleotide is conventionally followed by an oblique dash, e.g. 2′, to distinguish it from the number of an atom in the nitrogenous base.) The purine bases **adenine** and **guanine**, together with the pyrimidine bases **cystosine** and **uracil** (Fig. 3.1), are found in RNA. A small number of methyl-derivatives of these compounds may also be present. In DNA, only the purine bases **adenine** and **guanine** and the pyrimidine bases **cytosine** and **thymine** (Fig. 3.1) are present.

Covalent glycosidic bonds link the C1′ atom of each pentose with either the N9 atom of a purine or the N1 atom of a pyrimidine, the base-pentose unit being termed a **nucleoside**.

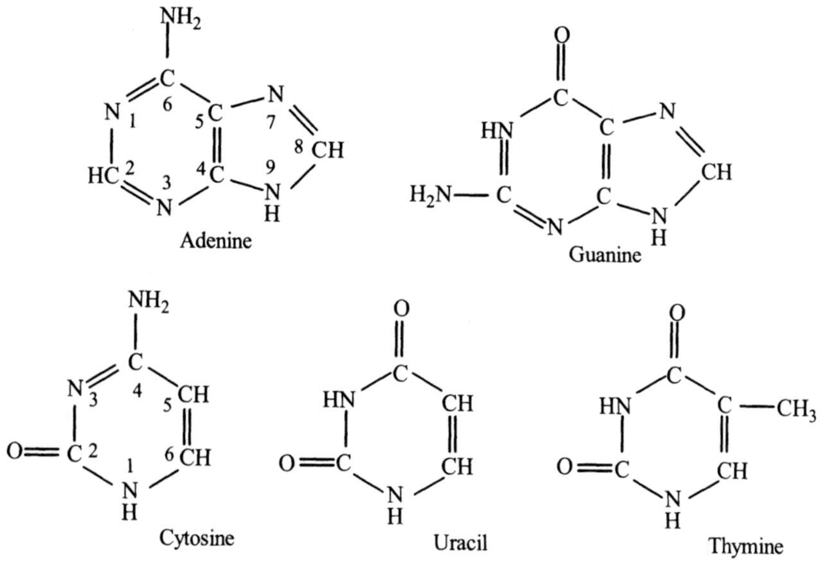

Fig. 3.1 – The structures of purine and pyrimidine bases found in nucleic acids.

The common nucleosides forming part of RNA are adenosine, guanosine, cytidine and uridine, and of DNA are deoxyadenosine, deoxyguanosine, deoxycytidine and deoxythymidine, the names being derived from those of the bases present. Thus mononucleotides may be regarded as nucleoside-phosphates and named accordingly, e.g. adenosine 5′-monophosphate (AMP). The structures of the polynucleotide chains found in RNA and DNA are indicated in Fig. 3.2.

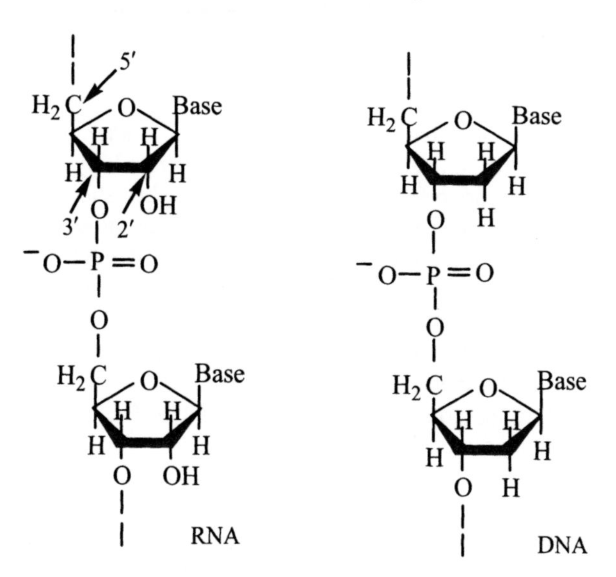

Fig.3.2 – The structures of the polynucleotide chains in RNA and DNA.

It can be seen that the linkages of the phosphodiester backbone involve atoms C3′ and C5′. The only C3′ atom not involved in such a linkage is at one end of the chain (termed the 3′ end) and the only free 5′ atom is at the other end (the 5′ end).

It was clearly established by about 1950 that DNA is a store of inherited information, since the total amount and the base composition of DNA in a cell are constant and are not affected by environmental factors. In **eukaryotic cells,** e.g. cells of higher animals and plants, DNA is located in the nucleus where, in association with protein, it forms the chromosomes (see Fig. 14.3). **Prokaryotic cells,** e.g. simple micro-organisms, are much smaller than eukaryotic cells and possess no membranous organelles or nucleus, the single molecule of DNA, largely free from protein, being tightly coiled in a nuclear zone.

The **central dogma of molecular genetics,** formulated by Francis Crick in 1953, is that the genetic information in a cell is the sequence of bases in DNA. The DNA in a cell can **replicate** itself exactly, prior to cell division, so that each daughter cell has the identical genetic information to the parent cell. RNA is synthesized using DNA as a template, so the information contained in the base sequence of DNA is **transcribed** into the base sequence of RNA. Three different types of RNA molecule are known: messenger RNA (mRNA), transfer RNA (tRNA) and ribosomal RNA (rRNA), the last-mentioned being associated with protein in the structures called **ribosomes,** which are found in the cytoplasm of cells. Protein biosynthesis involves all three types of RNA, including mRNA, whose base sequence is **translated** into the amino acid sequence of the polypeptide chain. Thus, the flow of genetic information towards its expression as protein structure, according to the central dogma of molecular genetics, may be summarized as follows:

$$\overset{\text{replication}}{\circlearrowleft}\ \text{DNA} \xrightarrow{\text{transcription}} \text{RNA} \xrightarrow{\text{translation}} \text{protein} \qquad (3.1)$$

3.1.2 The double-helix structure of DNA

In 1953 Francis Crick and James Watson proposed a structure for the **DNA** molecule consistent with experimental evidence obtained by Erwin Chargaff, Rosalind Franklin and Maurice Wilkins. Chargaff (1950) had shown that the amount of adenine (A) present in a DNA molecule was exactly the same as that of thymine (T), while the guanine (G) content was exactly equal to the cytosine (C) content; Franklin and Wilkins (1952), in X-ray diffraction studies on pure DNA fibres, had demonstrated a 3.4 Å periodicity along the axis. According to Crick and Watson, these features could be explained if the DNA molecule consists of two antiparallel polynucleotide chains arranged in a double-helix about a common axis: the hydrophilic sugar-phosphate backbones form the outer surface of the molecule, the hydrophobic bases being directed towards the inside.

The adenine bases in each chain are always paired with thymine bases in the other chain, guanine and cytosine bases being similarly paired (Fig. 3.3a). The distance between the backbones is just sufficient to accommodate a purine-pyrimidine unit linked by hydrogen bonds, only A–T and G–C pairs having suitable complementary groups to accomplish this (Fig. 3.3b).

Pyrimidine molecules are planar and purines approximately so, and the hydrogen-bonded base-pair arrangement is such that their planes are perpendicular to the axis of the double helix and occur at a repeat distance of 3.4 Å along it.

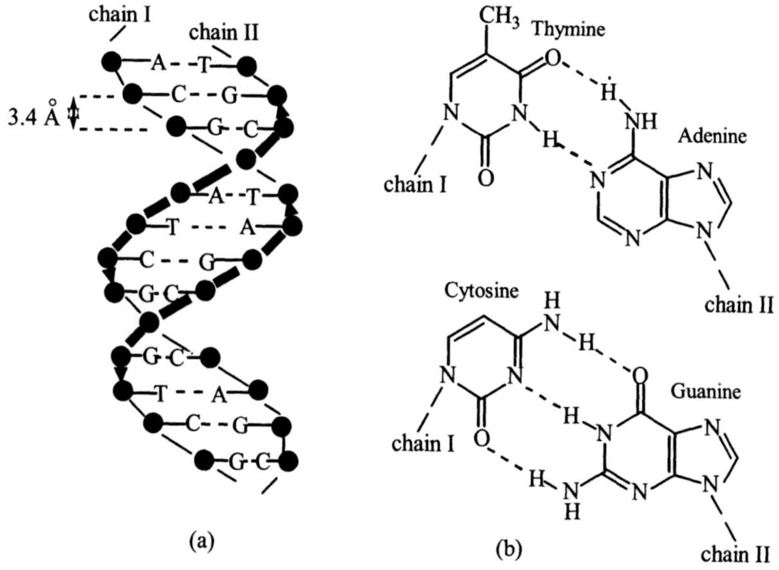

chain I
chain II
3.4 Å
A - - T
C - - G
G - C
A - - T
T - - - A
C - - G
G - C
G - C
T - - A
C - - G
G - C
(a)

CH_3 Thymine
Adenine
chain I
chain II
Cytosine
Guanine
chain I
chain II
(b)

Fig. 3.3 – (a) The double-helix structure of DNA, proposed by Crick and Watson. The structure is stabilized by hydrogen bonding between complementary purine and pyrimidine bases in the two helices. Hydrogen bonds can be formed as shown in (b).

This model quickly gained general acceptance and Crick, Watson and Wilkins were awarded the Nobel Prize in 1962.

The central dogma of molecular genetics (section 3.1.1) can easily be explained in terms of the double-helix structure of DNA. In DNA replication, each strand of double-helix acts as a template, an entire new complementary strand being synthesized alongside. Thus, two identical DNA molecules result from the molecule initially present, each possessing in its double-helix one of the original strands and one newly synthesized strand. In transcription, a section of one of the DNA strands acts as a template for the synthesis of a complementary strand of RNA, which is then released from the template. The synthesis of RNA is sequential, one nucleotide at a time adding to the 5′ end, the process being catalysed by **RNA polymerase**.

Although this was (and still remains) entirely consistent with the evidence, it soon emerged that the transcription process is far from simple, particularly in eukaryotic cells. Here (as shown by Robert Roeder and William Rutter) there are three separate RNA polymerases, termed I, II and III, each responsible for the synthesis of different types of RNA (mRNA being produced by the action of RNA polymerase II). For his role in elucidating the detailed mechanism of transcription in the yeast, *Saccharomyces cerevisiae*, where Mg^{2+} ions appear to help bind nucleotides to RNA polymerase II, and different stages of the transcription process are marked by conformational changes in the enzyme, Roger Kornberg was awarded a Nobel Prize in 2006.

Translation, which is of particular relevance to the subject of this book, is discussed in the next section.

3.1.3 The translation of genetic information into protein structure

Before amino acids can be incorporated into a protein, they must first be activated by linkage to a molecule of tRNA.

$$\text{amino acid} + \text{tRNA} + \text{ATP} \rightleftharpoons \text{aminoacyl–tRNA} + \text{AMP} + (\text{PP})_i \quad (3.2)$$

The reaction is catalysed by an **aminoacyl-tRNA synthase** and proceeds via an enzyme-bound AMP-amino acid complex. These enzymes are highly specific, governing which amino acids can be activated (only L-amino acids) and to which tRNA molecule they can be linked. As a result of this, each tRNA molecule is specific for a particular amino acid.

The base sequences of several tRNA molecules were elucidated by Robert Holley (1965) and others, using different phosphodiesterases to remove residues from the 3′ and 5′ ends of the RNA, and identifying them by laborious procedures involving chromatography and electrophoresis. X-ray crystallography studies were also carried out, the results showing that tRNA molecules have a characteristic structure maintained by hydrogen bonding between complementary bases (A···U, G···C) at certain points (Fig. 3.4). All molecules have a –C–C–A sequence at the 3′ terminus, which is added in the cytoplasm after transport of the rest of the molecule from the nucleus, where it is synthesized.

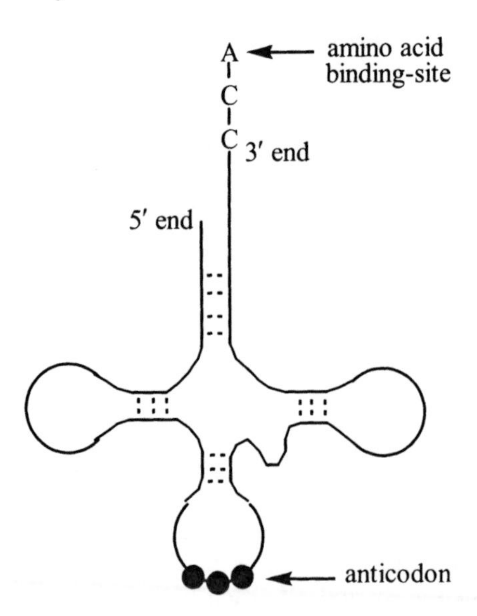

Fig. 3.4 – A simplified, two-dimensional representation of tRNA.

The amino acid binds to the terminal 3′–OH of the tRNA by an ester linkage. Diametrically opposite this, in an exposed position, is a sequence of three bases termed the **anticodon.**

Each molecule of tRNA has a unique anticodon and, as we have seen, will bind only one particular amino acid. Here then is the clue to the **genetic code,** the link between the base sequence of nucleic acids and the amino acid sequence of proteins.

The three bases constituting the anticodon are complementary to a triplet of bases, called a **codon,** in mRNA. The structure of mRNA, governed by the structure of a section of DNA, consists of a sequence of such codons and thus specifies the amino acid sequence of a polypeptide chain. By adding synthetic polyribonucleotides of known composition to cell-free protein-synthesizing systems from *E. coli* and investigating the structure of the polypeptide chains synthesized, Marshall Nirenberg, Gobind Khorana and Severo Ochoa (1961-1966) were able to decipher the entire genetic code (Fig. 3.5). This is very much the same for all known forms of life, only extremely minor variations having been discovered.

First position	Second position				Third position
(5′ end)	U	C	A	G	(3′ end)
U	Phe	Ser	Tyr	Cys	U
	Phe	Ser	Tyr	Cys	C
	Leu	Ser	Terminate	Terminate	A
	Leu	Ser	Terminate	Trp	G
C	Leu	Pro	His	Arg	U
	Leu	Pro	His	Arg	C
	Leu	Pro	Gln	Arg	A
	Leu	Pro	Gln	Arg	G
A	Ile	Thr	Asn	Ser	U
	Ile	Thr	Asn	Ser	C
	Ile	Thr	Lys	Arg	A
	fMet/Met	Thr	Lys	Arg	G
G	Val	Ala	Asp	Gly	U
	Val	Ala	Asp	Gly	C
	Val	Ala	Glu	Gly	A
	Val	Ala	Glu	Gly	G

Fig. 3.5 – The genetic code

Protein synthesis is **initiated** by the presence of mRNA, whose initiating codon binds to a **peptidyl binding site (P-site)** of a ribosome sub-unit in the presence of various factors, stimulating assembly of the intact ribosome.

The initiating codon is AUG or, less commonly, GUG. In prokaryotic cells, where protein synthesis has been investigated in the greatest detail, these code for *N*-formylmethionine (fMet) at the start of a sequence but for methionine and valine respectively elsewhere, different tRNA molecules being involved. In the cytoplasm of eukaryotic cells the initial amino acid is methionine (without the formyl group), which again is not carried by the same tRNA molecule as other methionine residues in the chain.

In prokaryotic cells, initiation results in the P-site being occupied by *N*-formylmethionine-tRNA, whose anticodon pairs with the initiating codon of the mRNA. The overall process is accompanied by the hydrolysis of GTP to GDP and P_i, which provides the energy required. Although the code is always read in the $5' \rightarrow 3'$ direction, the initiating codon is not necessarily at the $5'$ terminus of the mRNA. The intervening bases possibly act as a further component of the initiation process.

When this process is complete, the next codon after the initiating one (reading towards the $3'$ end) lies in close proximity to an **aminoacyl binding site (A-site)** in the ribosome. The **elongation cycle** commences when the appropriate aminoacyl-tRNA enters this site, its anticodon pairing with the codon of the mRNA (Fig. 3.6a); this process also is linked to the hydrolysis of a molecule of GTP. A peptide bond is then formed between the α-carboxyl group of the amino acid in the P-site and the α-amino group of that in the A-site, the dipeptide formed being held by the tRNA in the A-site (Fig. 3.6b). This reaction is catalysed by a **peptidyl transferase** enzyme, which is an integral part of the ribosome. Various elongation factors are also required.

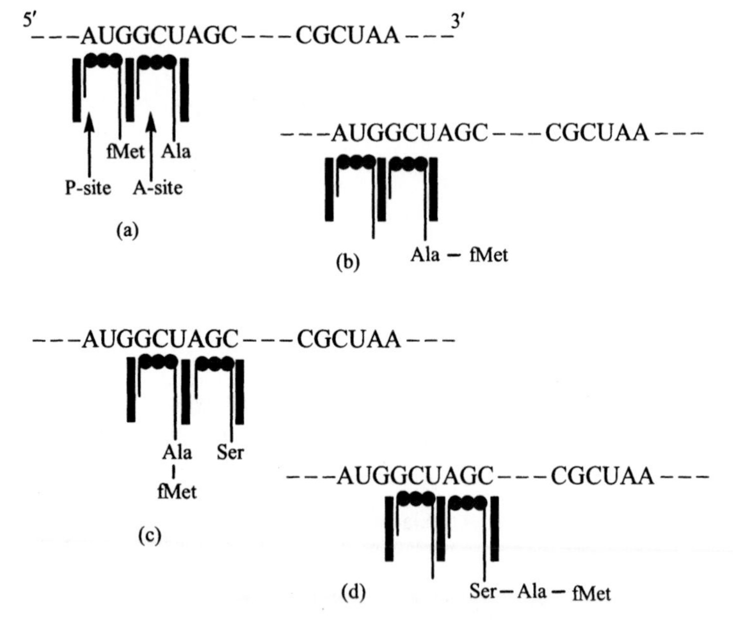

Fig. 3.6 – Stages during the synthesis of a polypeptide chain at a prokaryotic ribosome (see text for details). In the cytoplasm of eukaryotic cells, the initiating codon codes for Met rather than fMet. In either case, removal of the N-terminal residues by deformylase and/or aminopeptidase enzymes after translation would result in a polypeptide chain of outline structure Ala-Ser----Arg.

Translocation then takes place, the uncharged tRNA leaving the A-site and being replaced by the entire contents of the P-site, the mRNA also moving to maintain its pairing with the tRNA anticodon. Once again, energy for this process is provided by the hydrolysis of GTP. Translocation brings a new codon into association with the vacated A-site, which is then filled by the appropriate aminoacyl-tRNA (Fig. 3.6c). A peptide bond is formed as before, the growing peptide chain being held by the tRNA in the A-site (Fig. 3.6d) prior to a further translocation step.

This cyclic sequence of events continues, an appropriate amino acid being added to the peptide chain for each codon in the mRNA. During each turn of the cycle, the peptidyl-tRNA waits in the P-site for the next aminoacyl-tRNA to fill the A-site, this being the basis for the naming of these sites.

Eventually a **termination** codon is reached, when release factors cause the mRNA and the complete polypeptide chain to leave the ribosome. The last amino acid to be added to the peptide chain is the C-terminus. When no longer associated with a mRNA molecule, each ribosome breaks down into its sub-units. However, as long as the mRNA remains intact, it can recombine with a ribosome sub-unit and initiate the process once again.

3.1.4 Modification of protein structure after translation

Each polypeptide chain synthesized in bacteria by translation of the message carried by mRNA has *N*-formylmethionine as the N-terminus. The formyl group does not appear in the final protein, being removed by the action of a **deformylase.** The resulting methionine residue may also be quickly removed, by an **aminopeptidase,** as may several other amino acid residues near the N-terminus. Even more modifications, e.g. the attachment of prosthetic groups, may take place before the polypeptide chain folds to take up its correct tertiary structure. It is likely that generally similar processes take place in all organisms, including eukaryotic ones, although the details may differ.

The section of DNA which carries the information for the synthesis of a single protein is called a **gene**; that for the synthesis of a single polypeptide chain may be termed a **cistron.** Eukaryotic genes include untranslated sequences **(introns)** interspersed with expressed sequences **(exons)**, the RNA complementary to introns being discarded at the transcription stage, when that complementary to the exons is linked together **(spliced)**. Thus, in eukaryotic cells, mRNA is formed by the removal of certain sections from a longer strand, termed **heteronuclear (hn) RNA**. The discovery, by Richard Roberts and Phillip Sharp in 1977, that eukaryotic genes consist of introns and exons, was eventually rewarded by a Nobel Prize in 1993.

It might be assumed that, if a protein consists of more than one polypeptide chain, more than one cistron is responsible for its synthesis. However, this is not necessarily so, particularly where the polypeptide chains are all identical. Even if they are not, it is possible that the protein could be synthesized originally as a single polypeptide chain and subsequently cleaved in one or more places. For example, the hormone **insulin** is synthesized as the inactive polypeptide **preproinsulin,** which is converted to **proinsulin** by removal of 24 N-terminal amino acid residues, and subsequently activated by further enzymic cleavage. The final stage of the process may be represented diagrammatically as shown in Fig. 3.7.

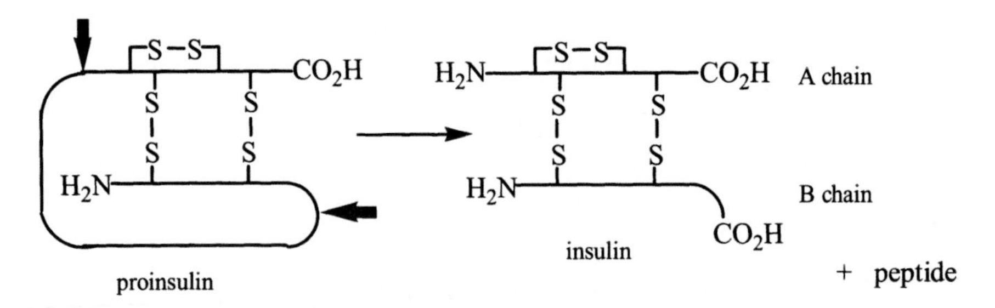

Fig. 3.7 – Diagrammatic representation of the enzymic cleavage of proinsulin (points of cleavage indicated by thick arrows). The A chain of insulin is 21 residues in length, the B chain 30 residues, and the discarded peptide 30 residues.

Some proteolytic enzymes, e.g. **chymotrypsin** and **trypsin**, may similarly be synthesized as inactive polypeptides, the active enzyme being produced by cleavage of peptide bonds (see section 5.1.2).

When a protein has to be transported to another part of the cell after translation, it is identified to the appropriate transport system by a **signal** or **leader sequence** of predominantly hydrophobic amino acid residues, which is removed at a specific point during the transport process. One such signal sequence is the 23-residue section of preproinsulin referred to above (excluding the N-terminal methionine residue). Preproinsulin is synthesized as an unbridged polypeptide at a ribosome attached to a **docking protein** on the surface of the endoplasmic reticulum (see Fig. 14.3) in pancreatic β-cells, and as it passes through a protein pore (termed a **translocon**) into the lumen of the endoplasmic reticulum for transport, the signal peptide is removed by a membrane-bound peptidase. The proinsulin molecule thus created then folds, enabling the cross-linking disulphide bridges to form.

3.1.5 Control of protein synthesis

Various mechanisms are present within the living cells to prevent the unnecessary synthesis of protein. Molecules of mRNA have a relatively short life, being broken down by ribonucleases. Therefore, a continuing requirement for the protein in question necessitates further synthesis of mRNA. On the other hand, if there is no further requirement for the protein, **transcriptional control** may prevent RNA synthesis.

The first control system to be investigated in detail was that of the **lactose operon** (*lac* operon) of *Escherichia coli.* This micro-organism can utilize lactose as its sole source of glucose, the enzymes **β-galactosidase, galactoside permease** and **galactoside transacetylase** (also known as galactoside *o*-acetyltransferase) being produced in large amounts. Galactoside permease is required for the transport of lactose across the bacterial cell membrane, while β-galactosidase catalyses its subsequent hydrolysis to glucose and galactose. The precise physiological function of the third enzyme, galactoside transacetylase, is not known. If the organism is grown in the presence of the source material, lactose, synthesis of these enzymes may be **induced**; if grown in presence of the catabolic end-product, glucose, synthesis is **repressed**.

In 1961 Francois Jacob and Jacques Monod, after a study of mutants unable to produce one or more of these enzymes, came to the conclusion that all three are normally synthesized together under the control of a **regulator gene** (*i*) and an **operator site** (*o*); these are both involved in the induction process. Catabolite repression is achieved at a promoter site (*p*), which otherwise stimulates the binding of RNA polymerase to the DNA template in readiness for transcription to take place. The *i, p* and *o* sites are adjacent to the actual structural genes for the synthesis of β-galactosidase (*z*), permease (*y*)) and transacetylase (*a*), the whole unit being termed the **lactose operon** (Fig. 3.8). Later work, especially that of Walter Gilbert and Beno Müller-Hill, confirmed this model and enabled many of the details of the mechanism to be elucidated. The promoter is the site of binding of a complex formed between 3′, 5′-cyclic AMP (cAMP) and a **catabolite gene activator protein** (*CAP*), which stimulates the binding of RNA polymerase to the adjacent *o* site in this and other catabolite gene systems in *E. coli.* The synthesis of the *CAP* is the responsibility of yet another gene. Glucose inhibits the synthesis of cAMP so, in the presence of glucose, the cAMP-*CAP* complex is not formed, and RNA polymerase cannot be bound.

The *i* gene is responsible for the synthesis of a **repressor protein** which binds tightly to the *o* site, thus preventing binding of RNA polymerase, and hence transcription of the *z, y* and *a* genes. This repressor protein has four sub-units, each with a binding site for allolactose, a metabolite of lactose. The presence of lactose results in the formation of some allolactose, which becomes attached to the repressor protein and prevents it from binding to the operator. For transcription to take place, the cAMP-*CAP* complex must bind to the *p* site, but the repressor protein must *not* bind to the *o* site.

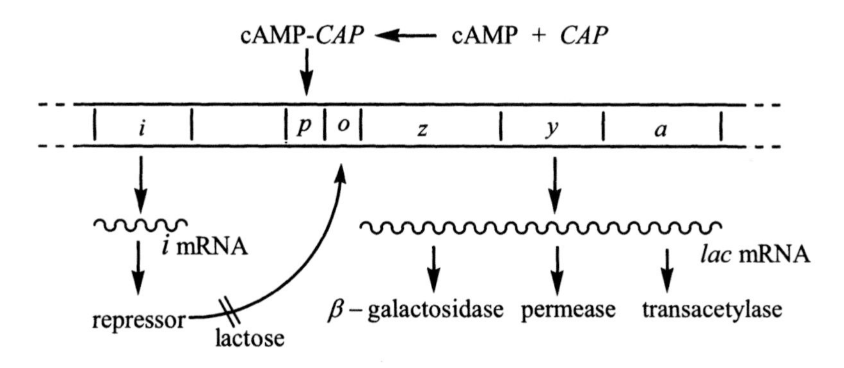

Fig. 3.8 – The lactose operon of *E. coli.*

In the presence of both lactose and glucose, repressor protein binding does not take place but neither does promoter activation. In the absence of both lactose and glucose, the promoter is activated but the operator is repressed. In the absence of lactose and the presence of glucose, both of the negative influences are at work. In none of these situations does transcription of the lactose operon occur: it takes place only when the organism really requires the enzymes in question, i.e. when lactose is present and glucose is not otherwise available. Jacob and Monod were awarded the Nobel Prize in 1965, partly for their work in the elucidation of the lactose operon.

A similar control mechanism is found with the **tryptophan operon** of *E. coli,* the enzymes for tryptophan synthesis not being produced in the presence of this amino acid. In this example, the metabolic end-product, tryptophan, acts as a **co-repressor**: only a tryptophan-repressor complex can bind to the operator to prevent transcription.

In general, transcriptional control of protein synthesis is very important in micro-organisms, but some degree of **translational control,** i.e. control of the rate of the translation process, may also be present. For example, it seems that some ribosomal proteins can prevent the synthesis of certain other ribosomal proteins by action at the translational level.

A similar situation exists in eukaryotic cells. Thus, in the yeast *Saccharomyces cerevisiae,* the galactokinase, galactose-1-P uridylyltransferase and UDP-glucose 4-epimerase enzymes essential for galactose utilization have genes (*GAL1, GAL7* and *GAL10* respectively) located on the same chromosome, although they are not transcribed into the same RNA molecule, each having its own binding site for RNA polymerase. The synthesis of these enzymes is induced by galactose, which apparently binds to a *GAL80* protein, preventing it binding to a *GAL4* protein, as shown by investigations during the 1980s. In the absence of galactose, it seems that the *GAL4-GAL80* protein complex binds to the chromosome and prevents transcription of the *GAL1, GAL7* and *GAL10* genes.

Regardless of whether galactose is present, synthesis of the enzymes is repressed in the presence of glucose by a mechanism which does not involve cAMP, in contrast to the situation in *E. coli.* Another difference is that, in general, yeast regulatory proteins can bind a considerable distance away from the genes they influence.

Even more differences from prokaryotic systems are found in multicellular eukaryotes. The cells of multicellular organisms interact with each other to a considerable extent and, unlike unicellular organisms, do not generally have a dependence on nutrients from the surrounding medium. Hence the induction of enzyme synthesis in response to the presence of a particular exogenous nutrient is not a common feature in complex eukaryotes. All cells in the same organism are the carriers of identical genetic information, but only a very small proportion of this is expressed, or indeed can be expressed, at any given time. The identities of the genes expressed vary according to the location and function of a cell, and also change as the organism matures. Thus, regulation of gene expression is important in **differentiation,** i.e. in determining which types of protein may or may not be synthesized by cells of a particular kind, and in **development.** Most genes in any cell type are located in highly condensed **heterochromatic** regions of chromosomes and these remain inactive. Only genes in expanded **euchromatic** regions, which vary according to cell type, can be expressed, and the actual expression of these genes is subject to further control. Such regulation generally takes place at the transcriptional level, by the binding of specific regulator proteins to the chromosome, either at **promoters,** adjacent to binding sites for RNA polymerase, or at **enhancers,** located elsewhere on the chromosome. The mechanism of action of regulation at enhancers may involve chromosomal folding, to bring the regulatory protein into the vicinity of the appropriate gene.

In both prokaryotes and eukaryotes, a **helix-turn-helix** motif (see section 2.5.2) is a common feature of DNA-binding domains of regulatory proteins. In eukaryotes, this forms part of a characteristic **homeobox sequence,** which binds to a I80-base segment near a starting point for transcription. Other types of common DNA-binding domains have been termed **zinc fingers** and **leucine zippers.** Many regulatory proteins also have binding sites for other proteins.

Some **hormones,** which are intercellular messengers, operate through transcriptional control. For example, there is evidence that **glucocorticoids**, e.g. cortisol, link with proteins to form complexes which bind to promoter sites on chromosomes in mammalian liver, stimulating the binding of RNA polymerase. Glucocorticoids have long been known to mobilize amino acids for glucose synthesis in the liver, and it seems that the mechanism of action involves increasing the production of the enzymes involved, e.g. tyrosine transaminase, tryptophan-2,3-dioxygenase and glutamate-ammonia ligase, by increasing the rate of transcription. Glucocorticoids act only on cells which contain a specific receptor protein in the cytoplasm (i.e. on liver cells and those of peripheral tissue such as muscle, adipose and lymphoid tissue). This receptor protein normally binds to **heat shock protein 90 (hsp 90)**, which prevents functional interaction with DNA, but the binding of a corticosteroid displaces hsp 90, allowing the activated receptor to dimerize, travel to the nucleus and stimulate transcription. Other steroid hormones operate in a similar fashion, as do thyroid hormones although, in the latter case, the receptors never leave the nucleus.

3.1.6 Sequence determination

The base sequence of a single strand of DNA may be determined by the **chain-terminator method**, the development of which led to a second Nobel Prize for Fred Sanger (see section 2.4.5). In the presence of a DNA polymerase and the four deoxynucleoside triphosphates (dATP, dCTP, dGTP and dTTP), which in this procedure are labelled with ^{32}P, the DNA acts as a template for the synthesis of a complementary strand. This synthesis starts at the 3′-end of the original strand, the new one growing in the 5′ to 3′ direction. The same process is carried out in four separate vessels, under identical conditions, except that a dideoxy-analogue (lacking a 3′-OH group) of a different one of the four deoxynucleoside triphosphates is present in each. This brings about competition in each in each flask for incorporation into the growing DNA chains of a normal deoxynucleoside triphosphate or its dideoxy-analogue, the latter event preventing further elongation of a particular chain. When the process is complete, the DNA fragments in each flask are denatured and loaded next to each other on a polyacrylamide gel, to be separated according to size by electrophoresis, utilizing the principles of molecular sieving (see section 16.3). Because of the incorporation of ^{32}P, the bands may then be detected by autoradiography. From the results, the DNA sequence can be deduced. For example, if the smallest fragment came from the flask containing the dideoxy-ATP, the first base in the sequence must be adenine. If the second smallest fragment came from the flask containing dideoxy-CTP, the next base must be cytosine, and so on. Of course, this sequence will be complementary to that of the original DNA strand (see problem 3.1).

Automated systems (e.g. the Applied Biosystems Prism 310 Genetic Analyzer) have been developed from this basic idea, adapted to incorporate fluorescence-based technologies.

Once the sequence of the original DNA strand has been determined, the amino acid sequence corresponding to this can be deduced from a knowledge of the genetic code, as outlined in section 3.1.3. Thus, the primary structure of a protein may be determined indirectly by use of this procedure. In practice, complications arise because some sections of genes are not translated into proteins (see section 3.1.4), and also because the chain-terminator method requires the use of an oligodeoxynucleotide primer. Nevertheless, such problems can be overcome (see section 21.1.2), enabling the primary structure of proteins to be determined more quickly by indirect means than by the more traditional methods. To some extent, the two approaches may be used in complementary fashion.

An alternative to the chain-terminator procedure for sequencing DNA, the **chemical cleavage method**, was developed by Alan Maxam and Walter Gilbert, the latter, like Sanger, being the recipient of a Nobel Prize. In this technique, a sample of single-strand DNA is labelled at the 5′-end with ^{32}P. The sample is then divided into four portions, each being subsequently subjected to different treatments to cleave particular types of linkage. However, conditions are chosen so that not all such linkages are broken, just a few throughout each chain. So, dimethyl sulphate (DMS) is used to methylate guanine residues, allowing piperidine to cleave chains in a random fashion at the 5'-side of these. Thus, for example, if a DNA strand has a sequence (starting from the 5′-end) ^{32}P-CATGAGTCGA, and, if a sample is treated with DMS/piperidine, the products should include ^{32}P-CAT, ^{32}P-CATGA and ^{32}P-CATGAGTC. These can be separated according to size by polyacrylamide gel electrophoresis (PAGE) and then, because of the ^{32}P-label, detected by autoradiography.

In fact, DMS methylates both of the purines, guanine and adenine, but under the alkaline conditions used in the above procedure, methylated guanine residues are much more susceptible to hydrolysis, a prerequisite for chain cleavage by piperidine. However, under mildly acidic conditions, chain cleavage could equally well take place next to guanine or adenine, allowing a different set of fragments to be separated according to size. The finding of a fragment of the same size obtained under the two different conditions (alkaline and acidic) must mean that the fragment had been produced by cleavage of the chain at the 5′-end of a guanine residue, whereas a fragment obtained under acidic conditions which did not appear under alkaline ones must have resulted from cleavage at the 5′-end of an adenine residue.

In similar fashion, chains may be cleaved at the 5'-end of pyrimidine (cytosine or thymine) residues by treatment with hydrazine followed by piperidine. However, in the presence of 2M HCl, the reaction of hydrazine with thymine is suppressed, so cleavage takes place only adjacent to cytosine residues.

So, four procedures are available, which can cleave the DNA chain at the 5′-ends of guanine; guanine or adenine; cytosine; or cytosine or thymine residues. By subjecting a sample to cleavage by each of the four treatments, in separate containers, followed by PAGE and autoradiography, the complete base sequence of the DNA chain can be deduced (see problem 3.2).

Note that the sequence obtained is that of the sample itself, not, as in the chain-terminator procedure, that of a complementary strand. If the 3′-end of the DNA chain is known to be the start of an expressed sequence, the amino acid sequence of the peptide which would be formed can be deduced by using the genetic code, as outlined above.

3.2 THE PROPERTIES OF PROTEINS

3.2.1 Chemical properties of proteins

The chemical properties of proteins are largely those of the side chains of the constituent amino acids. Thus **arginine** side chains, each containing a **guanidine** group, can react with α-naphthol in the presence of an oxidizing agent such as sodium hypochlorite to produce a red colour: this is the **Sakaguchi** reaction. Similarly, **tryptophan** side chains, being **indoles,** can react with glyoxylic acid in the presence of concentrated sulphuric acid to produce a purple colour: this is the **Hopkins-Cole** reaction. Positively-charged **amino** groups in the side chains of **arginine** and **lysine** can bind to **Coomassie Blue** dye by means of electrostatic interactions, giving rise to a colour-change from red-brown to blue.

Tyrosine side chains, each possessing a **phenolic** group, can undergo a variety of reactions. If treated with mercuric sulphate and sodium nitrate and then heated, a red complex is produced by the **Millon** reaction. They also undergo the **Folin-Ciocalteu** reaction if treated with tungstate and molybdate, a blue colour being formed.

All of these procedures, particularly the last mentioned, can be used for the **quantitative estimation of proteins,** the intensity of the colour produced being dependent on the number of reacting groups present. However, it is usually necessary to assume that the protein being estimated has an average distribution of amino acid residues, and no reacting groups other than those in the protein must be present.

An alternative method for the quantitative estimation of protein makes use of the peptide bond rather than an amino acid side chain as the reacting unit: this is the **biuret** reaction, which involves treatment with cupric sulphate in alkali. The copper ions form tetradentate complexes with opposite pairs of peptide-bonded nitrogen, giving rise to a purple colour. The widely used **Lowry** method for protein determination combines the **biuret** and **Folin-Ciocalteu** procedures to give greater sensitivity than can be achieved by the biuret technique alone, the enhanced colour resulting from transfer of electrons from the tetradentate copper complexes to the Folin-Ciocalteu reagent. **Bicinchoninic acid (BCA)** may also be used to increase the sensitivity of the biuret technique, the reaction between peptide bonds and cupric ions (Cu^{2+}) producing cuprous ions (Cu^{+}) each of which can link with two molecules of BCA, an intense purple complex being formed. A further method for protein analysis makes use of the fact that **tyrosine** and **tryptophan** side chains absorb light at 280 nm.

As will be seen later (Chapters 10 and 11), functional groups in amino acid side chains play an important role in the catalytic activity of enzymes. Many agents can inactivate enzymes by binding to these functional groups.

So, **heavy metal ions** (e.g. Ag^{+}) bind strongly to the **sulphydryl** groups of **cysteine** residues and thus may act as poisons to a great many enzymes.

$$\underset{\text{enzyme}}{\text{E–SH}} + Ag^{+} \rightarrow \text{E–S–Ag} + H^{+} \tag{3.3}$$

3.2.2 Acid-base properties of proteins

According to the definitions given by Johannes Brőnsted and Thomas Lowry in 1923, which are particularly applicable to acid-base reactions in dilute aqueous solution, an **acid** is a **proton donor** and a **base** is a **proton acceptor**. The equation for the dissociation of an acid, e.g. acetic acid, is as follows:

$$CH_3COOH \rightleftharpoons H^{+} + CH_3COO^{-} \tag{3.4}$$

The equation for the protonation of a base, e.g. ammonia (NH_3), can be written in the same form, if looked at from the point of view of the dissociation of the corresponding conjugate acid (NH_4^{+}):

$$NH_4^{+} \rightleftharpoons H^{+} + NH_3 \tag{3.5}$$

Thus, *any* acid-base reaction can be written in the form:

$$HA \rightleftharpoons H^{+} + A^{-} \tag{3.6}$$

where HA is an acid, or proton donor, and A^{-} is a base, or proton acceptor. A **weak acid** has a high affinity for its proton and dissociates only partially at most pH values, whereas a **strong acid** has a low affinity for its proton and dissociates readily. Thus, the **strength** of an acid (which has nothing to do with its overall **concentration)** is indicated by its **dissociation constant** K_s. For the acid HA, the dissociation constant K_a at a particular temperature is given by:

$$K_a = [H^{+}][A^{-}]/[HA] \tag{3.7}$$

K_a should really be called the *apparent* dissociation constant, since the terms in square brackets are usually taken to be the experimentally determined concentrations. To obtain the true thermodynamic dissociation constant, allowance would have to be made for departures from ideal behaviour. Also, to be strictly accurate, the expression K_s should involve $[H_2O]$ and $[H_3O^{+}]$. However, if investigations are performed in dilute aqueous solution, water will always be present in great excess, its concentration not being measurably affected by the reactions taking place. Therefore, these terms may be ignored, provided the aim is simply to compare systems which are all in dilute aqueous solution.

The expression for K_a (equation 3.7) can be rearranged to give:

$$[H^{+}] = K_a[HA]/[A^{-}] \tag{3.8}$$

Taking the negative logarithm of both sides:

$$-\log_{10}[H^+] = -\log_{10}(K_a[HA]/[A^-])$$

$$= -\log_{10}K_a - \log_{10}([HA]/[A^-])$$

$$= -\log_{10}K_a + \log_{10}([A^-]/[HA])$$

By definition, $-\log_{10}[H^+] = pH$, and $-\log_{10}K_a = pK_a$.

Hence,
$$pH = pK_a + \log_{10}\left(\frac{[A^-]}{[HA]}\right). \qquad (3.9)$$

This is known as the **Henderson-Hasselbalch equation** (after Lawrence Henderson and Karl Hasselbalch). It gives the relationship between the pH and the degree of ionization of any ionizable species HA (Fig. 3.9).

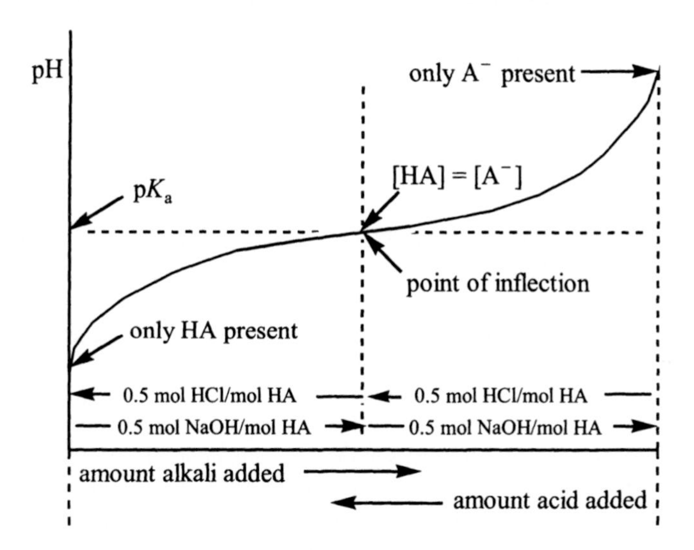

Fig. 3.9 – Titration curve for the ionisable species HA, as predicted by the Henderson-Hasselbalch equation.

When the pH is equal to the pK_a value, exactly half of the molecules present are in the dissociated form, i.e. $[HA] = [A^-]$. If strong acid (e.g. HCl) is added to this solution, one mole of A^- is converted to HA for every mole of acid added. Similarly one mole of HA is converted to A^- for every mole of strong base (e.g. NaOH) added. Therefore, if a known amount of strong acid or base is added to a known amount of HA, the new value of [HA] and $[A^-]$ can easily be calculated, and the resulting pH obtained from the Henderson-Hasselbalch equation.

If the solution where pH = pK_a is treated with 0.5 moles of HCl for every mole of HA present (including that present in the dissociated form), then the HA will be fully protonated. Addition of 0.5 moles of NaOH per mole of HA will return the system to its initial position, and addition of a further 0.5 moles of NaOH per mole of HA will result in complete deprotonation. To return once more to the initial position, 0.5 moles of HCl are required for every mole of HA present. The pH changes taking place throughout this process are indicated in Fig. 3.9.

It can be seen that the pK_a value is characterized by a point of inflection on the curve. Therefore, any ionizable group can be titrated against strong acid or alkali and, from the graph of pH against titre obtained, the pK_a value of the group can be determined. If two groups are present which ionize over completely different pH ranges (e.g. the α-carboxyl and α-amino groups of free amino acids), a graph of pH against titre will show two separate ionization curves, enabling the pK_a value of each group to be determined with ease. However, if two groups are present which ionize over approximately the same pH range, then the graph of pH against titre obtained will be a composite of two ionization curves and accurate determination of the pK_a values will be difficult. If several ionizable groups are present, the situation will be even more complicated.

The groups contributing to the acid-base properties of **proteins** are the ionizable side chains of aspartate, glutamate, histidine, cysteine, tyrosine, lysine and arginine residues, together with N-terminal α-amino groups and C-terminal α-carboxyl groups. All other α-amino and α-carboxyl groups present are, of course, involved in peptide bonds and thus not free to ionize. The pK_a values of all these groups are given in Fig. 3.10. However, it must be pointed out that the values given are only approximate ones, since the actual values depend to a considerable extent upon the environment within the protein (see section 10.1.6).

The overall titration curve for a protein will show the superimposed effects of all the ionizations present. For example, if a solution of **ribonuclease** is taken to acid pH and then titrated against NaOH, the observed graph of pH against titre can be analysed to show that all the ionizable groups known to be present, i.e. one α-carboxyl, ten side chain carboxyl, four histidine imidazole, six tyrosyl hydroxyl, one α-amino, ten side chain amino and four arginine guanidine groups, have contributed to the overall effect.

If we now return to Fig. 3.9 we can see that the change in pH brought about by the addition of a given amount of strong acid or base reaches a minimum value where pH = pK_a: in other words, the **buffering capacity** of an ionizable group is greatest where the pH is near its pK_a value. Proteins (excluding fibrous proteins which are largely insoluble) can act as **buffers**, their buffering capacity being greatest near the pK_a value of their most abundant ionizable side chain. Thus, haemoglobin is able to play an important buffering role in erythrocytes (red blood cells), since it contains a relatively large amount of an amino acid (histidine) whose side chain has a pK_a value near the intracellular pH.

Figure 3.9 also shows that the total electrical charge present depends on the degree of dissociation of ionizable groups and hence on pH. Therefore, most proteins will have a net positive charge at low pH values and will travel towards the cathode on electrophoresis. Conversely, at high pH values most proteins will have a net negative charge and travel towards the anode.

Ionizable group	Dissociation reaction	pK_a
α-carboxyl	$-COOH \rightleftharpoons H^+ + -COO^-$	3.0
aspartyl carboxyl	$-CH_2COOH \rightleftharpoons H^+ + -CH_2COO^-$	3.9
glutamyl carboxyl	$-CH_2CH_2COOH \rightleftharpoons H^+ + -CH_2CH_2COO^-$	4.1
histidine imadazole	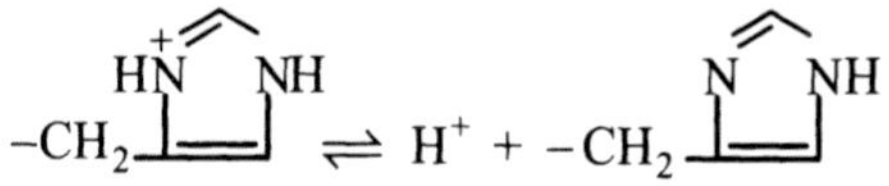 $-CH_2$–(imidazole ring, $H\overset{+}{N}$, NH) $\rightleftharpoons H^+ + -CH_2$–(imidazole ring, N, NH)	6.0
α-amino	$-NH_3^+ \rightleftharpoons H^+ + -NH_2$	8.0
cysteine sulphydryl	$-CH_2SH \rightleftharpoons H^+ + -CH_2S^-$	8.4
tyrosyl hydroxyl	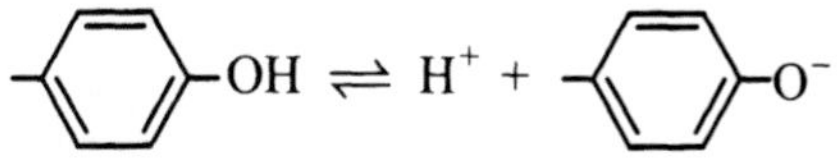 –(benzene ring)–OH $\rightleftharpoons H^+ +$ –(benzene ring)–O^-	10.1
lysyl amino	$-(CH_2)_4NH_3^+ \rightleftharpoons H^+ + -(CH_2)_4NH_2$	10.8
arginine guanidine	$-(CH_2)_3NH.C(={}^+NH_2).NH_2 \rightleftharpoons H^+ + -(CH_2)_3NH.C(=NH).NH_2$	12.5

Fig. 3.10 – The ionisable groups which contribute to the acid-base properties of proteins, shown with their approximate pK_a values. These can vary by several pH units according to their environment within the protein.

The pH at which there is no net charge on the molecule (i.e. at which there is an equal balance between positive and negative charges) is termed the **isoelectric point**. A protein having a relatively high content of basic amino acids will have a relatively high isoelectric point, one having a high content of acidic amino acids will have a low isoelectric point.

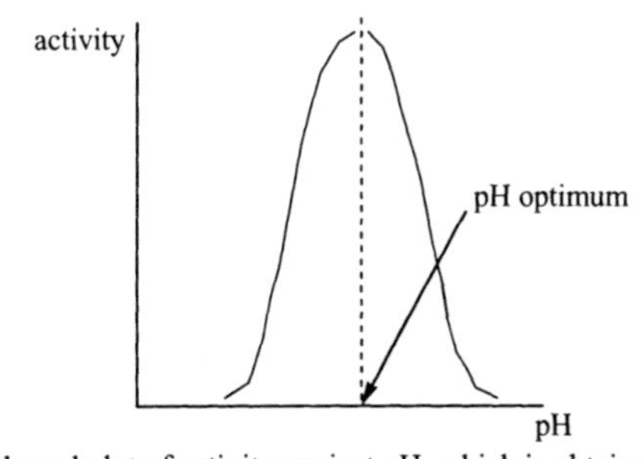

Fig. 3.11 – A bell-shaped plot of activity against pH, which is obtained for many enzymes.

Globular proteins often function correctly only when certain ionizable side chains are in a specified form, making their usefulness pH-dependent. Each enzyme, therefore, has a characteristic pH optimum and is active over a relatively small pH range. In many cases a bell-shaped plot of activity against pH is obtained (Fig. 3.11). This is discussed in more detail in section 10.1.6.

3.2.3 The solubility of globular proteins

Solubility in aqueous solvents is enhanced by the formation of weak ionic interactions, including hydrogen bonds, between solute molecules and water. Therefore, any factor which interferes with this process must influence solubility. Electrical interactions between solute molecules will also affect solubility, since repulsive forces will hinder the formation of insoluble aggregates. In general, the solubility of a globular protein in an aqueous solvent is influenced by four main factors: salt concentration, pH, the organic content of the solvent and the temperature.

Addition of a small amount of **neutral salt** to a solution can increase the solubility of a protein. The added ions can cause small changes in ionization of amino acid side chains and can also interfere with interactions between protein molecules, the overall effect being to increase interactions between solute and solvent. This phenomenon, known as **salting-in**, depends solely on the **ionic strength** of the salt solution. (The **ionic strength** is the value obtained by multiplying the concentration of each ion by the square of its charge, adding together the results obtained for the different ions present and dividing by two: i.e. ionic strength = $\frac{1}{2}\Sigma[A]Z_A^2$.) Thus, divalent ions such as Mg^{2+} and SO_4^{2-} are relatively more effective than monovalent ions such as Na^+ and Cl^-. At very high salt concentrations, the abundance of interactions between the added ions and water decreases the possibilities for protein-water interactions, often resulting in the protein being precipitated from solution (Fig. 3.12). This is termed **salting-out**.

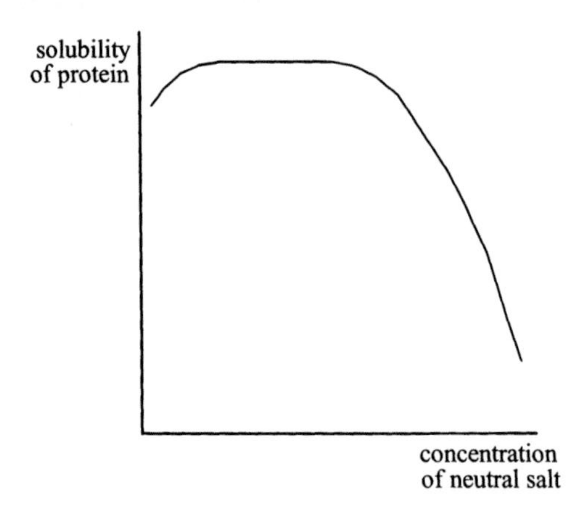

Fig. 3.12 – The effect of neutral salt concentration on protein solubility. The phenomena of salting-in, at low salt concentration, and salting-out, at high salt concentration, may be observed

Some cations, e.g. Zn^{2+} and Pb^{2+}, have a more direct action on protein solubility, linking with protein anions to form **insoluble complexes.** Proteins may also be precipitated by treatment with various acids, e.g. trichloroacetic, perchloric, picric or sulphosalicylic acids, which form **acid-insoluble salts** with protein cations. Such techniques are often used to remove proteins from solutions prior to the analysis of other substances.

At **extremes of pH,** the pattern of charges carried by the ionizable side chains will be very different from that under normal conditions, so the compact tertiary structure will usually be disrupted, a more random structure being formed: this process is termed **denaturation.** Since the tertiary structure of an active globular protein is characterized by the majority of hydrophobic groups being hidden inside the molecule, disruption will bring these into contact with the aqueous solvent and the solubility of the protein will decrease considerably. Reversion to the original conditions will sometimes, but not always, result in the protein refolding into the tertiary structure required for functional activity: although this structure is likely to be the most favourable one from the point of view of energy, its re-establishment may be extremely slow because of the tangles produced during denaturation.

The solubility of each globular protein will also decrease markedly over a very narrow pH range around its **isoelectric point** (Fig. 3.13). Molecules of the protein are electrically neutral at this particular pH, so there is no net repulsion between them to prevent the formation of insoluble aggregates.

The introduction of a **water-miscible component,** e.g. ethanol, into the solvent lowers its **dielectric constant.** This increases the attractive forces between groups of opposite charge within a protein molecule and thus diminishes their linkages with surrounding water molecules (section 2.3.2). The solubility of the protein will decrease in consequence.

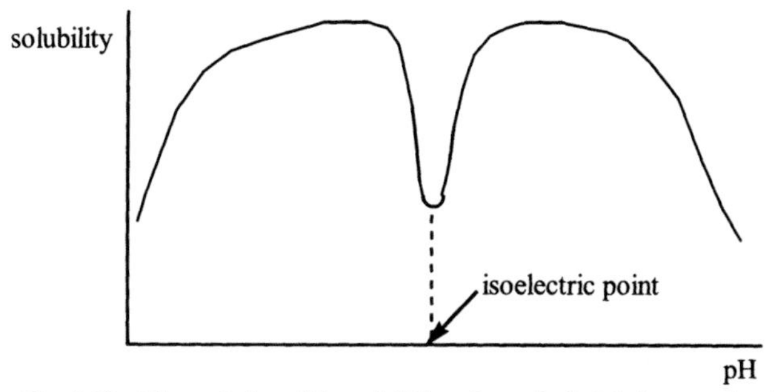

Fig. 3.13 – The variation of the solubility of a typical globular protein with pH.

Techniques involving varying pH, salt concentration and organic solvent concentration are widely used to separate mixtures of proteins by differential precipitation (see section 16.1.4).

The solubility of globular proteins increases with the **temperature** up to about 40-50°C. Above this temperature, thermal agitation tends to disrupt tertiary structure, leading to denaturation and a sharp decrease in solubility. In enzymes, this effect is paralleled to some extent by changes in activity: rates of enzyme-catalysed reactions increase with increasing temperature, since collisions between molecules become more frequent, until the enzyme is denatured and catalytic activity is lost (Fig. 3.14). A few proteins are also denatured if the temperature is reduced to below 10°C. Other conditions which lead to protein denaturation include the presence of high concentrations (e.g. 6 M) of urea or guanidine HCl, which disrupt the non-covalent interactions stabilizing tertiary structure.

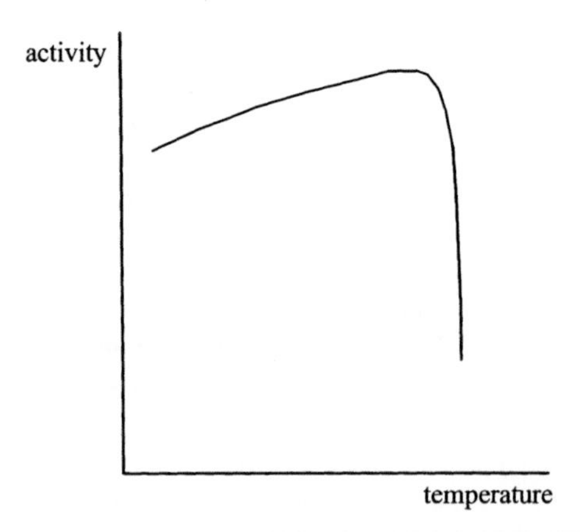

Fig. 3.14 – The variation of the activity of a typical enzyme with temperature

SUMMARY OF CHAPTER 3

Genetic information is stored in cells as the base sequence of DNA. This can be replicated prior to cell division, so that each cell produced contains the same genetic information as the original cell. The information may be transcribed into RNA structure and then translated into protein structure in such a way that the amino acid sequence of each protein synthesized is determined by the base sequence of a section of DNA known as a gene. Hence, the primary structure of a protein may be deduced if the structure of the corresponding gene is known.

Various control mechanisms exist to prevent unnecessary protein synthesis taking place. In simple micro-organisms, control of protein synthesis by transcriptional control is of particular importance, but some degree of translational control may also occur. In higher organisms, it seems likely that transcriptional control can determine which proteins are produced by a particular type of cell.

The chemical and acid-base properties of proteins are largely those of the side chains of the amino acid residues present. The solubility of a protein in an aqueous solvent is influenced by salt concentration, pH, organic solvent content and temperature.

FURTHER READING

Alberts, B., Johnson, A. *et al* (2002), *Molecular Biology of the Cell*, 4th edn., Garland (Chapters 4-7).

Baldwin, R. L. and Rose, G. D. (1999), Is protein folding hierarchic?, *Trends in Biochemical Sciences,* **24**, 26-33, 77-83.

Berg, J. M., Tymocsko, J. L and Stryer, L. (2006), *Biochemistry*, 6th edn., Freeman (Chapters 4, 5, 28-31).

Chapman, K. E. and Higgins, S. J. (eds.) (2001), *Regulation of Gene Expression: Essays in Biochemistry*, **37**, Portland Press.

Dinner, A. R., Sali, A. *et al* (2000), Understanding protein folding via free-energy surfaces from theory to experiment, *Trends in Biochemical Sciences*, **25**, 331-339.

Frydman, J. (2001), Folding of newly translated proteins in vivo, *Annual Review of Biochemistry*, **70**, 603-647.

Green, R. and Noller, H. F. (1997), Ribosomes and translation, *Annual Review of Biochemistry*, **66**, 679-716.
Kapp, L. D. and Lorsh, J. R. (2004), The molecular mechanisms of eukaryotic translation, *Annual Review of Biochemistry*, **73**, 657-704.
Lodish, H., Berk, A. *et al* (2004), *Molecular Cell Biology,* 5th edn., Freeman (Chapters 915).
Malik, S. and Roeder, R. G. (2000), Transcriptional regulation through mediator-like coactivators in yeast and metazoan cells, *Trends in Biochemical Sciences,* **25**, 277-283.
Mata, J., Marguerat, S. and Bahler, J. (2005), Post-transcriptional control of gene expression, *Trends in Biochemical Sciences*, **30**, 506-514.
Montminy, M. (1997), Transcriptional regulation by cyclic AMP, *Annual Review of Biochemistry,* **66**, 807-822.
Moore, P. B. and Steitz, T. A. (2003), The structural basis of large ribosomal subunit function, *Annual Review of Biochemistry*, **72,** 813-850.
Morimoto, R. I., Kline, M. P. *et al* (1997), The heat-shock response - regulation and function of heat-shock proteins and molecular chaperones, *Essays in Biochemistry,* **32**, 17-29, Portland Press.
Naar, A. M., Lemon, B. D. Tjian, R. (2001), Transcriptional coactivator complexes, *Annual Review of Biochemistry*, **70**, 475-501.
Nelson, D. L. and Cox, M. M. (2004), *Lehninger Principles of Biochemistry,* 4th edn., Worth (Chapters 8, 24-28).
Parodi, A. J. (2000), Protein glycosylation and its role in protein folding, *Annual Review of Biochemistry*, **69**, 69-93.
Proudfoot, N. (2000), Connecting transcription to messenger RNA processing, *Trends in Biochemical Sciences*, **25**, 290-293.
Stetefeld, J. and Ryan, M. A. (2005), Structural and functional diversity generated by alternative mRNA splicing, *Trends in Biochemical Sciences*, **30**, 515-521.
Wimberley, B. T., Brodersen, D. E. *et al* (2000), Structure of the 30S ribosomal subunit, *Nature,* **407**, 327-339 (see also 340-348).
Voet, D. and Voet, J. G. (2004), *Biochemistry*, 3rd edn., Wiley (Chapters 5, 29-34).

PROBLEMS

3.1 The chain-terminator sequencing procedure was applied to a single strand of DNA, and a diagram of the resulting polyacrylamide gel electrophoresis pattern is shown below, the largest fragment being on the right. Lane A represents the contents of the flask to which dideoxy-ATP has been added, lane C that containing dideoxy-CTP, and so on.

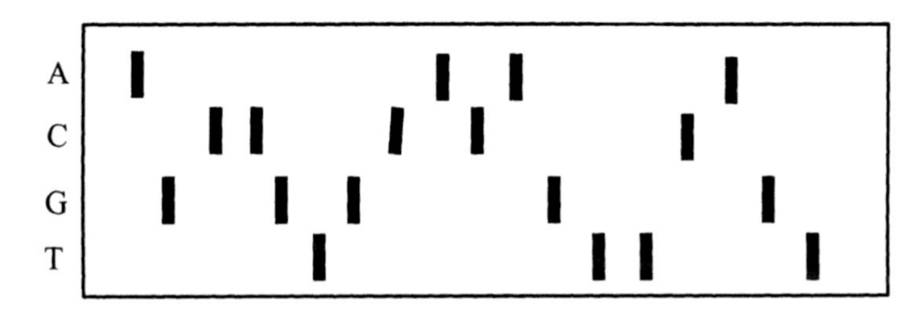

From this information, deduce the structure of the DNA strand to which the procedure has been applied.

3.2 The chemical cleavage sequencing procedure was applied to a single strand of DNA, and a diagram of the resulting polyacrylamide gel electrophoresis pattern is shown below, the largest fragment being on the right. Lane G represents the contents of the flask where cleavage took place to the 5′-side of a guanine residue; lane A + T that where cleavage took place to the 5′-side of either a guanine or an adenine residue; lane C that where cleavage took place to the 5′-side of a cytosine residue; and C + T that where cleavage took place to the 5′-side of either a cytosine or a thymine residue

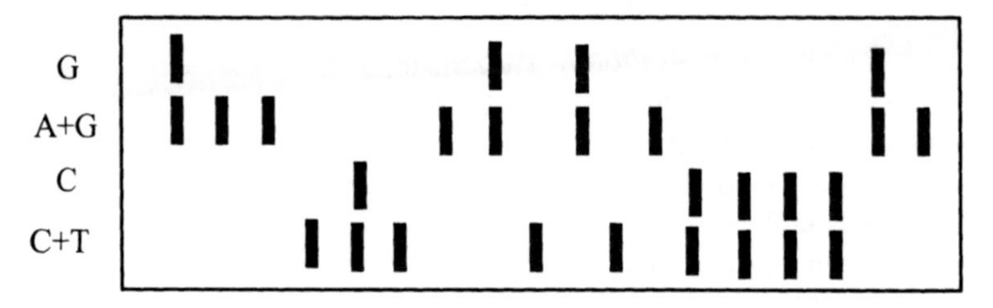

(a) From this information, deduce the structure of the DNA strand to which the procedure had been applied.

(b) Assuming that the base at the 3′-end of this is the first in a coding sequence, deduce the primary structure of the specified peptide.

3.3 (a) What are the pH values of the solutions obtained by mixing:

(i) 0.2 mol dm^{-3} sodium acetate with an equal volume of 0.1 mol dm^{-3} HCl, and

(ii) 0.2 mol dm^{-3} sodium acetate with an equal volume of 0.1 mol dm^{-3} acetic acid?

(b) Serial dilutions of an acetic acid solution were prepared so that tube number 1 contained 5 cm^3 of 0.32 mol dm^{-3} acetic acid, tube number 2 contained 5 cm^3 of 0.16 mol dm^{-3} acetic acid, tube number 3 contained 5 cm^3 of 0.08 mol dm^{-3} acetic acid, and so on. To each was added 1 cm^3 of a solution of the protein casein in 0.1 mol dm^{-3} sodium acetate. The greatest degree of protein precipitation was observed to occur in tube number 4. What does this indicate about the isoelectric point of casein? (Assume the pK_a of acetic acid is 4.74.)

3.4 The imidazole side chain of histidine is ionizable. Calculate the percentage present in the protonated form at (a) pH 3, (b) pH 5, (c) pH 7 and (d) pH 9. (Assume the pK_a of the imidazole group is 6.0.)

4

Specificity of Enzyme Action

4.1 TYPES OF SPECIFICITY

A characteristic feature of enzymes is that they are specific in action. Some enzymes exhibit **group specificity**, i.e. they may act on several different, though closely related, substrates to catalyse a reaction involving a particular chemical group. An example of this kind of enzyme is alcohol dehydrogenase, which will catalyse the oxidation of a variety of alcohols. Another is hexokinase, which will assist the transfer of phosphate from ATP to several different hexose sugars. Other enzymes will act only on one particular substrate, when they are said to exhibit **absolute specificity**. For example, glucokinase catalyses the transfer of phosphate from ATP to glucose and to no other sugar (although the specificity of the so-called 'glucokinase' found in liver is less clear-cut, in contrast to the true glucokinases of bacteria and invertebrates).

Uncatalysed reactions often give rise to a wide range of products, but enzyme-catalysed reactions are **product-specific** as well as being **substrate-specific**. Also, in addition to showing chemical specificity, enzymes exhibit **stereochemical specificity**: if a substrate can exist in two stereochemical forms, chemically identical but with a different arrangement of atoms in three-dimensional space (section 2.2.2), then only one of the isomers will undergo reaction as a result of catalysis by a particular enzyme. For example, L-amino acid oxidase mediates the oxidation of L-amino acids to oxo acids. A separate enzyme, D-amino acid oxidase, is required for the corresponding oxidation of D-amino acids.

Even greater specificity is shown by the fungal enzyme glucose oxidase, which catalyses the reaction:

$$\beta\text{-D-glucose} + O_2 \rightleftharpoons \text{D-gluconolactone} + H_2O_2 \qquad (4.1)$$

No other naturally-occurring sugar, including α-D-glucose and β-D-galactose, can be acted upon to any appreciable extent.

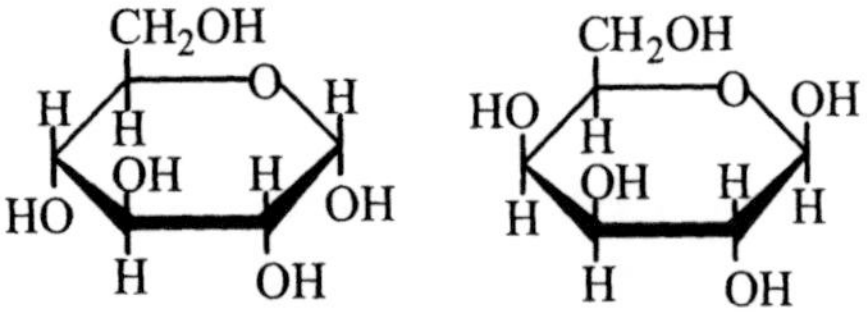

α – D – glucose $\qquad\qquad$ β – D – galactose

The only enzymes which act on both stereoisomeric forms of a substrate are those whose function is to interconvert L- and D-isomers. An example is alanine racemase, which catalyses the reaction:

$$\text{L-alanine} \rightleftharpoons \text{D-alanine} \qquad (4.2)$$

Enzyme-catalysed reactions may yield stereospecific products even when the substrate possesses no asymmetric carbon atom. For example, the action of glycerol kinase on glycerol always results in the production of L-glycerol-3-phosphate (sn-glycerol-3- phosphate):

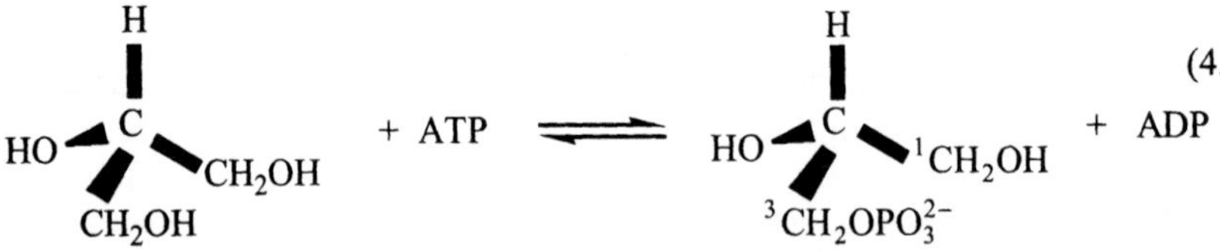

$$\text{glycerol} + \text{ATP} \rightleftharpoons \text{L-glycerol-3-phosphate} + \text{ADP} \qquad (4.3)$$

No L-glycerol-1-phosphate is formed, even though the two $-CH_2OH$ groups of glycerol are chemically identical.

4.2 THE ACTIVE SITE

In order to explain the stereochemical specificity of enzymes (section 4.1), Alexander Ogston (1948) pointed out that there must be *at least three different points of interaction* between enzyme and substrate (Fig. 4.1).

These interactions can have either a binding or a catalytic function: **binding sites** link to specific groups in the substrate, ensuring that the enzyme and substrate molecules are held in a fixed orientation with respect to each other, with the reacting group or groups in the vicinity of **catalytic sites**.

For example, sites A″ and A‴ (Fig. 4.1) might represent binding sites for R″ and R‴ respectively, and A′ a catalytic site for a reaction involving R′. Thus, even if R′ and R″ are chemically identical (as with glycerol in the glycerol kinase reaction mentioned in section 4.1), the asymmetry of the enzyme-substrate complex means that only R′ can react, providing binding site A‴ is specific for R‴. R″ can never undergo reaction under these conditions, since it is not brought into the vicinity of site A′ even when R′ binds to site A″.

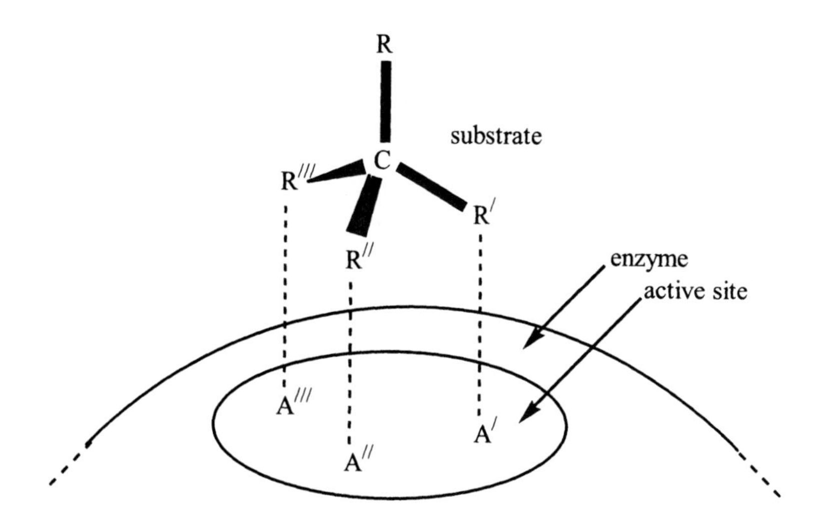

Fig. 4.1 – Diagrammatic representation of three-point interaction between enzyme and substrate. A′, A″ and A‴ are sites on the enzyme which interact with groups R′, R″ and R‴, respectively, of the substrate. Each point of interaction may have a binding or a catalytic function.

Generally similar considerations apply to enzymes catalysing reactions involving more than one substrate. In this case, the reacting groups of each substrate are brought together in the vicinity of one or more catalytic sites.

The region which contains the binding and catalytic sites is termed the **active site**, or **active centre**, of the enzyme. This comprises only a small proportion of the total volume of the enzyme and is usually at or near the surface, since it must be accessible to substrate molecules. In some cases, X-ray diffraction studies have revealed a clearly-defined pocket or cleft in the enzyme molecule into which the whole or part of each substrate can fit (see Fig. 2.10).

Although the active site is given a planar representation in Fig. 4.1, it should be realised that it has, in fact, a three-dimensional structure since it consists of portions of a polypeptide chain. The amino acid residues involved may be widely separated in the primary structure, being brought together in space because of the twists and turns within the molecule. The binding and catalytic sites must be either amino acid residues or cofactors, the latter being themselves bound to amino acid side chains. Substrate binding may involve a variety of linkages (see section 2.3.2), but the bonds formed are usually relatively weak (i.e. non-covalent).

Those amino acid residues in the active site which do not have a binding or catalytic function may nevertheless contribute to the specificity of the enzyme. Their side chains must be of suitable size, shape and character not to interfere with the binding of the substrate, but they might interfere with the binding of other, chemically similar, substances.

The active site often includes both polar and non-polar amino acid residues, creating an arrangement of hydrophilic and hydrophobic **microenvironments** not found elsewhere on an enzyme molecule. Hence, the function of an enzyme may depend not only on the spatial arrangement of binding and catalytic sites, but also on the **environment** in which these sites occur.

Thus it can be seen that the three-point interaction theory provides only a limited explanation of enzyme specificity, a more complete view coming from consideration of a whole range of interactions in three-dimensional space.

4.3 THE FISCHER 'LOCK-AND-KEY' HYPOTHESIS

As early as 1890, Emil Fischer suggested that enzyme specificity implied the presence of complementary structural features between enzyme and substrate: a substrate might fit into its complementary site on the enzyme as a key fits into a lock. This is entirely consistent with the more detailed aspects of active site structure discussed in section 4.2. According to the **lock-and-key** model, all structures remain fixed throughout the binding process (Fig. 4.2).

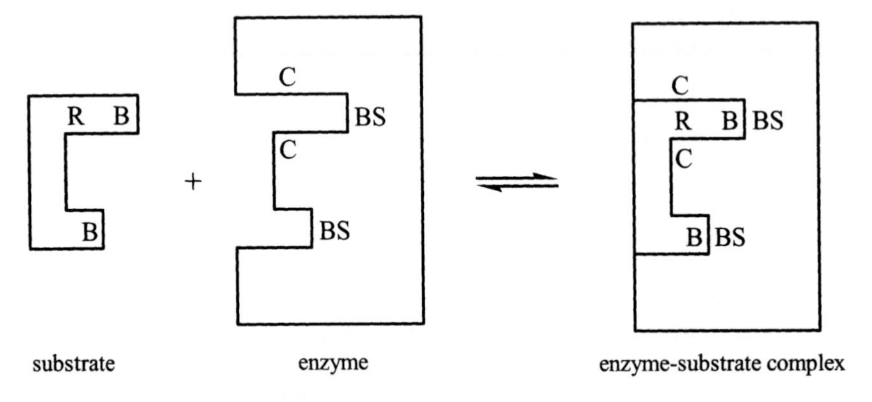

Fig. 4.2 – Diagrammatic representation of the interaction between an enzyme and its substrate, according to the lock-and-key model. In the example illustrated, the single substrate is bound at two points, bringing the reacting group into the vicinity of two different catalytic sites. (BS = a binding site on the enzyme, C = a catalytic site, B = a binding group on the substrate and R = a reacting group, i.e. a group undergoing enzyme-catalysed reaction.)

4.4 THE KOSHLAND 'INDUCED-FIT' HYPOTHESIS

The lock-and-key hypothesis explains many features of enzyme specificity, but takes no account of the known flexibility of proteins.

X-ray diffraction analysis and data from several forms of spectrometry, including nuclear magnetic resonance (NMR), have revealed differences in structure between free and substrate-bound enzymes. Thus, the binding of a substrate to an enzyme may bring about a **conformational change**, i.e. a change in three-dimensional structure but not in primary structure. This is not necessarily surprising, for the bonds formed between a substrate and its binding sites may have replaced previously existing linkages between each binding site and neighbouring groups on the enzyme. Also, the presence of a substrate at the active site may exclude water molecules and thus make the region more non-polar. Both of these factors could be responsible for some degree of change in tertiary structure taking place.

Daniel Koshland, in his **induced-fit** hypothesis of 1958, suggested that the structure of a substrate may be complementary to that of the active site in the enzyme-substrate complex, but not in the free enzyme: a conformational change takes place in the enzyme during the binding of substrate which results in the required matching of structures (Fig. 4.3). The induced-fit hypothesis essentially requires the active site to be floppy and the substrate to be rigid, allowing the enzyme to wrap itself around the substrate, in this way bringing together the corresponding catalytic sites and reacting groups. In some respects, the relationship between a substrate and an active site is similar to that between a hand and a woollen glove: in each interaction the structure of one component (substrate or hand) remains fixed and the shape of the second component (active site or glove) changes to become complementary to that of the first.

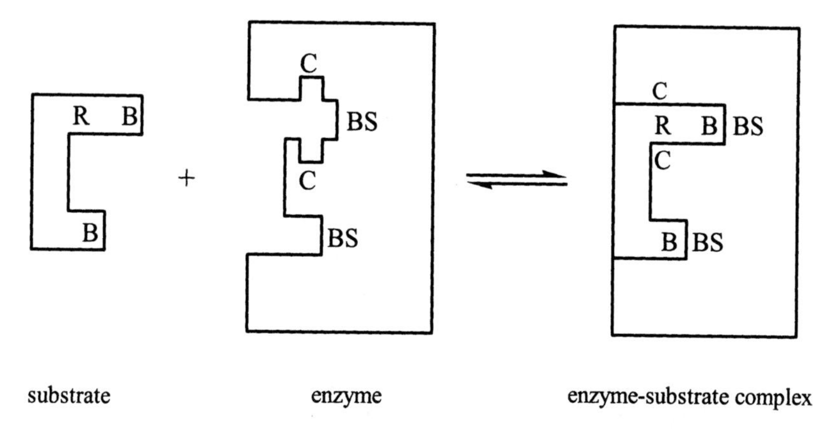

Fig. 4.3 – Diagrammatic representation of the interaction between an enzyme and its substrate, according to the induced-fit model (abbreviations as for Fig. 4.2).

Such a mechanism could help to achieve a high degree of specificity for the enzyme. In the lock-and-key mechanism, the active site is always structurally intact, with the catalytic sites aligned and freely accessible. Thus a suitable reacting group, whether part of an appropriately-bound substrate or not, can come into contact with the region of catalytic activity and some degree of reaction take place. In the induced-fit mechanism, on the other hand, different catalytic components might be separated by a considerable margin in the free enzyme, minimising the risk of a chance collision of a reactive group with both of them (see Fig. 4.3). It is also possible that access to the catalytic groups of the free enzyme might be blocked.

Only when a binding group of the substrate is recognized by the corresponding site of the enzyme and the binding process proceeds does the conformational change take place, which results in all the relevant groups in substrate and enzyme coming together. Of course, a similar binding group in a substance other than the substrate might trigger off a conformational change but, in general, this would not result in catalytic groups being brought together in the vicinity of an appropriate reacting group, so no reaction would take place (Fig. 4.4). This would be termed **non-productive binding**.

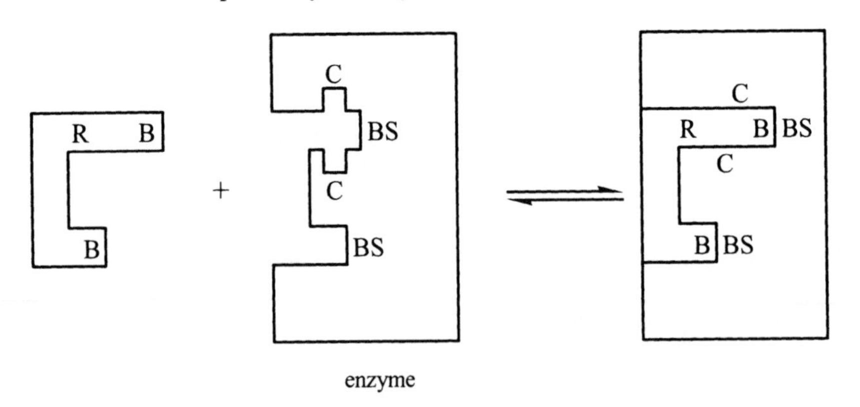

Fig. 4.4 – Diagrammatic representation of the non-productive interaction between an enzyme and a compound resembling a substrate in some respects but not others, according to the induced-fit model (abbreviations as for Fig. 4.2). In the example shown, the arm containing R is longer than the corresponding arm in the substrate.

An example of a reaction which appears to proceed via an induced-fit mechanism is that catalysed by yeast hexokinase:

$$\text{D-hexose} + \text{ATP} \rightleftharpoons \text{D-hexose-6-P} + \text{ADP} \tag{4.4}$$

In the absence of hexose, bound ATP is hydrolysed extremely slowly, even though, in chemical terms, this hydrolysis could be brought about by the action of a water molecule in the solvent just as well as by an –OH group in the hexose. This, together with X-ray diffraction evidence, suggests that the binding of the hexose causes a conformational change in the enzyme which activates the ATP.

Conformational changes of the type discussed in this section have been shown to play a part in the mechanism of action of several other enzymes, e.g. carboxypeptidase A (section 11.4.4). They have also been useful in explaining the behaviour of allosteric enzymes (section 13.3.1).

4.5 HYPOTHESES INVOLVING STRAIN OR TRANSITION-STATE STABILIZATION

Although the lock-and-key and induced-fit models can explain enzyme specificity, neither suggests any direct mechanism by which the catalysed reaction may be driven forward. Substrate-binding often involves the expenditure of a considerable amount of energy and, although it serves a very useful purpose in bringing reacting and catalytic groups together, further energy must be supplied before the reaction can proceed.

John (J. B. S.) Haldane, in 1930, pointed out that if the binding energy was used to distort the substrate in such a way as to facilitate the subsequent reaction, then less energy would be required for the reaction to take place. This concept was developed further by Linus Pauling, in 1948.

Let us assume, for example, that the structure of the active site is almost complementary to that of a substrate, but not exactly so. If the structure of the active site is rigid, the substrate must be distorted slightly in order to bind to the enzyme. This distortion might result in the stretching, and thus weakening, of a bond which is subsequently to be cleaved, thus assisting the forward reaction (Fig. 4.5).

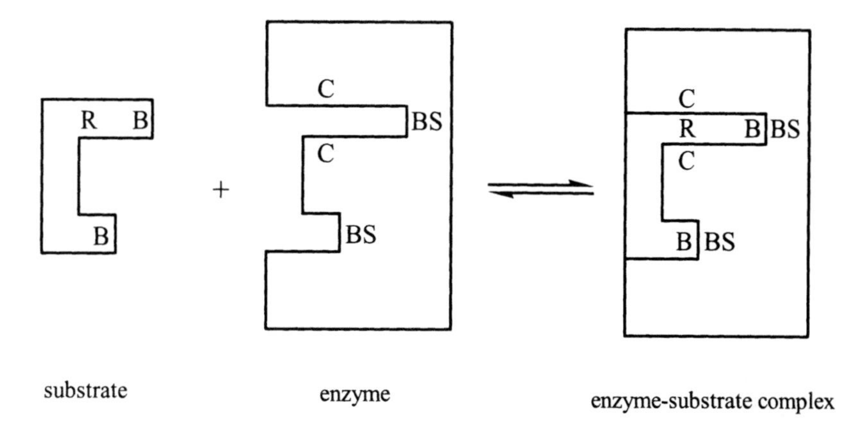

Fig. 4.5 – Diagrammatic representation of the interaction between an enzyme and its substrate, incorporating a 'strain' effect, indicated by the stretching of the section containing R during binding (abbreviations as for Fig. 4.2).

In fact, little clear-cut evidence has been obtained for the occurrence of distorted binding. An alternative, and possibly more likely, mechanism for driving the reaction forward is **transition-state stabilization.** This assumes that the substrate is bound in an undistorted form, but the enzyme-substrate complex possesses various unfavourable interactions. These tend to distort the substrate in such a way as to favour the following reaction sequence: enzyme-substrate complex → transition-state → products (see sections 6.2 and 6.6). As the reaction proceeds, the unfavourable interactions diminish, and are absent from the transition-state.

Thus, the overall effects of **strain** and **transition-state stabilization** are very similar, but the sequence of events is slightly different in the two cases. An example of an enzyme-catalysed reaction proceeding via a transition-state stabilization mechanism is the hydrolysis of peptides by **chymotrypsin**. **Lysozyme** (section 11.3.4) is often cited as an example of an enzyme which operates by a strain mechanism, but even in this case the true mechanism may be transition-state stabilization.

4.6 FURTHER COMMENTS ON SPECIFICITY

Specificity of enzyme action is determined by two separate factors: the relative ability of a potential substrate to bind to the enzyme and, once bound, its relative ability to undergo a reaction to form products. Only the overall rate of product formation indicates whether the enzyme can utilize a particular potential substrate.

No single one of the hypotheses discussed in sections 4.3-4.5 is able to account for the features of catalysis and specificity observed in all enzyme-catalysed reactions. Moreover, in some cases at least, a contribution from more than one of these factors appears to be present.

The induced-fit and strain/transition-state stabilization mechanisms are not necessarily mutually exclusive. If the active site of an enzyme has a floppy structure and moulds itself around the substrate during the binding process, then distortion of the substrate would be unlikely to occur. However, a more precise conformational change taking place in the protein during the binding of substrate could result in some degree of strain being present in the latter. In either case, the enzyme-substrate complex formed could possess internal stress and thus assist the subsequent reaction by transition-state stabilization. The mechanism for the hydrolysis of peptides by **papain** appears to involve a conformational change in the enzyme in addition to either a strain or transition-state stabilization factor.

In general, irrespective of the mechanism of an enzyme-catalysed reaction, the major factor governing specificity is the stability of the enzyme-bound transition-state which exists during the conversion of enzyme-bound substrate to products. A potential substrate which can form a relatively stable transition-state when bound to the enzyme will be converted to products at an appreciable rate. This was first pointed out by Pauling in 1948 and has since been confirmed by experiments with **transition-state analogues**. These are stable compounds which resemble the transition-state compounds thought to be formed as part of a reaction sequence. Such analogues have been shown to bind very tightly to the active sites of the appropriate enzymes, more tightly in fact than the corresponding substrate or products.

The investigation of transition-state structure is difficult because it occurs only transiently under normal conditions. However, if a reaction is carried out at low temperatures (e.g. at –21°C in aqueous dimethyl sulphoxide), the lifetimes of intermediates are extended and their structures may be studied by techniques such as NMR. This is termed cryoenzymology.

SUMMARY OF CHAPTER 4

Enzymes exhibit chemical and stereochemical specificity with respect both to substrates and products. Such specificity requires at least three different points of interaction between enzyme and substrate. Each substrate is bound to the enzyme at specific sites to form an enzyme-substrate complex in which reacting groups are held in close proximity to each other and to catalytic sites. That region of the enzyme's three-dimensional structure which contains the substrate-binding sites and the catalytic sites is termed the active site.

According to the Fischer lock-and-key hypothesis, the active site has rigid structural features which are complementary to those of each substrate. In contrast, the Koshland induced-fit hypothesis suggests that at least some active sites are flexible, possessing a structure complementary to that of a substrate only when the latter is bound to the enzyme.

These models can explain some aspects of enzyme specificity, but do not suggest any mechanism for driving forward the enzyme-catalysed reaction. The Haldane and Pauling concept of strain, and the transition-state stabilization modification of this, explains how distortion of the substrate during or after binding can facilitate the subsequent reaction. This is not necessarily inconsistent with an induced-fit mechanism.

The most important factor in determining whether an enzyme will act on a particular substrate to produce a product appears to be the stability of the enzyme-bound transition-state which would have to be formed.

FURTHER READING

Berg, J. M., Tymoczko, J. L and Stryer, L. (2006), *Biochemistry*, 6th edn., Freeman (Chapter 8).

Cornish-Bowden, A. and Cardenas, M. L. (1991), Hexokinase and 'glucokinase' in liver metabolism, *Trends in Biochemical Sciences*, **16**, 281-282.

Fersht, A. (1999), *Structure and Mechanism in Protein Science,* Freeman (Chapters 7-9, 13).

Khan, S. and Sheetz, M. P. (1997), Force effects on biochemical kinetics, *Annual Review of Biochemistry*, **66**, 785-805.

Lolis, E. and Petsko, G. A. (1990), Transition-state analogues in protein crystallography - probes of the structural source of enzyme catalysis, *Annual Review of Biochemistry,* **59**, 597-630.

Nelson, D. L. and Cox, M. M. (2004), *Lehninger Principles of Biochemistry,* 4th edn., Worth (Chapter 6).

Price, N. C. and Stevens, L. (1999), *Fundamentals of Enzymology,* 3rd edn., Oxford University of Press (Section 5.3).

Schramm, V. L. (1998), Enzyme transition-states and transition-state analog design, *Annual Review of Biochemistry,* **67**, 693-720.

Voet, D. and Voet, J. G. (2004), *Biochemistry*, 3rd edn., Wiley (Chapters 13, 15).

5

Monomeric and Oligomeric Enzymes

5.1 MONOMERIC ENZYMES

5.1.1 Introduction

Monomeric proteins are those which consist of only a single polypeptide chain, so they cannot be dissociated into smaller units. Very few monomeric enzymes are known, and all of these catalyse hydrolytic reactions. In general, they contain between 100 and 300 amino acid residues and have molecular weights in the range 13 000 - 35 000. Some, e.g. **carboxypeptidase A**, are associated with a metal ion, but most act without the help of any cofactor.

A number of monomeric enzymes are **proteases** (or **proteolytic enzymes**), i.e. they catalyse the hydrolysis of peptide bonds in other proteins. In order to prevent them doing generalized damage to all cellular proteins, they are often synthesized in an inactive form known as a **proenzyme** or **zymogen**, and activated as required. Such enzymes include the **serine proteases**, so called because of the presence in the active site of an essential serine residue, i.e. a serine residue whose presence is essential for enzymic activity (see section 10.1.2).

5.1.2 The serine proteases

The serine proteases **chymotrypsin, trypsin** and **elastase**, which are produced in an inactive form by the mammalian pancreas, form a closely related group of enzymes. Although only about 40% of the primary structure is common to all three enzymes, most of the catalytically important amino acid residues correspond exactly. X-ray crystallography studies have also shown that their tertiary structures are very similar.

They are believed to function by an identical mechanism (see section 11.3.2) and show a similar pH optimum of about pH 8. All are **endopeptidases**, hydrolysing peptide bonds in the middle of polypeptide chains, but their specificities are different. **Chymotrypsin** has a large hydrophobic binding pocket which will bind **phenylalanine, tryptophan** and **tyrosine** side chains, enabling cleavage of the peptide bond at the carbonyl side of one of these residues.

In **trypsin**, aspartate replaces serine at the bottom of the binding pocket, giving this enzyme a specificity for cleaving bonds adjacent to amino acid residues with basic side chains, i.e. **lysine** or **arginine**. In the case of **elastase,** two glycine residues at the mouth of the binding pocket in chymotrypsin or trypsin are replaced by valine and threonine, whose bulky side chains block the pocket and result in the enzyme specifically cleaving bonds adjacent to residues with small non-polar side chains, e.g. **alanine.**

Chymotrypsin is synthesized in the pancreas as the zymogen chymotrypsinogen (or pre-chymotrypsin). This is a single polypeptide chain of 245 residues containing five intra-chain disulphide bridges. On passing into the intestine, where proteolytic enzymes are required to digest dietary proteins, chymotrypsinogen is attacked by trypsin. This breaks the peptide bond between arginine-15 and isoleucine-16, producing π-chymotrypsin. Already the molecule has full enzymatic activity, but further changes then take place: a dipeptide is removed from positions 14 and 15 by the action of another molecule of π-chymotrypsin, producing δ-chymotrypsin, and further chymotrypsin digestion removes a dipeptide from positions 147 and 148 to give the final product, α-chymotrypsin (Fig. 5.1).

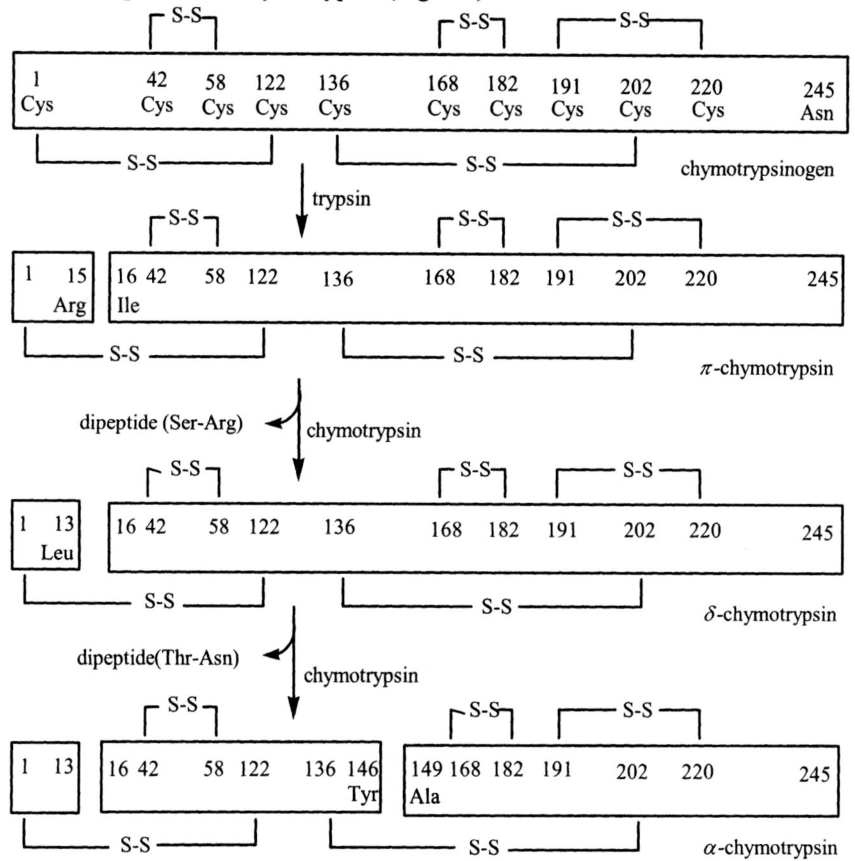

Fig. 5.1 – Diagrammatic representation of the formation of bovine α-chymotrypsin from the inactive zymogen chymotrypsinogen, as elucidated by the work of Blow and colleagues (1968).

Chymotrypsin contains three polypeptide chains linked by disulphide bridges, so it is not strictly a monomeric enzyme, but the sequential numbering system of the original chymotrypsinogen molecule is usually maintained. In contrast, **trypsin** *is* a genuine monomeric enzyme. Trypsinogen lacks nine amino acid residues at the N-terminus, by comparison with chymotrypsinogen, so cannot form the equivalent of the 1-122 disulphide bridge. The action of enteropeptidase (or trypsin itself) in the intestine removes a hexapeptide from the N-terminus of trypsinogen to produce the active trypsin, which is equivalent to the main chain of π-chymotrypsin and has the same N- and C-termini. **Elastase** is similarly produced from its corresponding zymogen by the action of trypsin.

The similar primary structures and almost identical tertiary structures of these three enzymes suggest that they evolved from a common ancestor by **divergent evolution.** The gene for the ancestral enzyme may have been duplicated several times, enabling different **mutations** (accidental changes of base sequence) in these duplicated genes to result in the production of slightly different enzymes.

Serine proteases involved in **blood clotting,** e.g. thrombin, and some **bacterial serine proteases** (e.g. from *S. griseus*) may also have evolved from this common ancestor, since their structures are similar to those of the enzymes from mammalian pancreas. In contrast, other bacterial serine proteases, e.g. **subtilisin** from *Bacillus amyloliquifaciens,* have very different primary and tertiary structures from those of the mammalian serine proteases. However, the active site structures and mechanism of action of all these enzymes are almost identical. This may suggest **convergent evolution,** i.e. the acquisition of similar characteristics from different starting materials by independent evolutionary pathways.

5.1.3 Some other monomeric enzymes

Pepsin A, like the pancreatic serine proteases, plays a role in the digestion of proteins eaten by mammals. It is called an **acid protease** because it functions at the low pH values found in the stomach. Peptide fragments are removed from the inactive form, pepsinogen, by the action of acid or other pepsin molecules to produce the active enzyme. This has a preference for cleaving bonds with a non-polar amino acid residue on either side. Another acid protease found in the stomach is **chymosin (rennin)**. Others are found in micro-organisms.

A group of **thiol proteases**, similar in structure to each other, are found in plants. These include **papain**, from the papaya fruit, and **ficain** (formerly **ficin**), from figs. Other thiol proteases, of different structure, are found in bacteria and mammalian lysosomes. The essential cysteine residue in each of these enzymes plays a similar role to that of serine in the serine proteases.

Several **exopeptidases**, which remove terminal amino acid residues from polypeptide chains, are well known. Bovine pancreatic **carboxypeptidase A**, a monomeric enzyme containing one zinc ion per molecule, will break the peptide bonds linking C-terminal non-polar amino acids to the rest of the chain. It is produced when trypsin removes peptide fragments from the zymogen, procarboxypeptidase A. A very similar enzyme, **carboxypeptidase B**, which has a specificity for C-terminal amino acids with basic side chains, is also secreted as a zymogen by bovine pancreas.

Not all monomeric enzymes act on proteins. Well-known ones which hydrolyse other substrates include **ribonuclease** (section 11.3.3) and **lysozyme** (section 11.3.4).

5.2 OLIGOMERIC ENZYMES

5.2.1 Introduction

Oligomeric proteins consist of two or more polypeptide chains, which are usually linked to each other by non-covalent interactions and never by peptide bonds. The component polypeptide chains are termed **sub-units** and may be identical to or different from each other. If they are identical, they are sometimes called **protomers. Dimeric proteins** consist of **two, trimeric proteins** of **three** and **tetrameric proteins** of **four** sub-units. The molecular weight is usually in excess of 35 000.

The vast majority of known enzymes are oligomeric: for example, all of the enzymes involved in glycolysis possess either two or four sub-units. It is, therefore, reasonable to assume that the sub-units of oligomeric proteins gain properties in association that they do not have in isolation. Such enzymes are not synthesized as inactive zymogens, but their activities may be regulated in a far more precise way by **feed-back inhibition**. This is possible because many oligomeric proteins exhibit **allostery**, i.e. their different binding sites interact (see Chapters 12 and 13). Some examples of oligomeric enzymes are considered below, in order to see what other advantages may result from the association of sub-units.

5.2.2 Lactate dehydrogenase

Vertebrate **lactate dehydrogenase (LDH)** is an example of an oligomeric enzyme where each sub-unit has the same function, in this case to catalyse the conversion of lactate to pyruvate (see equation 1.4 in section 1.3.4).

The enzyme, as found in many species, is a tetramer of molecular weight 140 000. However, although each sub-unit has a molecular weight of about 35 000, two types, of different amino acid composition, are found within each species: they are the **M-form**, which predominates in **skeletal muscle** and other largely anaerobic tissues; and the **H-form**, the predominant sub-unit in the **heart**. The two types of sub-unit are produced by separate genes. Each monomer is catalytically inactive, but it can combine with others of the same or different type to produce the active tetrameric enzyme. All combinations of H and M sub-units are equally possible, so five **isoenzymes** of LDH can exist: H_4, H_3M, H_2M_2, HM_3 and M_4 (termed LDH_1– LDH_5 respectively).

Although these catalyse the same reaction, they do so with different characteristics (the properties of H_3M, H_2M_2 and HM_3 being intermediate between those of H_4 and M_4) which enables the different isoenzymes to play different physiological roles. The precise mechanisms involved are far from clear, but Nathan Kaplan and his colleagues (1964) proposed an **aerobic-anaerobic hypothesis** whose major features may be seen if we consider LDH in its metabolic context (Fig. 5.2). Pyruvate may be produced from carbohydrates (by glycolysis) or from amino acids. Under anaerobic conditions, it may undergo LDH-mediated conversion to lactate.

When oxygen is freely available, pyruvate is metabolized to enter the tricarboxylic acid cycle. The tricarboxylic acid cycle and the pathway of glycolysis are important from the point of view of making energy available in a suitable form within the cell. Both lead to the synthesis of ATP, an important intermediate in energy metabolism (section 11.5.4). Lactate can only be produced from pyruvate and can only be metabolized back to pyruvate.

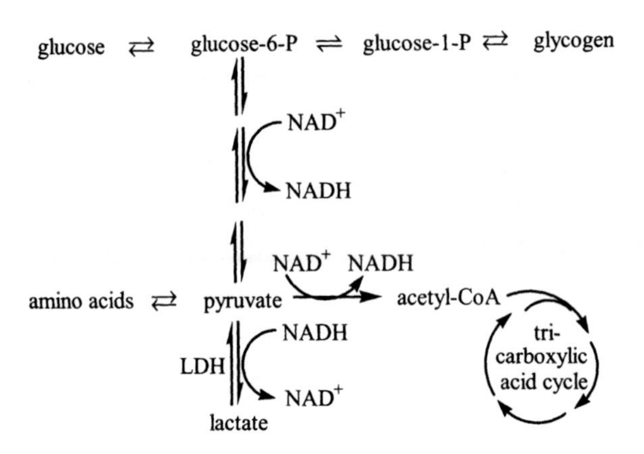

Fig. 5.2 – The main features of pyruvate metabolism in vertebrates (simplified for clarity).

Under aerobic conditions, most of the pyruvate formed is channelled into the tricarboxylic acid cycle to ensure maximum ATP production. Thus, tissues which have a plentiful and constant oxygen supply do not normally require much lactate production to take place and tend to be rich in the H_4 isoenzyme: this converts pyruvate to lactate at a relatively low rate, the process possibly even being inhibited by pyruvate. However, such tissues may require to utilize lactate (produced elsewhere) as substrate and, when lactate levels are high, forcing the reaction in favour of net pyruvate formation, the inhibiting effect of pyruvate on H_4 is removed.

Under anaerobic conditions, the tricarboxylic acid cycle cannot operate, leaving the cell dependent upon glycolysis for ATP production. Without a constant supply of NAD^+, this too would break down, but the LDH-mediated conversion of pyruvate to lactate can ensure that NAD^+ levels are maintained. Therefore, in tissues which can become oxygen-starved, an LDH isoenzyme (M_4) with a high capacity for converting pyruvate to lactate is required. The lactate produced eventually finds its way to the heart or liver, via the blood stream, and pyruvate is reformed.

The Kaplan hypothesis, as outlined above, seems quite plausible, and is consistent with much of the available experimental evidence, but not all of it. For example, although the M_4 isoenzyme has a higher turnover number (see section 7.1.3) than H_4 for pyruvate to lactate conversion, it apparently has a *lower* affinity than H_4 for the substrate. Also, the predominant isoenzyme in human liver is M_4, whereas a predominance of H_4 might have been expected from the aerobic-anaerobic hypothesis (see also section 14.2.1). Moreover, this hypothesis does not explain why five (and not just two) different isoenzymes are required, presumably enabling each tissue to have the type (or types) appropriate to its needs.

Nevertheless, whatever the precise reasons for the presence of five isoenzymes of lactate dehydrogenase within an organism, it is apparent that an arrangement enabling this to be achieved with only two different sub-units (and two genes) is more efficient than the possible alternative of having five different active monomeric isoenzymes (requiring five genes).

5.2.3 Lactose synthase

Mammary gland **lactose synthase** is an example of an oligomeric enzyme where a non-functional sub-unit modifies the behaviour of a functional sub-unit. This enzyme, as isolated from milk, consists of two sub-units: one of these is a catalytically inactive protein, **α-lactalbumin**, found only in mammary gland; the other is ***N*-acetyllactosamine synthase**, better known as **β-1,4-galactosyltransferase**, an enzyme present in most tissues. In the absence of α-lactalbumin, β-1,4-galactosyltransferase catalyses the reaction:

$$\text{UDP-galactose} + N\text{-acetylglucosamine} \rightleftharpoons \text{UDP} + N\text{-acetyllactosamine} \quad (5.1)$$

This is important in the synthesis of the carbohydrate components of glycoproteins. The enzyme is also produced and stored in the mammary gland during pregnancy, when levels of α-lactalbumin are low. After the birth of the baby, reduced synthesis of the hormone progesterone in the mother leads to increased synthesis of the luteotrophic hormone (prolactin), stimulating the production of α-lactalbumin in the mammary gland. This combines with the stored β-1,4-galcosyltransferase to form lactose synthase, an enzyme which facilitates production of the lactose component of the milk required for the new-born baby. Lactose synthase catalyses the reaction: UDP-galactose + glucose $\rightleftharpoons$ UDP + lactose. Thus it can be seen that the presence of the α-lactalbumin sub-unit changes the specificity of the enzyme, causing it to transfer galactose to glucose rather than to *N*-acetylglucosamine. The overall process involves conformational changes in the region of the active site of the enzyme (see section 14.2.6).

5.2.4 Tryptophan synthase

Bacterial **tryptophan synthase** is an example of an oligomeric enzyme which contains two different functional sub-units. The enzyme catalyses the reaction:

$$\text{indole-3-glycerol-P} + \text{L-serine} \rightarrow \text{L-tryptophan} + \text{glyceraldehyde-3-P} \quad (5.2)$$

It can be dissociated into two α sub-units, each of molecular weight 29 000, and a β_2 sub-unit, of molecular weight 90 000. The β_2 sub-unit further dissociates in the presence of 4 M urea to give two β sub-units, each of which has a binding site for the coenzyme pyridoxal phosphate. The isolated α sub-units will catalyse the reaction:

$$\text{indole-3-glycerol phosphate} \rightleftharpoons \text{indole} + \text{glyceraldehyde-3-phosphate} \quad (5.3)$$

The isolated β_2 sub-unit also has catalytic activity, but for the reaction:

$$\text{indole} + \text{L-serine} \xrightarrow{\text{pyridoxal-P}} \text{L-tryptophan} \quad (5.4)$$

So, the different sub-units of tryptophan synthase can be seen to catalyse separate halves of the overall reaction. However, the rates of these partial reactions are less than 5% the rate of the reaction catalysed by the intact $\alpha_2\beta_2$ enzyme. Also, significantly, indole is not released from the intact enzyme. Clearly, the oligomer has a degree of organization not possessed by the isolated sub-units. In fact, as shown by Edith Miles and colleagues (1990), the intermediate compound, indole, is passed down a short tunnel, directly from the active site of an α sub-unit to that of a β sub-unit, greatly increasing the efficiency of the overall process.

Another example of an enzyme where different sub-units catalyse different, though linked, reactions is discussed in section 11.5.8.

5.2.5 The pyruvate dehydrogenase multienzyme complex

The **pyruvate dehydrogenase multienzyme complex** of bacteria and animal cells shows the same type of organization as tryptophan synthase (section 5.2.4), but on an even larger scale. The Enzyme Commission (section 1.3) recommended that such a complex should be regarded as a system of separate enzymes rather than as a single enzyme.

The pyruvate dehydrogenase complex enables pyruvate to enter the tricarboxylic acid cycle, by catalysing its overall conversion to acetyl-CoA:

$$\text{pyruvate} + \text{CoASH} + \text{NAD}^+ \rightarrow \text{acetyl-CoA} + \text{CO}_2 + \text{NADH} \quad (5.5)$$

Lester Reed and colleagues (1968) showed that the *E. coli* enzyme consists of 60 polypeptide chains and has a molecular weight of about 4 600 000. Three separate catalytic activities are present: **pyruvate dehydrogenase (E_1)** (which also catalyses a decarboxylation); **dihydrolipoyl acetyltransferase (E_2)**; and **dihydrolipoyl dehydrogenase (E_3)**. The reaction sequence is shown in Fig. 5.3.

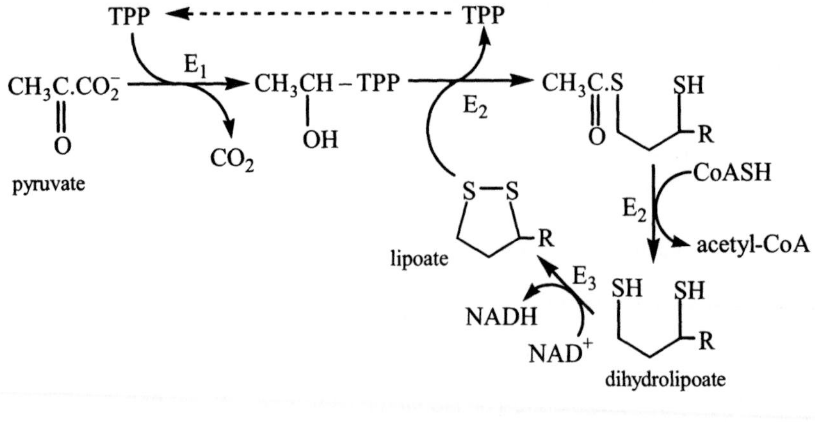

Fig. 5.3 – The reactions catalysed by the pyruvate dehydrogenase complex of *E. coli*. E_1, E_2 and E_3 are the separate enzymes making up the complex (see text for details).

The whole process takes place with the substrate bound to the enzyme, either directly or via the cofactors **thiamine pyrophosphate (TPP)** and **lipoate**.

TPP is associated with E_1 while the side chain of lipoate is covalently bound, by an amide linkage, to a lysyl residue of E_2. Hence the cofactor is actually **lipoamide** rather than **lipoate.** Protein E_3 also contains a prosthetic group, FAD. The reaction mechanism is discussed in section 11.5.6.

The enzyme complex is about 300 Å in diameter and its features have been observed by electron microscopy. It has a polyhedral structure, with each of the sub-units appearing approximately spherical. The complex is held together by non-covalent forces and may easily undergo dissociation. At alkaline pH, the sub-units of the E_1 protein can be separated from those of the E_2 and E_3 proteins. At neutral pH and high urea concentration, the E_2 and E_3 proteins can be separated from each other. If the various sub-units are mixed together at neutral pH in the absence of urea, the multienzyme complex will spontaneously reform, but E_1 and E_3 sub-units will not re-associate unless E_2 is present. The core of the complex consists of 24 sub-units of E_2, associated as trimers, with a symmetrical arrangement of E_1 and E_3 sub-units around this cubical core. Along each of the twelve edges of the cube is a dimer of E_1, and on each of the six faces of the cube is a dimer of E_3. It appears that, although the side chain of each lipoamide cofactor is attached to E_2, its length and flexibility enables the lipoyl head to make contact with the active groups on adjacent E_1 and E_3 molecules and thus link the various processes taking place.

SUMMARY OF CHAPTER 5

Monomeric proteins consist of a single polypeptide chain; oligomeric enzymes have two or more such chains. Only a few enzymes, mainly hydrolases, are monomeric. These are often synthesized as inactive zymogens and activated by the removal of peptide fragments.

Oligomeric enzymes are often allosteric, enabling their activities to be regulated by feed-back inhibition. Varying combinations of different sub-units making up an oligomeric enzyme can enable a wide range of expression to be obtained. The association of different sub-units can also increase the efficiency of an enzyme, since a sequence of reactions can take place without the intermediate products being released. Such organization is most notable in multienzyme complexes.

FURTHER READING

Berg, J. M., Tymocsko, J. L. and Stryer, L. (2006), *Biochemistry*, 6th edn., Freeman (Chapters 9, 10, 15-17, 27).

Huang, X., Holden, H. M.. and Raushel, F. M. (2001), Channelling of substrates and intermediates in enzyme-catalysed reactions, *Annual Review of Biochemistry*, **70**, 149-180.

Manchester, K. L. (1994), LDH: plus ça change, plus c'est la même chose - continuing problems with textbook presentations of the kinetic properties of the isoenzymes of lactate dehydrogenase, *Biochemical Education,* **22**, 91-93.

Nelson, D. L. and Cox, M. M., *Lehninger Principles of Biochemistry*, 4th edn., Worth (Chapters 6, 14-16).

Price, N. C. and Stevens, L. (1999), *Fundamentals of Enzymology*, 3rd edn., Oxford University Press (Chapters 3, 7).

Voet, D. and Voet, J. G. (2004), *Biochemistry,* 3rd edn., Wiley (Chapters 14-17, 21)

Part 2

Kinetic and Chemical Mechanisms of Enzyme-Catalysed Reactions

6

An Introduction to Bioenergetics, Catalysis and Kinetics

6.1 SOME CONCEPTS OF BIOENERGETICS

6.1.1 The first and second laws of thermodynamics

Bioenergetics, a branch of thermodynamics, is concerned with the changes in energy and similar factors as a biochemical process takes place, and not with the mechanism or speed of the process.

The **first law of thermodynamics** states that energy can neither be created nor destroyed, but can be converted into other forms of energy or used to perform work.

The **second law of thermodynamics** states that the entropy, or degree of disorder, of the universe is always increasing. However, it says nothing about a particular system under study, be it a mechanical engine, a living cell or a chemical reaction. Life, a state of high organization or low entropy, can be maintained for a while by the consumption of a highly organized form of chemical energy (food) and, in the case of photosynthetic organisms, light energy. This energy is either converted to a less organized form of energy (heat) or utilized to perform work. However, the approach towards ultimate thermodynamic equilibrium is certain: the death and subsequent decomposition of every organism.

Heat cannot be used to perform work by systems operating at constant temperature and constant pressure, such as the living cell. Thus, we have the concept of two forms of energy: that which can be used to perform work (called free energy) and that which cannot. For any system under study, processes can only take place spontaneously which result in a decrease in the free energy of the system, i.e. a transfer of free energy to the surroundings, unless there is interference from outside the system.

6.1.2 Enthalpy, entropy and free energy

In the terminology of thermodynamics, a **closed system** is one which can exchange energy but not matter with its surroundings.

The exchange of energy must involve thermal transfer and/or the performance of work. If, in a closed system at constant temperature and pressure, a process takes place which involves a transfer of heat to or from the surroundings and a change in volume of the system, then from the first law of thermodynamics, $\Delta E = \Delta H - P\Delta V$, where ΔE is the increase in **intrinsic energy** of the system, ΔH is the increase in **enthalpy** and $P\Delta V$ is the **work done** on the surroundings by increasing the volume of the system by ΔV at constant pressure P and temperature T. The enthalpy change is defined simply as the quantity of heat absorbed by the system under these conditions and can be determined by calorimetric experiments.

Under these conditions, the increase in **entropy** of the surroundings is $-\Delta H/T$. If the process took place under conditions of thermodynamic reversibility, i.e. infinitely slowly, the increase in entropy of the system, ΔS, would be $\Delta H/T$. For the process to take place spontaneously under thermodynamically *irreversible* conditions, ΔS must be greater than $\Delta H/T$, giving an overall increase in the entropy of the system plus surroundings, as required by the second law of thermodynamics. Hence, $\Delta S - (\Delta H/T) > 0$, and $\Delta H - T\Delta S < 0$.

J. Willard Gibbs (1878) defined the increase in **free energy** of the system, ΔG, as $\Delta G = \Delta H - T\Delta S$. So, for any spontaneous process at constant temperature and pressure, $\Delta G < 0$.

6.1.3 Free energy and chemical reactions

For a chemical reaction, the change in the **Gibbs free energy function (ΔG)** is the energy which is available to do **work** (for example, osmotic work, muscular work or biosynthetic work) as the reaction proceeds from given concentrations of reactants and products (the **'initial conditions'**) to **chemical equilibrium.** Consider a reaction:

$$A \rightleftharpoons B$$

If, at given concentrations of A and B, the sign of ΔG is negative, then the system would lose free energy to its surroundings as the reaction proceeded to equilibrium in the direction written, i.e. it would be energetically favourable for the reaction to proceed in that direction. Assuming no interference with the system, the concentration of B would increase and that of A would decrease until chemical equilibrium was reached.

If, on the other hand, at given concentrations of A and B the sign of ΔG is positive, then the system could only make free energy available to its surroundings if the reaction proceeded to equilibrium in the *opposite* direction to that written, and this is the only process that could happen spontaneously. Again, assuming no interference with the system from outside, the concentration of B would decrease and that of A would increase until chemical equilibrium was reached.

If, under given conditions, the value of ΔG is zero, then chemical equilibrium has been attained.

It should be realised that, at each point, ΔG for the forward reaction is of equal value but opposite sign to ΔG for the reverse reaction.

The relationship between ΔG and the concentration of reactants and products may be seen by considering the reaction:

$$A + B \rightleftharpoons C + D$$

At a particular temperature (T K)

$$\Delta G = -RT\log_e\left(\frac{[C_{eq}][D_{eq}]}{[A_{eq}][B_{eq}]}\right) + RT\log_e\left(\frac{[C_0][D_0]}{[A_0][B_0]}\right) \tag{6.1}$$

where R is the gas constant

$[C_{eq}]$ is the equilibrium concentration of C (and similarly for $[D_{eq}]$ etc.)

$[C_0]$ is the initial concentration of C (and similarly for $[D_0]$ etc.)

$\log_e(\ ) = \ln(\ ) = 2.303 \log_{10}(\ )$

However, by definition, the **equilibrium constant K_{eq}** $= \left(\dfrac{[C_{eq}][D_{eq}]}{[A_{eq}][B_{eq}]}\right)$ (6.2)

$$\therefore \Delta G = -RT\log_e K_{eq} + RT\log_e\left(\frac{[C_0][D_0]}{[A_0][B_0]}\right) \tag{6.3}$$

Under the special conditions where the initial concentration of all reactants and products are 1.0 mol l^{-1},

$\Delta G = -RT \log_e K_{eq} = \Delta G^{\ominus}$ (called the **standard free energy change**). (6.4)

Thus, in general,

$$\Delta G = \Delta G^{\ominus} + RT\log_e\left(\frac{\text{product of initial concentrations of products}}{\text{product of inital concentrations of reactants}}\right) \tag{6.5}$$

6.1.4 Standard free energy

The **standard free energy change, $\Delta G^{\ominus}$**, is a useful concept. As we have seen above, it can be calculated if the equilibrium constant of a reaction is known:

$$\Delta G^{\ominus} = -RT\log_e K_{eq} = -2.303\ RT\log_{10} K_{eq} \tag{6.6}$$

It is also the difference between the standard free energy of formation of the reactants and the standard free energy of formation of the products, each term being adjusted to the stoichiometry of the reaction equation. For a reaction where a molecules of A react with b molecules of B: $a\text{A} + b\text{B} \rightleftharpoons c\text{C} + d\text{D}$

$$\Delta G^{\ominus} = (c.G^{\ominus}_C + d.G^{\ominus}_D) - (a.G^{\ominus}_A + b.G^{\ominus}_B) \tag{6.7}$$

where $G^{\ominus}_C$ is the **standard free energy of formation** of C, etc.

The **standard free energy of formation** is a measure of the total amount of free energy a compound can yield on complete decomposition. Values for many compounds and groups have been calculated and are available in the literature. Standard free energies of formation are additive, and so are standard free energy changes.

Therefore, if the standard free energy change for A $\rightleftharpoons$ B is $\Delta G^{\ominus}_1$ and that for B $\rightleftharpoons$ C is $\Delta G^{\ominus}_2$, then the standard free energy change for A $\rightleftharpoons$ C can be determined by addition and is $\Delta G^{\ominus}_1 + \Delta G^{\ominus}_2$:

Reaction	*Standard free energy change*
A $\rightleftharpoons$ B	$\Delta G^{\ominus}_1$
B $\rightleftharpoons$ C	$\Delta G^{\ominus}_2$
A $\rightleftharpoons$ C	$\Delta G^{\ominus}_1 + \Delta G^{\ominus}_2$

Standard free energies for redox reactions can be calculated from the **standard electrode potential**, if this is known:

$$\Delta G^{\ominus} = -nF\Delta E_{\ominus} \tag{6.8}$$

where n is the number of electrons transferred; F is the Faraday number; and $\Delta E_{\ominus}$ is the standard electrode potential $= E_{\ominus}$(oxidising couple) – $E_{\ominus}$(reducing couple).

So, $\Delta G^{\ominus}$ can be calculated by a variety of methods, of which that from K_{eq} is particularly useful. Strictly speaking, $\Delta G^{\ominus}$ refers to the standard free energy change at pH 0. Biochemists, therefore, prefer to use instead the term $\Delta G^{\ominus\prime}$, calculated at some defined physiological pH, usually pH 7. Apart from the pH difference, the meanings of $\Delta G^{\ominus}$ and $\Delta G^{\ominus\prime}$ are the same. If ΔG at pH 7 is being calculated for a reaction involving H^+, the $[H^+]$ term has already been taken into account if $\Delta G^{\ominus\prime}$ is used in the calculation, but needs to be included if $\Delta G^{\ominus}$ is used instead.

For reactions where water is a reactant or product, its concentration is set by convention at 1.0 mol l^{-1} to remove it from the calculations, since water is always present in excess.

6.1.5 Bioenergetics and the living cell

Although it is ΔG and not $\Delta G^{\ominus\prime}$ which actually determines how a reaction will proceed, the relationship between the two terms means that the latter will give some idea of whether the reaction is likely to go largely in the forward or the reverse direction over most concentrations of reactant and product, or whether it will be a finely balanced reaction. However, it should be understood that this relies on the assumption that the reaction is allowed to proceed without interference from outside, which would generally be the case in a test-tube experiment.

In the living cell, in contrast, such an assumption is unlikely to be valid, for free energy made available from one reaction can be used to drive another in an energetically unfavourable direction, provided the two reactions have a **common intermediate** (this is termed the **principle of common intermediates**). Such coupled reactions often involve ATP, for this can be readily synthesized within the cell by either **substrate-level phosphorylation** or **oxidative phosphorylation**, yet is readily hydrolysed to ADP or AMP with the release of much free energy. For example, for the reaction, ribulose-5-phosphate + $P_i \rightleftharpoons$ ribulose-1,5-bisphosphate, $\Delta G^{\ominus\prime}$ is + 13.4 kJ mol^{-1}. This is a large enough positive value to suggest that the reaction should go in the direction of ribulose-5-phosphate formation at most concentrations of reactant and product, but in practice it goes almost exclusively in the opposite direction, due to coupling with ATP breakdown.

Reaction	*Standard free energy change*
ribulose-5-P + $P_i \rightleftharpoons$ ribulose-1,5-bisP	+13.4 kJ mol^{-1}
ATP $\rightleftharpoons$ ADP + P_i	−31.0 kJ mol^{-1}
ribulose-5-P + ATP $\rightleftharpoons$ ribulose-1,5-bisP + ADP	−17.6 kJ mol^{-1}

Another consideration in the living cell is that concentrations of reactants and products are constantly changing, due to the effects of other reactions and to the intake of food. All this has the effect of keeping most reactions away from chemical equilibrium. If they were all allowed to proceed to equilibrium, there would be no free energy available to perform work, and life would be impossible.

6.2 FACTORS AFFECTING THE RATES OF CHEMICAL REACTIONS

6.2.1 The collision theory

Molecules can react only if they come into contact with each other. Therefore, any factor which increases the rate of collisions, e.g. increased concentration of the reactants or increased temperature, will increase the reaction rate (according to the principles of Svante Arrhenius and Jacobus van't Hoff).

However, not all colliding molecules react. This can be partly explained by steric reasons, for not all collisions will result in the appropriate groups of molecules coming into contact, particularly if the reactants are complex. A further and more important reason is that not all colliding molecules possess between them sufficient energy to undergo a reaction.

6.2.2 Activation energy and the transition-state theory

Not all molecules of the same type will possess the same amount of energy, taking all forms of energy into account. The energy of an individual molecule will depend, for example, on what collisions that molecule has recently been involved in.

In order for a reaction to take place, colliding molecules must have sufficient energy to overcome a **potential-barrier** known as the **energy of activation**. This is true even of energetically favourable reactions, i.e. those which can proceed spontaneously with liberation of free energy.

The fact that a reaction *can*, in general, proceed spontaneously does not mean it *will* necessarily do so *in all circumstances.* By way of analogy, consider a ball on a hillside. The ball would tend to roll downhill; it would not spontaneously roll uphill. Nevertheless, if a stone was in its path, the ball would stay where it was. The stone would be the **potential-barrier** which would have to be overcome before the ball could roll down the hill. It is a similar potential-barrier which stops human beings bursting spontaneously into flame, for we burn with a great liberation of energy, if heated to a high-enough temperature to begin the process.

The best explanation for the requirement for activation energy in a chemical reaction is the **transition-state theory**, developed by Henry Eyring. This postulates that every chemical reaction proceeds via the formation of an unstable intermediate between reactants and products. Consider the hydrolysis of an ester:

$$R'C(=O)OR'' + H_2O \longrightarrow R'C(=O)OH + R''OH \tag{6.9}$$

It is believed that the reaction mechanism involves the addition of water to the ester to form a transition-state compound having regions of positive and negative charge which cannot be stabilized:

$$R'C(=O)OR'' + H_2O \longrightarrow \overset{\delta-}{O}{\cdots}C(R')(OR''){\cdots}\overset{\delta+}{O}H_2 \longrightarrow R'C(=O)OH + R''OH \tag{6.10}$$

An unstable compound, by definition, will break down to give a more stable one and so must possess more free energy than the stable compound. A free energy profile of a reaction involving an unstable transition-state, such as the one above, is given in Fig. 6.1.

The activation energy is, therefore, the energy needed to form the transition-state from the reactants. The transition-state is unstable, and will very quickly break down to form the products (or back to reactants), but no products can be formed from reactants unless the transition-state has been formed. The free energy of activation thus acts as a potential-barrier to the reaction taking place.

An estimate of the activation energy may be obtained from the reaction rate at different temperatures, as will now be discussed.

The probability that a particular molecule possesses energy in excess of a value E is $e^{-E/RT}$ (where R is the gas constant and T the absolute temperature).

For a molecule of energy E to react, it must collide with a molecule of energy at least $E^{\neq} - E$, where $E^{\neq}$ is the **activation energy**. (To be more precise, the free energy of activation, $\Delta G^{\neq}$, should be used instead of $E^{\neq}$, but we will disregard that fine distinction.) The probability of a colliding pair having sufficient energy to react is:

$$e^{-E/RT} . e^{(E^{\neq}-E)/RT} = e^{-E^{\neq}/RT} \tag{6.11}$$

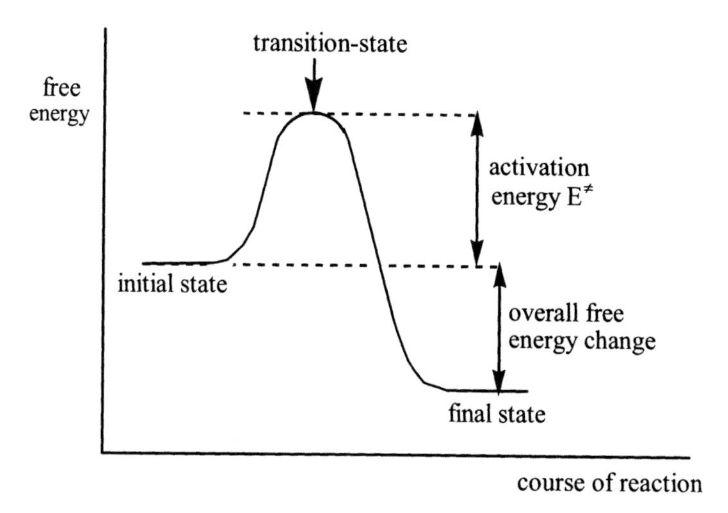

Fig. 6.1 – Free energy changes for an energetically favourable reaction proceeding via the formation of a transition-state.

This is consistent with the **Arrhenius equation**, first derived in experimentally in 1889:

$$k = \text{constant} \times \mathrm{e}^{-E^{\neq}/RT} \tag{6.12}$$

(where k is the **rate constant**, a characteristic of the reaction at temperature T; the other constants have been shown to be equal to PZ, where P is a **steric factor** and Z the **collision frequency**). Taking logs,

$$\log_e k = \log_e PZ - \left(\frac{E^{\neq}}{RT}\right) \tag{6.13}$$

If the rate constant $= k_{T_1}$ at absolute temperature T_1 and
$= k_{T_2}$ at absolute temperature T_2,

$$\log_e\left(\frac{k_{T_1}}{k_{T_2}}\right) = \log_e k_{T_1} - \log_e k_{T_2} = \left(\log_e PZ - \frac{E^{\neq}}{RT_1}\right) - \left(\log_e PZ - \frac{E^{\neq}}{RT_2}\right)$$

$$= \frac{E^{\neq}}{R}\left(\frac{1}{T_2} - \frac{1}{T_1}\right) = \frac{E^{\neq}}{R}\left(\frac{T_1 - T_2}{T_1 T_2}\right)$$

$$\therefore E^{\neq} = R\left(\frac{T_1 T_2}{T_1 - T_2}\right)\log_e\left(\frac{k_{T_1}}{k_{T_2}}\right) = 2.303R\left(\frac{T_1 T_2}{T_1 - T_2}\right)\log_{10}\left(\frac{k_{T_1}}{k_{T_2}}\right) \tag{6.14}$$

Assuming that the rate of reaction is proportional to the rate constant,

$$E^{\neq} = 2.303R\left(\frac{T_1T_2}{T_1 - T_2}\right)\log_{10}\left(\frac{v_{T_1}}{v_{T_2}}\right) \tag{6.15}$$

(where v_{T_1} is the rate of reaction at temperature T_1 and v_{T_2} is the rate of reaction at temperature T_2).

If the reaction rate (or the rate constant) has been determined at more than two temperatures, $E^{\neq}$ can be calculated by plotting $\log_{10}$ k (or $\log_{10}$ v) against $1/T$. The slope of the graph is $-E^{\neq}/2.303R$.

It is important to realize that the activation energy calculated depends on the assumptions made and cannot be used to decide between possible reaction mechanisms.

6.2.3 Catalysis

A catalyst accelerates a chemical reaction without changing its extent and can be removed unchanged from amongst the end-products of the reaction. It has no overall thermodynamic effect: the amount of free energy liberated or taken up when a reaction has been completed will be the same whether a catalyst is present or not.

In most cases, a catalyst acts by reducing the energy of activation. The catalyst, or part of it, combines with the reactants to form a different transition-state from that involved in the uncatalysed reaction; one which is more stable and, therefore, of lower energy (Fig. 6.2).

Note that the initial and final states are at the same free energy levels for the catalysed and uncatalysed reactions, and the overall free energy change as the reaction proceeds is also the same.

As a rough generalization, therefore, an uncatalysed reaction might proceed as follows:

$$A + B \xrightarrow{\text{very slow}} A \cdots B \xrightarrow{\text{very fast}} \text{products}$$

The same reaction, but in the presence of a catalyst C, might take place like this:

$$A + B + C \xrightarrow{\text{fast}} A \cdots B \cdots C \xrightarrow{\text{fast}} \text{products} + C$$

Because the catalyst stabilizes the transition-state, the breakdown of the intermediate to the products might be slower than in the uncatalysed reaction. Nevertheless, the overall rate is determined by the slowest step in the sequence, so the overall rate of the catalysed reaction will be faster than that for the uncatalysed reaction.

Catalysts are often **acids or bases**: acids stabilize the transition-state by donating a proton; bases by accepting a proton. The hydrolysis of an ester, which we considered earlier, may be catalysed by a base (X–O$^-$) as follows:

$$\underset{\text{OR}''}{\overset{\text{R}'}{\text{O=C}}} + \text{O}\begin{smallmatrix}\text{H}\\ \text{H}\end{smallmatrix} + \text{X-O}^- \rightarrow \text{O}\!\equiv\!\underset{\text{OR}''}{\overset{\text{R}'}{\text{C}}}\text{---}\overset{\text{H}}{\text{O}}\text{----H}\cdot\text{---}\overset{\delta-}{\text{O}}\text{—X} \rightarrow \underset{\text{OH}}{\overset{\text{R}'}{\text{O=C}}} + \text{R}''\text{OH} + \text{X-O}^- \tag{6.16}$$

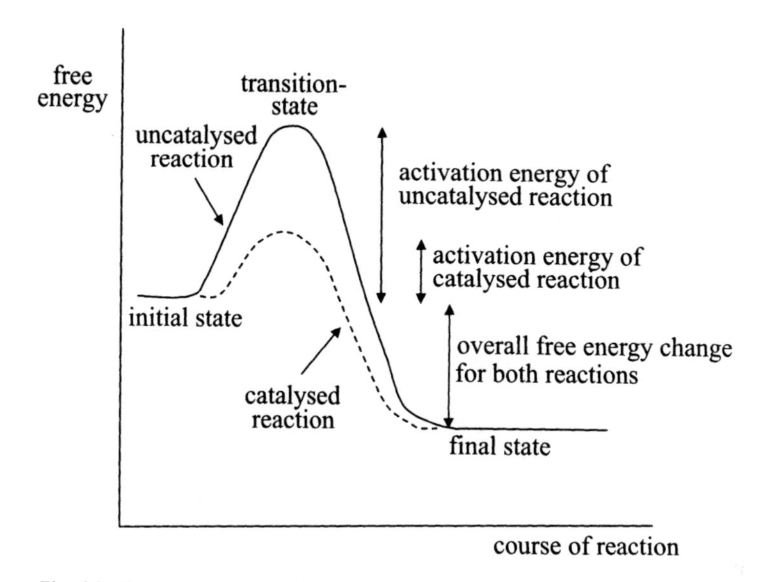

Fig. 6.2 – Free energy changes for an energetically favourable reaction, showing the effects of a catalyst

Other common forms of catalysis include **covalent catalysis**, where the transition-state is stabilized by changes involving covalent bonds, and **metal ion catalysis**, where the transition-state is stabilized by electrostatic interactions with a metal ion.

In this section we have looked at the free energy profiles of energetically favourable reactions, considering the changes taking place when the molecules of the reactant are converted to molecules of the product. Of course, most reactions are reversible, but note that the activation energy required for the reverse, energetically unfavourable, reaction in Figs 6.1 and 6.2 is the energy required to form the transition-state from the products and is thus much greater than the activation energy in the energetically favourable direction. The proportion of colliding product molecules possessing between them sufficient energy to overcome the potential-barrier will be even less than for the reactants. However, the greater the product concentration, the greater the number of molecules present capable of undergoing the reverse reaction. As we have seen, the overall trend, indicated by ΔG, towards chemical equilibrium depends on the actual concentrations of reactants and products present. The position of chemical equilibrium is not affected by the presence of a catalyst: the catalyst merely accelerates its attainment.

6.3 KINETICS OF UNCATALYSED CHEMICAL REACTIONS

6.3.1 The Law of Mass Action and the order of reaction

Kinetics is the study of reaction rates and the factors influencing them. It is not concerned with the chemical nature of the changes taking place.

All kinetic work is based on the **Law of Mass Action** proposed by Cato Maximilian Guldberg and Peter Waage in 1867. This states that the rate of a reaction is proportional to the product of the **activities** of each reactant, each **activity** being raised to the power of the number of molecules of that reactant taking part, as indicated by the reaction equation. For a reaction:

$$a\text{A} + b\text{B} \rightarrow \text{products}$$

the rate is proportional to (activity of A)^a × (activity of B)^b. For practical purposes, the term **activity** is usually replaced by **concentration**, although strictly speaking it is only equal to the concentration in ideal gases and in very dilute solutions.

The Law of Mass Action led to the concept of the **order of reaction.**

A **first-order reaction** is one which proceeds at a rate proportional to the concentration of *one* reactant. Thus, for a first-order reaction A→P taking place at a constant temperature and pressure in a dilute solution, reaction rate v at any time t is given by

$$v = -\text{d[A]}/\text{d}t = +\text{d[P]}/\text{d}t = k\text{[A]} \tag{6.17}$$

where v_0 = reaction rate at time t
k = rate constant
[A] = concentration of reactant A at time t
[P] = product concentration at time t
$-\text{d[A]}/\text{d}t$ = rate of decrease of [A]
$+\text{d[P]}/\text{d}t$ = rate of increase of [P].

A **second-order reaction** is one which proceeds at a rate proportional to the concentration of *two* reactants or to the *second power* of a single reactant. So, for a second-order reaction A + B → P, taking place at a constant temperature and pressure in a dilute solution,

$$v = -\text{d[A]}/\text{d}t = -\text{d[B]}/\text{d}t = +\text{d[P]}/\text{d}t = k\text{[A][B]} \tag{6.18}$$

the terms having the same general meaning as above. Similarly, for a second-order reaction of the form 2A → P,

$$v = -\text{d[A]}/\text{d}t = +\text{d[P]}/\text{d}t = k\text{[A]}^2 \tag{6.19}$$

A two-reactant reaction can, under certain circumstances, be regarded as a pseudo single-reactant one. For instance, reactions taking place in an aqueous medium and involving water and one other reactant are usually considered pseudo single-reactant since water will be in vast excess and its concentration will not change significantly during the course of the reaction.

It should also be noted that a **zero-order reaction** is possible. This is one whose rate is *independent* of the concentration of any of the reactants.

6.3.2 The use of initial velocity

For any reaction, starting with reactants only and measuring the appearance of product with time, a graph of the form represented in Fig. 6.3 is obtained.

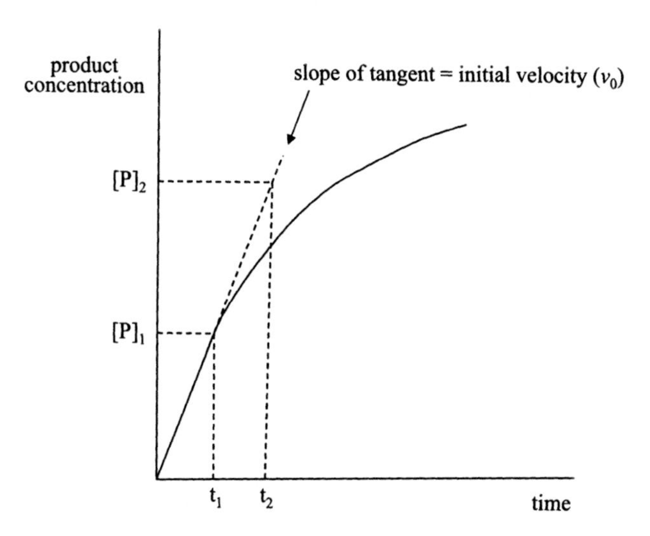

Fig. 6.3 – Graph of product concentration against time for a chemical reaction.

The rate of the reaction at any time t is the slope of the curve at that point. This may be constant for a short time at the start of the reaction and then decreases with the decreasing concentration of the reactant(s) as the reaction proceeds, finally falling to zero. At this point, either all the reactants have been converted into products or, much more commonly, a chemical equilibrium has been set up, with the rate of the forward reaction now being equal to the rate of the back reaction.

The **initial velocity** (v_0) of the reaction is the reaction rate at $t = 0$ and may be determined by drawing a tangent to the graph, as shown in Fig. 6.3. From this tangent,

$$v_0 = \frac{[P]_2 - [P]_1}{t_2 - t_1} \tag{6.20}$$

The units for v_0 are those used for product concentration, divided by those used for time. The importance of initial velocity is that it is a kinetic parameter determined for the reaction in a situation which can be easily specified. The concentrations of each reactant are known from the amounts actually added. This would not be true at any other point during the reaction. Also, since there are no products present at $t = 0$, no back reaction will be taking place.

It will be apparent that the **initial velocity** will depend on the **initial concentrations** of the reactants. For a single-reactant reaction, it has been seen in the previous section that, for the reaction to be first-order, $v = k[A]$. Thus, at time $t = 0$, $v_0 = k[A_0]$, where $[A_0]$ is the initial concentration of A.

Similarly, for the single-reactant second-order reaction, $v_0 = k[A_0]^2$ and for a single-reactant zero-order reaction, $v_0 = k[A_0]^0 = k$. The graphs corresponding to these equations are shown in Fig. 6.4. This approach can also be applied to reactions involving more than one reactant. If all the reactants but one are maintained at fixed concentrations, the order of reaction with respect to the variable reactant can be determined. The overall order of reaction will be the sum of the orders with respect to each of the individual reactants.

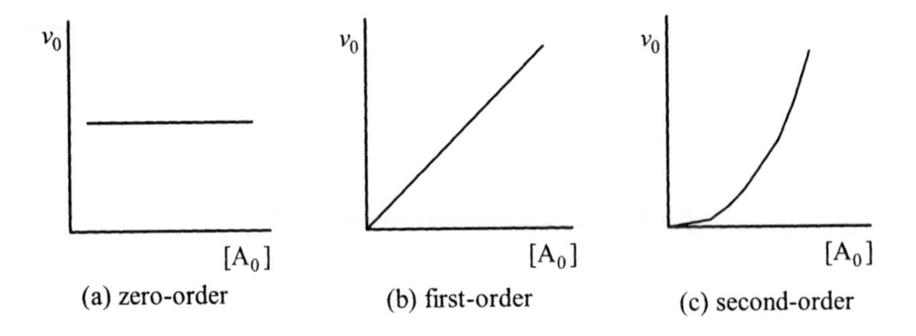

Fig. 6.4 – Graphs of initial velocity against initial reactant concentration for a single-reactant reaction.

6.4 KINETICS OF ENZYME-CATALYSED REACTIONS: AN HISTORICAL INTRODUCTION

Ludwig Wilhelmy, in 1850, showed that the rate of acid-hydrolysis of sucrose was proportional to the sucrose concentration, at constant acid concentration. From the terms defined in the previous section, therefore, we would say that this reaction was first-order with respect to sucrose.

When the identical reaction, but catalysed by the enzyme **invertase**, was similarly investigated, different results were obtained. Adrian Brown (1902) showed that, at low sucrose concentrations, the reaction was again first-order with respect to the substrate, but at higher concentrations it became zero-order.

This was found to be generally true for all single-substrate enzyme-catalysed reactions and for multi-substrate reactions where the concentrations of all the substrates but one were kept constant. A graph of initial velocity (v_0) against initial substrate concentration ($[S_0]$) at constant total enzyme concentration ($[E_0]$) was found to be a rectangular hyperbola, as shown in Fig. 6.5.

Such a graph has the general equation

$$v_0 = \frac{a[S_0]}{[S_0] + b} \qquad (6.21)$$

where a and b are constants. The constant a represents the maximum value of v_0 (called V_{max}) and b is the value of $[S_0]$ where $v_o = ½V_{max}$.

Although some enzymes do not give hyperbolic graphs (see Chapters 12 and 13), the attainment of a maximum initial velocity with increasing substrate concentration at constant total enzyme concentration is characteristic of all enzymes.

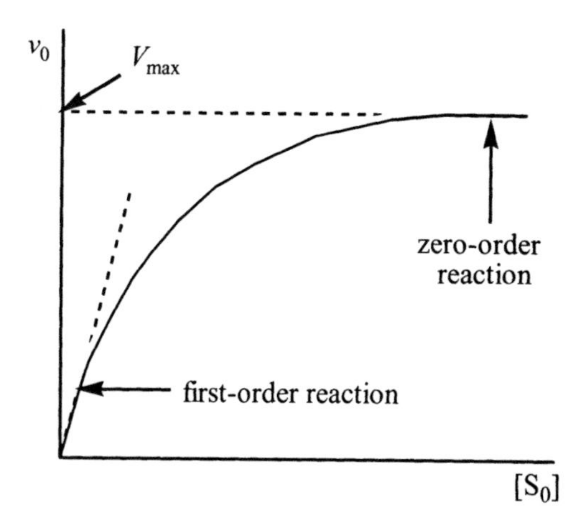

Fig. 6.5 – Graph of initial velocity against initial substrate concentration at constant total enzyme concentration for a single-substrate enzyme-catalysed reaction. Note that the curve is distorted to demonstrate the approach to V_{max}, which in fact is only achieved at an infinite substrate concentration.

The explanation for this feature was first given by Brown, with particular reference to the hydrolysis of sucrose, but subsequently found to be of general application. Enzyme and substrate combine to form an **enzyme-substrate complex**, which undergoes a further reaction to break down to enzyme and products:

$$E + S \xrightarrow{\text{rate constant } k_1} ES \xrightarrow{\text{rate constant } k_2} E + P$$

The overall rate of reaction (the rate of formation of P) must be limited by the amount of enzyme available and by the rate of breakdown of the enzyme-substrate complex. If the substrate concentration is sufficiently high, it will '**saturate**' the enzyme, i.e. force an immediate reaction with each available enzyme molecule to form an enzyme-substrate complex. Under these conditions, therefore, there will be no free enzyme present and the concentration of enzyme-substrate complex ([ES]) will be the total enzyme concentration present ($[E_0]$), making the overall rate of reaction $k_2[E_0]$ (from the **Law of Mass Action**). This is independent of substrate concentration and so cannot be increased by using still higher substrate concentrations. It is, therefore, the maximum initial velocity possible at this enzyme concentration. Hence

$$V_{max} = k_2[E_0] \tag{6.22}$$

In contrast, at very low substrate concentrations, the enzyme will be far from saturated and the overall rate of reaction will be limited by the rate at which enzyme and substrate molecules react to form an enzyme-substrate complex. At constant enzyme concentration this will be proportional to the substrate concentration and, therefore, a first-order reaction will result.

At intermediate substrate concentrations, the enzyme will be partially saturated with substrate and the order of reaction will be somewhere between zero-order and first -order.

Note that the terms 'saturated' and 'partially saturated' refer to the population of enzyme molecules, and not to each individual molecule. An enzyme molecule which binds a single substrate molecule cannot in itself be partially saturated at any given moment, but a population of such molecules can be, some being substrate-bound and some free.

The existence of an enzyme-substrate complex was first demonstrated experimentally in 1936 by spectroscopic studies on the enzymes **peroxidase** and **catalase**. Since then, the presence of such complexes has been amply documented by X-ray crystallography, ESR, NMR, fluorimetry and other techniques.

6.5 METHODS USED FOR INVESTIGATING THE KINETICS OF ENZYME-CATALYSED REACTIONS

6.5.1 Initial velocity studies

Initial velocity studies have been found particularly useful for investigating the kinetics of enzyme-catalysed reactions. In addition to the general advantages of such studies, mentioned earlier, there is the extra reason here that enzymes are often unstable in solution, so the restriction of investigations to v_0 determinations give the best chance of avoiding errors caused by loss of enzyme activity with time.

The initial velocity of a reaction is usually determined from a graph of product concentration against time, as shown in Fig. 6.3. Alternatively, a direct instrumental reading known to be proportional to product concentration (e.g. absorbance units) may be plotted against time. The rate of reaction could also be determined from the disappearance of substrate as the reaction proceeds, but in general it is considered a better technique to measure the appearance of something from an initial value of nothing rather than the disappearance of something from an initial large value, particularly when it is the initial rate which is all-important.

The experiments are performed at constant temperature and pH using a method which enables the course of the reaction to be monitored continuously, and, preferably, automatically. The actual technique used will depend on the reaction being investigated.

If there is a difference between substrate and product in the absorbance of light of a particular wavelength, **spectrophotometric** or **colorimetric techniques** (preferably the former) may be used. The **molar absorption coefficient** of a substance, if known, will enable the actual concentration of that substance to be calculated from an absorbance reading, since absorbance = molar absorption coefficient x concentration (mol l^{-1}) x cell light path (cm).

Of particular suitability for investigation by spectrophotometric techniques are the wide range of oxidation/reduction reactions involving NAD^+/NADH or $NADP^+$/NADPH interconversion.

The reduced forms of these dinucleotides have an absorbance peak at 340 nm, whereas the oxidized forms do not (Fig. 6.6), thus making it easy to follow the course of their interconversion by monitoring absorbance at this wavelength.

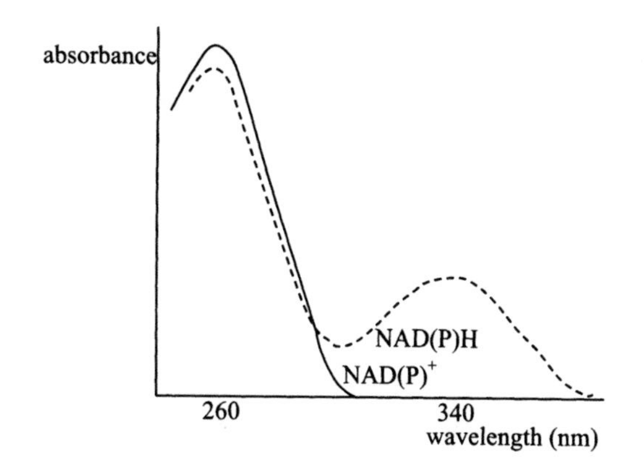

Fig. 6.6 – The ultraviolet absorbance spectra of $NAD(P)^+$ and NAD(P)H.

This interconversion of these substances may also be followed by **fluorimetric techniques,** which give greater sensitivity than spectrophotometric ones.

Reactions where gases are evolved or taken up as the reaction proceeds can be investigated by **manometric** techniques. The reaction is performed in an airtight vessel coupled to a manometer, which records changes in volume or pressure.

Ion-selective electrodes (e.g. the oxygen electrode) can be used where relevant to a particular reaction. More generally, other **electrochemical methods** (e.g. conductometry) can be applied wherever there are differences in electrical properties between substrate and product. These and other techniques are discussed in more detail in section 18.1.

Methods which are not immediately applicable to the investigation of a particular reaction may nevertheless be used if the reaction is coupled to a suitable **indicator reaction.** The principle here is to add excess of the indicator enzyme, so the rate of the indicator reaction will be a measure of the rate of the reaction under investigation. For example, **glucose oxidase** catalyses:

$$\beta\text{-D-glucose} + O_2 \rightarrow \text{gluconolactone} + H_2O_2 \qquad (6.23)$$

This reaction can be followed by manometric or oxygen electrode techniques, but is not suitable for investigation by spectrophotometric methods. However, if excess **peroxidase** is added to break down the hydrogen peroxide as it is formed, this reaction may be linked to the production of a coloured dye from a suitable chromogen, e.g. guaiacum:

$$H_2O_2 + \text{chromogen} \rightarrow H_2O + \tfrac{1}{2}O_2 + \text{dye} \qquad (6.24)$$

The rate of appearance of dye, determined spectrophotometrically, will be an indicator of the rate of reaction catalysed by glucose oxidase.

An alternative to the use of coupled reactions is to introduce an artificial substrate which, when reacting, results in a measurable change not provided by the natural substrate. An example is *p*-nitrophenyl phosphate, a colourless compound, which is hydrolysed to the yellow *p*-nitrophenol by the action of **alkaline phosphatase.**

6.5.2 Rapid-reaction techniques

In order to test some of the hypotheses made above, e.g. to investigate the formation and breakdown of the enzyme-substrate complex or complexes, it is necessary to use techniques capable of detecting changes taking place over time scales of the order of magnitude of 1 millisecond.

Of particular importance is a detailed kinetic study of the changes taking place over the first fraction of a second of the reaction (termed **transient kinetics).** This is usually performed using rapid mixing techniques of the **continuous-flow or stopped-flow** variety, the reaction being monitored by some suitable detector coupled to a display unit and computer.

With **continuous-flow** systems, introduced by Hamilton Hartridge and Francis (F. J. W.) Roughton, streams of enzyme and substrate converge and are pumped together at a fixed speed down capillary-bore tubing to ensure rapid mixing and elimination of dead-space. From a knowledge of the dimensions of the tube and the flow rate, the time taken from mixing to the arrival at the detector can be calculated. By altering the flow rate, this time of reaction can be changed. Also, several detectors can be placed at different points along the flow tube. The main disadvantage of continuous-flow methods is that they tend to be wasteful of reagents.

With **stopped-flow** techniques, developed by Britton Chance and by Quentin Gibson, solutions of enzymes and substrate are rapidly injected together into an observation chamber and the course of reaction monitored continuously.

The main limitation of both these techniques is the time taken to mix enzyme and substrate(s). This problem does not apply to the alternative approach, the investigation of **relaxation kinetics**. Here the enzyme and substrate(s) are mixed and the system allowed to come to equilibrium; then the position of the equilibrium is rapidly altered by a sudden temperature or pressure change, and the approach to the new position of equilibrium constantly monitored. This technique, pioneered by Manfred Eigen (1963), is particularly suitable for the investigation of reactions which are readily reversible, e.g. those catalysed by **aminotransferases**.

6.6 THE NATURE OF ENZYME CATALYSIS

Enzymes behave like any other catalysts in forming with the reactants a **transition-state** of lower **free energy** than that which would be found in the uncatalysed reaction. As we have seen, there is kinetic and spectroscopic evidence for the existence of **enzyme-substrate complexes**; however, these are not synonymous with transition-states. For a single-substrate reaction, the enzyme initially binds the substrate at a specific binding-site to form a relatively stable enzyme-substrate complex, this process taking place via the formation of an unstable transition-state. In the enzyme-substrate complex, the reacting groups are held in close proximity to each other and to the catalytic-site of the enzyme.

The catalysed reaction can now take place, via the formation of another unstable transition-state, to give the product. However, at this point the product may well still be bound to the enzyme, in which case another relatively-stable reaction intermediate, the **enzyme-product complex**, would exist before the free product was liberated.

The free energy profile of a reaction of this type is depicted in Fig. 6.7. For reactions involving several substrates and products, the process would be even more complicated, but the principles remain the same.

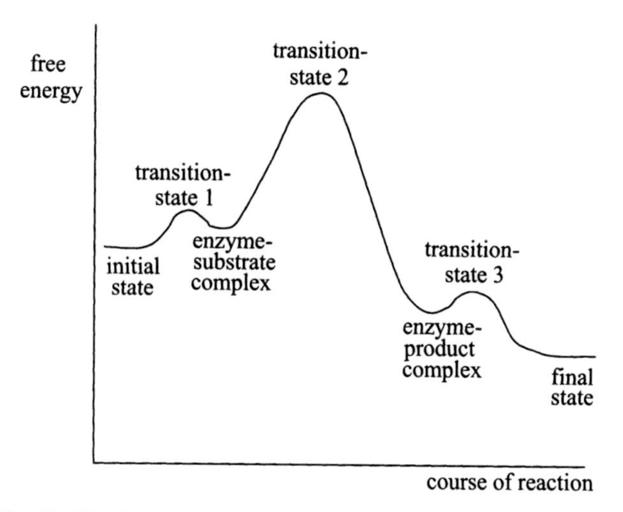

Fig. 6.7 – The free energy profile of an enzyme-catalysed reaction involving the formation of an enzyme-substrate and an enzyme-product complex.

The activation energy for the rate-limiting step of enzyme-catalysed reactions may be estimated as for uncatalysed reactions by investigating the rate of reaction at different temperatures.

For a reaction of the form:

$$\mathrm{E+S} \xrightarrow{k_1} \mathrm{ES} \xrightarrow{k_2} \mathrm{E+P} \text{ , or}$$

$$\mathrm{E+S} \xrightarrow{k_1} \mathrm{ES} \xrightarrow{k_2} \mathrm{EP} \xrightarrow{k_3} \mathrm{E+P} \ (\text{where ES} \rightarrow \text{EP is the rate-limiting step})$$

the overall rate of reaction is given by $k_2[\mathrm{ES}]$ and the best estimate of k_2 comes from V_{max} , since $V_{max} = k_2[\mathrm{E_0}]$.

For such a reaction investigated at absolute temperatures T_1 and T_2 , at the same initial enzyme concentration:

$$\begin{aligned}\text{Activation energy} &= 2.303R\left(\frac{T_1 T_2}{T_1 - T_2}\right)\log_{10}\left(\frac{k_2 \text{ at } T_1}{k_2 \text{ at } T_2}\right) \\ &= 2.303R\left(\frac{T_1 T_2}{T_1 - T_2}\right)\log_{10}\left(\frac{V_{max} \text{ at } T_1}{V_{max} \text{ at } T_2}\right)\end{aligned} \quad (6.25)$$

(the terms being as defined in section 6.2.2).

Relatively small changes in activation energy can greatly alter the rate of reaction: an enzyme which reduces the activation energy from 100 kJ mol^{-1} to 60 kJ mol^{-1}, perfectly reasonable figures, increases the reaction rate by about 10 million.

SUMMARY OF CHAPTER 6

Biochemical processes can only proceed spontaneously in such a direction that the free energy of the system, i.e. the energy that can be used to perform work, decreases. However, even energetically-favourable chemical reactions have to overcome a potential-barrier, known as the activation energy, before the reaction can take place. This is explained by the need to form unstable transition-states. Catalysts, including enzymes, act by allowing the formation of different, more stable, transition-states and, thus, reduce the activation energy. The position of chemical equilibrium is unchanged but is reached much faster than in the equivalent uncatalysed reaction.

The initial velocity of enzyme-catalysed reactions has a limiting value at each total enzyme concentration; this occurs when the enzyme is saturated with substrate. Enzymes react with substrates to form enzyme-substrate complexes. These are quite distinct from the transition-states which also occur as part of the process of enzyme catalysis.

FURTHER READING

Cornish-Bowden, A. (1995), *Fundamentals of Enzyme Kinetics,* 2nd edn., Portland Press (Chapters 1-3).

Lodish, H., Berk, A. *et al* (2004), *Molecular Cell Biology,* 5th edn., Freeman (Chapter 2).

Metiu, H. (2006), *Physical Chemistry: Thermodynamics* and *Physical Chemistry: Kinetics*, Garland.

Nelson, D. L. and Cox, M. M. (2004), *Lehninger Principles of Biochemistry,* 4th edn., Worth (Chapters 6, 13).

Nichols, D. G. and Ferguson, S. J. (1992), *Bioenergetics* 2, Academic Press.

Voet, D. and Voet, J. G. (2004), *Biochemistry,* 3rd edn., Wiley (Chapters 3, 14).

Wright, M. R. (1999), *Fundamental Chemical Kinetics*, Horwood.

PROBLEMS

(Assume $R = 8.314$ J K^{-1} mol^{-1}, $F = 96487$ J V^{-1}, activity = concentration.)

6.1 (a) The following reaction characteristics have been demonstrated at pH 7.5 and 310 K:

$$\text{aspartate} + \text{citrulline} \rightleftharpoons \text{argininosuccinate} + H_2O \qquad K'_{eq} = 1.6 \times 10^{-6}$$
$$\text{arginine} + \text{fumarate} \rightleftharpoons \text{argininosuccinate} \qquad K'_{eq} = 93$$
$$\text{arginine} + H_2O \rightleftharpoons \text{citrulline} + NH_4^+ \qquad K'_{eq} = 1.4 \times 10^{5}$$
$$\text{aspartate} + H_2O \rightleftharpoons \text{malate} + NH_4^+ \qquad K'_{eq} = 7.5 \times 10^{-3}$$

Calculate $\Delta G^{\ominus\prime}$ at 310 K and pH 7.5 for the reaction:

$$\text{fumarate} + H_2O \rightleftharpoons \text{malate}$$

If $\Delta H^{\ominus\prime}$ for this reaction is -16.6 kJ mol^{-1}, what will be the value of $\Delta S^{\ominus\prime}$ under these conditions?

(b) If the concentration of fumarate is 10^{-4} mol l^{-1} and of malate 9×10^{-4} mol l^{-1}, calculate ΔG for the formation of malate from fumarate at 310 K and pH 7.5. If a system with these initial concentrations was allowed to proceed to equilibrium, deduce the final concentrations of fumarate and malate.

6.2 For the reaction:

$$\text{malate} + NAD^+ \rightleftharpoons \text{oxaloacetate} + NADH + H^+,$$

$\Delta G^{\ominus\prime} = 27.96$ kJ mol^{-1} at pH 7.0 and 25°C.

(a) Calculate the difference in $E'_\ominus$ between the oxaloacetate/malate couple and the NAD^+/$NADH + H^+$ couple at pH 7.0 and 25°C.

(b) Calculate the value of ΔG for the reaction at pH 7.0 and 25°C when malate and NAD^+ are both present at 0.01 mol l^{-1} and oxaloacetate and NADH are both present at 0.02 mol l^{-1}.

(c) Calculate the concentration of oxaloacetate equal to concentration of NADH present in equilibrium with 0.05 mol l^{-1} each of malate and NAD^+ at pH 7.0 and 25°C.

6.3 A reaction, in the presence of a certain amount of enzyme, was found to have a V_{max} of 61.7 nmol s^{-1} at 25°C. In the presence of the same amount of enzyme at 37°C, the V_{max} was 120.8 nmol s^{-1}. Using sensitive detection techniques, it was possible to follow the course of the uncatalysed reaction, which had a rate of 30.2×10^{-3} nmol s^{-1} at 25°C and 88.4×10^{-3} nmol s^{-1} at 37°C. Calculate the activation energy for the catalysed and uncatalysed reactions.

6.4 For an uncatalysed reaction A + B $\rightleftharpoons$ C + D, the progress of the reaction was followed spectrophotometrically for different initial concentrations of A, the initial concentration of B always being 10 mmol l^{-1}. No products were present initially in any experiment. The following results were obtained:

$[A_0]$ (mmol l^{-1})	Absorbance at time t (min)					
	t = 0.5	1.0	1.5	2.0	2.5	3.0
10.0	0.073	0.127	0.180	0.234	0.272	0.292
8.0	0.062	0.105	0.148	0.191	0.224	0.247
6.0	0.051	0.085	0.116	0.149	0.175	0.191
4.0	0.040	0.061	0.083	0.104	0.123	0.135
2.0	0.030	0.041	0.051	0.062	0.070	0.075

Determine the order of reaction with respect to A.

6.5 (a) The following data were obtained at pH 7.5 and 37°C:
for the reaction glucose-6-P $\rightleftharpoons$ glucose-1-P, $K'_{eq} = 6.12 \times 10^{-2}$;
for the reaction fructose-6-P + UTP $\rightleftharpoons$ UDP-glucose + $2P_i$, $K'_{eq} = 4.52 \times 10^3$;
for the reaction glucose-1-P + UTP $\rightleftharpoons$ UDP-glucose + $(PP)_i$, $K'_{eq} = 1.00$;
for the reaction $2P_i \rightleftharpoons (PP)_i$, $K'_{eq} = 3.06 \times 10^{-5}$.

From this, determine the K'_{eq} at pH 7.5 and 37°C for the reaction:

$$\text{glucose-6-P} \rightleftharpoons \text{fructose-6-P}.$$

If the initial concentration of glucose-6-P is 9.0 mmol l^{-1} and of fructose-6-P is 2.0 mmol l^{-1}, what is the value of ΔG under these conditions?

(b) The reaction in the presence of hexose phosphate isomerase was found to have a V_{max} of 3.6 μmol min^{-1} at 25°C and of 7.0 μmol min^{-1} at 37°C, the conditions being identical in all other respects. The reaction in the absence of enzyme proceeded at a rate of 5.4 nmol min^{-1} at 37°C, and of 15.5 nmol min^{-1} at 50°C, the conditions being identical in all other respects. Determine the activation energies of the catalysed and uncatalysed reactions. What assumptions are being made in the above calculations?

7

Kinetics of Single-Substrate Enzyme-Catalysed Reactions

7.1 THE RELATIONSHIP BETWEEN INITIAL VELOCITY AND SUBSTRATE CONCENTRATION

7.1.1 The Henri and Michaelis-Menten equations

In section 6.4 we discussed experimental evidence which showed that, for many single-substrate and pseudo single-substrate enzyme-catalysed reactions, there was a hyperbolic relationship between initial velocity v_0 and initial substrate concentration $[S_0]$ (see Fig. 6.5 and equation 6.21) so that:

$$v_0 = \frac{V_{max}[S_0]}{[S_0]+b} \tag{7.1}$$

where V_{max} is the maximum v_0 at a particular $[E_0]$ and b is another constant.

Kinetic models to explain these findings were proposed by Victor Henri (1903) and by Leonor Michaelis and Maude Menten (1913). These were essentially similar, but Michaelis and Menten did a great deal of experimental work to give their treatment a sounder basis, e.g. unlike Henri they recognized the importance of using initial velocity v_0 rather than any velocity v.

Let us consider a single-substrate enzyme-catalysed reaction where there is just one substrate-binding site per enzyme. The simplest general equation for such a reaction would be:

$$E + S \underset{k_{-1}}{\overset{k_1}{\rightleftharpoons}} ES \underset{k_{-2}}{\overset{k_2}{\rightleftharpoons}} E + P$$

If investigations are restricted to the initial period of the reaction, the product concentration is negligible and the formation of ES from product can be ignored.

Under these conditions, therefore, the reaction simplifies still further to:

$$E + S \underset{k_{-1}}{\overset{k_1}{\rightleftharpoons}} ES \xrightarrow{k_2} E + P$$

The rate of formation of ES at any time t (within the initial period when the product concentration is negligible) = k_1[E][S], where [E] is the concentration of free enzyme and [S] the concentration of free substrate at time t.

Also at time t, the rate of breakdown of ES back to E and S = k_{-1}[ES], where [ES] is the concentration of enzyme-substrate complex at this time.

The **Michaelis-Menten assumption** was that an **equilibrium** between enzyme, substrate and enzyme-substrate complex was almost instantly set up and maintained, the breakdown of enzyme-substrate complex to products being too slow to disturb this equilibrium. Using this assumption, therefore:

$$k_1[E][S] = k_{-1}[ES] \tag{7.2}$$

The constants may then be separated from the variables, giving:

$$\frac{[E][S]}{[ES]} = \frac{k_{-1}}{k_1} = K_s \tag{7.3}$$

where K_s is the dissociation constant of ES. The total concentration of enzyme present [E_0] must be the sum of the concentration of free enzyme [E] and the concentration of bound enzyme [ES]. Therefore, in order to involve total enzyme concentration (a quantity which can easily be determined and specified) into the above equation, the following substitution is made:

$$[E] = [E_0] - [ES]$$

$$\therefore ([E_0] - [ES])[S]/[ES] = K_s$$

$$\therefore K_s[ES] = ([E_0] - [ES])[S]$$

$$= [E_0][S] - [ES][S]$$

$$\therefore [ES][S] + K_s[ES] = [E_0][S]$$

$$\therefore [ES]([S] + K_s) = [E_0][S]$$

$$\therefore [ES] = [E_0][S]/([S]+K_s) \tag{7.4}$$

[ES] has been included in this way because, as discussed in section 6.4, this term governs the rate of formation of products (the overall rate of reaction) according to the relationship $v_0 = k_2$[ES]

If we substitute the expression for [ES] derived above, we obtain:

$$v_0 = \frac{k_2[E_0][S]}{[S] + K_s} \tag{7.5}$$

Moreover, we know that when the substrate concentration is very high, all the enzyme is present as the enzyme-substrate complex and the limiting initial velocity V_{max} is reached. Under these conditions, $V_{max} = k_2[E_0]$.

Therefore, we can substitute V_{max} for $k_2[E_0]$ in the expression for v_0 and get:

$$v_0 = \frac{V_{max}[S]}{[S]+K_s} \tag{7.6}$$

It was further assumed by Michaelis and Menten that the substrate was usually present in much greater concentrations than the enzyme. This is generally true, for enzymes, like all catalysts, are often present at very low concentrations. It is important to realize that we can talk about the substrate concentration being 'low' and giving first-order kinetics with a very small degree of **saturation** of the enzyme while at the same time the substrate concentration may be a thousand times that of the enzyme. This is valid because the formation of the enzyme-substrate complex from the enzyme and substrate is a reversible process.

If we make the assumption that the initial substrate concentration $[S_0]$ is very much greater than the initial enzyme concentration $[E_0]$, then the formation of the enzyme-substrate complex will result in an insignificant change in free substrate concentration. Hence, in the expression for v_0 derived above, we can substitute $[S_0]$ (a quantity easily specified) for [S], giving:

$$v_0 = \frac{V_{max}[S_0]}{[S_0]+K_s} \tag{7.7}$$

This is an equation of the form required to explain the experimental findings. However, it is unsatisfactory in that the Michaelis-Menten equilibrium-assumption cannot be generally applicable: some, possibly many, enzyme-catalysed reactions are likely to proceed at rates fast enough to disturb such an equilibrium.

7.1.2 The Briggs-Haldane modification of the Michaelis-Menten equation

The equation derived by Michaelis and Menten (section 7.1.1) was modified by George (G. E.) Briggs and John (J. B. S.) Haldane (1925), who introduced a more generally valid assumption, that of the **steady-state**. They argued that since the concentration of enzyme, and thus enzyme-substrate complex, was usually very small compared with the substrate concentration, then the rate of change of [ES] would be negligible compared to the rate of change of [P] over the initial period of the reaction, except during the very brief period at the beginning while ES was first being formed. In the absence of product, the concentration of ES would be determined by the total enzyme concentration, which remains constant throughout, and by the substrate concentration, which changes by a negligible amount, as a percentage of its initial value, over the period of interest. So, once this complex had been produced, it would be maintained in a steady-state, i.e. it would be broken down as fast as it was being formed, [ES] remaining constant.

Rapid reaction studies have shown that this is a reasonable assumption for most enzyme-catalysed reactions under the conditions indicated above. The steady-state is usually established within milliseconds of the start of the reaction and is maintained for a few minutes until the product concentration, and hence the rate of the reverse reaction, becomes significant.

If we again consider a single-substrate single-binding-site reaction:

$$E + S \underset{k_{-1}}{\overset{k_1}{\rightleftharpoons}} ES \xrightarrow{k_2} E + P$$

The rate of formation of ES at any time t (within the initial period when the product concentration is negligible) = $k_1[E][S]$.

The rate of breakdown of ES at this time = $k_{-1}[ES] + k_2[ES]$, since ES can break down to form products or reform reactants.

Using the steady-state assumption:

$$k_1[E][S] = k_{-1}[ES] + k_2[ES] = [ES](k_{-1} + k_2) \tag{7.8}$$

Separating the constants from the variables:

$$\frac{[E][S]}{[ES]} = \frac{k_{-1} + k_2}{k_1} = K_m \tag{7.9}$$

where K_m is another constant. Substituting $[E] = ([E_0] - [ES])$ as before:

$$\frac{([E_0] - [ES])[S]}{[ES]} = K_m$$

from which

$$[ES] = \frac{[E_0][S]}{[S] + K_m}$$

Again, since $v_0 = k_2[ES]$

$$v_0 = \frac{k_2[E_0][S]}{[S] + K_m} \tag{7.10}$$

and since $V_{max} = k_2[E_0]$

$$v_0 = \frac{V_{max}[S]}{[S] + K_m} \tag{7.11}$$

Finally, since the substrate concentration is usually much greater than the enzyme concentration, $[S] \approx [S_0]$, so

$$v_0 = \frac{V_{max}[S_0]}{[S_0]+K_m} \quad \text{at constant } [E_0] \tag{7.12}$$

This has the same form as the equation derived by Michaelis and Menten (7.7). The definition of the constant in the denominator is all that has changed. Hence equation 7.12 has retained the name, **Michaelis-Menten equation**, and K_m is called the **Michaelis constant**. The **equilibrium-assumption** is regarded as a special case of the more general **steady-state-assumption**, occurring where $k_{-1} \gg k_2$.

A graph of v_0 against $[S_0]$ will have the form of a rectangular hyperbola (Fig. 7.1), consistent with experimental findings for many enzyme-catalysed reactions.

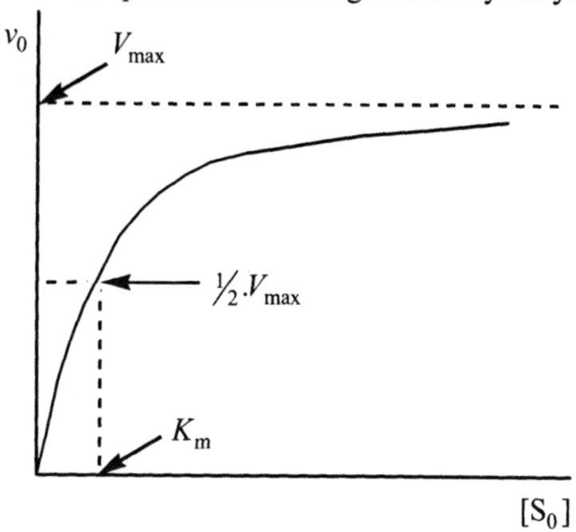

Fig. 7.1 – Graph of v_0 against $[S_0]$ at constant $[E_0]$ for a single- substrate enzyme-catalysed reaction, from the Michaelis-Menten equation.

V_{max}, the maximum initial velocity at a particular $[E_0]$, can be obtained from the graph as shown in Fig. 7.1. It will have the same units as v_0.

K_m can also be obtained from the graph, making use of the following reasoning. When $v_0 = \frac{1}{2} V_{max}$, $\frac{1}{2} V_{max} = V_{max}[S_0]/([S_0] + K_m)$

$$\therefore V_{max}([S_0] + K_m) = 2V_{max}[S_0]$$

$$\therefore K_m = [S_0]$$

Therefore, K_m is the value of $[S_0]$ which gives an initial velocity equal to $\frac{1}{2} V_{max}$, and K_m will have the same units as $[S_0]$. (However, note that the K_m measured in this way may be different from the true K_m, which is defined in terms of rate constants, if there are any inhibitors or other complicating factors present, as discussed in sections 8.2, 14.2 and 20.2.2: in these circumstances, the measured value may be termed the **apparent K_m** and given the symbol K'_m)

7.1.3 The significance of the Michaelis-Menten equation

The Michaelis-Menten equation, both in its original form and as modified by Briggs and Haldane, is derived with respect to a single-substrate enzyme-catalysed reaction with one substrate-binding site per enzyme and involving the formation of a single intermediate complex.

Such reactions are very rare indeed. However, the requirement for a single-substrate reaction can be taken to include pseudo single-substrate reactions and also variations with respect to one substrate in a multi-substrate reaction, provided the other substrates are maintained at a constant concentration. Similarly the requirement for an enzyme with a single substrate-binding site can include enzymes with more than one binding-site for the substrate in question, provided there is no interaction between the binding sites (see Chapters 12 and 13). Finally, the requirement for the formation of a single intermediate complex can include reactions of the type:

$$E + S \underset{k_{-1}}{\overset{k_1}{\rightleftharpoons}} ES \xrightarrow{k_2} EP \xrightarrow{k_3} E + P$$

provided ES→EP is the rate-limiting step of the overall reaction (see section 6.6).

The constant k_{cat} , called the **turnover number**, is often applied to enzyme-catalysed reactions. This is obtained from the general expression $V_{max} = k_{cat}[E_0]$. It represents the maximum number of substrate molecules which can be converted to products per molecule enzyme per unit time. For reactions of the simple type discussed above, $k_{cat} = k_2$; for more complex reactions, k_{cat} will be a function involving several individual rate constants. The turnover number for most enzymes lies in the range $1–10^4$ per second.

For the simple reaction, we have seen that $v_0 = k_2[E_0][S]/([S] + K_m)$. We can substitute for $[E_0]$ using $[E_0] = [E] + [ES]$, and then for [ES] using $[E][S]/[ES] = K_m$ (see section 7.1.2). That gives $v_0 = (k_2/K_m)[E][S]$ or, more generally, $v_0 = (k_{cat}/K_m)[E][S]$. The term k_{cat}/K_m is the **catalytic efficiency**. A high value (approaching that of k_1) indicates that the limiting factor for the overall reaction is the frequency of collisions between E and S molecules; a low value would be more in keeping with the equilibrium-assumption. A comparison of k_{cat}/K_m for alternative substrates can also be used as a measure of the specificity of an enzyme.

The Michaelis-Menten equation has been found to be applicable to a great many enzyme-catalysed reactions, and the constants V_{max} and K_m determined. V_{max} varies with the total concentration of enzyme present, but K_m is independent of enzyme concentration and is characteristic of the system being investigated. It can thus be used to identify a particular enzyme. In most cases, K_m values lie in the range 10^{-2}–10^{-6} mol l^{-1}. Also, since the catalytic step (ES→E + P, or ES→EP) is often the rate-limiting step, k_1 and k_{-1} are frequently much larger than k_2. Where this is the case, the Michaelis-Menten equilibrium assumption is valid and $K_m \approx K_s$. In general, K_s gives an indication of the **affinity** of the enzyme for the substrate: a low K_s value indicates a high affinity of enzyme for substrate, whereas a high K_s value indicates a low affinity.

If we turn our attention to processes taking place within the living cell, one of the factors which determines which of the several alternative metabolic pathways a substance enters is the K_m value of the first enzyme in each pathway. Consider, for example, the fate of the hexose-phosphates, glucose-6-phosphate, glucose-1-phosphate and fructose-6-phosphate, which are readily interconvertible in the cell and can be regarded as a single unit. The next step in the direction of glycolysis is mediated by **phosphofructokinase**, which has a much lower K_m for its substrate fructose-6-phosphate than the first enzyme in the direction of glycogen synthesis has for its substrate glucose-1-phosphate.

At most concentrations of hexose-phosphate, therefore, phosphofructokinase will be more saturated with substrate than the enzymes of glycogen synthesis and so most of the hexose-phosphate will be metabolized via glycolysis. Only at high hexose-phosphate concentrations, when phosphofructokinase is fully saturated with substrate, will glycogen synthesis become significant. Of course in general the fate of a metabolite depends on other factors besides K_m , including the concentration of each enzyme, the value of V_{max} at this enzyme concentration and the effects of activators and inhibitors (see section 14.2). Another enzyme in the glycolytic pathway, **triose phosphate isomerise**, has a catalytic efficiency near the upper limit of between 10^8 and 10^9 $(mol\ l^{-1})^{-1}s^{-1}$, so a reaction will take place just about every time it collides with a substrate molecule.

It should be realized that steady-state kinetic constants are determined in highly purified solutions by *in vitro* laboratory experiments, usually at the pH optimum of the enzyme under investigation. This is not necessarily the same pH as that *in vivo* in the living organism where the enzyme functions, since many enzymes are found in the same environment and not all will have an identical pH optimum, even though of course they must all have some activity at the physiological pH. Similarly the temperature at which the *in vitro* experiment is conducted may not be the *in vivo* temperature. Also, steady-state experiments are usually performed at enzyme concentrations much lower than those found *in vivo*. (This is not necessarily the case with rapid-reaction methods, discussed in section 7.2). Even more importantly, factors absent from *in vitro* experiments, e.g. membranes, may contribute to the action of an enzyme *in vivo*. Thus, despite the usefulness of *in vitro* kinetic studies, it cannot be assumed that they determine exactly how the enzyme behaves in the living organism.

From a practical viewpoint, a knowledge of the K_m of an enzyme is invaluable when assaying that enzyme (section 15.2): enzyme assays should be performed with the enzyme fully saturated with substrate. From the Michaelis-Menten equation, saturation is approached tangentially and only actually achieved at a substrate concentration of infinity. Nevertheless, if the K_m is known, the equation enables an initial substrate concentration to be calculated which, for all practical purposes, can be regarded as saturating the enzyme. For example, if $[S_0] = 100\ K_m$, then $v_0 = 0.99\ V_{max}$, irrespective of the actual enzyme concentration but provided it is much less than the substrate concentration.

7.1.4 The Lineweaver-Burk plot

The graph of the Michaelis-Menten equation, v_0 against $[S_0]$ (Fig. 7.1), is unsatisfactory as a means of determining V_{max} and K_m. The value of v_0 approaches V_{max} in tangential fashion, only actually attaining it, according to the Michaelis-Menten equation, at infinite substrate concentration. Hence, it is very difficult to use a plot of v_0 against $[S_0]$ to obtain an accurate value of V_{max} and hence of K_m. The graph, being a curve, cannot be accurately extrapolated upwards from values of v_0 at non-saturating concentrations.

Hans Lineweaver and Dean Burk (1934) overcame this problem without making any fresh assumptions. They simply took the Michaelis-Menten equation (7.12), $v_0 = V_{max}[S_0]/([S_0] + K_m)$, and inverted it:

$$\frac{1}{v_0} = \frac{[S_0]+K_m}{V_{max}[S_0]} = \frac{[S_0]}{V_{max}[S_0]} + \frac{K_m}{V_{max}[S_0]}$$

$$\therefore \frac{1}{v_0} = \frac{K_m}{V_{max}} \cdot \frac{1}{[S_0]} + \frac{1}{V_{max}} \quad \text{(the **Lineweaver-Burk equation**)} \qquad (7.13)$$

This is of the form $y = mx + c$, which is the equation of a straight line graph; a plot of y against x has a slope m and intercept c on the y-axis.

A plot of $1/v_0$ against $1/[S_0]$ (the **Lineweaver-Burk plot)** for systems obeying the Michaelis-Menten equation is shown in Fig. 7.2. The graph, being linear, can be extrapolated even if no experiment has been performed at anything approximating to a saturating substrate concentration, and from the extrapolated graph the values of K_m and V_{max} can be determined as shown in Fig. 7.2. Departure from linearity for a particular enzyme-catalysed reaction indicates that the assumptions inherent in the Michaelis-Menten equation are not valid in this instance.

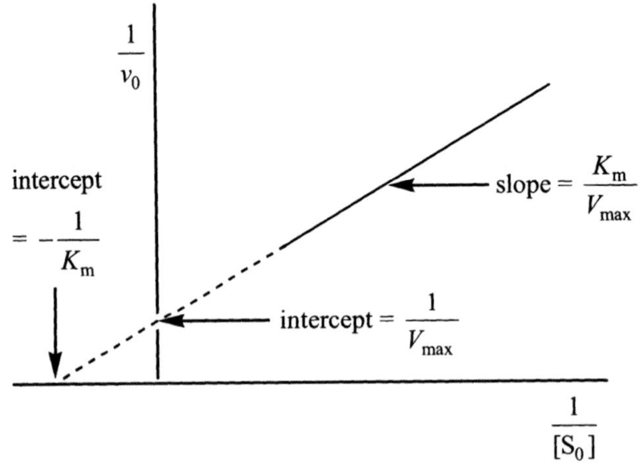

Fig. 7.2 – The Lineweaver-Burk plot.

7.1.5 The Eadie-Hofstee and Hanes plots

The Lineweaver-Burk plot has been criticized on several grounds. Firstly, and of least importance, the extrapolation across the $1/v_0$ axis to determine $-1/K_m$ sometimes reaches the edge of the graph paper before reaching the $1/[S_0]$ axis, possibly resulting in the graph having to be redrawn with altered axes. Secondly, it is said to give undue weight to measurements made at low substrate concentrations, when results are likely to be most inaccurate (this criticism should be borne in mind when a Lineweaver-Burk plot is being drawn). Thirdly, departures from linearity are less obvious than in some other plots, particularly the Eadie-Hofstee and Hanes plots. This could be very important if a reaction mechanism was being investigated.

The **Eadie-Hofstee plot** (devised by George Eadie and Barend Hofstee) takes as its starting point the Lineweaver-Burk equation (7.13), based in turn on the Michaelis-Menten equation (7.12). Both sides of the equation are multiplied by the factor $v_0 V_{max}$:

$$\frac{1}{(v_0)}(v_0)V_{max} = \frac{K_m}{(V_{max})}\cdot\frac{1}{[S_0]}v_0(V_{max}) + \frac{1}{(V_{max})}v_0(V_{max}\}$$

$$\therefore\ v_0 = -K_m\frac{v_0}{[S_0]} + V_{max} \tag{7.14}$$

Again, this is the equation of a straight line graph, from which V_{max} and K_m can be determined, as shown in Fig. 7.3a.

The **Hanes plot** (devised by Charles Hanes) similarly starts with the Lineweaver-Burk equation (7.13), which in this instance is multiplied throughout by $[S_0]$:

$$\frac{1}{v_0}[S_0] = \frac{K_m}{V_{max}}\cdot\frac{1}{[S_0]}[S_0] + \frac{1}{V_{max}}[S_0]$$

$$\therefore\ \frac{[S_0]}{v_0} = \frac{1}{V_{max}}[S_0] + \frac{K_m}{V_{max}} \tag{7.15}$$

Once more this gives a linear plot, from which V_{max} and K_m can be determined, as in Fig. 7.3b.

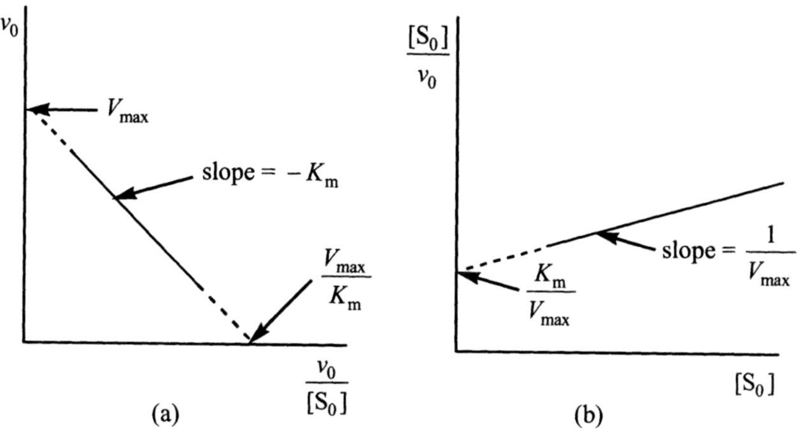

Fig. 7.3 – (a) The Eadie-Hofstee plot; (b) the Hanes plot.

The Eadie-Hofstee and Hanes plots are favoured by enzyme kineticists, but the Lineweaver - Burk plot continues to be widely used by enzymologists in general. The important thing is to obtain good experimental data covering a wide range of substrate concentrations, which are chosen so that the points will be evenly spread over the plot being used.

Computerized data processing, usually based on the least-squares approach to curve fitting and sometimes incorporating automatic rejection of points outside an arbitrarily set limit, gives a quick and convenient method of obtaining an estimate of K_m and V_{max}.

However, the results are not entirely reliable, and should never be accepted blindly: it is important to consider the actual graphs obtained, particularly to see if there is any evidence of a departure from linearity.

7.1.6 The Eisenthal and Cornish-Bowden plot

In view of the difficulties in obtaining reliable estimates of K_m and V_{max}, even with the aid of statistical analysis, from the plots discussed above, Robert Eisenthal and Athel Cornish-Bowden (1974) suggested a different approach, though one still based on the Michaelis-Menten equation (7.12). The reciprocal form of this equation, at constant $[E_0]$, gives

$$\frac{1}{v_0} = \frac{K_m + [S_0]}{V_{max}[S_0]}$$

$$\therefore \frac{V_{max}}{v_0} = \frac{K_m + [S_0]}{[S_0]} = \frac{K_m}{[S_0]} + 1 \tag{7.16}$$

Therefore, at constant v_0 and $[S_0]$, a plot of V_{max} against K_m is linear. The reader can be forgiven for feeling uneasy about this suggestion, for, at first sight, it seems nonsense to plot a constant (V_{max}) against another constant (K_m); nevertheless, the relationship is mathematically sound.

When $K_m = 0$, $V_{max} = v_0$, and when $V_{max} = 0$, $K_m = -[S_0]$. Therefore, each v_0, $[S_0]$ pair can be used to generate a line by marking v_0 on the V_{max} axis and $-[S_0]$ on the K_m axis. These two points are then joined up and the line extrapolated. Lines for all v_0, $[S_0]$ pairs, at constant $[E_0]$, must pass through the true values of K_m and V_{max}, so these values must be given by the point of intersection of all the lines. However, because of experimental error, the lines are likely to intersect over a range of values (Fig. 7.4). Nevertheless, in the opinion of many enzymologists, the best available estimates of K_m and V_{max} are the medians of the values obtained from the intersections of these lines, provided there is no evidence from other plots of departures from linearity.

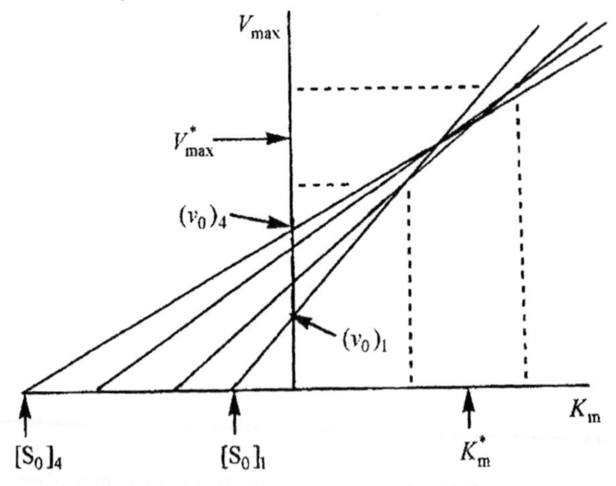

Fig. 7.4 – Typical Eisenthal–Cornish-Bowden plot for an enzyme-catalysed reaction at constant $[E_0]$. (V^*_{max} is the best estimate of V_{max}, K^*_m the best estimate of K_m.) See text for discussion.

7.1.7 The Haldane relationship for reversible reactions

All reactions are to some degree reversible, and many enzyme-catalysed reactions can function in either direction within the cell. It is therefore of interest to compare the kinetics of the forward and back reactions.

Consider a single-substrate reaction, $S \rightleftharpoons P$, proceeding via the formation of a single intermediate complex. In the forward direction this would be regarded as an ES complex, but in the reverse direction the same complex would be called EP.

$$S + E \underset{k_{-1}}{\overset{k_1}{\rightleftharpoons}} ES/EP \underset{k_{-2}}{\overset{k_2}{\rightleftharpoons}} P + E$$

The Michaelis-Menten equation (7.12) in the forward direction, at fixed $[E_0]$, gives:

$$v_f = \frac{V_{max}^s [S_0]}{[S_0] + K_m^s}$$

where v_f is the initial velocity in the forward direction,

V_{max}^s is the maximum initial velocity in the forward direction and

$$K_m^s = \frac{k_{-1} + k_2}{k_1}$$

The Michaelis-Menten equation for the reverse reaction (at constant $[E_0]$ gives:

$$v_b = \frac{V_{max}^p [P_0]}{[P_0] + K_m^p}$$

where v_b is the initial velocity in the back direction,

V_{max}^p is the maximum initial velocity for the back reaction and

$$K_m^p = \frac{k_1 + k_2}{k_{-2}}$$

J. B. S. Haldane derived a useful relationship between the kinetic constants and the equilibrium constants of the reaction. At equilibrium, the rate of the forward reaction equals the rate of the back reaction. Under these conditions:

$$k_{-1}[ES] = k_1[ES][S]$$

$$\therefore \frac{[ES]}{[E]} = \frac{k_1}{k_{-1}}[S]$$

Also under these conditions:

$$k_2[\mathrm{ES}] = k_{-2}[\mathrm{E}][\mathrm{P}]$$

$$\therefore \frac{[\mathrm{ES}]}{[\mathrm{E}]} = \frac{k_{-2}}{k_2}[\mathrm{P}] = \frac{k_1}{k_{-1}}[\mathrm{S}]$$

$$\therefore K_{\mathrm{eq}} = \frac{[\mathrm{P}]}{[\mathrm{S}]} = \frac{k_1}{k_{-1}}.\frac{k_2}{k_{-2}}$$

But $V_{\mathrm{max}}^{\mathrm{S}} = k_2[\mathrm{E}_0]$

and $V_{\mathrm{max}}^{\mathrm{P}} = k_{-1}[\mathrm{E}_0]$

Also,

$$\frac{K_{\mathrm{m}}^{\mathrm{s}}}{K_{\mathrm{m}}^{\mathrm{P}}} = \frac{(k_{-1}+k_2)}{k_1}\frac{k_{-2}}{(k_{-1}+k_2)} = \frac{k_{-2}}{k_1}$$

$$\therefore K_{\mathrm{eq}} = \frac{k_1 k_2}{k_{-1}k_{-2}} = \frac{V_{\mathrm{max}}^{\mathrm{s}} K_{\mathrm{m}}^{\mathrm{P}}}{V_{\mathrm{max}}^{\mathrm{P}} K_{\mathrm{m}}^{\mathrm{s}}} \tag{7.17}$$

This is known as the **Haldane relationship**.

If the equilibrium constant is known, this relationship can be used to check the validity of the kinetic constants which have been determined.

In general, it is likely that the K_{m} for the reaction in the metabolically important direction will be less than that for the reaction in the opposite direction, since K_{m} often varies inversely with affinity (see section 7.1.3). However, it should be realized that the direction of metabolic flow depends also on the concentrations of S and P present within the cell.

7.2 RAPID-REACTION KINETICS

7.2.1 Pre-steady-state kinetics

Although investigations of steady-state kinetics are of great importance to the enzymologist, as discussed in section 7.1, they have severe limitations. They allow the calculation of the kinetic constants K_{m} and k_{cat} , but the meaning of these constants depends on the assumptions made, such as the number of intermediates involved and the rates of their interconversion. These assumptions cannot be tested by ordinary 'test-tube' methods. The turnover number, k_{cat} , of many enzymes is in the order of 100 s^{-1}, i.e. 100 molecules of product can be produced per second per molecule of enzyme, which means that the slowest step in the mechanism is likely to have a half-life of only a few milliseconds. However, continuous-flow and stopped-flow rapid-mixing methods (section 6.5.2) can be designed to follow the build-up and decay of reaction intermediates over such a time-scale and they have been widely used for the study of enzyme-catalysed reactions.

The experimental results can be compared with the theoretical results for a particular mechanism, which are now generally produced with the help of a computer. Herbert Gutfreund and others have obtained simulated results for a variety of reaction mechanisms.

The simplest possible reaction is a first-order one of the form;

$$\mathrm{A} \xrightarrow{k} \mathrm{B}$$

The rate of reaction at time t is given by:

$$-\mathrm{d[A]/d}t = +\mathrm{d[B]/d}t = k\mathrm{[A]}$$

Integration of this gives:

$$\log_e[\mathrm{A_0}] - \log_e[\mathrm{A}] = kt \qquad (7.18)$$

(where $[\mathrm{A_0}]$ is the initial concentration of A); or

$$[\mathrm{A}] = [\mathrm{A_0}]e^{-kt} \qquad (7.19)$$

The course of this reaction is depicted in Fig. 7.5.

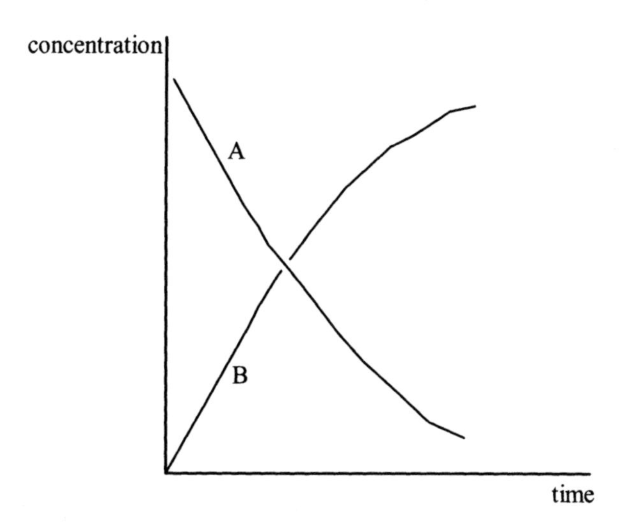

Fig. 7.5 – Theoretical progress curves for a reaction of the form A → B.

The integrated equation for most reaction mechanisms are far more complicated than this. Let us examine the course of reaction for a single-substrate enzyme-catalysed reaction involving a single intermediate complex, where the initial substrate concentration is much greater than the initial enzyme concentration:

$$\mathrm{E + S} \underset{k_{-1}}{\overset{k_1}{\rightleftharpoons}} \mathrm{ES} \underset{k_{-2}}{\overset{k_2}{\rightleftharpoons}} \mathrm{E + P}$$

The rate of increase of [ES] at time t (within the initial period when [P] is negligible) is given by:

$$\begin{aligned} d[ES]/dt &= k_1[E][S] - k_{-1}[ES] - k_2[ES] \\ &= k_1([E_0] - [ES])([S_0] - [ES] - [P]) - k_{-1}[ES] - k_2[ES] \end{aligned}$$

Since $[S_0] \gg [E_0]$, then $([S_0] - [ES] - [P]) \approx [S_0]$

$$\therefore \quad d[ES]/dt = k_1([E_0] - [ES])[S_0] - k_{-1}[ES] - k_2[ES] \qquad (7.20)$$

This may be integrated to give an expression showing the change in [ES] with time. The integration is beyond the scope of this book, but the results are illustrated in Fig. 7.6. The changes in concentration with time for all the other participants in the reaction can be calculated in similar fashion and these are also shown in Fig. 7.6.

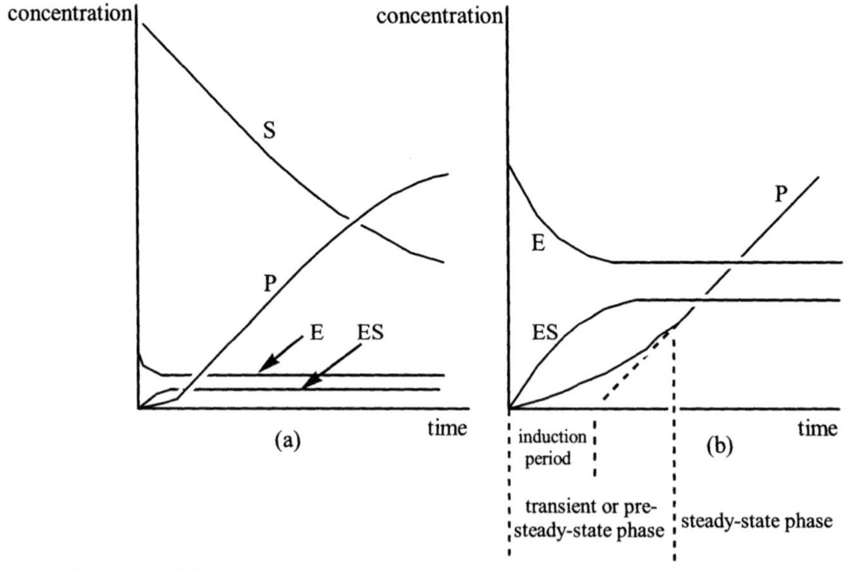

Fig. 7.6 – (a) Theoretical progress curve for a reaction of the form $E + S \rightleftharpoons ES \rightarrow E + P$, where $[S_0] \gg [E_0]$; (b) the same, showing the initial period in greater magnification.

The linear part of the [P] against t graph represents the steady-state phase of the reaction and has a slope = $(k_2[E_0][S_0])/([S_0] + K_m)$ (obtained by substituting K_m for $(k_{-1} + k_2)/k_1$ in the integrated equation). If the linear, steady-state portion of this graph is extrapolated downwards it intercepts the t-axis where $t = 1/(k_1[S_0] + K_m)$. This is called the **induction period**. If the experimental results are consistent with such a mechanism (e.g. no other reaction intermediate is implicated) and if K_m is known from steady-state investigations, then k_1, k_2 and hence k_{-1} can be calculated.

Because of the finite time taken for mixing, the induction period must be greater than about 5 ms for meaningful results to be obtained. Enzymes with a turnover number greater than a few hundred per second catalyse reactions proceeding too quickly to be analysed by such methods.

The calculated progress curves for a slightly more complicated reaction:

$$\mathrm{E+S} \underset{k_{-1}}{\overset{k_1}{\rightleftharpoons}} \mathrm{ES} \underset{k_{-2}}{\overset{k_2}{\rightleftharpoons}} \mathrm{EP} \underset{k_{-3}}{\overset{k_3}{\rightleftharpoons}} \mathrm{E+P}$$

where the rate-limiting step is EP→E + P, are as shown in Fig. 7.7.

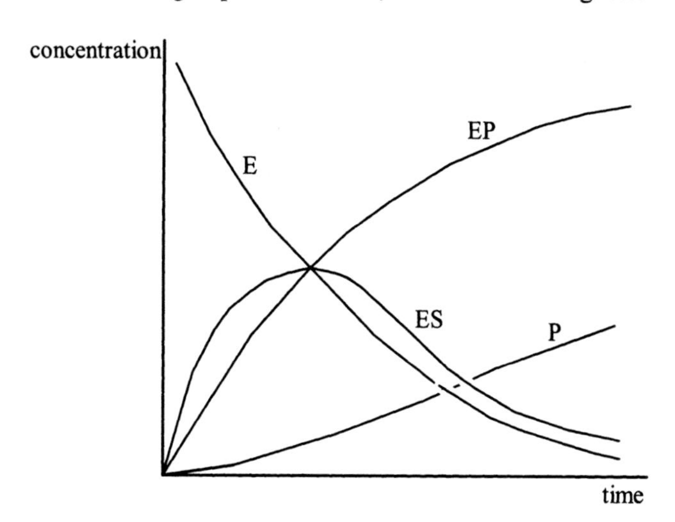

Fig. 7.7 – Theoretical progress curves for a reaction of the form E + S ⇌ ES ⇌ EP → E + P, EP → E + P being the rate-limiting step.

As an example of the use of rapid-reaction methods in the investigation of reaction mechanisms, let us consider some findings for the enzyme **chymotrypsin**. This will catalyse the hydrolysis of *p*-nitrophenylacetate, giving the coloured *p*-nitrophenol as one of the products. Brian Hartley and B. A. Kilby (1952) found that immediately after mixing the reagents there was an **initial burst** when one mole of *p*-nitrophenol per mole of enzyme was liberated much more rapidly than the subsequent steady-state liberation of *p*-nitrophenol. This suggested rapid acylation of the enzyme, the rate-limiting step of the overall reaction being the hydrolysis of this acyl-enzyme:

$$\text{E} + p\text{-nitrophenylacetate} \longrightarrow \text{E-}p\text{-nitrophenylacetate} \longrightarrow \text{E-acetate} \xrightarrow[\text{step}]{\text{rate-limiting}} \text{E} + \text{acetate}$$
$$\downarrow$$
$$p\text{-nitrophenol} \qquad (7.21)$$

Herbert Gutfreund (1955) followed the liberation of *p*-nitrophenol using stopped-flow methods and confirmed these conclusions, as well as determining some rate constants.

From a study of many enzymes, Gutfreund concluded that the rate-limiting step was usually one of the following: a change in conformation producing a reactive complex from the initial ES complex; the chemical interconversion of substrate to product; or the release of product from enzyme.

7.2.2 Relaxation kinetics

Rapid reaction kinetics can also be investigated by relaxation methods and, since these start with the solutions already mixed and at equilibrium, they can be applied to processes taking place too quickly for investigation by pre-steady-state methods.

It can be shown that, if a chemical reaction at equilibrium is subjected to any extremely rapid pressure or temperature change (e.g. 5°C in 5 μs) and the position of equilibrium changes slightly in consequence, the system will **relax** towards the new equilibrium position according to the relationship:

$$[\Delta A] = [\Delta A]_0 e^{-t/\tau} \qquad (7.22)$$

where $[\Delta A]$ is the difference between the concentration of A at time t and at the new equilibrium position, i.e. $[\Delta A] = [A] - [A_{eq}]$. $[\Delta A]_0$ is the difference between the concentration of A immediately after the temperature or pressure change and at the new equilibrium position; and τ is the **relaxation time**.

From the above equation (7.22):

$$\log_e[\Delta A] = \log_e[\Delta A]_0 - t/\tau \qquad (7.23)$$

$\therefore$ the slope of a graph of $\log_e[\Delta A]$ against t gives $-1/\tau$

This relationship holds for all chemical reactions where the perturbation in equilibrium is small, but the meaning of τ in terms of reaction constants depends on the actual reaction mechanism

Consider the simplest reversible reaction $A \rightleftharpoons B$, with k_1 and k_{-1} being, as usual, the rate constants for the forward and back reactions. Where the total concentration of A and B present is $[B_0]$, i.e. $[A] + [B] = [B_0]$, the rate of the forward reaction at time t is given by:

$$-d[A]/dt = +d[B]/dt = k_1[A] - k_{-1}[B]$$

$$= k_1[A] - k_{-1}([B_0] - [A])$$

$$= k_1[A] - k_{-1}[B_0] + k_{-1}[A]$$

$$= [A](k_1 + k_{-1}) - k_{-1}[B_0] \qquad (7.24)$$

At equilibrium,

$$k_1[A_{eq}] = k_{-1}[B_{eq}]$$

$$\therefore\ k_1[\mathrm{A_{eq}}] = k_{-1}([\mathrm{B_0}] - [\mathrm{A_{eq}}])$$

$$= k_{-1}[\mathrm{B_0}] - k_{-1}[\mathrm{A_{eq}}]$$

$$\therefore\ [\mathrm{A_{eq}}](k_1 + k_{-1}) = k_{-1}[\mathrm{B_0}]$$

Substituting for $k_{-1}[\mathrm{B_0}]$ in the rate equation derived above (equation 7.24):

$$-\mathrm{d}[\mathrm{A}]/\mathrm{d}t = +\mathrm{d}[\mathrm{B}]/\mathrm{d}t = [\mathrm{A}](k_1 + k_{-1}) - [\mathrm{A_{eq}}](k_1 + k_{-1})$$

$$= (k_1 + k_{-1})([\mathrm{A}] - [\mathrm{A_{eq}}])$$

$$= (k_1 + k_{-1})[\Delta\mathrm{A}]$$

However, the rate of change of $[\Delta\mathrm{A}]$ must be the same as the rate of change of [A]:

$$\therefore -\mathrm{d}[\mathrm{A}]/\mathrm{d}t = -\mathrm{d}[\Delta\mathrm{A}]/\mathrm{d}t = (k_1 + k_{-1})[\Delta\mathrm{A}]$$

$$\therefore\ \mathrm{d}[\Delta\mathrm{A}]/\mathrm{d}t = -(k_1 + k_{-1})[\Delta\mathrm{A}]$$

Integrating, $\log_e[\Delta\mathrm{A}] = \text{constant} - (k_1 + k_{-1})t$

At $t = 0$, $[\Delta\mathrm{A}] = [\Delta\mathrm{A}]_0$, so constant $= \log_e[\Delta\mathrm{A}]_0$

$$\therefore\ \log_e[\Delta\mathrm{A}] = \log_e[\Delta\mathrm{A}]_0 - (k_1 + k_{-1})t \qquad (7.25)$$

This is of the same form as the general relaxation equation so, by comparison with this, it can be seen that $\tau = 1/(k_1 + k_{-1})$ for the reaction $\mathrm{A} \rightleftharpoons \mathrm{B}$.

Similarly, for the reaction $\mathrm{E} + \mathrm{S} \rightleftharpoons \mathrm{ES}$, again with rate constants k_1 and k_{-1} for the forward and back reactions, it can be deduced that $1/\tau = k_1([\mathrm{E_{eq}}] + [\mathrm{S_{eq}}]) + k_{-1}$.

Reactions of this form include the binding of one substrate to an enzyme in the absence of other substrates, preventing the reaction from proceeding further. This has been used to investigate the binding of NADH to lactate dehydrogenase. A plot of $([\mathrm{E_{eq}}] + [\mathrm{S_{eq}}])$ against $1/\tau$ results in a straight line with the intercept equal to k_{-1} and the slope equal to k_1, so these constants can be calculated.

The meaning of τ for more complex reactions is, needless to say, very much more complicated and only in special cases can it be identified with individual steps and particular rate constants.

7.3 THE KING AND ALTMAN PROCEDURE

When more than one or two intermediates are present in a reaction sequence, the derivation of an overall rate equation becomes extremely complex, involving the solving of several simultaneous equations.

However, Edward King and Carl Altman (1956) devised empirical rules which allow the rate equation for a particular mechanism to be written down by inspection as a function of individual rate constants, and this procedure was developed by Tze-Fei Wong and Charles Hanes (1962). The theoretical basis of these rules involves matrix algebra and is beyond the scope of this book. We will merely indicate the approach, using some simple examples. First let us consider again the simplest possible enzyme-catalysed reaction under steady-state conditions:

$$\mathrm{E}+\mathrm{S} \underset{k_{-1}}{\overset{k_1}{\rightleftharpoons}} \mathrm{ES} \xrightarrow{k_2} \mathrm{E}+\mathrm{P}$$

The King and Altman procedure requires the reaction to be written as a cyclic process, showing the interconversion of the enzyme-forms involved (in this case E and ES). Each step must be described by a κ (kappa), which is the product of the rate constant and the concentration of free substrate involved in that step, or just the rate constant if no substrate is involved. Hence for the above sequence we have:

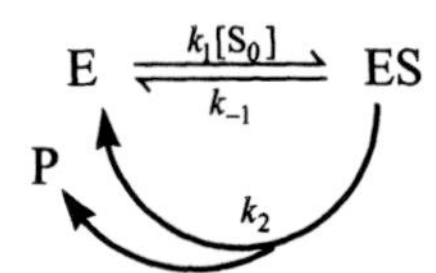

For each enzyme species we must then work out all the pathways by which that species may be synthesized. Each pathway must contain $n-1$ steps, where n is the number of enzyme species present (in this case 2) and the product of the kappas for the steps in each pathway is determined; each kappa product will contain $n-1$ terms. Pathways where two arrows arrive at a single enzyme species are permitted, but pathways where two arrows leave a single enzyme species are forbidden, so there are no closed loops. For the simple system being considered here (where $n = 2$) we have:

Enzyme species	*Pathways forming enzyme species*	*Sum of kappa products*
E	$\mathrm{E} \xleftarrow{k_{-1}}$ and E $\overset{k_2}{\curvearrowleft}$	$k_{-1}+k_2$
ES	$\xrightarrow{k_1[\mathrm{S_0}]} \mathrm{ES}$	$k_1[\mathrm{S_0}]$

For each enzyme species the following relationship then holds:

$$\frac{[\text{enzyme species}]}{[\mathrm{E_0}]} = \frac{\text{sum of kappa products of that species}}{\text{sum of all kappa products}}$$

Hence, for the mechanism under question,

$$\frac{[\mathrm{E}]}{[\mathrm{E_0}]} = \frac{k_{-1}+k_2}{k_{-1}+k_2+k_1[\mathrm{S_0}]} \quad \text{and} \quad \frac{[\mathrm{ES}]}{[\mathrm{E_0}]} = \frac{k_1[\mathrm{S_0}]}{k_{-1}+k_2+k_1[\mathrm{S_0}]} \tag{7.26}$$

The overall rate equation is given, as in section 7.1.2, by $v_0 = k_2[\mathrm{ES}]$. Substituting the expression for [ES] determined by the King and Altman procedure (7.26):

$$v_0 = \frac{k_2 k_1 [\mathrm{S_0}][\mathrm{E_0}]}{k_{-1} + k_2 + k_1[\mathrm{S_0}]} = \frac{k_2[\mathrm{S_0}][\mathrm{E_0}]}{\left(\dfrac{k_{-1} + k_2}{k_1}\right) + [\mathrm{S_0}]} \tag{7.27}$$

which, given the definition of K_m and the fact that $V_{max} = k_2[\mathrm{E_0}]$, is identical to the equation derived in section 7.1.2 (i.e. equation 7.12).

Let us now use the King and Altman procedure to determine the rate equation for a slightly more complicated reaction sequence, that involving three enzyme species suggested for reactions catalysed by chymotrypsin (section 7.2.1). Under steady-state conditions this may be written:

$$\mathrm{AX + E} \underset{k_{-1}}{\overset{k_1}{\rightleftharpoons}} \mathrm{EAX} \xrightarrow[\searrow \mathrm{A}]{k_2} \mathrm{EX} \xrightarrow{k_3} \mathrm{E + X}$$

In cyclical form this is:

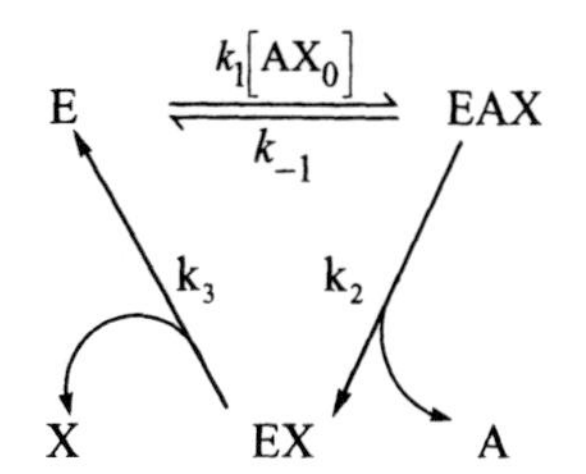

Enzyme species	*Pathways forming enzyme species*	*Sum of kappa products*
E	E ← k_3, ↙ k_2 and E ← k_{-1}, ← k_3	$k_2 k_3 + k_{-1} k_3$
EAX	$k_1[\mathrm{AX_0}]$ → EAX, ← k_3	$k_1 k_3[\mathrm{AX_0}]$
EX	$k_1[\mathrm{AX_0}]$ →, ↙ k_2 EX	$k_1 k_2[\mathrm{AX_0}]$

From this, for example,

$$\frac{[\mathrm{EX}]}{[\mathrm{E_0}]} = \frac{k_1 k_2[\mathrm{AX_0}]}{k_2 k_3 + k_{-1} k_3 + k_1 k_3[\mathrm{AX_0}] + k_1 k_2[\mathrm{AX_0}]} \tag{7.28}$$

Under steady-state conditions the overall rate of reaction is given by:

$$v_0 = -\mathrm{d[A]/d}t = k_2[\mathrm{EAX}] = \mathrm{d[X]/d}t = k_3[\mathrm{EX}] \qquad (7.29)$$

If we substitute for either [EAX] or [EX] in the expression obtained by the King and Altman procedure, we obtain the equation:

$$v_0 = \frac{k_1 k_2 k_3 [\mathrm{AX_0}][\mathrm{E_0}]}{k_2 k_3 + k_{-1} k_3 + k_1 k_3 [\mathrm{AX_0}] + k_1 k_2 [\mathrm{AX_0}]}$$

$$= \frac{k_1 k_2 k_3 [\mathrm{AX_0}][\mathrm{E_0}]}{k_3 (k_2 + k_{-1}) + k_1 [\mathrm{AX_0}](k_2 + k_3)} \qquad (7.30)$$

SUMMARY OF CHAPTER 7

The hyperbolic graphs of v_0 against $[S_0]$ obtained experimentally for many enzyme-catalysed reactions can be obtained by the Michaelis-Menten equation. The original equilibrium-assumption of Michaelis and Menten is now seen as a special case of the steady-state-assumption of Briggs and Haldane. Steady-state kinetics can be used to calculate K_m and k_{cat} , which are characteristic of a particular enzyme. These calculations are facilitated by the use of linear plots derived from the Michaelis-Menten equation.

The kinetic mechanism of a reaction can be investigated in more detail by the use of rapid-reaction techniques. Continuous-flow, stopped-flow and relaxation methods give information as to the intermediates involved in the reaction and enable individual rate constants to be calculated.

A procedure which enables the rate equation for a complex reaction sequence to be written down by inspection has been described by King and Altman.

FURTHER READING

As for Chapter 6, plus:

Bisswanger, H. (2002), *Enzyme Kinetics: Principles and Methods,* Wiley–VCH.

Cornish-Bowden, A. (1995), *Fundamentals of Enzyme Kinetics,* 2nd edn., Portland Press (Chapters 4, 11, 12).

Cornish-Bowden, A. (1995), *Analysis of Enzyme Kinetic Data,* Oxford University Press.

Gutfreund, H. (1995), *Kinetics for the Life Sciences - Principles, Transmitters and Catalysts,* Cambridge University Press.

Gutfreund, H. (1999), Rapid-flow techniques and their contribution to enzymology, *Trends in Biochemical Sciences,* **24,** 457-460.

Marangoni, A. G. (2002), *Enzyme Kinetics: A Modern Approach*, Wiley (Chapters 1-3, 6, 14-15).

Methods in Enzymology, **63** (1979), **249** (1995), **308** (1999), **354** (2002): Enzyme kinetics and mechanism, Academic Press.

Price, N. C. and Stevens, L. (1999), *Fundamentals of Enzymology,* 3rd edn., Oxford University Press (Chapter 4).

Schulz, A. R. (1994), *Enzyme Kinetics,* Cambridge University Press (Chapters 1-4).

PROBLEMS

7.1 The following results were obtained for an enzyme-catalysed reaction

Substrate concentration (mmol l^{-1}):	5.0	6.67	10.0	20.0	40.0
Initial velocity (μmol l^{-1} min^{-1}):	147	182	233	323	400

Calculate K_m and V_{max}

7.2 Malate dehydrogenase catalyses the reaction:

$$\text{(S)-malate} + NAD^+ \rightleftharpoons \text{oxaloacetate} + NADH + H^+$$

The rate of the forward reaction was investigated in the presence of saturating concentrations of malate and a fixed concentration of enzyme. The following results were obtained:

	Absorbance (at 340 nm) at time t (minutes)					
[NAD^+] (mmol l^{-1})	t = 0.5	1.0	1.5	2.0	2.5	3.0
1.5	0.033	0.056	0.079	0.102	0.122	0.138
2.0	0.036	0.063	0.089	0.116	0.138	0.154
2.5	0.040	0.069	0.099	0.128	0.150	0.168
3.33	0.043	0.075	0.108	0.140	0.163	0.175
5.0	0.047	0.084	0.121	0.158	0.177	0.184
10.0	0.052	0.095	0.137	0.180	0.192	0.200

Calculate K_m and V_{max}.

7.3 Ficain is a proteolytic enzyme which catalyses the hydrolysis of a variety of substrates. One such reaction was followed by rapid reaction techniques, measuring the change of product concentration with time, and the following results were obtained:

Time (s):	0	0.1	0.2	0.4	0.5
[Product] (μmol l^{-1}):	0	4	8	17	37

Assuming that the reaction proceeds by the simplest possible mechanism, calculate the rate constants involved, given that the initial substrate concentration was 1.67 mmol l^{-1} and the K_m 15 mmol l^{-1}.

7.4 Use the King and Altman procedure to obtain a rate equation for the reaction:

$$E + S \underset{k_{-1}}{\overset{k_1}{\rightleftharpoons}} ES \underset{k_{-2}}{\overset{k_2}{\rightleftharpoons}} EP \xrightarrow{k_3} E + P$$

8

Enzyme Inhibition

8.1 INTRODUCTION

Inhibitors are substances which tend to decrease the rate of an enzyme-catalysed reaction. Although some act on a substrate or cofactor, we will restrict our discussion here to those which combine directly with an enzyme. **Reversible inhibitors** bind to an enzyme in a reversible fashion and can be removed by dialysis (or simply dilution) to restore full enzymic activity, whereas **irreversible inhibitors** cannot be removed from an enzyme by dialysis. Sometimes it may be possible to remove an irreversible inhibitor from an enzyme by introducing another component to the reaction mixture, but this would not affect the classification of the original interaction.

Reversible inhibitors usually rapidly form an equilibrium system with an enzyme to show a definite degree of inhibition (depending on the concentration of enzyme, inhibitor and substrate) which remains constant over the period when initial velocity studies are normally carried out. In contrast, the degree of inhibition by irreversible inhibitors may increase over this period of time.

In this chapter we shall be concerned mainly with the inhibition of simple single-substrate enzyme-catalysed reactions. This group includes most single-substrate reactions obeying Michaelis-Menten kinetics (section 7.1.3). The inhibition of two-substrate enzyme-catalysed reactions will be discussed in section 9.3.2.

8.2 REVERSIBLE INHIBITION

8.2.1 Competitive inhibition

Competitive inhibitors often resemble the substrates whose reactions they inhibit, and because of this structural similarity they may compete for the same binding site on the enzyme. The enzyme-bound inhibitor then either lacks the appropriate reactive group or it is held in an unsuitable position with respect to the catalytic site of the enzyme or to other potential substrates for a reaction to occur.

In either case, a **dead-end complex** is formed, and the inhibitor must dissociate from the enzyme and be replaced by a molecule of substrate before a reaction can take place at that particular enzyme molecule.

For example, **malonate** ($^-O_2C.CH_2.CO_2^-$) is a competitive inhibitor of the reaction catalysed by **succinate dehydrogenase**:

$$\underset{\text{succinate}}{^-O_2C.CH_2CH_2.CO_2^-} \rightleftharpoons \underset{\text{fumarate}}{^-O_2C.CH{=}CH.CO_2^-} \qquad (8.1)$$

Malonate has two carboxyl groups, like the substrate, succinate, and can fill the succinate-binding site on the enzyme. However, the subsequent reaction involves the formation of a double bond, and since malonate, unlike succinate, has only one carbon atom between the carboxyl groups, it cannot react.

The effect of a competitive inhibitor depends on the inhibitor concentration, the substrate concentration and the relative affinities of the substrate and the inhibitor for the enzyme. In general, at a particular inhibitor and enzyme concentration, if the substrate concentration is low, the inhibitor will compete favourably with the substrate for the binding sites on the enzyme and the degree of inhibition will be great. However, if, at this same inhibitor and enzyme concentration, the substrate concentration is high, then the inhibitor will be much less successful in competing with the substrate for the available binding sites and the degree of inhibition will be less marked. At very high substrate concentrations, molecules of substrate will greatly outnumber molecules of inhibitor and the effect of the inhibitor will be negligible. Hence V_{max} for the reaction is unchanged (Fig. 8.1). However the apparent K_m, the substrate concentration when $v_0 = \frac{1}{2}V_{max}$, is clearly increased as a result of the inhibition, and is given the symbol K'_m.

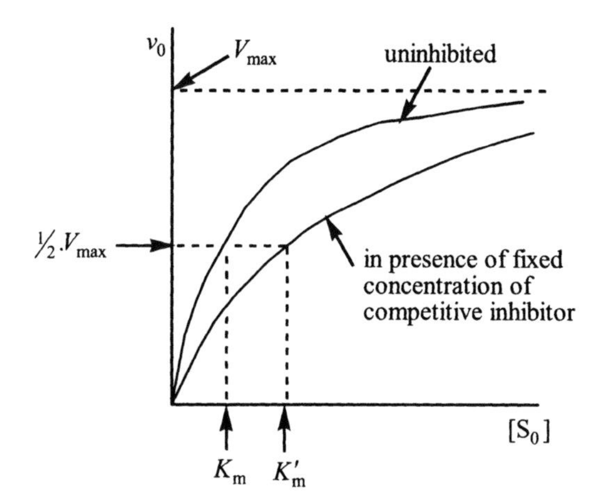

Fig. 8.1 – Michaelis-Menten plot (at fixed $[E_0]$) showing the effect of a competitive inhibitor.

Let us investigate the steady-state kinetics of a simple single-substrate single-binding-site single-intermediate enzyme-catalysed reaction in the presence of a competitive inhibitor, I.

$$\begin{array}{ccc} E + S & \underset{k_{-1}}{\overset{k_1}{\rightleftharpoons}} ES \xrightarrow{k_2} E + P \\ -I \big\Updownarrow +I & \\ EI & \end{array} \qquad (8.2)$$

The dissociation constant for the reaction between E and I is K_i , where K_i = [E][I]/[EI]. In this context, K_i is called the **inhibitor constant**. Equilibrium between enzyme and inhibitor will normally be established almost instantaneously on mixing.

If we begin to derive the initial velocity equation using the steady-state assumption exactly as in section 7.1.2, we reach equation 7.9:

$$\frac{[E][S]}{[ES]} = \frac{k_{-1} + k_2}{k_1} = K_m$$

As before, we wish to make a substitution for [E] in this expression, and here we have to take the inhibitor into account, since:

$$[E_0] = [E] + \{ES] + [EI]$$

$$= [E] + [ES] + [E][I]/K_i$$

$$= [E]\left(1 + \frac{[I]}{Ki}\right) + [ES]$$

$$\therefore [E] = \frac{[E_0] - [ES]}{\left(1 + \frac{[I]}{K_i}\right)}$$

If we now make the substitution for [E]:

$$\frac{([E_0] - [ES])[S]}{\left(1 + \frac{[I]}{K_i}\right)[ES]} = K_m$$

$$\therefore \frac{([E_0] - [ES])[S]}{[ES]} = K_m\left(1 + \frac{[I]}{K_i}\right)$$

If we continue to develop the argument exactly as in section 7.1.2, we reach the expression:

$$v_0 = \frac{V_{max}[S_0]}{[S_0] + K_m\left(1 + \frac{[I_0]}{K_i}\right)} \quad (8.3)$$

This is an equation of the same form as the Michaelis-Menten equation, the only difference being that K_m has been increased by a factor $(1 + ([I_0]/K_i))$. (Note that the inhibitor concentration will usually be of the same order of magnitude as the substrate concentration and thus much greater than the enzyme concentration, so $[I] \approx [I_0]$ just as $[S] \approx [S_0]$.) Therefore, for simple competitive inhibition, V_{max} is unchanged but K_m is altered so that $K'_m = K_m(1 + ([I_0]/K_i))$, where K'_m is the apparent K_m in the presence of an initial concentration $[I_0]$ of competitive inhibitor. It can be seen that K_i is equal to the concentration of competitive inhibitor which apparently doubles the value of K_m.

The Lineweaver-Burk equation in the presence of a competitive inhibitor will be:

$$\frac{1}{v_0} = \frac{K'_m}{V_{max}} \cdot \frac{1}{[S_0]} + \frac{1}{V_{max}} \quad (8.4)$$

and Lineweaver-Burk plots showing the effect of competitive inhibition are shown in Fig. 8.2.

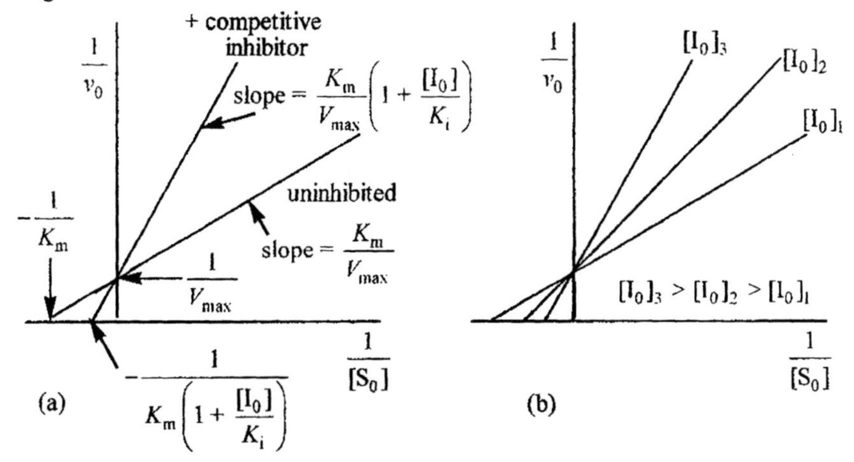

Fig. 8.2 – (a) Lineweaver-Burk plot showing the effect of competitive inhibition; (b) the same, showing plots for several inhibitor concentrations at fixed enzyme concentrations.

It must be pointed out that this identical expression would be obtained if the inhibitor-binding site was separate from the substrate-binding site, provided the binding of the substrate to the enzyme resulted in the blockage of the inhibitor-binding site by a conformational change or other mechanism. In this situation the inhibitor could bind to E but not to ES, exactly as discussed above. For this reason it has become common to classify an inhibitor as competitive if, in its presence, a Lineweaver-Burk plot is obtained with changed K_m but unchanged V_{max}, *irrespective of the actual mechanism involved* (see Fig. 8.3a/b).

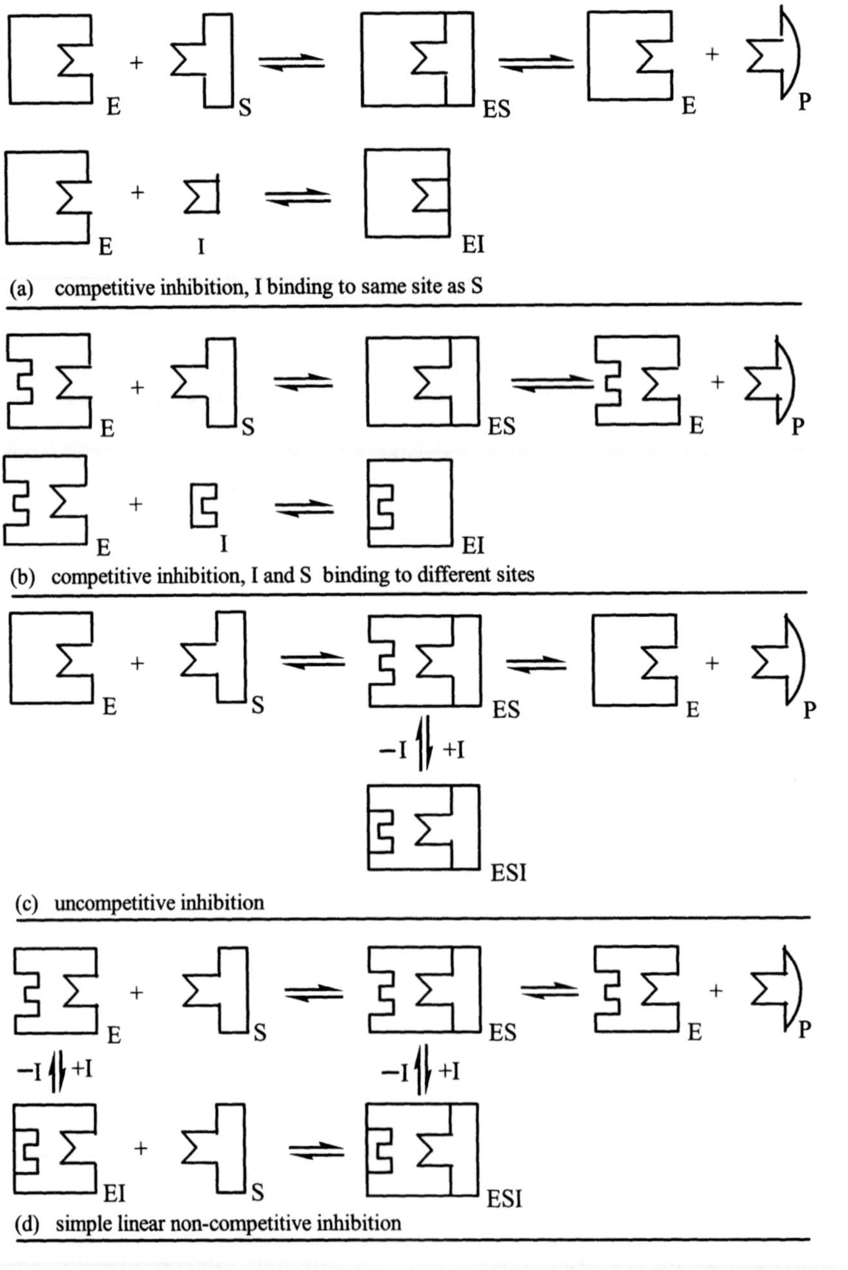

(a) competitive inhibition, I binding to same site as S

(b) competitive inhibition, I and S binding to different sites

(c) uncompetitive inhibition

(d) simple linear non-competitive inhibition

Fig. 8.3 – Diagrammatic representation of some possible examples of reversible inhibition.

Once competitive inhibition has been identified, it is desirable to determine the inhibitor constant K_i. It is obtained from the expression $K'_m = K_m(1 + ([I_0]/K_i))$, but a graphical method is preferred to a direct substitution of numbers, to allow errors in individual determinations to be averaged out. From the above expression:

$$K'_m = \frac{K_m}{K_i}[I_0] + K_m \tag{8.5}$$

so a plot of K'_m (determined from the intercept on the $1/[S_0]$ axis of the primary Lineweaver-Burk plot) against $[I_0]$ will be linear, with the intercept on the $[I_0]$ axis giving $-K_i$ (Fig. 8.4a). Similarly, since the slope of the Lineweaver-Burk plot in the presence of competitive inhibitor is

$$\frac{K_m}{V_{max}}\left(1 + \frac{[I_0]}{K_i}\right) \tag{8.6}$$

a graph of slope of the **primary** (Lineweaver-Burk) plot against $[I_0]$ will also be linear, the intercept on the $[I_0]$ axis giving $-K_i$ (Fig. 8.4b). The **secondary** plots of K'_m against $[I_0]$ and of slope against $[I_0]$ must, of course, be constructed from data obtained at fixed $[E_0]$.

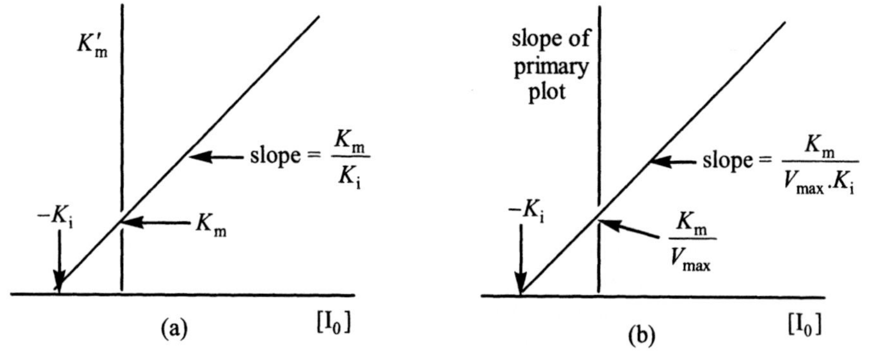

Fig. 8.4 – Secondary plots for competitive inhibition.

An alternative graphical means of calculating K_i was suggested by Malcolm Dixon (1953). The Lineweaver-Burk equation in the presence of a competitive inhibitor is:

$$\frac{1}{v_0} = \frac{K_m\left(1 + \frac{[I_0]}{K_i}\right)}{V_{max}} \cdot \frac{1}{[S_0]} + \frac{1}{V_{max}}$$

$$= \frac{K_m}{V_{max}[S_0]} \cdot \frac{[I_0]}{K_i} + \frac{K_m}{V_{max}[S_0]} + \frac{1}{V_{max}} \tag{8.7}$$

Therefore, at fixed $[S_0]$, a plot of $1/v_0$ against $[I_0]$ (the **Dixon plot**) is linear. When $[I_0] = -K_i$, it can be seen that $1/v_0 = 1/V_{max}$ and is therefore independent of $[S_0]$, so Dixon plots for different values of $[S_0]$ (but fixed $[E_0]$) intersect where $[I_0] = -K_i$ (Fig. 8.5).

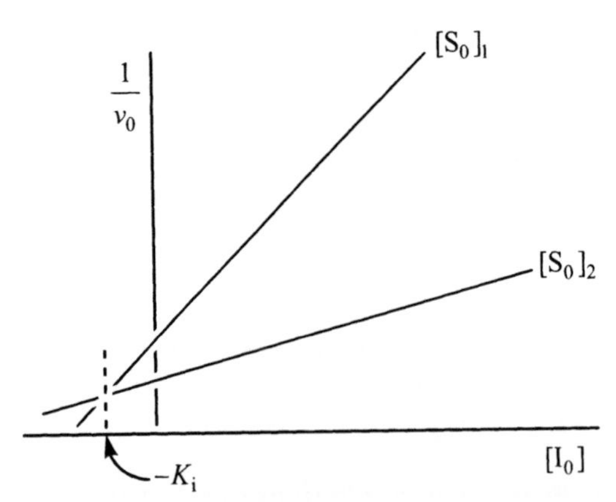

Fig. 8.5 – Dixon plot for competitive inhibition.

The simplest forms of competitive inhibition considered above are sometimes termed **linear competitive inhibition** because both primary and secondary plots are linear. In more complicated systems, the primary plots may be linear but the secondary plots non-linear. For example, if not one but two molecules of inhibitor can bind to the substrate binding site, then **parabolic competitive inhibition** is said to occur, because of the shape of the secondary and Dixon plots. Similarly, if the inhibitor binds to a different site from the substrate and reduces the affinity of the enzyme for the substrate without altering the reaction characteristics of that substrate which does bind, then **hyperbolic competitive inhibition** results. In each case the primary plots are linear and the inhibition patterns are indistinguishable from those for **linear competitive inhibition**.

Competitive inhibitors, like other types of inhibitors, may be used to help elucidate metabolic pathways by causing accumulation of intermediates. In this way Hans Krebs and colleagues used inhibition by malonate to investigate the tricarboxylic acid cycle, of which succinate dehydrogenase is a component. Provided they are not dangerous to humans, competitive inhibitors may be used in medicine or agriculture as chemotherapeutic drugs, insecticides or herbicides, to destroy or prevent growth of unwanted organisms. For example, sulphonamides such as sulphanilamide, once widely used in medicine, are competitive inhibitors of bacterial enzymes involved in the biosynthesis of the coenzyme tetrahydrofolate (see section 11.5.9) from *p*-aminobenzoic acid, the structure of which is similar to that of sulphanilamide:

H_2N–⟨ring⟩–SO_2NH_2 sulphanilamide

H_2N–⟨ring⟩–CO_2H *p*-aminobenzoic acid

The metabolic pathway for the formation of tetrahydrofolate from *p*-aminobenzoic acid is not found in humans, so sulphonamides can be used to limit the growth of bacteria with relatively little risk to the patient.

A detailed investigation of the binding characteristics of various competitive inhibitors which bind at the same site as a natural substrate can give useful information about factors governing the binding of the substrate. In two-substrate enzyme-catalysed reactions, competitive inhibition studies can help elucidate the reaction mechanism (section 9.3.2). In this context and in general, it should be noted that the product of a reaction often resembles the substrate and therefore may act as a competitive inhibitor. There are a few instances where competitive inhibition by a product may play an important role in metabolic regulation within the living cell. For example, **2,3-bisphosphoglycerate** inhibits its own formation from **3-phosphoglyceroyl phosphate**, a reaction catalysed by **bisphosphoglycerate mutase**.

8.2.2 Uncompetitive inhibition

Uncompetitive inhibitors bind only to the enzyme-substrate complex and not to the free enzyme. Substrate-binding could cause a conformational change to take place in the enzyme and reveal an inhibitor binding site (Fig. 8.3c), or the inhibitor could bind directly to the enzyme-bound substrate. In neither case does the inhibitor compete with the substrate for the same binding site, so the inhibition cannot be overcome by increasing the substrate concentration. Both K_m and V_{max} are altered, but a distinctive kinetic pattern emerges under steady-state conditions.

Once again let us consider the simplest situation:

$$\begin{array}{c} \mathrm{E + S \rightleftharpoons ES \rightarrow E + P} \\ -\mathrm{I} \updownarrow +\mathrm{I} \\ \mathrm{ESI} \end{array} \tag{8.8}$$

where ESI is a **dead-end complex** (see Fig. 8.3c). The inhibitor constant K_i = [ES][I]/[ESI]. For this system,

$$\begin{aligned} [\mathrm{E_0}] &= [\mathrm{E}] + [\mathrm{ES}] + [\mathrm{ESI}] \\ &= [\mathrm{E}] + [\mathrm{ES}] + \frac{[\mathrm{ES}]\{\mathrm{I}]}{K_i} \\ &= [\mathrm{E}] + \mathrm{ES}\left(1+\frac{[\mathrm{I}]}{K_i}\right) \end{aligned}$$

$$\therefore [\mathrm{E}] = [\mathrm{E_0}] - [\mathrm{ES}]\left(1+\frac{[\mathrm{I}]}{K_i}\right)$$

Under steady-state conditions, as previously noted, [E][S]/[ES] = K_m (equation 7.9).

Substituting for [E] and continuing as in section 7.1.2, the outcome is:

$$v_0 = \frac{V_{max}[S_0]}{[S_0]\left(1+\frac{[I_0]}{K_i}\right)+K_m} \tag{8.9}$$

Dividing throughout by $(1 + ([I_0]/K_i))$ gives:

$$v_0 = \frac{\frac{V_{max}}{\left(1+\frac{[I_0]}{K_i}\right)}[S_0]}{[S_0]+\frac{K_m}{\left(1+\frac{[I_0]}{K_i}\right)}} \tag{8.10}$$

This is an equation of the same form as the Michaelis-Menten equation, the constants K_m and V_{max} both being divided by a factor $(1 + ([I_0]/K_i))$. Thus, for uncompetitive inhibition:

$$V'_{max} = \frac{V_{max}}{\left(1+\frac{[I_0]}{K_i}\right)} \quad \text{and} \quad K'_m = \frac{K_m}{\left(1+\frac{[I_0]}{K_i}\right)} \tag{8.11}$$

where V'_{max} is the value of V_{max} in the presence of an initial concentration $[I_0]$ of uncompetitive inhibitor and K'_m is the apparent value of K_m under the same conditions. An inhibitor concentration equal to K_i will halve the values of both V_{max} and K_m.

The Lineweaver-Burk equation in the presence of an uncompetitive inhibitor is:

$$\frac{1}{v_0} = \frac{K'_m}{V'_{max}}.\frac{1}{[S_0]}+\frac{1}{V_{max}} \tag{8.12}$$

and the slope of a Lineweaver-Burk plot is equal to:

$$\frac{K'_m}{V'_{max}} = \frac{K_m}{V_{max}}.\frac{\left(1+\frac{[I_0]}{K_i}\right)}{\left(1+\frac{[I_0]}{K_i}\right)} = \frac{K_m}{V_{max}} \tag{8.13}$$

In other words, the slope of a Lineweaver-Burk plot is not altered by the presence of an uncompetitive inhibitor, but both intercepts change (Fig. 8.6). As before, the inhibitor constant K_i can be determined using **secondary plots**. For uncompetitive inhibition,

$$\frac{1}{V'_{max}} = \frac{1}{V_{max}}\left(1+\frac{[I_0]}{K_i}\right) \quad \text{and} \quad \frac{1}{K'_m} = \frac{1}{K_m}\left(1+\frac{[I_0]}{K_i}\right) \tag{8.14}$$

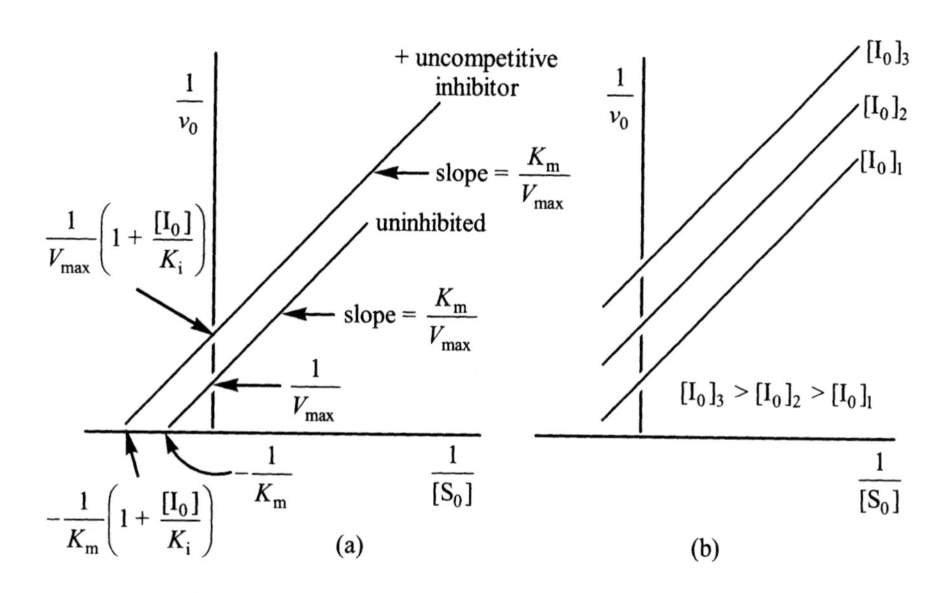

Fig. 8.6 – (a) Lineweaver-Burk plot showing the effect of uncompetitive inhibition; (b) the same, showing plots for several inhibitor concentrations at fixed enzyme concentration.

Hence, plots of $1/V'_{max}$ or $1/K'_m$ (obtained from intercepts on $1/v_0$ and $1/[S_0]$ axes respectively of the primary plot) against $[I_0]$ are linear, the intercept on the $[I_0]$ axes giving K_i (Fig. 8.7).

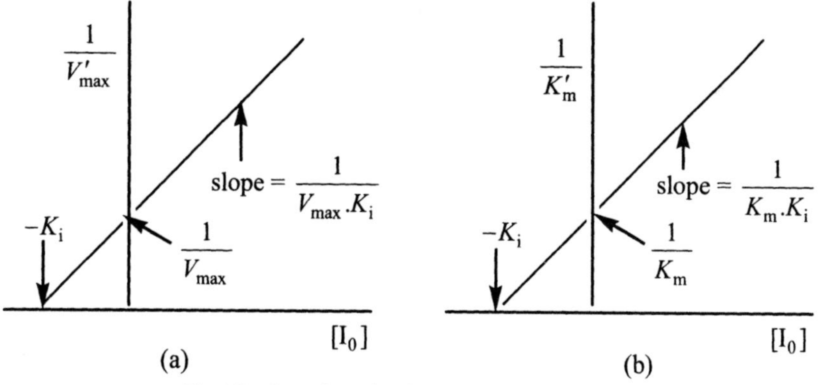

Fig. 8.7 – Secondary plots for uncompetitive inhibition.

Uncompetitive inhibition of single-substrate enzyme-catalysed reactions is a rare phenomenon, one of the few possible examples known being the inhibition of aryl sulphatase by hydrazine, and another the inhibition of intestinal alkaline phosphatase by phenylalanine. However, uncompetitive inhibition patterns are seen with two-substrate reactions and this may help in the elucidation of the reaction mechanism (see section 9.3.2).

8.2.3 Non-competitive inhibition

A non-competitive inhibitor can combine with an enzyme molecule to produce a **dead-end complex,** regardless of whether a substrate molecule is bound or not. Hence the inhibitor must bind at a different site from the substrate. We shall consider only the case where the inhibitor destroys the catalytic activity of the enzyme, either by binding to the catalytic site or as a result of a conformational change affecting the catalytic site, but does not affect substrate-binding (see Fig. 8.3d). The situation for a simple single-substrate reaction will be as follows:

$$\begin{array}{ccccc} E & \underset{-S}{\overset{+S}{\rightleftharpoons}} & ES & \rightarrow & P \\ {\scriptstyle -I}\Big\uparrow\Big\downarrow{\scriptstyle +I} & & {\scriptstyle -I}\Big\uparrow\Big\downarrow{\scriptstyle +I} & & \\ EI & \underset{-S}{\overset{+S}{\rightleftharpoons}} & ESI & & \end{array} \qquad (8.15)$$

Even this is a complex situation, for ES can be arrived at by alternative routes, making it impossible for an expression of the same form as the Michaelis-Menten equation to be derived using the general **steady-state-assumption**. However, types of non-competitive inhibition consistent with a Michaelis-Menten-type equation and a linear Lineweaver-Burk plot can occur if the **equilibrium-assumption** (section 7.1.1) is valid.

In the simplest possible model, **simple linear non-competitive inhibition**, the substrate does not affect inhibitor-binding. Under these conditions, the reactions E + I $\rightleftharpoons$ EI and ES + I $\rightleftharpoons$ ESI have an identical dissociation constant K_i , again called the **inhibitor constant**. The total enzyme concentration is effectively reduced by the inhibitor, decreasing the value of V_{max} but not altering K_m , since neither inhibitor nor substrate affects the binding of the other.

Let us once again derive an initial velocity equation, bearing in mind that on this occasion we are considering only the special case where $K_m \approx K_s$. Equation 7.9 will still be valid, despite the condition that $k_{-1} \gg k_2$, so [E][S]/[ES] = K_m

In the presence of a non-competitive inhibitor which will bind equally well to E or to ES, i.e. where K_i = [E][I]/[ES] = [ES][I]/[ESI]:

$$[E_0] = [E] + [ES] + [EI] + [ESI]$$

$$= [E] + [ES] + \frac{[E][I]}{K_i} + \frac{[ES][I]}{K_i}$$

$$= ([E] + \{ES]) \left(1 + \frac{[I]}{K_i}\right)$$

$$\therefore\ [E] + [ES] = \frac{[E_0]}{\left(1 + \frac{[I]}{K_i}\right)}$$

$$\therefore [E] = \frac{[E_0]}{\left(1+\frac{[I]}{K_i}\right)} - [ES]$$

If we continue exactly as in section 7.1.2 we conclude that, under these conditions:

$$v_0 = \frac{V_{max}}{\left(1+\frac{[I_0]}{K_i}\right)} \cdot \frac{[S_0]}{([S_0]+K_m)} \tag{8.16}$$

This is of the form of the Michaelis-Menten equation, with V_{max} being divided by a factor $(1 + ([I_0]/K_i))$. Thus, for simple linear non-competitive inhibition, K_m is unchanged and V_{max} is altered so that:

$$V'_{max} = \frac{V_{max}}{\left(1+\frac{[I_0]}{K_i}\right)}, \quad \text{or} \quad \frac{1}{V'_{max}} = \frac{1}{V_{max}}\left(1+\frac{[I_0]}{K_i}\right) \tag{8.17}$$

where V'_{max} is the value of V_{max} in the presence of a concentration $[I_0]$ of non-competitive inhibitor. It follows that K_i for such a system is the inhibitor concentration which halves the value of V_{max}. The Lineweaver-Burk equation for simple linear non-competitive inhibition is:

$$\frac{1}{v_0} = \frac{K_m}{V'_{max}} \cdot \frac{1}{[S_0]} + \frac{1}{V_{max}} \tag{8.18}$$

and Lineweaver-Burk plots showing the effect of such inhibition are shown in Fig. 8.8.

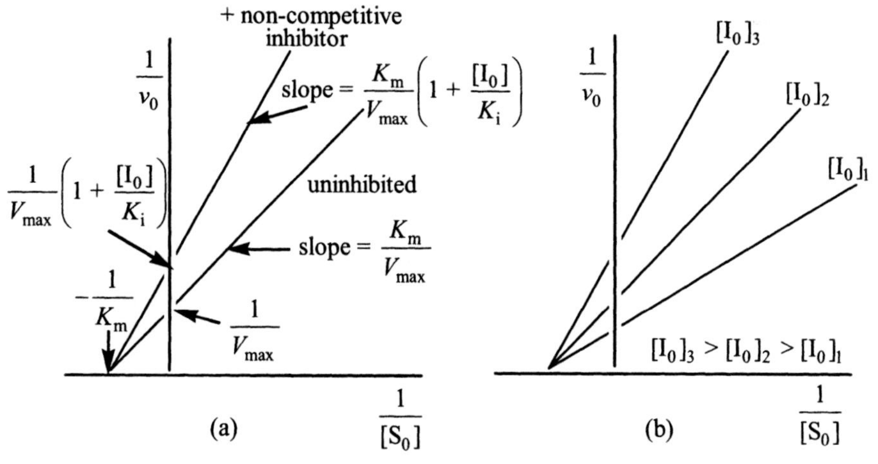

Fig. 8.8 – (a) Lineweaver-Burk plot for simple linear non-competitive inhibition; (b) the same, showing plots for several inhibitor concentrations at fixed enzyme concentration.

Once this type of inhibition has been established, the inhibitor constant K_i may be determined using secondary plots. Since

$$\frac{1}{V'_{max}} = \frac{1}{V_{max}}\left(1+\frac{[I_0]}{K_i}\right) \tag{8.19}$$

and the slope of the primary Lineweaver-Burk plot is

$$\frac{K_m}{V_{max}}\left(1+\frac{[I_0]}{K_i}\right) \tag{8.20}$$

it follows that plots of $1/V'_{max}$ against $[I_0]$ and of slope of the primary plot against $[I_0]$ are linear, the intercept on the $[I_0]$ axis giving $-K_i$ (Fig. 8.9).

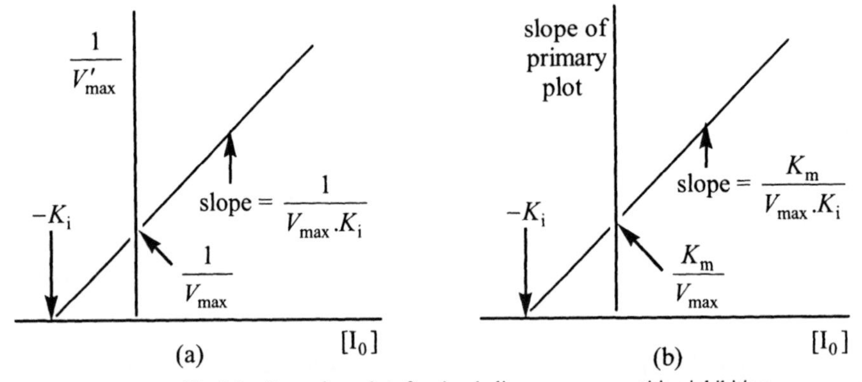

Fig. 8.9 – Secondary plots for simple linear non-competitive inhibition.

A Dixon plot may also be used to determine K_i . From the Lineweaver-Burk equation:

$$\frac{1}{v_0} = \frac{K_m}{V_{max}}\left(1+\frac{[I_0]}{K_i}\right).\frac{1}{[S_0]} + \frac{1}{V_{max}}\left(1+\frac{[I_0]}{K_i}\right)$$

$$= \frac{K_m}{V_{max}[S_0]} + \frac{K_m[I_0]}{V_{max}K_i[S_0]} + \frac{1}{V_{max}} + \frac{[I_0]}{V_{max}K_i}$$

$$= \left(\frac{K_m}{V_{max}[S_0]} + \frac{1}{V_{max}}\right)\frac{[I_0]}{K_i} + \frac{K_m}{V_{max}[S_0]} + \frac{1}{V_{max}} \tag{8.21}$$

Thus a Dixon plot of $1/v_0$ against $[I_0]$ will be linear at fixed $[E_0]$ and $[S_0]$ for simple linear non-competitive inhibition. When $[I_0] = -K_i$, it can be seen that $1/v_0 = 0$. Hence the intercept on the $[I_0]$ axis will give $-K_i$ (Fig. 8.10).

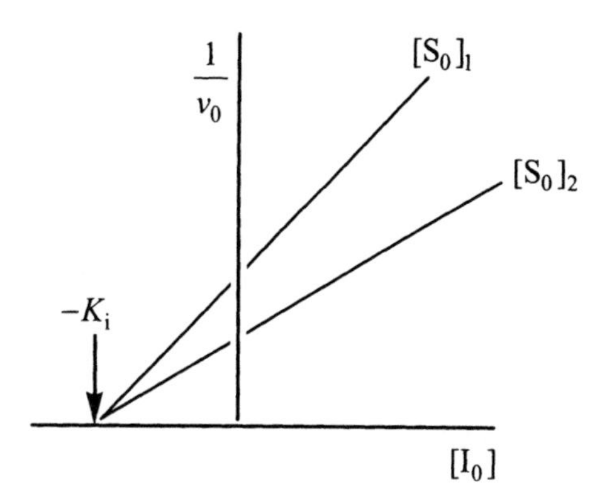

Fig. 8.10 – Dixon plot for simple linear non-competitive inhibition.

Some enzymologists point out, with reason, that simple linear non-competitive inhibition is a special case of general non-competitive inhibition, and other forms of non-competitive inhibition may show different characteristics. However, since it is far easier to determine the inhibitor pattern than the actual mechanism of inhibition, and since the same pattern may be seen for different mechanisms (see section 8.2.1), it has become established that inhibitors are classified according to the overall pattern observed. On this basis, non-competitive inhibition is only said to be present when the characteristics of simple linear non-competitive inhibition (i.e. linear Lineweaver-Burk plot, altered V_{max} but unchanged K_m) are demonstrated. If a linear Lineweaver-Burk plot is obtained but inhibition patterns characteristic of simple competitive inhibition (Fig. 8.2), uncompetitive inhibition (Fig. 8.6) or simple linear non-competitive inhibition (Fig. 8.8) are *not* observed, then **mixed inhibition** (section 8.2.4) is said to occur, irrespective of the mechanism involved. In any discussion of enzyme inhibition, the basis for classification used should be made clear.

Few clear-cut instances of non-competitive inhibition of single-substrate enzyme-catalysed reactions are known, one possible case being the inhibition of fructose-bisphosphatase by AMP in the direction of fructose-6-phosphate formation. Hydrogen ions may be regarded as providing one of the simplest examples of non-competitive inhibition. Some enzymes, e.g. chymotrypsin, where the catalytic site includes a proton acceptor, may be inhibited by increasing hydrogen ion concentration (i.e. decreasing pH), Lineweaver-Burk plots at different pH values over a relatively narrow range showing the characteristics of non-competitive inhibition. However, it should be borne in mind that the effects of changing pH on enzyme activity are complex (see sections 3.2.2 and 10.1.6).

Heavy-metal ions and organic molecules which bind to –SH groups of cysteine residues in the enzyme are sometimes quoted as being examples of non-competitive inhibitors, as are groups such as cyanide which bind to the metal ions of metallo-enzymes and destroy enzyme activity. However, in many cases such effects are irreversible, ruling out non-competitive inhibition, which must be reversible. The confusion may arise because irreversible inhibition can give kinetic patterns apparently characteristic of non-competitive inhibition, even though the two types of inhibition are otherwise quite distinct (section 8.3).

Nevertheless, it should be understood that the toxicity of substances such as cyanide, carbon monoxide, hydrogen sulphide and heavy metals is due to their action as enzyme inhibitors, whatever the precise mechanism in each case.

8.2.4 Mixed inhibition

In section 8.2.3 we obtained an expression for simple linear non-competitive inhibition which depended on the equilibrium-assumption (section 7.1.1) being valid, and further assumed that substrate-binding and inhibitor-binding were completely independent. Let us now consider the situation where the second assumption is *not* made.

There are two processes by which inhibitor may bind to the enzyme:

$$\mathrm{E} + \mathrm{I} \rightleftharpoons \mathrm{EI} \text{ (inhibitor constant } K_\mathrm{i}\text{) ; and}$$

$$\mathrm{ES} + \mathrm{I} \rightleftharpoons \mathrm{ESI} \text{ (inhibitor constant } K_\mathrm{I}\text{ .}$$

Hence, $K_\mathrm{i} = [\mathrm{E}][\mathrm{I}]/[\mathrm{EI}]$ and $K_\mathrm{I} = [\mathrm{ES}][\mathrm{I}]/[\mathrm{ESI}]$

As in section 8.2.3, $[\mathrm{E}][\mathrm{S}]/[\mathrm{ES}] = K_\mathrm{m}$; and,

$$[\mathrm{E_0}] = [\mathrm{E}] + [\mathrm{ES}] + [\mathrm{EI}] + [\mathrm{ESI}].$$

If we develop the argument as before, but this time without assuming that K_i and K_I are identical:

$$[\mathrm{E_0}] = [\mathrm{E}] + [\mathrm{ES}] + \frac{[\mathrm{E}][\mathrm{I}]}{K_\mathrm{i}} + \frac{[\mathrm{ES}][\mathrm{I}]}{K_\mathrm{I}}$$

$$= [\mathrm{E}]\left(1+\frac{[\mathrm{I}]}{K_\mathrm{i}}\right) + [\mathrm{ES}]\left(1+\frac{[\mathrm{I}]}{K_\mathrm{I}}\right)$$

$$\therefore [\mathrm{E}] = \frac{[\mathrm{E_0}]-[\mathrm{ES}]\left(1+\frac{[\mathrm{I}]}{K_\mathrm{I}}\right)}{\left(1+\frac{[\mathrm{I}]}{K_\mathrm{i}}\right)}$$

Substituting for [E] in the expression for K_m:

$$\frac{\left([\mathrm{E_0}]-[\mathrm{ES}]\left(1+\frac{[\mathrm{I}]}{K_\mathrm{I}}\right)\right)[\mathrm{S}]}{\left(1+\frac{[\mathrm{I}]}{K_\mathrm{i}}\right)[\mathrm{ES}]} = K_\mathrm{m}$$

$$\therefore [E_0][S] - [S][ES]\left(1+\frac{[I]}{K_I}\right) = K_m[ES]\left(1+\frac{[I]}{K_i}\right)$$

$$\therefore [ES]\left([S]\left(1+\frac{[I]}{K_I}\right)+K_m\left(1+\frac{[I]}{K_i}\right)\right) = [E_0][S]$$

$$\therefore [ES] = \frac{[E_0][S]}{[S]\left(1+\frac{[I]}{K_I}\right)+K_m\left(1+\frac{[I]}{K_i}\right)}$$

Continuation as before gives:

$$v_0 = \frac{V_{max}[S_0]}{[S_0]\left(1+\frac{[I_0]}{K_I}\right)+K_m\left(1+\frac{[I_0]}{K_i}\right)} \quad (8.22)$$

If numerator and denominator are both divided by $(1 + ([I_0]/K_I))$

$$v_0 = \frac{\frac{V_{max}}{\left(1+\frac{[I_0]}{K_I}\right)}.[S_0]}{[S_0]+\frac{K_m\left(1+\frac{[I_0]}{K_i}\right)}{\left(1+\frac{[I_0]}{K_I}\right)}} \quad (8.23)$$

This is of the same form as the Michaelis-Menten equation and can be written:

$$v_0 = \frac{V'_{max}[S_0]}{[S_0]+K'_m} \quad (8.24)$$

where

$$V'_{max} = \frac{V_{max}}{\left(1+\frac{[I_0]}{K_I}\right)} \quad \text{and} \quad K'_m = K_m\frac{\left(1+\frac{[I_0]}{K_i}\right)}{\left(1+\frac{[I_0]}{K_I}\right)} \quad (8.25)$$

Similarly, the Lineweaver-Burk equations is:

$$\frac{1}{v_0} = \frac{K'_m}{V'_{max}} \cdot \frac{1}{[S_0]} + \frac{1}{V'_{max}} \tag{8.26}$$

and a Lineweaver-Burk plot will be linear. However, in general, K_m , V_{max} and slope, which equals

$$\frac{K'_m}{V'_{max}} = \frac{K_m}{V_{max}}\left(1 + \frac{[I_0]}{K_i}\right) \tag{8.27}$$

are all affected by the inhibitor. Thus, plots at different inhibitor concentrations (at fixed $[E_0]$) will not intersect on either axis, nor will the slope be the same, so the pattern will be different from those characteristic of competitive, non-competitive and uncompetitive inhibition, and is given the name mixed inhibition. It must be realized that this describes the overall pattern observed and does not imply that more than one type of inhibitor is present.

In the situation where $K_i > K_I$, the plots cross to the left of the $1/v_0$ axis but above the $1/[S_0]$ axis (Fig. 8.11a). This situation has been termed **competitive-non-competitive inhibition**, because the pattern observed lies between those for competitive (Fig. 8.2) and non-competitive (Fig. 8.8) inhibition.

In the situation where $K_i < K_I$, the plots cross to the left of the $1/v_0$ axis and below the $1/[S_0]$ axis (Fig. 8.11b). This form of mixed inhibition has been termed **non-competitive-uncompetitive inhibition** because the pattern is intermediate between those for non-competitive (Fig. 8.8) and uncompetitive (Fig. 8.6) inhibition.

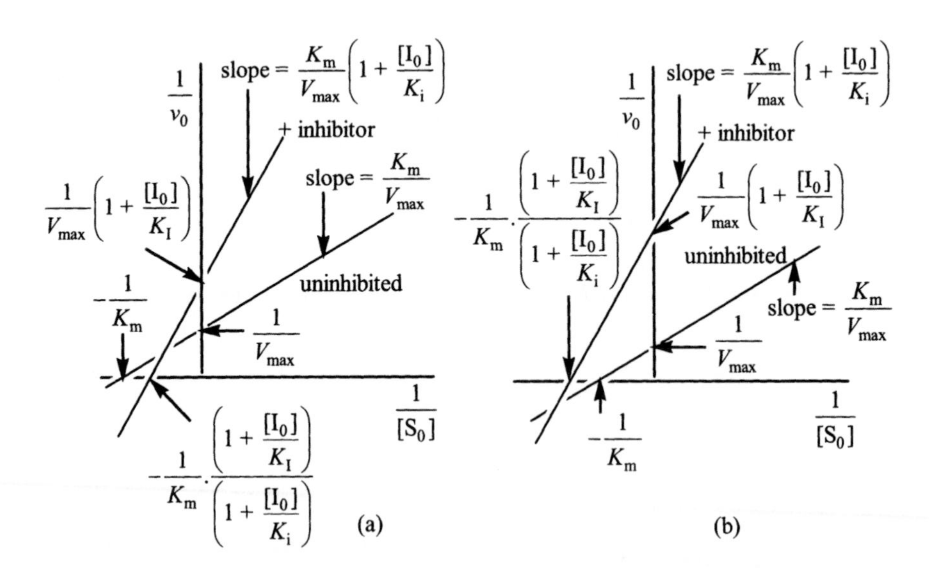

Fig. 8.11 – Lineweaver-Burk plots showing the effect of mixed inhibition: (a) $K_I > K_i$; (b) $K_I < K_i$.

In either case, K_i and K_I can be determined using secondary plots. For mixed inhibition:

$$\frac{1}{V'_{max}} = \frac{1}{V_{max}}\left(1+\frac{[I_0]}{K_I}\right) \tag{8.28}$$

and slope for the inhibited reaction = slope for the uninhibited reaction × $(1+([I_0]/K_i))$. Hence a secondary plot of $1/V'_{max}$ against $[I_0]$ will be linear, the intercept on the $[I_0]$ axis giving $-K_i$ (Fig. 8.12a). A graph of slope of primary plot against $[I_0]$ will also be linear, the intercept on the $[I_0]$ axis giving $-K_i$ (Fig. 8.12b).

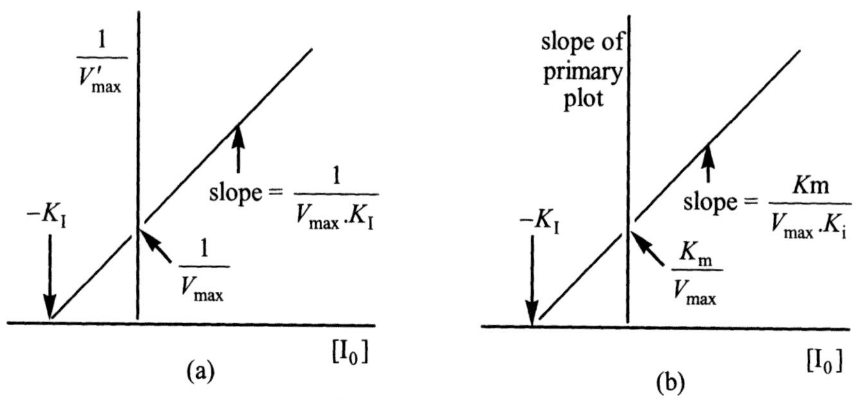

Fig. 8.12 – Secondary plots for mixed inhibition.

However, $$K'_m = K_m \frac{\left(1+\frac{[I_0]}{K_i}\right)}{\left(1+\frac{[I_0]}{K_I}\right)} \tag{8.29}$$

which means that a graph of K'_m against $[I_0]$ will *not* be linear.

The equation for v_0 derived above is a relatively general one, since no assumptions were made about the values of K_i and K_I, and it can be simplified for special cases. If ESI cannot be formed, then $K_I = \infty$ and the equation becomes that for competitive inhibition (section 8.2.1), regardless of whether the substrate and inhibitor bind to the same or different sites. If the complex ESI can occur but not EI, then $K_i = \infty$ and the equation simplifies to that for uncompetitive inhibition (section 8.2.2). When $K_i = K_I$, the equation reduces to that for simple linear non-competitive inhibition (section 8.2.3).

8.2.5 Partial inhibition

Hitherto we have considered only situations where enzyme-inhibitor complexes are dead-end ones, i.e. where no product can be formed from them. Let us now return to the general system described in section 8.2.4 and consider what would happen if the inhibition was only partial and the ESI complex could break down (with rate constant = k'_2) to yield product according to the equation:

$$\text{ESI} \xrightarrow{k'_2} \text{E} + \text{P} + \text{I}$$

Under these conditions, the overall initial velocity is given by:

$$v_0 = k_2[\text{ES}] + k'_2[\text{ESI}]$$

$$= k_2[\text{ES}] + k'_2\frac{[\text{ES}][\text{I}]}{K_\text{I}}$$

$$= k_2[\text{ES}]\left(1+\frac{k'_2[\text{I}]}{k_2K_\text{I}}\right)$$

Using the same procedure as before, an expression of the same form as the Michaelis-Menten equation can be obtained:

$$v_0 = \frac{V_{\max}[\text{S}_0]\dfrac{\left(1+\dfrac{k'_2[\text{I}_0]}{k_2K_\text{I}}\right)}{\left(1+\dfrac{[\text{I}_0]}{K_\text{I}}\right)}}{[\text{S}_0]+\dfrac{K_\text{m}\left(1+\dfrac{[\text{I}_0]}{K_\text{i}}\right)}{\left(1+\dfrac{[\text{I}_0]}{K_\text{I}}\right)}} \qquad (8.30)$$

Hence a Lineweaver-Burk plot would be linear. However, **secondary plots** of intercept or slope against $[\text{I}_0]$ will not be linear, enabling **partial inhibition** to be distinguished from inhibition involving **dead-end complexes**.

8.2.6 Substrate inhibition

A characteristic of enzyme-catalysed reactions, as we have seen, is that for a given enzyme concentration, the initial reaction velocity increases with increasing initial substrate concentration to a limiting value, $V_{\max}$ (section 6.4). At still higher substrate concentrations, the initial velocity is sometimes found to be less than the maximum value. In some instances the observations may be explained away on the basis of interaction between the detecting system and excess substrate, but in other cases it appears that the substrate, in very high concentrations, really can inhibit its own conversion to product.

Let us consider one possible mechanism for substrate inhibition at high substrate concentration, in relation to the reaction catalysed by **succinate dehydrogenase**. For a reaction to take place, both carboxyl groups of the substrate have to bind to the enzyme:

$$\text{E}\left|\begin{array}{l} \text{- - }{}^-O_2C \\ \quad\;\;\, | \\ \quad\;\; CH_2 \\ \quad\;\;\, | \\ \quad\;\; CH_2 \\ \quad\;\;\, | \\ \text{- - }{}^-O_2C \end{array}\right. \text{succinate} \quad \longrightarrow \quad \text{E}\left|\begin{array}{l} \text{- - }{}^-O_2C \\ \quad\;\;\, | \\ \quad\;\; CH \\ \quad\;\;\, \| \\ \quad\;\; CH \\ \quad\;\;\, | \\ \text{- - }{}^-O_2C \end{array}\right. \text{fumarate} \; + \; 2H \qquad (8.31)$$

At very high substrate concentrations there is an increased possibility of carboxyl groups from two separate substrate molecules binding to the same enzyme:

$$\text{E}\left|\begin{array}{l} \text{- - }{}^-O_2C.CH_2.CH_2.CO_2^- \\ \\ \\ \text{- - - }{}^-O_2C.CH_2.CH_2.CO_2^- \end{array}\right.$$

If this happens, a reaction cannot take place until one of them has dissociated away again. The characteristic features of substrate inhibition are shown in Fig. 8.13.

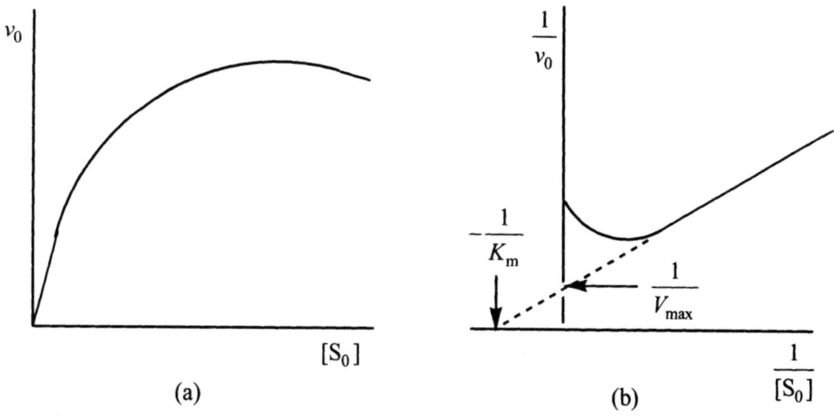

Fig. 8.13 – (a) Michaelis-Menten and (b) Lineweaver-Burk plots, showing the effects of substrate inhibition.

In general, substrate inhibition occurs when a molecule of substrate binds to one site on the enzyme and then another molecule of substrate binds to a separate site on the enzyme to form a dead-end complex. This can be regarded as a form of uncompetitive inhibition, the extra substrate molecule being the inhibitor. As shown in section 8.2.2, the initial velocity equation for uncompetitive inhibition (8.10) is:

$$v_0 = \frac{\dfrac{V_{\text{max}}}{\left(1+\dfrac{[\text{I}_0]}{K_\text{i}}\right)}[\text{S}_0]}{[\text{S}_0]+\dfrac{K_\text{m}}{\left(1+\dfrac{[\text{I}_0]}{K_\text{i}}\right)}}$$

If the inhibitor is identical to the substrate, this becomes:

$$v_0 = \frac{\dfrac{V_{\text{max}}}{\left(1+\dfrac{[\text{S}_0]}{K_\text{i}}\right)}[\text{S}_0]}{[\text{S}_0]+\dfrac{K_\text{m}}{\left(1+\dfrac{[\text{S}_0]}{K_\text{i}}\right)}} = \frac{V_{\text{max}}[\text{S}_0]}{[\text{S}_0]\left(1+\dfrac{[\text{S}_0]}{K_\text{i}}\right)+K_\text{m}} \tag{8.32}$$

This is consistent with the plots in Fig. 8.12. At very low $[\text{S}_0]$ the term $[\text{S}_0]/K_\text{i}$ is negligible and the expression reduces to the normal Michaelis-Menten equation. When $[\text{S}_0]$ is very high, then

$$[\text{S}_0]\left(1+\frac{[\text{S}_0]}{K_\text{i}}\right)+K_\text{m} \approx [\text{S}_0]\left(1+\frac{[\text{S}_0]}{K_\text{i}}\right)$$

and the equation simplifies to:

$$v_0 = \frac{V_{\text{max}}}{\left(1+\dfrac{[\text{S}_0]}{K_\text{i}}\right)} \tag{8.33}$$

Under these circumstances, v_0 decreases as $[\text{S}_0]$ increases, as observed for substrate inhibition at high substrate concentrations.

8.2.7 Allosteric inhibition

The forms of inhibition considered previously in this chapter have many applications (section 8.2.1) but in general they do not play a major role in the normal functioning of the living cell. In contrast, allosteric inhibition plays a vital role in metabolic regulation. Consider a biosynthetic pathway:

$$\text{A} \rightarrow \text{B} \rightarrow \text{C} \rightarrow \text{D} \rightarrow \text{E} \rightarrow \text{F}$$

Unnecessary production of excess F may be prevented, and supplies of A conserved, by feedback inhibition, where the end-product F acts as an allosteric inhibitor of an early enzyme in the pathway, e.g. that catalysing the reaction A $\rightarrow$ B.

An allosteric inhibitor, by definition (see section 13.2.3), binds to the enzyme at a site distinct from the substrate-binding site. Therefore, some of the types of inhibition we have considered previously may be regarded as forms of allosteric inhibition. However, the term allosteric inhibition is usually reserved for the situation where the inhibitor, rather than forming a **dead-end complex** with the enzyme, influences **conformational changes** which may alter the **binding characteristics** of the enzyme for the substrate or the subsequent **reaction characteristics** (or both). The Michaelis-Menten plot becomes less hyperbolic and more sigmoidal (S-shaped) (Fig. 13.1), which means that the rate of reaction is reduced at low substrate concentrations but not necessarily at others.

If the binding characteristics alone are affected, V_{max} will usually remain unchanged, so the inhibition pattern could be regarded as competitive. Similarly, other forms of allosteric inhibition, where V_{max} is altered, could be regarded as giving non-competitive or mixed inhibition, depending on whether K_m (the substrate concentration where $v_0 = \frac{1}{2}V_{max}$) is changed or not. However, in most cases the Michaelis-Menten equation is not obeyed in the presence of allosteric inhibitors, nor are linear Lineweaver-Burk plots obtained, so the terms competitive, non-competitive and mixed inhibition are, in general, not considered applicable.

Almost all enzymes known to be subject to end-product (feed-back) inhibition are oligomeric proteins, so the mechanism of allosteric inhibition may involve interactions between the enzyme sub-units. In support of this, allosteric control, but not catalytic activity, of some enzymes has been shown to be lost when the oligomer is separated into its monomeric units.

Allosteric inhibition is discussed in more detail in Chapters 13 and 14.

8.3 IRREVERSIBLE INHIBITION

An irreversible inhibitor binds to the active site of the enzyme by an irreversible reaction, $E + I \rightarrow EI$, and hence cannot subsequently dissociate from it. A covalent bond is usually formed between inhibitor and enzyme. The inhibitor may act by preventing substrate-binding or it may destroy some component of the catalytic site. Compounds which irreversibly denature the enzyme protein or cause non-specific inactivation of the active site are not usually regarded as irreversible inhibitors.

Of course in practice no process is totally irreversible, but an inhibitor which shows great affinity for the enzyme (dissociation constant in the order of 10^{-9} mol l^{-1}) is regarded as irreversible. Unlike reversible inhibition, where an equilibrium is quickly set up between inhibitor and enzyme, making the system suitable for investigation by initial velocity studies, irreversible inhibition is progressive and will increase with time until either all the inhibitor or all the enzyme present has been used up in forming enzyme-inhibitor complex.

Regardless of whether the reaction between enzyme and irreversible inhibitor has gone to completion before initial velocity studies are commenced, these will give little or no information about the characteristics of the inhibitor (see below).

It is much more useful to describe the system in terms of the rate constant for the binding of inhibitor to substrate. If these react on a 1:1 basis, then the time taken to reduce the enzyme activity to 50% of its original value will be inversely proportional to the initial inhibitor concentration (at fixed total enzyme concentration).

If a molar excess of the inhibitor is present, all molecules of enzyme present will eventually become bound to inhibitor and catalytic activity will be reduced to a residual level or completely lost. (In general, the binding of such an inhibitor to a catalytic site will totally destroy catalytic activity, but binding to a substrate-binding site may not prevent a small amount of catalysis taking place, provided the catalytic site remains active and accessible to substrate molecules approaching it by random movement in free solution.) Hence irreversible inhibitors may be titrated against an enzyme, particularly where the reaction between enzyme and inhibitor proceeds rapidly. In contrast, reversible inhibitors, like substrates (section 7.1.1), can be present in great excess without saturating the available enzyme.

Irreversible inhibitors effectively reduce the concentration of enzyme present. An inhibitor of initial concentration $[I_0]$ will reduce the concentration of active enzyme from an initial value of $[E_0]$ to $[E_0] - [I_0]$, assuming the inhibitor is not in excess. If a substrate is introduced after the reaction between inhibitor and enzyme has gone to completion, a system which obeys the Michaelis-Menten equation in the absence of inhibitor will still do so. The value of K_m will be the same as for the uninhibited reaction, but V_{max} will be reduced (to V'_{max}).

In the absence of inhibitor, $V_{max} = k_{cat}[E_0]$

In the presence of inhibitor, $V'_{max} = k_{cat}([E_0] - [I_0])$

$$\therefore \frac{V'_{max}}{V_{max}} = \frac{[E_0]-[I_0]}{[E_0]}$$

$$\therefore V'_{max} = V_{max}[E_0]\left(1-\frac{[I_0]}{[E_0]}\right) \qquad (8.34)$$

Similar results would be obtained even if the reaction between enzyme and inhibitor had not gone to completion, provided the degree of inhibition was relatively constant over the period when initial velocity studies were being carried out.

Therefore, patterns resembling those for reversible non-competitive inhibition, with unchanged K_m and reduced V_{max}, may be obtained with irreversible inhibitors, even when the inhibitor binds to the same site as the substrate. However, the very real differences between the two forms of inhibition should be clear from the above discussion, and any attempt to calculate K_i from initial velocity measurements would be a totally meaningless exercise, since the relationship between V'_{max} and V_{max} does not involve K_i in this case. Hence, if a pattern of non-competitive inhibition is obtained in the investigation of a system, it is important to establish whether the inhibition is reversible or irreversible before the results can be interpreted.

Many irreversible inhibitors attack –SH groups (in cysteine side chains) which are often found at the active sites of enzymes. Important examples are alkylating agents, such as iodoacetate and iodoacetamide, which form covalent linkages with essential –SH groups:

$$\underset{\text{enzyme}}{E-SH} + \underset{\text{iodoacetate}}{ICH_2.CO_2^-} \rightarrow E-S-CH_2CO_2^- + HI \qquad (8.35)$$

Another well-known group are the organophosphorus compounds which react with essential –OH groups (in serine side chains) of some enzymes. An example is **diisopropylphosphofluoridate (DFP),** which is a nerve poison since one of the enzymes it inactivates is acetylcholinesterase, important in nerve function:

$$\underset{\text{enzyme}}{E-OH} + \underset{\text{DFP}}{F-\overset{\displaystyle OCH(CH_3)_2}{\underset{\displaystyle OCH(CH_3)_2}{P{=}O}}} \rightarrow E-O-\overset{\displaystyle OCH(CH_3)_2}{\underset{\displaystyle OCH(CH_3)_2}{P{=}O}} + HF \qquad (8.36)$$

Irreversible inhibitors are useful in the investigation of the active site of an enzyme, since the inhibitor, unlike the substrate, will remain firmly bound to one of the amino acids of the enzyme and thus act as a marker to enable it to be identified (Chapter 10). Some organophosphorus compounds are also used as insecticides.

SUMMARY OF CHAPTER 8

Competitive inhibitors usually compete with the substrate for the same binding site on the enzyme. In the characteristic form, Michaelis-Menten kinetics are obeyed, K_m is increased and V_{max} unchanged.

Uncompetitive inhibitors bind to a site other than the substrate-binding site on the enzyme-substrate complex, altering the K_m and V_{max} but not the slope of the Lineweaver-Burk plot.

Non-competitive inhibitors bind to a site other than the substrate-binding site on the enzyme and enzyme-substrate complex. In the characteristic form, Michaelis-Menten kinetics are obeyed, K_m is unchanged and V_{max} decreased.

Forms of inhibition obeying Michaelis-Menten kinetics but not giving patterns characteristic of competitive, uncompetitive or non-competitive inhibition are usually termed mixed inhibition, irrespective of the actual mechanism.

Secondary plots and Dixon plots enable the inhibitor constant K_i to be calculated and help distinguish between mechanisms which give identical primary Lineweaver-Burk plots.

These types of inhibition are used in biochemical research and have applications in medicine and agriculture. Allosteric inhibition, which results in more sigmoidal reaction characteristics, plays an important role in metabolic regulation in the living cell. All these forms of inhibition are reversible, but irreversible inhibition is also known. Irreversible inhibitors have been used to identify amino acids in the active sites of enzymes.

FURTHER READING

As for Chapter 7, plus:

Birk, Y. (2003), *Plant Protease Inhibitors: Significance in Nutrition, Plant Protection, Cancer Prevention and Genetic Engineering*, Springer-Verlag.

Choudhary, M. I. (ed.) (1996), *Biological Inhibitors,* Harwood Academic Publishers.

Cornish-Bowden, A. (1995), *Fundamentals of Enzyme Kinetics,* 2nd edn., Portland Press (Chapter 5).

Huber, R. (1990), Serine and cysteine proteases and their natural inhibitors - structures and implications for function and drug design, *Proceedings of the Association of Clinical Biochemists National Meeting,* 1990, 2-4.

Kubota, E., Dean, R. G. *et al* (2002), Inhibition of peptidases in the control of blood pressure, *Essays in Biochemistry*, **38**, 129-139, Portland Press.

Marangoni, A. G. (2002), *Enzyme Kinetics: A Modern Approach*, Wiley (Chapters 4-5, 13).

Ogden, J. E. and Moore, P. K. (1995), Inhibition of nitric oxide synthase - a novel class of therapeutic agent?, *Trends in Biotechnology,* **13**, 70-78.

Schudt, C. Dent, C. and Rabe, K. F. (eds.) (1996), *Phosphodiesterase Inhibitors,* Academic Press.

Smith, H. J. (2004), *Enzymes and their Inhibition: Drug Development*, Taylor and Francis.

Zollner, H. (1993), *Handbook of Enzyme Inhibitors,* 2nd edn., VCH.

PROBLEMS

8.1 An enzyme-catalysed reaction was found to be affected by two inhibitors, A and B. The following results were obtained at fixed total enzyme concentration:

[substrate] (mmol l^{-1})	Initial velocity (absorbance units per minute)		
	Uninhibited	With 1 mmol l^{-1} A	With 1 mmol l^{-1} B
50	0.684	–	–
20	1.08	–	–
10	1.43	1.01	0.653
5	1.02	0.649	0.468
3.3	0.798	0.476	0.363
2.5	0.657	0.374	0.296
2.0	0.549	0.311	0.250

Comment on the results.

8.2 The system investigated in problem 7.1 was investigated again under identical conditions but in the presence of an inhibitor, giving the following data:

[Substrate] (mmol l^{-1})	5.0	6.67	10.0	20.0	40.0
Initial velocity (μmol $l^{-1}min^{-1}$)	100	122	156	222	278

Determine the type of inhibition. If K_i for this system is 2.9 mmol l^{-1}, calculate the inhibitor concentration present.

8.3 The system investigated in problem 7.2 was investigated again under identical conditions but in the presence of oxaloacetate (initial concentration 2.0 mmol l^{-1} in each case). These results were obtained:

		Absorbance (at 340 nm) at time t (minutes)					
[NAD^+] (mmol l^{-1})	t =	0.5	1.0	1.5	2.0	2.5	3.0
1.5		0.026	0.042	0.058	0.074	0.088	0.102
2.0		0.028	0.045	0.062	0.080	0.096	0.112
2.5		0.029	0.047	0.066	0.084	0.102	0.117
3.33		0.030	0.050	0.070	0.090	0.108	0,123
5.0		0.032	0.053	0.074	0.097	0.115	0.130
10.0		0.033	0.057	0.080	0.103	0.124	0.140

What type of inhibition is exhibited? What would be the value of V_{max}' in presence of 3.0 mmol l^{-1} oxaloacetate?

8.4 A single-substrate enzyme-catalysed reaction was investigated in the presence of 1.0 mmol l^{-1} inhibitor and in the absence of inhibitor, the initial enzyme concentration being constant throughout. The following results were obtained:

		[Product] (μmol l^{-1}) at time t (s)					
[Substrate] (mmol l^{-1})	t =	0	60	120	180	240	300
5.0 inhibited		0	110	221	333	430	480
5.0 uninhibited		0	161	320	482	598	662
6.67 inhibited		0	142	281	420	531	598
6.67 uninhibited		0	194	388	581	745	796
10.0 inhibited		0	183	367	549	705	752
10.0 uninhibited		0	263	525	789	998	1120
20.0 inhibited		0	279	558	837	1050	1170
20.0 uninhibited		0	400	798	1200	1520	1760
50.0 inhibited		0	398	798	1190	1520	1760
50.0 uninhibited		0	576	1150	1730	2170	2460

Determine the type of inhibition and calculate the values of K_m' and V_{max}' in presence of 3.0 mmol l^{-1} inhibitor.

8.5 The following results were obtained for a single-substrate enzyme-catalysed reaction (at fixed initial enzyme concentration) in the presence of an inhibitor, I:

[substrate] (mmol l^{-1})	v_0 (absorbance units min^{-1}) at various values of $[I_0]$ (mmol l^{-1})				
	$[I_0]$ = 1.0	2.0	3.0	4.0	5.0
2.0	0.432	0.396	0.365	0.339	0.317
2.5	0.485	0.448	0.417	0.389	0.365
3.33	0.552	0.516	0.485	0.457	0.432
5.0	0.642	0.609	0.579	0.552	0.528

What can you deduce about the nature of the inhibition?

8.6 A single-substrate enzyme catalysed reaction was investigated at fixed initial enzyme concentration in the presence of an inhibitor, I, the following data being obtained:

[substrate] (mmol l^{-1})	v_0 (absorbance units min^{-1}) at various values of $[I_0]$ (mmol l^{-1})				
	$[I_0]$ = 1.0	2.0	3.0	4.0	5.0
2.0	0.400	0.351	0.317	0.291	0.271
2.5	0.447	0.394	0.356	0.328	0.306
3.33	0.506	0.447	0.405	0.374	0.350
5.0	0.584	0.521	0.475	0.440	0.413

What can be deduced from these results about the nature of the inhibition?

9

Kinetics of Multi-Substrate Enzyme-Catalysed Reactions

9.1 EXAMPLES OF POSSIBLE MECHANISMS

9.1.1 Introduction

Most biochemical reactions involve at least two substrates, so it is necessary to consider the kinetics of such reactions. This is a vast and extremely complex topic and, for reasons of simplicity, we will restrict our discussion to some specific examples of **two-substrate two-product (bi-bi)** reactions. These are often transfer reactions of one type or another (including oxidation/reduction reactions) and can best be represented as: $AX + B \rightleftharpoons BX + A$.

The reaction mechanism may be a **sequential** one, where both substrates bind to the enzyme to form a **ternary complex** before the first product is formed, or it may be **non-sequential**.

9.1.2 Ping-pong bi-bi mechanism

An example of a non-sequential mechanism is the ping-pong bi-bi or double-displacement mechanism:

$$AX + E \rightleftharpoons E.AX \rightleftharpoons EX.A \rightleftharpoons EX + A \qquad (9.1)$$

$$EX + B \rightleftharpoons EX.B \rightleftharpoons E.BX \rightleftharpoons E + BX \qquad (9.2)$$

AX first binds to the enzyme E, forming a binary complex E.AX (X is usually a small group and does not participate in the reaction as a free molecule, so it is not regarded as a separate reactant). An intramolecular reorganization takes place, the bond E–X being formed and the X–A bond being broken. The first product, A, then leaves before the second substrate arrives. B cannot bind to the enzyme E but can bind to the modified enzyme EX.

Since only one substrate is present on the enzyme at any one time, there may only be a **single binding site**. Another intramolecular rearrangement takes place, the bond B–X being formed and the bond E–X being broken. The second product, BX, is then liberated, leaving the enzyme in its original form. A diagrammatic representation devised by Wallace (W. W.) Cleland shows this sequence of events as follows:

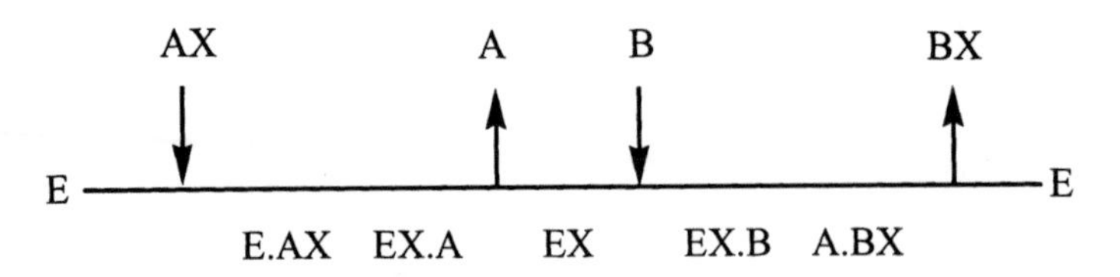

9.1.3 Random-order mechanism

A random-order mechanism is one in which any substrate can bind first to the enzyme and any product can leave first. It is a sequential mechanism and for a two-substrate reaction involves the formation of a **ternary complex** (one involving enzyme and both substrates):

$$
\begin{array}{lcccl}
E + AX \rightleftharpoons E.AX & & & & EA \rightleftharpoons E + A \\
\quad\searrow^{+B} & & & \nearrow^{-BX} & \\
 & E.AX.B & \rightleftharpoons & E.A.BX & \quad (9.3) \\
\quad\nearrow^{+AX} & & & \searrow^{-A} & \\
E + B \rightleftharpoons EB & & & & E.BX \rightleftharpoons E + BX
\end{array}
$$

There will be **two binding sites** on the enzyme, one for A/AX and one for B/BX:

9.1.4 Compulsory-order mechanism

A compulsory-order (or simply ordered) mechanism is a sequential mechanism where the order of binding to and leaving the enzyme is compulsory. For a two-substrate reaction, a **ternary complex** will be involved. The precise order must be specified, e.g.

$$E + AX \rightleftharpoons E.AX \xrightleftharpoons{+B} E.AX.B \rightleftharpoons E.A.BX \xrightleftharpoons{-BX} EA \rightleftharpoons E + A \quad (9.4)$$

or

$$E + B \rightleftharpoons EB \xrightleftharpoons{+AX} E.B.AX \rightleftharpoons E.BX.A \xrightleftharpoons{-A} E.BX \rightleftharpoons E + BX \quad (9.5)$$

As before, the enzyme will have separate binding sites for A/AX and B/BX.

9.2 STEADY-STATE KINETICS

9.2.1 The general rate equation of Alberty

Many two-substrate enzyme-catalysed reactions obey the Michaelis-Menten equation with respect to one substrate at constant concentrations of the other substrate. This applies both to reactions catalysed by enzymes with just one binding site per substrate and to those with several binding sites per substrate, provided there is no interaction between the binding sites. For such reactions, Robert Alberty (1953) derived the general equation:

$$v_0 = \frac{V_{max}[AX_0][B_0]}{K_m^B[AX_0]+K_m^{AX}[B_0]+[AX_0][B_0]+K_s^{AX}K_m^B} \quad (9.6)$$

where V_{max} is the maximum possible v_0, when AX and B are both saturating,

K_m^{AX} is the concentration of AX which gives ½V_{max} when B is saturating,

K_m^B is the concentration of B which gives ½V_{max} when AX is saturating,

K_s^{AX} is the dissociation constant for $E + AX \rightleftharpoons E.AX$

The total enzyme concentration is constant and much smaller than the concentrations of the two substrates.

At very large $[B_0]$, the general equation simplifies to:

$$v_0 = \frac{V_{max}}{\left(1+\frac{K_m^{AX}}{[AX_0]}\right)} = \frac{V_{max}[AX_0]}{[AX_0]+K_m^{AX}} \quad (9.7)$$

(which is the Michaelis-Menten equation, 7.12). Similarly at very large $[AX_0]$,

$$v_0 = \frac{V_{max}}{\left(1+\frac{K_m^B}{[B_0]}\right)} = \frac{V_{max}[B_0]}{[B_0]+K_m^B} \quad (9.8)$$

At constant but not saturating $[B_0]$, the general equation can be rearranged to give:

$$v_0 = \frac{V_{max}K_1[AX_0]}{[AX_9]+K_2} \quad (9.9)$$

(which is of the form of the Michaelis-Menten equation) where

$$K_1 = \frac{[B_0]}{[B_0]+K_m^B} \quad \text{and} \quad K_2 = \frac{K_s^{AX}K_m^B+K_m^{AX}[B_0]}{[B_0]+K_m^B} \quad (9.10)$$

At constant but non-saturating [AX_0], a similar expression can be obtained, also of the form of the Michaelis-Menten equation.

The reason for the mixed $K_s^{AX}K_m^B$ term rather than $K_m^{AX}K_m^B$ can be seen if we consider a compulsory-order mechanism of the form:

$$A + AX \rightleftharpoons EAX \rightleftharpoons EAXB \rightarrow\rightarrow \text{products}$$

As [B_0] tends to zero, there will be very little formation of EAXB from EAX, so $E + AX \rightleftharpoons EAX$ will be very close to equilibrium and K_m^{AX} tends to K_s^{AX}. This is consistent with the Alberty equation, for at a constant but very low value of [B_0],

$$K_1 \approx [B_0]/K_m^B \quad \text{and} \quad K_2 \approx K_s^{AX}(K_m^B)/(K_m^B)$$

giving

$$v_0 = \frac{V_{max}[B_0][AX_0]}{K_m^B([AX_0]+K_s^{AX})} \qquad (9.11)$$

an expression involving K_s^{AX} but not K_m^{AX}.

For a compulsory-order mechanism where B binds first to the enzyme, the $K_s^{AX}K_m^B$ term would be replaced in the general equation by $K_s^BK_m^{AX}$.

For a random-order mechanism, either term could be used, and would be justified on grounds similar to the above.

Many random- and compulsory-order reactions involving ternary complexes obey the general rate equation of Alberty, particularly where the rate-limiting step is the interconversion of the ternary complexes (E.AX.B $\rightleftharpoons$ E.A.BX, the main chemical reaction taking place, involving the breakage of the A–X bond and the formation of the B–X bond in the forward direction). A similar observation was made for single-substrate reactions in section 7.1.3, for these obeyed the Michaelis-Menten equation if ES→EP was the rate-limiting step. A random-order mechanism where all the steps except the interconversion of the ternary complexes are rapid is sometimes said to have a **random-order rapid-equilibrium bi-bi mechanism**.

For a ping-pong bi-bi mechanism, the liberation of A from the enzyme in the initial period of the reaction will be irreversible because the concentration of product A present will be negligible. Hence $K_s^{AX} = 0$ and so $K_s^{AX}.K_m^B = 0$, giving this mechanism a simpler rate equation:

$$v_0 = \frac{V_{max}[AX_0][B_0]}{K_m^B[AX_0]+K_m^{AX}[B_0]+[AX_0][B_0]} \qquad (9.12)$$

9.2.2 Plots for mechanisms which follow the general rate equation

Two-substrate reactions obeying the general rate equation of Alberty also obey the Michaelis-Menten equation with respect to one substrate, provided the concentration of the other substrate is maintained fixed. It follows, therefore, that the corresponding double-reciprocal (i.e. Lineweaver-Burk) plot should be linear. Such **primary plots** are drawn for reactions investigated with the fixed substrate concentration being in one of two categories: either saturating the enzyme, or approximately half-saturating it. The theoretical results are shown in Fig. 9.1, together with some of the characteristics of the graphs.

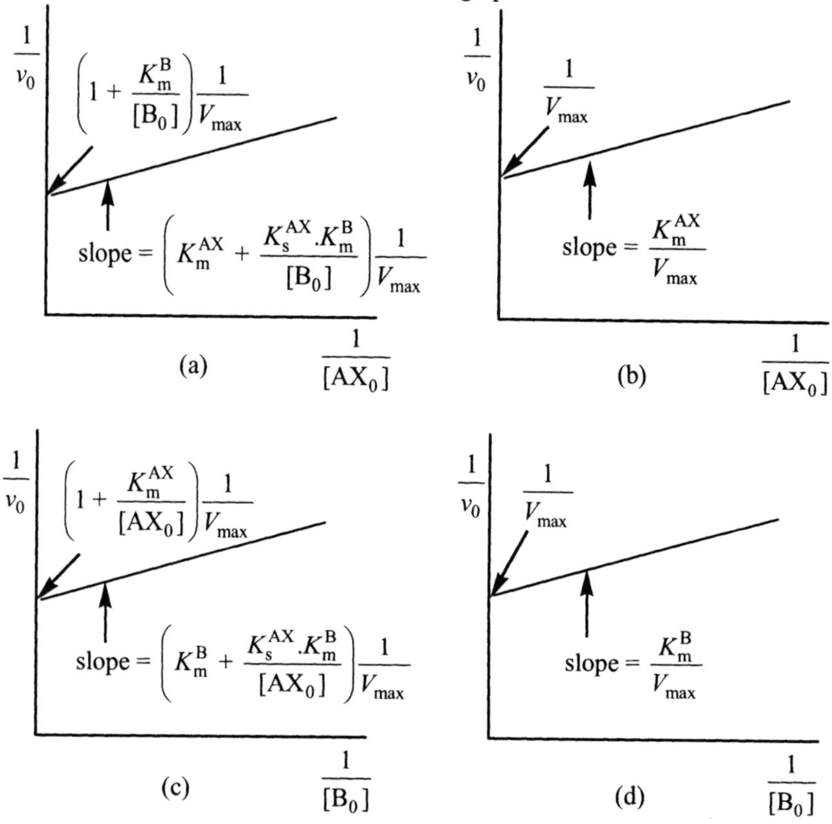

Fig. 9.1 – Primary plots for reactions with mechanisms following the general rate equation: (a) $1/v_0$ against $1/[AX_0]$, where $[B_0]$ is constant and $\approx K^B{}_m$; (b) $1/v_0$ against $1/[AX_0]$, where {B_0] is saturating (i.e. pseudo single-substrate with regard to AX; (c) $1/v_0$ against $1/[B_0]$, where $[AX_0]$ is constant and $\approx K^{AX}{}_m$; and (d) $1/v_0$ against $1/[B_0]$, where $[AX_0]$ is saturating (i.e. pseudo single-substrate with respect to B).

The values of the various constants may be determined by means of **secondary plots**. So, for example, if primary plots of $1/v_0$ against $1/[AX_0]$ (Fig. 9.1a) are drawn for a series of values of $[B_0]$, then a secondary plot of intercept (on the $1/v_0$ axis) against $1/[B_0]$ will have an intercept of $1/V_{max}$ and slope of K_m^B / V_{max}, enabling both of these constants to be determined.

Furthermore, a secondary plot of primary plot slope against $1/[B_0]$ will have an intercept of K_m^{AX}/V_{max} and a slope of $K_s^{AX}K_m^B/V_{max}$, enabling the values of K_m^{AX} and K_s^{AX} to be determined (see problem 9.1).

9.2.3 The general rate equation of Dalziel

Keith Dalziel (1957) gave the general rate equation in the form:

$$\frac{[E_0]}{v_0} = \phi_0 + \frac{\phi_{AX}}{[AX_0]} + \frac{\phi_B}{[B_0]} + \frac{\phi_{AXB}}{[AX_0][B_0]} \tag{9.13}$$

The ϕ terms, called **kinetic coefficients**, can be obtained by drawing primary and secondary plots as follows. Primary plots of $[E_0]/v_0$ against $1/[AX_0]$ at constant $[B_0]$ are drawn for a series of different values of $[B_0]$. The slopes and the intercepts on the $[E_0]/v_0$ axis for each graph are determined, and then **secondary replots** are drawn as in Fig. 9.2. From the secondary replots, the values of ϕ_0, ϕ_{AX}, ϕ_B and ϕ_{AXB} can be determined. The whole procedure could be repeated from primary plots of $[E_0]/v_0$ against $1/[B_0]$ at constant $[AX_0]$, but the same results should be obtained.

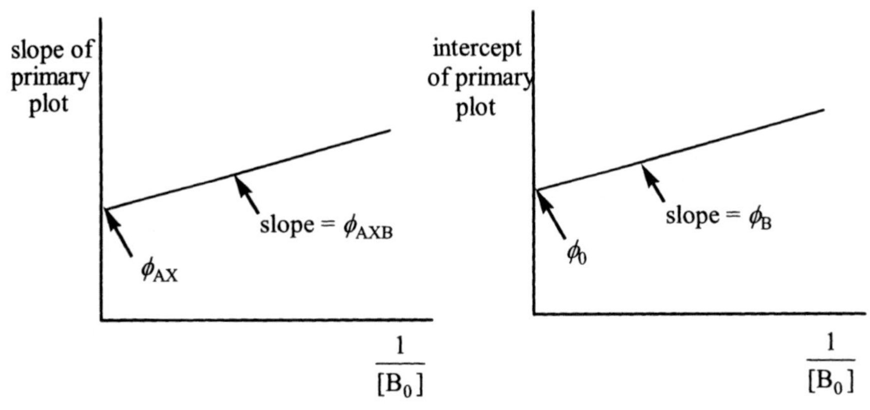

Fig. 9.2 – Secondary replots to enable calculation of the kinetic coefficients of Dalziel.

The Dalziel equation can be arranged in the same form as the Alberty equation and the relationship between the respective constants determined. Thus, V_{max} in the Alberty equation is $[E_0]/\phi_0$ in the Dalziel equation, whereas $K_m^{AX} = \phi_{AX}/\phi_0$, $K_m^B = \phi_B/\phi_0$ and $K_s^{AX} = \phi_{AXB}/\phi_B$.

9.2.4 Rate constants and the constants of Alberty and Dalziel

The general rate equation of Alberty bears an obvious resemblance to the Michaelis-Menten equation and so serves as a useful introduction to the subject of two-substrate kinetics. However, the Dalziel equation may be of more value to an enzyme kineticist.

The kinetic constants in the general rate equations are, of course, functions involving the individual rate constants of the steps involved in the reaction. Relationships can be derived for particular mechanisms and it is found that the Dalziel kinetic coefficients give more straightforward relationships between rate constants than do the Alberty constants. In either case, the mechanisms involved in calculating these relationships can be extremely laborious, so the **King and Altman procedures** (section 7.3) prove extremely useful.

Let us consider the ping-pong bi-bi mechanism (section 9.1.2) under steady-state conditions. Since the initial concentrations of both products are zero, the release of these products from the enzyme will be effectively irreversible. The reaction sequence, in cyclic form, is:

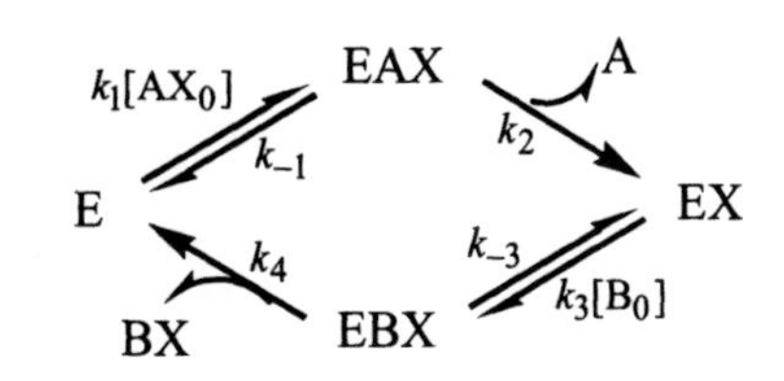

Enzyme species	*Pathways forming enzyme species*	*Sum of kappa products*
E	k_{-1}, $k_3[B_0]$, k_4 → E and k_2, $k_3[B_0]$, k_4 → E	$k_{-1}k_3k_4[B_0] + k_2k_3k_4[B_0]$ $= k_3k_4[B_0](k_{-1} + k_2)$
EAX	$k_1[AX_0]$, $k_3[B_0]$, k_4 → EAX	$k_1k_3k_4[AX_0][B_0]$
EX	$k_1[AX_0]$, k_2, k_4 → EX and $k_1[AX_0]$, k_2, k_{-3} → EX	$k_1k_2k_4[AX_0] + k_1k_2k_{-3}[AX_0]$ $= k_1k_2[AX_0](k_{-3} + k_4)$
EBX	$k_1[AX_0]$, k_2, $k_3[B_0]$ → EBX	$k_1k_2k_3[AX_0][B_0]$

Hence, for example:

$$\frac{[EAX]}{[E_0]} = \frac{k_1k_3k_4[AX_0][B_0]}{k_1k_3[AX_0][B_0](k_2 + k_4) + k_3k_4[B_0](k_{-1} + k_2) + k_1k_2[AX_0](k_{-3} + k_4)}$$

The overall rate of reaction $v_0 = k_2[EAX] = k_4[EBX]$.

Substituting for [EAX] and rearranging,

$$\frac{[E_0]}{v_0} = \frac{k_1k_3[AX_0][B_0](k_2 + k_4) + k_3k_4[B_0](k_{-1} + k_2) + k_1k_2[AX_0](k_{-3} + k_4)}{k_1k_2k_3k_4[AX_0][B_0]}$$

$$\therefore \frac{[\mathrm{E}_0]}{v_0} = \frac{k_2+k_4}{k_2k_4} + \frac{k_{-1}+k_2}{k_1k_2}\cdot\frac{1}{[\mathrm{AX}_0]} + \frac{k_{-3}+k_4}{k_3k_4}\cdot\frac{1}{[\mathrm{B}_0]} \quad (9.14)$$

This is in the form of the Dalziel equation. If the expression is rearranged into the form of the Alberty equation we obtain:

$$v_0 = \frac{\left(\dfrac{k_2k_4}{k_2+k_4}\right)[\mathrm{E}_0][\mathrm{AX}_0][\mathrm{B}_0]}{\dfrac{k_2}{k_3}\left(\dfrac{k_{-3}+k_4}{k_2+k_4}\right)[\mathrm{AX}_0]+\dfrac{k_4}{k_1}\left(\dfrac{k_{-1}+k_2}{k_2+k_4}\right)[\mathrm{B}_0]+[\mathrm{AX}_0][\mathrm{B}_0]} \quad (9.15)$$

Steady-state rate equations for other mechanisms can be obtained similarly.

9.3 INVESTIGATION OF REACTION MECHANISMS USING STEADY-STATE METHODS

9.3.1 The use of primary plots

Reaction mechanisms which obey the general rate equation of Alberty (and that of Dalziel) give linear primary plots of $1/v_0$ against $1/[\mathrm{AX}_0]$ at constant $[\mathrm{B}_0]$, and of $1/v_0$ against $1/[\mathrm{B}_0]$ at constant $[\mathrm{AX}_0]$. As shown in Fig. 9.1, the expressions for the intercept on the $1/v_0$ axis and the slopes of these graphs both include the concentration of the fixed substrate at non-saturating concentrations, so both intercept and slope change if the experiments are repeated with the fixed substrate concentration set at a different value. The primary plots for such reactions, e.g. those proceeding by compulsory-order ternary-complex mechanisms or by random-order rapid-equilibrium bi-bi mechanisms, will be of the form shown in Fig. 9.3a.

In contrast, the equation for reactions proceeding by a ping-pong bi-bi mechanism is simpler in that $K_{\mathrm{s}}^{\mathrm{AX}}K_{\mathrm{m}}^{\mathrm{B}} = 0$. From Fig. 9.1 it can be seen that this results in the slope of the primary plot being independent of the concentration of the fixed substrate. A series of primary plots obtained at different concentrations of the fixed substrate would thus be parallel if a ping-pong bi-bi mechanism operates (Fig. 9.3b).

Compulsory-order and random-order ternary-complex mechanisms can be distinguished from ping-pong bi-bi mechanisms, but not from each other, by the use of primary plots.

For example, Louis Hersh and William Jencks (1967) obtained a graph of the form of Fig. 9.3b for **3-oxoacid CoA-transferase** and concluded that the mechanism was ping-pong bi-bi, probably proceeding as follows:

(9.16)

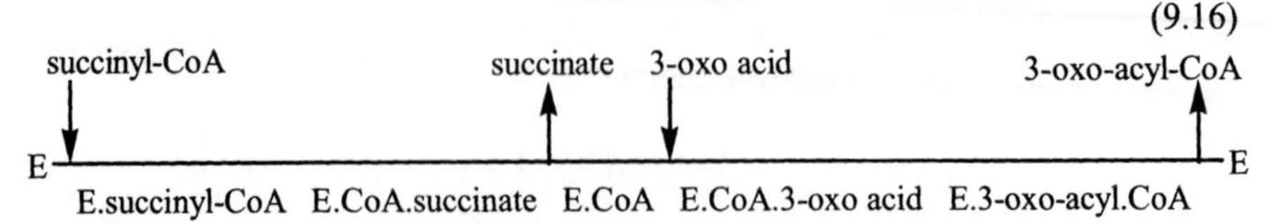

They confirmed this mechanism by isolating the E–CoA intermediate.

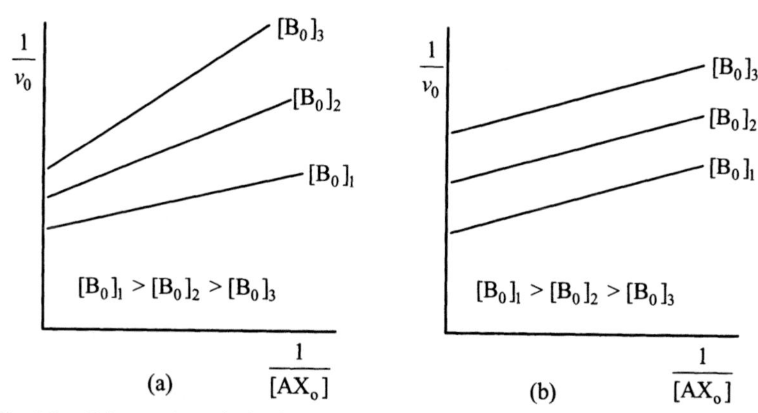

Fig. 9.3 – Primary plots obtained at non-saturating concentrations of both substrates: (a) for compulsory-order and random-order ternary-complex mechanisms; (b) for ping-pong bi-bi mechanisms. (In each case similar results would be obtained for plots of $1/v_0$ against $1/[B_0]$ at fixed $[AX_0]$.)

The drawing of primary plots is not in itself sufficient basis for concluding that a particular mechanism is ping-pong bi-bi, because it is difficult to be certain whether lines drawn from experimental results really are parallel. Herbert Fromm and co-workers (1962) concluded from primary plots that yeast **hexokinase** had a ternary-complex mechanism but brain hexokinase probably proceeded by a ping-pong bi-bi mechanism involving an E–P intermediate. Although isoenzymes do not necessarily act by identical mechanisms, later studies showed that the plots for brain hexokinase met far below the horizontal axis, so this enzyme too forms a ternary-complex.

9.3.2 The use of inhibitors which compete with substrates for binding sites

The use of inhibitors which compete with one of the substrates for a site on the enzymes can give useful information as to the mechanism of the reaction. A product of the reaction, for example, if present at the start may compete with a substrate for a binding site on the enzyme and thus slow down the rate of the forward reaction.

Because of the complex nature of two-substrate reactions, the fact that a particular inhibitor acts in a competitive way will not necessarily result in a characteristic competitive inhibition pattern (Fig. 8.2) being obtained. For double-reciprocal (Lineweaver-Burk) primary plots (Fig. 9.1) drawn for experiments performed at a single concentration of the fixed substrate but at a series of different inhibitor concentrations, the overall pattern will only be competitive if the intercept on the $1/v_0$ axis is unaffected by the inhibitor. If, on the other hand, the intercept depends on the inhibitor concentration but the slope does not, then a series of parallel lines will be obtained at different inhibitor concentrations and the overall inhibition pattern will be uncompetitive (Fig. 8.6). If both intercept and slope are influenced by inhibitor concentration and the plots at different inhibitor concentrations converge to the left of the $1/v_0$ axis, then the overall inhibition pattern is mixed (Fig. 8.11), except in the special case of non-competitive inhibition (Fig. 8.8) where the plots converge at the horizontal axis. All of these inhibition patterns may be seen with two-substrate reactions where the inhibition is essentially competitive in nature.

Cleland formulated a **series of rules** which enable the inhibition patterns for a particular mechanism to be predicted.

One rule states that **the intercept on the $1/v_0$ axis of a double-reciprocal plot is affected only by an inhibitor which binds reversibly to an enzyme-form other than that to which the variable substrate combines**. The explanation for this is straightforward. For there to be no change in intercept as a result of the presence of an inhibitor, the variable substrate in saturating concentrations must be able to prevent the inhibitor binding. *Therefore, for a characteristic pattern of competitive inhibition to occur, the inhibitor must be able to compete with the substrate whose concentration is being varied for a site on the same enzyme-form.*

Another rule is that **the slope of a double-reciprocal plot is affected by an inhibitor which binds to the same enzyme-form as the variable substrate, or to an enzyme-form which is connected by a series of reversible steps to that with which the variable substrate combines**. The explanation for this rule is as follows. Most enzyme-catalysed reactions obey the Michaelis-Menten equation (7.12):

$$v_0 = \frac{V_{max}[S_0]}{[S_0]+K_m^S}$$

where S is the only substrate, or the only substrate whose concentration is being varied. When $[S_0]$ is low, $K_m^S + [S_0] \approx K_m^S$ and therefore $v_0 = (V_{max}/K_m^S)[S_0]$. Hence the apparent first-order constant when $[S_0]$ is low is V_{max}/K_m^S, and this is the reciprocal of the slope of the Lineweaver-Burk plot (Fig. 7.2). The substrate S binds to the enzyme E according to the equation $E + S \rightleftharpoons ES$, so any factor which increases the concentration of ES relative to E will reduce V_{max}/K_m^S and hence increase the slope of the double-reciprocal (Lineweaver-Burk) plot. Similarly anything which decreases the concentration of ES relative to E will increase V_{max}/K_m^S and decrease the slope of the double-reciprocal plot. Hence, for there to be no change in slope in the presence of an inhibitor, these factors cannot be present: i.e. *for a pattern characteristic of uncompetitive inhibition to be seen, there must be no reversible link between inhibitor and variable substrate.* Most of the steps involved in enzyme-catalysed reactions are reversible, but irreversible steps which, in certain circumstances, may give rise to uncompetitive inhibition occur where a substrate is present in saturating concentrations or where an inhibitor forms a dead-end complex with the enzyme. It should be realized that, in the case of uncompetitive inhibition of a single-substrate reaction (section 8.2.2), there is no reversible link between inhibitor and substrate, because the inhibitor binds to ES to form a dead-end complex.

We will now consider product inhibition of the ping-pong bi-bi mechanism and ternary-complex mechanisms obeying the general rate equation (section 9.2.1). Non-saturating concentrations of the fixed-substrate will be assumed unless otherwise stated.

Let us look first at a **random-order ternary-complex mechanism** (equation 9.3):

$$
\begin{array}{l}
E + AX \rightleftharpoons E.AX \xrightleftharpoons{+B} E.AX.B \rightleftharpoons E.A.BX \xrightleftharpoons{-BX} EA \rightleftharpoons E + A \\
E + B \rightleftharpoons EB \xrightleftharpoons{+AX} E.AX.B \rightleftharpoons E.A.BX \xrightleftharpoons{-A} E.BX \rightleftharpoons E + BX
\end{array}
$$

If the forward reaction is investigated under steady-state conditions at fixed $[AX_0]$ and variable $[B_0]$ in the presence of the product BX, then a competitive inhibition pattern will result: the variable substrate B will compete with BX for the same site on the same enzyme-form (E). A graph of the form of Fig. 9.4a will result if the experiment is performed at a series of concentrations of the product BX. Similar results will be obtained for inhibition by the product A at fixed $[B_0]$ and variable $[AX_0]$. However, the other two combinations give less clear-cut predictions. Particularly in the case of a rapid-equilibrium mechanism, there may be competition between substrate and inhibitor for the same enzyme-form, even where the competition is not for the same *site*. Consider, for example, inhibition by the product A where the variable substrate is B. Saturating concentrations of B will remove all free E from the system, thus preventing A binding to it, a characteristic of **competitive** inhibition. However, excess B cannot prevent the possibility of A binding to EB to form a dead-end complex EBA. If such a complex can be formed, a **mixed** inhibition pattern would result (Fig. 9.4b).

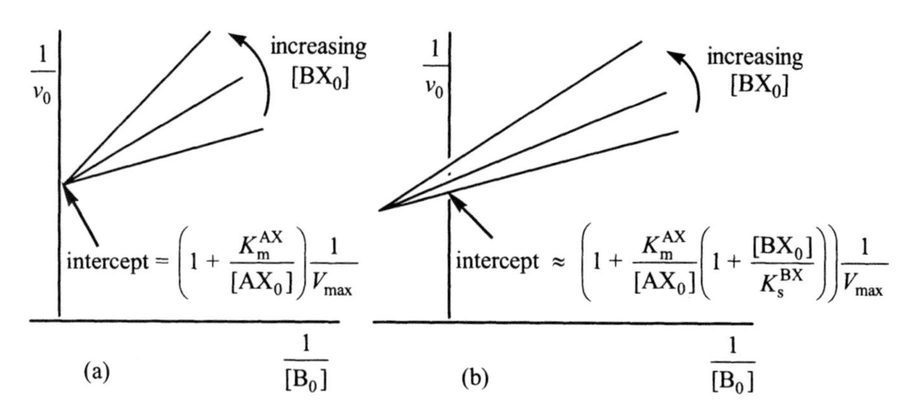

Fig. 9.4 – Plots of $1/v_0$ against $1/[B_0]$ at constant and non-saturating $[AX_0]$ showing : (a) a pattern characteristic of competitive inhibition: and (b) a pattern characteristic of mixed inhibition. The inhibitor is the product, BX.

Similarly, if BX is the inhibitor and AX the varying substrate, the inhibitor pattern will be characteristic of **competitive** inhibition unless the dead-end complex E.AX.BX can be formed (a less likely possibility than EBA because of congestion at the binding sites).

Product inhibition patterns of the type discussed above have been found for reactions with rapid-equilibrium random-order mechanisms (e.g. that catalysed by **creatine kinase**).

They have also been found for random-order mechanisms where the rate-limiting step is not solely the interconversion of the ternary complexes (e.g. reactions catalysed by **hexokinase** enzymes).

For a **compulsory-order mechanism** where AX binds first (equation 9.4):

$$AX + E \rightleftharpoons E.AX \rightleftharpoons E.AX.B \rightleftharpoons E.A.BX \rightleftharpoons EA \rightleftharpoons E + A$$

A and AX will compete for a site on the enzyme-form E. Therefore, overall **competitive** inhibition (as in Fig. 9.4a) will result from inhibition by the product A where AX is the variable substrate. No other combination of inhibitor and variable substrate will give a pattern characteristic of **competitive** inhibition for this mechanism. Inhibition by the product BX where B is the variable substrate will give **mixed** inhibition (as in Fig. 9.4b) since BX and B do not bind to the same enzyme-form, BX binding to EA and B to E.AX.

Similarly, for a **compulsory-order mechanism** of the form in equation 9.5:

$$B + E \rightleftharpoons EB \rightleftharpoons E.B.AX \rightleftharpoons E.BX.A \rightleftharpoons E.BX \rightleftharpoons E + BX$$

only inhibition by the product BX where B is the variable substrate will give a pattern characteristic of competitive inhibition.

For a **ping-pong bi-bi mechanism** equations 9.1 and 9.2):

$$AX + E \rightleftharpoons E.AX \rightleftharpoons EX.A \rightleftharpoons EX + A$$

$$EX + B \rightleftharpoons EX.B \rightleftharpoons E.XB \rightleftharpoons E + BX$$

AX and BX compete for a site on the enzyme-form E, whereas A and B compete for a site on the enzyme-form EX. Therefore inhibition by BX when AX is the variable substrate will give overall **competitive** inhibition, as will inhibition by A when B is the variable substrate. However, this only applies to ping-pong bi-bi mechanisms where there is a single binding site, e.g. many reactions catalysed by **transaminase (aminotransferase)** enzymes.

In all of the situations discussed above, there is a reversible link between inhibitor and variable substrate at non-saturating concentrations of the fixed substrates, so the possibility of uncompetitive inhibition does not arise. The overall inhibition patterns under these conditions may be summarized as follows:

Mechanism	*Product used as inhibitor*	*Inhibition pattern obtained*	
		with respect to varying $[AX_0]$	with respect to varying $[B_0]$
compulsory-order ternary complex (AX binding first)	A	competitive	mixed
	BX	mixed	mixed
compulsory-order ternary complex (B binding first)	A	mixed	mixed
	BX	mixed	competitive
random-order ternary complex	A	competitive	competitive or mixed
random-order ternary complex	BX	competitive or mixed	competitive
ping-pong bi-bi	A	mixed	competitive
ping-pong bi-bi	BX	competitive	mixed

Although these general conclusions may not apply in certain instances because of the effect of individual rate constants, a systematic study will in most cases give good evidence as to the mechanism of the reaction.

Further information may be obtained by performing inhibition studies in the presence of saturating concentration of the fixed substrate. Of particular interest is the situation where the saturating fixed substrate is the second substrate to be bound in a compulsory-order mechanism. For example, if $[B_0]$ is saturating for a mechanism where AX must bind first to the enzyme, as in the following sequence:

$$E \underset{}{\overset{+AX}{\rightleftharpoons}} E.AX \underset{}{\overset{+B}{\rightleftharpoons}} E.AX.B \rightleftharpoons E.A.BX \underset{+BX}{\rightleftharpoons} EA \rightleftharpoons E + A$$

inhibition by BX will give an **uncompetitive** pattern with respect to varying AX because there is no reversible link between inhibitor and variable substrate.

Dead-end inhibitors which are analogues of the products will give generally similar inhibitor patterns. However, there is a fundamental difference between the two types of inhibition. In the case of product inhibition, the enzyme-inhibitor complex is an intermediate normally present but which in the circumstances is non-productive for the reaction in the forward direction. In the case of dead-end inhibitors, the enzyme-inhibitor complex is not present in the uninhibited reaction and so effectively removes some of the enzyme from the reacting system. Therefore, there may be slight differences between the inhibition patterns in the two cases. For example, for a compulsory-order mechanism where AX binds first, inhibition by the product BX gives a **mixed** inhibition pattern with respect to varying AX at fixed but non-saturating concentrations of B. However, an analogue of BX which binds in the same way but forms a dead-end complex will give overall **uncompetitive** inhibition because there is no reversible link between inhibitor and variable substrate.

9.4 INVESTIGATION OF REACTION MECHANISMS USING NON-STEADY-STATE METHODS

9.4.1 Isotope exchange at equilibrium

Paul Boyer, in 1959, first suggested the investigation of isotope exchange at chemical equilibrium as a means of investigating reaction mechanisms, and he and his co-workers did much to develop the technique.

If isotope exchange can be demonstrated between a reactant and product in the absence of other reactants and products, then a **ping-pong mechanism** must be indicated. For example, isotope exchange takes place between the substrate, ortho-phosphate, and the product, glucose-1-phosphate, in the presence of the enzyme **sucrose phosphorylase**, but in the absence of the other substrate, sucrose, and the other product, fructose. Also, isotope can be exchanged between sucrose and fructose in the absence of orthophosphate and glucose-1-phosphate. This and other evidence suggests the following mechanism for the forward reaction:

(9.17)

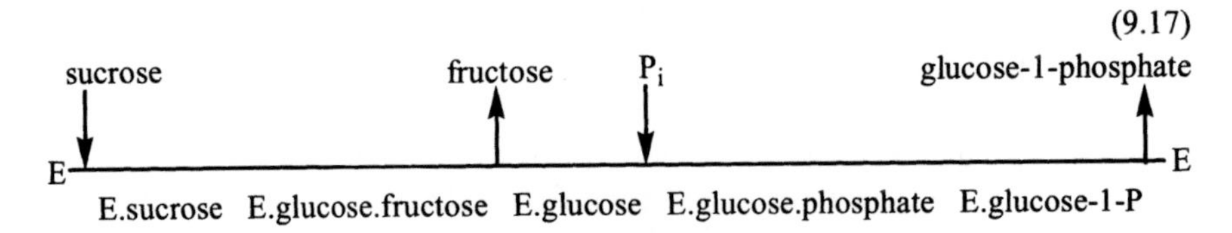

Of more general application is the investigation of the change in the rate of equilibrium isotope exchange when the concentration of reactants and products are altered without changing the position of equilibrium. Consider a reaction of the general form: AX + B $\rightleftharpoons$ BX + A. At equilibrium, the rate of forward reaction equals the rate of back reaction and $([BX][A])/([AX][B]) = K_{eq}$

A small amount of radioactively-labelled B is introduced (an amount too small to significantly affect the equilibrium) and the rate of formation of labelled BX is measured. Then the concentrations of A and AX are increased, keeping the ratio [A]:[AX] constant so as not to alter the position of equilibrium. The equilibrium concentrations of B and BX will remain unchanged, but the rate of isotope exchange between B and BX will be affected.

If the reaction has a compulsory-order mechanism of the form:

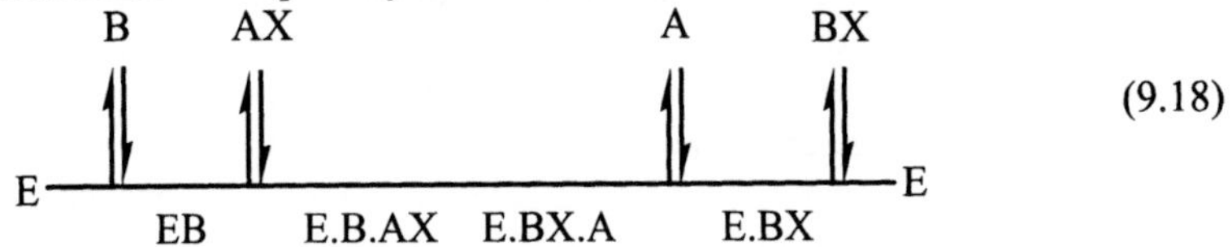

(9.18)

slight increases in the concentrations of A and AX may increase the rate of isotope exchange between B and BX. However, substantial increases in the concentrations of A and AX will force the enzyme towards formation of the ternary-complexes E.B.AX and E.BX.A, making it more difficult for B to dissociate from EB and BX to dissociate from EBX, and thus reducing the B→BX exchange (Fig. 9.5a).

Different results are found for other mechanisms. If the reaction has a compulsory-order mechanism with AX binding first:

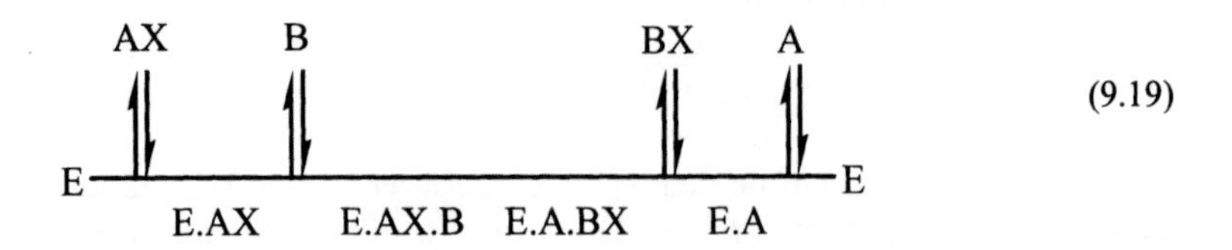

(9.19)

as the concentrations of AX and A increase, the free enzyme is forced into the EAX and EA forms. Since EAX is the form which reacts with B while EA does not affect the initial velocity of liberation of BX from E.A.BX, the rate of isotope exchange between B and BX will increase in a hyperbolic manner with increasing [AX] (Fig. 9.5b). Similar results will be obtained if the reaction has a random-order ternary-complex mechanism.

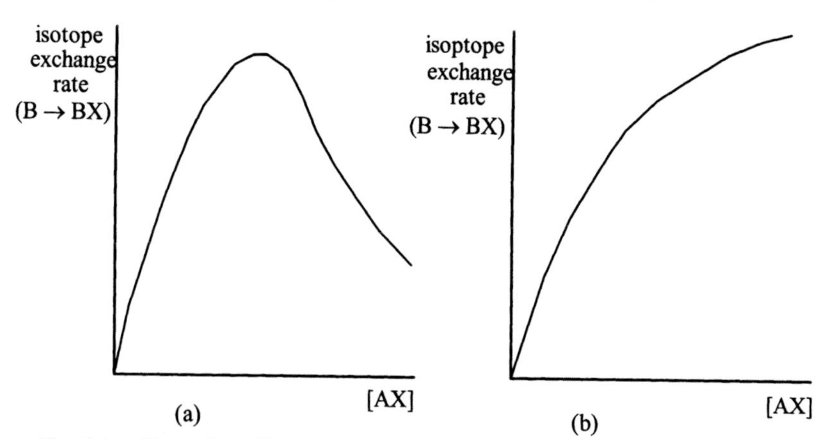

Fig. 9.5 – Plots of equilibrium isotope exchange from B to BX against [AX] at constant [AX]/[A]: (a) for a compulsory-order mechanism, B binding first; and (b) for a compulsory-order mechanism, AX binding first, or for a random-order mechanism.

A decreased rate of equilibrium isotope exchange with increased reactant and product concentration in an experiment of this type is diagnostic of a compulsory--order mechanism where the isotope exchange rate is being measured between the first substrate to be bound and the last product to be released.

An interesting example is the reaction catalysed by horse liver **alcohol dehydrogenase**: $CH_3CH_2OH + NAD^+ \rightleftharpoons CH_3CHO + NADH\ (+ H^+)$.

Increased concentrations of CH_3CH_2OH and CH_3CHO, keeping the ratio of one to the other constant, slows down the rate of isotope exchange between NAD^+ and NADH. It can therefore be concluded that this reaction has a compulsory-order ternary-complex mechanism, with NAD^+ binding first and NADH being the last product released. However, this is not the whole story, as will be seen in section 9.4.2.

9.4.2 Rapid-reaction studies

Rapid-reaction techniques may also be used to investigate the mechanism of two-substrate reactions. Hugo Theorell and Britton Chance (1951) investigated the reaction catalysed by horse liver **alcohol dehydrogenase** (section 9.4.1), using stopped-flow techniques with a double-beam detector. There is a difference between the absorbance of NADH and E.NADH at 350 nm but not at 328 nm. Therefore, monitoring the differences between the absorbances at 350 nm and 328 nm gives the rate of E.NADH → E + NADH conversion. Also, the rate of change of absorbance at 328 nm gives the rate of reduction of NAD^+ (i.e. $E.NAD^+ \rightarrow E.NADH$ conversion). Theorell and Chance found that the interconversion of the ternary-complexes took place extremely rapidly and that the rate-limiting step of the overall reaction was the dissociation of NADH from the E.NADH complex. This type of compulsory-order mechanism, characterized by very rapid interconversion of the ternary complexes, has been termed the **Theorell-Chance mechanism** and is represented as follows:

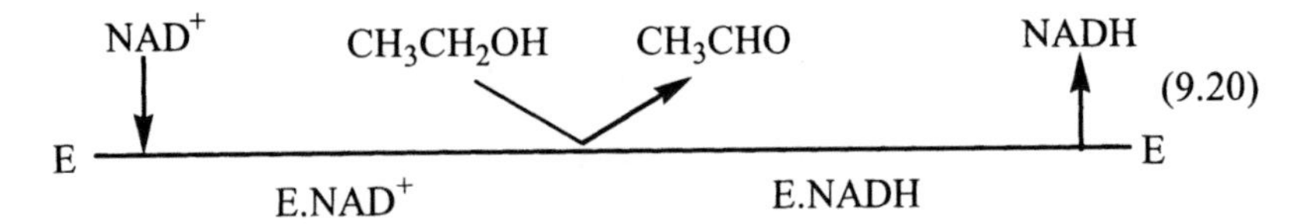

(9.20)

SUMMARY OF CHAPTER 9

Two-substrate enzyme-catalysed reactions may proceed by a variety of mechanisms, including the ping-pong bi-bi, compulsory-order ternary-complex and random-order ternary-complex mechanisms. General rate equations have been derived by Alberty and Dalziel for two-substrate reactions which obey the Michaelis-Menten equation with respect to one substrate at fixed concentrations of the other. The Alberty equation has the more obvious resemblance to the Michaelis-Menten equation but the constants of the Dalziel equation give a simpler relationship between the rate constants for the individual steps involved in the reaction. These relationships may be determined for a particular mechanism by the use of King and Altman procedures.

The mechanism of a two-substrate reaction may be investigated by steady-state methods, including product inhibition studies, and by non-steady-state methods, such as equilibrium isotope exchange and rapid-reaction techniques.

FURTHER READING

As for Chapter 7, plus:

Cornish-Bowden, A. (1995), *Fundamentals of Enzyme Kinetics,* 2nd edn., Portland Press (Chapters 6, 7).

Marangoni, A. G. (2002), *Enzyme Kinetics: A Modern Approach*, Wiley (Chapter 7).

Schulz, A. R. (1994), *Enzyme Kinetics,* Cambridge University Press (Chapters 5-10).

Wilson, K. and Walker, J. (2000), *Principles and Techniques of Practical Biochemistry,* 5th edn., Cambridge University Press (Chapter 7).

PROBLEMS

9.1 An enzyme-catalysed reaction of the form, AX + B $\rightleftharpoons$ BX + A, was investigated at fixed temperature, pH and enzyme concentration. The following results were obtained:

$[AX_0]$ (mmol l^{-1})	Initial velocity (μmol l^{-1}) at $[B_0]$ (mmol l^{-1}) $[B_0]$ = 1.0	1.5	2.0	3.0	5.0	10.0
1.0	77	103	125	158	200	250
1.5	97	130	158	200	254	319
2.0	111	150	182	231	294	370
3.0	130	176	214	273	349	441
5.0	152	205	250	319	410	521
10.0	172	234	286	366	472	602

Given that the reaction obeys the general rate equation, determine the values of K_s^{AX}, K_m^{AX} and K_m^{B}.

9.2 For the enzyme-catalysed transfer reaction, AX + B $\rightleftharpoons$ BX + A, a series of experiments were performed at fixed total enzyme concentration. The results are summarized in the following table, which shows the initial velocity of the reaction at different initial concentrations of AX, B and I (I being an inhibitor resembling BX in structure):

$[AX_0]$ (mmol l^{-1})	$[I_0]$ (mmol l^{-1})	Initial velocity (μmol l^{-1}) at $[B_0]$ (mmol l^{-1}) $[B_0]$ = 20.0	10.0	5.0	3.33	2.50	2.0
3.0	0	1460	1190	881	694	575	488
4.0	0	1720	1410	1030	806	667	564
6.0	0	2080	1690	1230	961	794	668
8.0	0	2380	1920	1360	1060	870	735
8.0	1.0	2250	1740	1200	913	738	617
8.0	2.0	2080	1590	1060	800	641	532

Further, product inhibition studies showed that the presence of A leads to competitive inhibition patterns with varying $[AX_0]$ at fixed $[B_0]$. What can you conclude about the mechanism of the reaction?

9.3 Malate dehydrogenase catalyses the reaction:

$$\text{malate} + \text{NAD}^+ \rightleftharpoons \text{oxaloacetate} + \text{NADH} + \text{H}^+$$

A malate dehydrogenase enzyme was investigated at pH 8 and 298 K as follows. The initial velocity of the reaction, in the direction of oxaloacetate formation, was determined spectrophotometrically by the appearance of NADH at 340 nm at different initial concentrations of malate and NAD^+. The effect of inhibition by the product oxaloacetate was also investigated. No NADH was present initially in any experiment, and the total enzyme concentration was the same in each case. The following results were obtained:

$[NAD^+]$ (mmol l^{-1})	[oxaloacetate] (mmol l^{-1})	v_0 (Δ absorbance $min^{-1} \times 10^3$) at [malate] (mmol l^{-1}) [malate] = 1.25	1.5	2.0	2.5	3.33	5.0	10.0
2.0	0	19	21	25	28	32	38	45
2.0	1.0	16	18	21	24	27	32	38
2.0	2.0	14	15	19	21	24	28	33
3.0	0	24	26	32	36	41	49	59
4.0	0	27	31	37	42	48	57	69
6.0	0	31	36	43	49	57	67	83

The equilibrium rate of conversion of NAD^+ to NADH was investigated by isotope exchange studies in the presence of different concentrations of malate, but keeping the ratio of [malate] to [oxaloacetate] constant at 100:1 and the enzyme concentrations the same in each case. The following results were obtained:

malate concentration (mmol l^{-1})	50	100	150	200	250	300
equilibrium reaction rate, $NAD^+ \rightarrow NADH$ (μmol l^{-1} min^{-1})	41	58	21	15	12	10

What can you conclude about the reaction mechanism from the above data?

9.4 An enzyme-catalysed transfer reaction of the form:

$$AX + B \rightleftharpoons BX + A$$

was investigated as follows.

The effect of a dead-end inhibitor I, similar in structure to BX, was investigated at varying initial concentrations of AX but fixed initial concentrations of B and enzyme. Temperature and pH were maintained constant and no BX or A were present at zero time. The following results were obtained:

		Initial velocity (μmol $l^{-1}min^{-1}$) at [I] (mmol l^{-1})		
$[AX_0]$ (mmol l^{-1})	[I] =	0	1.0	2.0
2.0		14.3	12.5	11.1
2.5		16.7	14.3	12.5
3.3		19.9	16.6	14.2
5.0		24.9	20.1	16.7
10.0		33.4	25.1	20.0

The effect of inhibition of the forward reaction by the product A at a concentration of 1.0 mmol l^{-1} was investigated at varying initial concentrations of AX under identical conditions to those above. The following results were obtained:

Initial concentration of AX (mmol l^{-1}):	2.0	2.5	3.3	5.0	10.0
Initial velocity (μmol $l^{-1}min^{-1}$):	10.5	12.5	15.3	20.1	28.6

The forward reaction was also investigated at varying initial concentrations of B and at a series of fixed (but non-saturating) initial concentrations of AX. Initial enzyme concentration, temperature and pH were the same for each experiment. No product A was present at zero time in any experiment, and no dead-end inhibitor was present at any time. In some experiments the product BX was also initially absent, but in others the effect of product inhibition by BX was investigated. These results were obtained:

		v_0 (μmol $l^{-1}min^{-1}$) at $[B_0]$ (mmol l^{-1})				
$[AX_0]$(mmol l^{-1})	$[BX_0]$(mmol l^{-1})	$[B_0]$ = 2.0	2.5	3.3	5.0	10.0
2.0	0	9.5	11.0	13.0	15.9	20.4
2.5	0	10.8	12.4	14.7	18.1	23.4
3.3	0	12.3	14.3	17.0	21.1	27.6
2.0	1.0	8.2	9.5	11.2	13.8	17.9
2.0	2.0	7.3	8.4	9.9	12.3	16.0

Equilibrium isotope exchange studies, performed at fixed enzyme concentration and constant [A]:[AX] ratio gave these results;

Concentration of AX (mmol l^{-1})	50	100	150	200	250	300
Equilibrium reaction rate, B→BX (μmol $l^{-1}min^{-1}$)	25	54	60	64	68	70

What can be concluded about the reaction mechanism from the above data?

9.5 Deduce the constants of the Dalziel equation in terms of rate constants for the reaction mechanism:

$$E \underset{k_{-1}}{\overset{k_1}{\rightleftharpoons}} E.AX \underset{k_{-2}}{\overset{k_2}{\rightleftharpoons}} E.AX.B \underset{k_{-3}}{\overset{k_3}{\rightleftharpoons}} E.A.BX \xrightarrow{k_4} EA \xrightarrow{k_5} E$$

9.6 Saccharopine dehydrogenase (also known as lysine-oxoglutarate reductase) catalyses the first step in lysine catabolism in mammalian liver:

$$\text{L-lysine} + \text{2-oxoglutarate} + \text{NADPH} \rightleftharpoons \text{L-saccharopine} + \text{NADP}^+$$

Purified bovine saccharopine dehydrogenase was subjected to product inhibition studies with saccharopine. In each experiment, the rate of reaction was determined by monitoring the conversion at 35°C of NADPH to $NADP^+$ at 340 nm.

In each case, no $NADP^+$ was present at zero time, and the concentrations of two of the three substrates were fixed. When the concentration of L-lysine was fixed, its value was 10 mmol l^{-1}; when the concentration of 2-oxoglutarate was fixed, its value was 4 mmol l^{-1}; and when the concentration of NADPH was fixed, its value was 0.4 mmol l^{-1}. The following data were obtained:

[saccharopine] (mmol l^{-1})	v_0 (absorbance units min^{-1}) at varying [L-lysine] (mmol l^{-1})				
	[L-lysine] = 1.0	1.25	1.67	2.5	5.0
1.1	0.143	0.167	0.200	0.250	0.333
0.80	0.167	0.193	0.232	0.288	0.385
0.28	0.246	0.286	0.338	0.417	0.541
0.	0.278	0.323	0.382	0.463	0.592

[saccharopine] (mmol l^{-1})	v_0 (abs. units min^{-1}) at varying [2-oxoglutarate] (mmol l^{-1})				
	[2-oxoglutarate] = 0.25	0.30	0.40	0.50	1.0
1.1	0.124	0.143	0.178	0.207	0.313
0.80	0.156	0.178	0.217	0.250	0.357
0.28	0.223	0.250	0.294	0.330	0.435
0	0.278	0.308	0.353	0.385	0.476

[saccharopine] (mmol l^{-1})	v_0 (absorbance units min^{-1}) at varying [NADPH] (mmol l^{-1})				
	[NADPH] = 0.10	0.125	0.167	0.25	0.50
1.1	0.177	0.195	0.217	0.246	0.284
0.80	0.189	0.208	0.235	0.270	0.313
0.28	0.227	0.257	0.299	0.357	0.439
0	0.242	0.278	0.327	0.398	0.500

What can be deduced from these results about the mechanism of the reaction catalysed by saccharopine dehydrogenase?

10

The Investigation of Active Site Structure

10.1 THE IDENTIFICATION OF BINDING SITES AND CATALYTIC SITES

10.1.1 Trapping the enzyme-substrate complex

The reversible character of the steps involved in enzyme-catalysed reactions makes the determination of each substrate-binding site less than straightforward. At steady-state, a constant amount of enzyme-substrate complex is known to be present, but if an attempt is made to isolate this and hydrolyse it so as to identify the amino acid to which the substrate is attached, the effort will not usually be rewarded with success. This is because the substrate will dissociate from the enzyme during the procedures involved. However, if the enzyme-substrate complex can be trapped in a modified form by some chemical process so that the substrate is no longer able to dissociate from the enzyme, then it may be possible to identify the substrate-binding site.

Consider, for example, the reversible reaction catalysed by **fructose-bisphosphate aldolase**:

$$\text{dihydroxyacetone-phosphate} + \text{glyceraldehyde-3-phosphate} \rightleftharpoons \text{fructose-1,6-bisphosphate} \qquad (10.1)$$

Bernard Horecker and colleagues showed in 1962 that if the reaction mixture was treated with sodium borohydride, a strong reducing agent, then an inactive complex was produced. On hydrolysis of this complex, ε-*N*-glyceryl lysine was found among the products. From this, it was concluded that dihydroxyacetone-phosphate normally binds to the ε (i.e. side chain) amino group of a lysine residue in the enzyme by a **Schiff base** (–N=CH–) linkage. The presumed sequences for the normal reaction and for the procedures used to identify the substrate-binding site are as follows:

$$
\underset{\text{enzyme}}{E-NH_2} + \underset{\text{dihydroxyacetone-P}}{\begin{matrix} CH_2OH \\ | \\ O{=}C \\ | \\ CH_2OPO_3^{2-} \end{matrix}} \rightleftharpoons \underset{\text{enzyme-substrate complex}}{E-N{=}\begin{matrix} CH_2OH \\ | \\ C \\ | \\ CH_2OPO_3^{2-} \end{matrix}} \xrightleftharpoons{\text{+ glyceraldehyde-3-P}} \text{fructose-1,6-bisP}
$$

$$
\downarrow \text{2H (borohydride)} \qquad (10.2)
$$

$$
\underset{\text{inactive complex}}{E-NH.\begin{matrix} CH_2OH \\ | \\ CH \\ | \\ CH_2OPO_3^{2-} \end{matrix}} \xrightarrow{\text{hydrolysis}} \underset{\varepsilon\text{-}N\text{-glyceryl lysine}}{\text{lysine}-NH.\begin{matrix} CH_2OH \\ | \\ CH \\ | \\ CH_2OH \end{matrix}}
$$

In general, once an enzyme-substrate complex has been trapped as an inactive complex, it may be subjected to partial hydrolysis and the amino acid sequence for a few residues on each side of the binding site determined. It may also be possible to determine the complete primary structure of the inactivated complex (as in section 2.4), and hence the substrate-binding site may be precisely located.

10.1.2 The use of substrate analogues

An alternative way of producing a complex more stable than the normal enzyme-substrate complex is to replace the natural substrate by an analogue which binds to the same site on the enzyme but is then less readily removed.

An example was discussed in section 7.2.1. The first step in the hydrolysis of *p*-nitrophenylacetate and other acyl esters by **chymotrypsin** is the rapid splitting of the ester to yield the first product (the alcohol) and form an acyl-enzyme. The subsequent liberation of the acyl group from the enzyme is extremely slow, allowing the structure of the complex to be investigated. It is found, as might be expected, that the acyl group binds by an ester linkage to the –OH group of a serine residue in the enzyme.

Enzymes form even stronger linkages with **irreversible inhibitors**, since in this case there is no subsequent reaction at all. Thus **DFP** (section 8.3) binds to the serine residue at the active site in chymotrypsin and other serine proteases to form very stable complexes. Partial hydrolysis of each enzyme-inhibitor complex gives a series of peptide fragments, which can be separated from each other and analysed. The amino acid sequence of any fragment containing DFP must be the primary structure of part of the active site. In this way, it can be shown that the amino acid sequence around the essential serine residue of chymotrypsin is -Gly-Asp-Ser-Gly-Gly-Pro-, and the serine residue in question identified as serine-195. An identical sequence is found for trypsin.

The assumption that an irreversible inhibitor binds at the active site of an enzyme is particularly valid when the inhibitor resembles a substrate. For example, tosyl (*N*-toluenesulphonyl)-L-phenylalanine chloromethyl ketone (TPCK) resembles esters which are hydrolysed by chymotrypsin, but TPCK itself acts as an irreversible inhibitor of this enzyme by alkylating the histidine-57 residue:

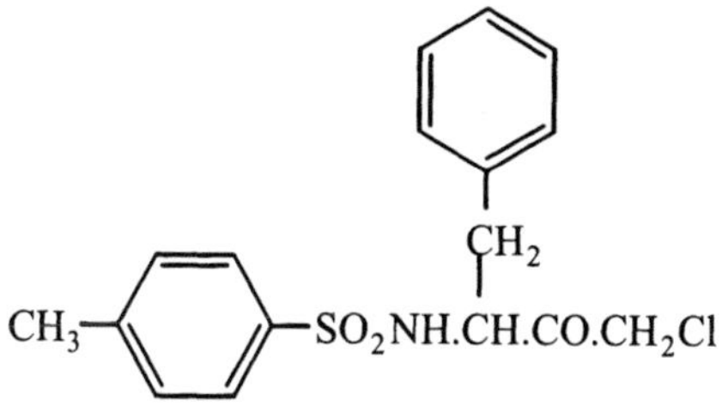

$$CH_3-C_6H_4-SO_2NH.CH(CH_2C_6H_5).CO.CH_2Cl$$

tosyl-L-phenylalanine chloromethyl ketone (TPCK)
(an inhibitor of chymotrypsin)

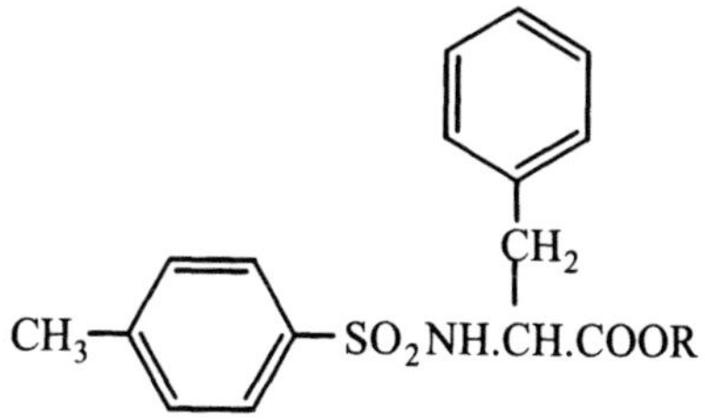

$$CH_3-C_6H_4-SO_2NH.CH(CH_2C_6H_5).COOR$$

ester of tosyl-L-phenylalanine

(a substrate for chymotrypsin)

Thus there is evidence that both serine-195 and histidine-57 are present at the active site of chymotrypsin. It would appear that the binding of TPCK to the enzyme brings the reactive –Cl group into close proximity to the histidine-57 residue and facilitates the formation of a covalent C–N bond between inhibitor and imidazole side chain.

The subject of enzyme inactivation by irreversible inhibitors and other agents is developed in the next section (10.1.3).

10.1.3 Enzyme modification by chemical procedures affecting amino acid side chains

If an enzyme is modified by the conversion of a particular amino acid side chain to a different form (e.g. by the action of an irreversible inhibitor) and this modification results in a loss of catalytic activity, then it is possible that the amino acid concerned is a component of the active site of the enzyme. However, it is also possible that the loss of activity is due to a change in tertiary structure resulting from a modification to an amino acid residue not present at the active site. The two situations may be distinguished by attempting to carry out the same modification in presence of excess amounts of substrate, which should protect the substrate-binding site and amino acid residues in the neighbouring region from being modified. Similar protection should be provided by excess amounts of reversible inhibitors which bind to the substrate-binding site, e.g. most competitive inhibitors. Thus, if an enzyme loses activity when one of its amino acid side chains (R) is modified by a particular treatment in the absence of substrate (or competitive inhibitor):

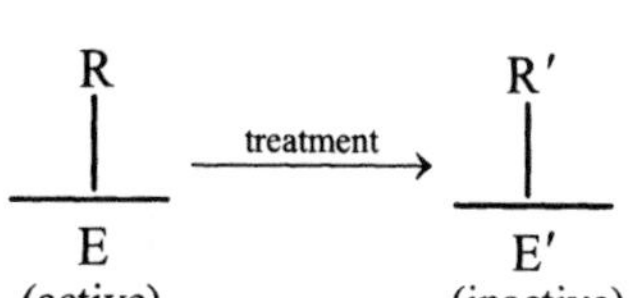

$$\underset{\substack{E \\ \text{(active)}}}{\overset{R}{|}} \xrightarrow{\text{treatment}} \underset{\substack{E' \\ \text{(inactive)}}}{\overset{R'}{|}} \tag{10.3}$$

but retains full activity if subjected to the same treatment in the presence of saturating amounts of the substrate (or competitive inhibitor):

$$\underset{\text{(active)}}{\underset{E}{R}} \xrightarrow{+S} \underset{ES}{\overset{S}{R}} \xrightarrow{\text{treatment}} \underset{ES}{\overset{S}{R}} \xrightarrow{\text{dialysis}} \underset{\text{(active)}}{\underset{E}{R}} \tag{10.4}$$

then the amino acid residue must be present at the active site of the enzyme. It should be realized that the presence of the substrate would protect not only the substrate-binding site (as illustrated above) but also neighbouring residues, which are likely to include the catalytic sites.

As a rough generalization, modification of a catalytic site would be expected to result in a complete loss of enzyme activity: the substrate would still be able to bind to the enzyme, but no subsequent catalysed reaction would occur. On the other hand, modification of a binding site might leave a residual catalytic activity: the substrate would no longer be able to bind to the enzyme, but might still be able to approach the catalytic site by random motion.

Enzyme modification studies carried out during the 1960s revealed **histidine, cysteine, serine, methionine, tyrosine, aspartate, glutamate, lysine** and **tryptophan** residues at the active sites of enzymes. Although the techniques used have long-since been superseded by others, a consideration of them serves as a useful introduction to more recent studies.

From the start, it seemed reasonable to think that the imidazole side chain of histidine could be an important contributor towards catalytic activity, because it is the only amino acid side chain to have a pK_a in the pH range at which most enzymes function: it can thus act as both a proton donor and a proton acceptor. This was soon confirmed. As mentioned in section 10.1.2, tosyl-L-phenylalanine chloromethyl ketone (TPCK) inhibits **chymotrypsin** irreversibly by alkylating histidine-57. This inactivation could be prevented by the presence of excess concentrations of the competitive inhibitor, benzamide, supporting the view that TPCK acts at the active site of the enzyme. TPCK does not inhibit **trypsin,** which has a different specificity to chymotrypsin, hydrolysing bonds adjacent to lysine rather than aromatic amino acid residues. This enzyme could be inactivated by tosyl-L-lysine chloromethyl ketone (TLCK), again by alkylation of essential histidine residues.

Alkylation of histidine residues by iodoacetate was shown (by William Stein, Stanford Moore and colleagues) to cause inactivation of pancreatic **ribonuclease**:

$$\underset{\text{enzyme}}{E{-}CH_2{-}\text{Im(N, NH)}} + \underset{\text{iodoacetate}}{ICH_2CO_2^-} \rightarrow \underset{\text{carboxymethyl-enzyme}}{E{-}CH_2{-}\text{Im(N, N}{-}CH_2CO_2^-)} + HI \tag{10.5}$$

Histidine-119 or, to a lesser degree, histidine-12, could be alkylated, but never both in the same ribonuclease molecule, suggesting that these two histidine residues were located close together at the active site.

Iodoacetamide, a similar alkylating agent, had no effect on the enzyme, so it appeared that the negative charge of the iodoacetate formed an electrostatic link with the protonated form of one of the histidine residues prior to the carboxymethylation of the other residue. In fact, iodoacetate itself did not usually attack histidine residues under the conditions used (pH 5.5), which suggested that histidine-12 and histidine-119 were in a special environment. The inactivation of ribonuclease by alkylation was prevented by the presence of excess phosphate, which binds to the active site.

Another technique which could be used to inactivate enzymes by modification of amino acid side chains was **photo-oxidation,** i.e. oxidation by activated oxygen in the presence of a photosensitizer such as methylene blue or rose bengal. This method was non-specific and might oxidize histidine, tryptophan, methionine and cysteines residues. However, some degree of specificity could be obtained by careful choice of photosensitizing dye and pH. With certain enzymes and under certain conditions, it was found that only a single amino acid residue was photo-oxidized. Thus Edward Westhead found that photo-oxidation of **enolase** in the presence of rose bengal affected only a single histidine residue, but this was sufficient to inactivate the enzyme. Magnesium ions, which are essential for the activation of enolase, could protect it from the effects of photo-oxidation. The oxidation product of a histidine residue depended on a variety of factors, including the environment of the residue in the protein. In some cases, aspartate was the major product.

The thiol group of **cysteine** residues could be alkylated by halogeno-compounds of the type which also alkylate histidine residues. For example, cysteine-25 of the proteolytic enzyme **papain** was alkylated by iodoacetamide, with resulting inactivation of the enzyme. Enzymes containing essential cysteine residues could also be inhibited by unsaturated compounds such as *N*-ethylmaleimide:

$$\underset{\text{enzyme}}{\begin{array}{c}SH\\|\\E\end{array}} + \underset{N\text{-ethylmaleimide}}{\begin{array}{l}CH.C(=O)\diagdown\\ \|\qquad\quad N.CH_2CH_3\\ CH.C(=O)\diagup\end{array}} \rightarrow \begin{array}{l}CH_2.C(=O)\diagdown\\ |\qquad\qquad N.CH_2CH_3\\ CH.C(=O)\diagup\\ |\\ E{-}S\end{array} \qquad (10.6)$$

Heavy-metal ions have a great affinity for thiol groups, and papain was found to be inactivated by *p*-chloromercuribenzoate (PCMB) and similar mercurial compounds. Such reagents tend to attack –SH groups in a non-specific fashion, so it could not be concluded that the cysteine residues in question were at the active site unless there was further evidence for this (as there was in the case of papain, as mentioned above).

The essential **serine** residue of **chymotrypsin** (serine-195) was modified by Daniel Koshland and co-workers (1963). Both the tosyl-derivative and its elimination product were found to be inactive:

$$\underset{\text{enzyme}}{E{-}CH_2{-}OH} + FSO_2{-}C_6H_4{-}CH_3 \rightarrow \underset{\substack{\text{tosyl-chymotrypsin}\\ \text{(inactive)}}}{E{-}CH_2{-}OSO_2{-}C_6H_4{-}CH_3} \xrightarrow{NaOH} \underset{\substack{\text{'anhydrochymotrypsin'}\\ \text{(inactive)}}}{E{=}CH_2} \quad (10.7)$$

Methionine residues, in common with those of other amino acids discussed in this section, could be modified by alkylation or photo-oxidation. Hence it was important to determine precisely which amino acids were affected in each modification experiment. Koshland and colleagues showed that photo-oxidation of **chymotrypsin** under certain conditions led to the oxidation of only a single methionine residue (methionine-192), which formed the corresponding sulphoxide. This modification led to only partial inactivation of the enzyme, as did alkylation of the same methionine residue. On the basis of these studies, it was concluded that methionine-192 was present at the active site of chymotrypsin and might play a part in substrate-binding, but was not otherwise involved in catalytic activity.

A **tyrosine** residue (tyrosine-248) was revealed to be at the active site of **carboxylase A** by a variety of modification techniques. Bert Vallee and colleagues showed that iodination or nitration of the benzene ring, or acylation of the phenolic –OH group, all inactivated the enzyme.

The presence of **aspartate** and **glutamate** at the active site of certain enzymes could be demonstrated by the production of esters of the side chain carboxyl groups. For example, Bernard Erlanger and others showed that *p*-bromophenacyl bromide could inactivate **pepsin** by forming an ester linkage with an aspartate residue:

$$\underset{\text{enzyme}}{E{-}CH_2{-}CO_2H} + \underset{p\text{-bromophenacyl bromide}}{BrCH_2CO{-}C_6H_4{-}Br} \rightarrow E{-}CH_2{-}CO{-}OCH_2CO{-}C_6H_4{-}Br + HBr \quad (10.8)$$

This modification could be prevented by the presence of excess substrate, showing that the aspartate residue was indeed at the activity site.

Koshland and colleagues treated **lysozyme** with aminomethanesulphonic acid and a carbodiimide, causing modification of all carboxyl groups present and a total loss of activity of the enzyme:

$$\underset{\text{enzyme}}{E-\overset{\overset{\displaystyle O}{\|}}{C}-OH} + \underset{\substack{\text{aminomethane-}\\\text{sulphonic acid}}}{H_2NCH_2SO_3H} \rightarrow E-\overset{\overset{\displaystyle O}{\|}}{C}-NHCH_2SO_3H + H_2O \qquad (10.9)$$

Modification in the presence of the substrate protected aspartate-52, with the result that 50% of total enzyme activity was retained, suggesting that this residue was a component of the active site.

A **lysine** residue (lysine-41) was shown to be present at the active site of **ribonuclease** by modification with dinitrofluorobenzene:

$$\underset{\text{enzyme}}{\overset{NH_2}{\underset{E}{|}}} + \underset{\text{dinitrofluorobenzene}}{F-C_6H_3(NO_2)-NO_2} \rightarrow \overset{HN}{\underset{E}{|}}-C_6H_3(NO_2)-NO_2 + HF \qquad (10.10)$$

Photo-oxidation of **tryptophan** residues in **lysozyme** inactivated the enzyme, suggesting the presence of these residues at the active site.

10.1.4 Enzyme modification by treatment with proteases

Just as enzyme-proteins may be modified by chemical procedures which affect the structures of amino acid side chains (section 10.1.3), so their primary structure may be modified by the action of proteolytic enzymes. In certain circumstances, this has provided information as to the structure of the active site of the enzyme. For example, the removal of the C-terminal tyrosine residue of **fructose-bisphosphate aldolase** resulted in a reduced activity towards its fructose-1,6-bisphosphate substrate. This, and other evidence, suggested that the tyrosine residue binds the 6-phosphate group.

Another well-known example concerns the action of **subtilisin** on pancreatic **ribonuclease**. The molecule is split into two portions: one is called the **S-peptide** (comprising the first 20 residues in the primary structure of ribonuclease) and the other is the **S-protein** (the other 104 residues of the ribonuclease molecule). Separately, the S-peptide and S-protein have no catalytic activity, but Frederic Richards and colleagues (1959) showed that if they were mixed in equimolar proportions, full enzyme activity was restored. This suggested that both contributed to the active site of the enzyme, agreeing with evidence discussed in section 10.1.3 that residues 12, 41 and 119 are all required for enzymic activity.

10.1.5 Enzyme modification by site-directed mutagenesis

Significant advances over the more traditional techniques discussed in sections 10.1.1-10.1.4 have been brought about by site-directed mutagenesis (for which Michael Smith was awarded a Nobel Prize in 1993). Enzymes with specific amino acid substitutions may be synthesized, and the effect on activity observed. Such techniques, involving recombinant DNA technology (see section 21.1), were introduced during the 1980s.

For example, Charles Craik and colleagues (1987) replaced aspartate-102 by asparagine in **trypsin,** and found that, at neutral pH, activity with an ester substrate was reduced by a factor of 10 000. The binding of the inhibitor DFP (see section 10.1.2) was similarly affected, and the binding of the inhibitor TLCK (see section 10.1.3) was reduced 5-fold. This implicated aspartate-102 in the mechanism of action of trypsin, and of the very similar **chymotrypsin.**

Similarly, in **triose phosphate isomerase**, Ronald Raines and colleagues (1986) demonstrated the involvement of glutamate-165 in the catalytic process by showing that replacement of this residue by aspartate resulted in a 1000-fold loss of activity.

For other examples see sections 10.2, 11.3.4, 11.4.4 and 14.2.6.

10.1.6 The effect of changing pH

Since the characteristics of ionizable side chains of amino acids depend on pH, enzymic activity varies with pH changes (section 3.2.2). At extremes of pH, the tertiary structure of the protein may be disrupted and the protein denatured. Even at moderate pH values, where the tertiary structure is not disrupted, enzyme activity may depend on the degree of ionization of certain amino acid side chains, and the pH profile of an enzyme may suggest the identity of those residues (as investigated by Ulf von Euler and colleagues (1924) and Malcolm Dixon (1953)).

Consider an enzyme whose catalytic activity depends on a certain amino acid side chain (Y^-) being able to act as a proton acceptor, according to the general equation

$$\underset{\text{protonated enzyme}}{\text{E.YH}} \rightleftharpoons \text{H}^+ + \underset{\text{deprotonated enzyme}}{\text{E.Y}^-} \qquad (10.11)$$

The titration graph of pH against added NaOH for such a system, as predicted by the Henderson-Hasselbalch equation, is shown in Fig. 3.9. Since the amount of $E.Y^-$ present is directly proportional to the amount of NaOH added, a graph of pH against the fraction of the enzyme in the form $E.Y^-$ will be exactly the same shape as Fig. 3.9. This graph, with the axes reversed, is shown in Fig. 10.1a.

The initial reaction velocity is given by $v_0 = k_{cat}[ES]$, where $k_{cat} = V_{max}/[E_0]$ (section 7.1.3). At each given pH, k_{cat} (and hence $V_{max}/[E_0]$) will be proportional to the fraction of enzyme in the form $E.Y^-$, if this form is required for catalytic activity.

$$\therefore V_{max}/[E_0] = k[EY^-]/[E_0] \quad \text{(where } k \text{ is a constant)}$$

$$\therefore V_{max} = k[EY^-]$$

At the optimal pH, where all of the enzyme present will be in the required form and V_{max} will have reached its optimal value, $(V_{max})_{opt}$, it follows that

$$(V_{max})_{opt} = k[E_0]$$

So, in general

$$V_{max} = (V_{max})_{opt} \cdot \frac{[EY^-]}{[E_0]} \qquad (10.12)$$

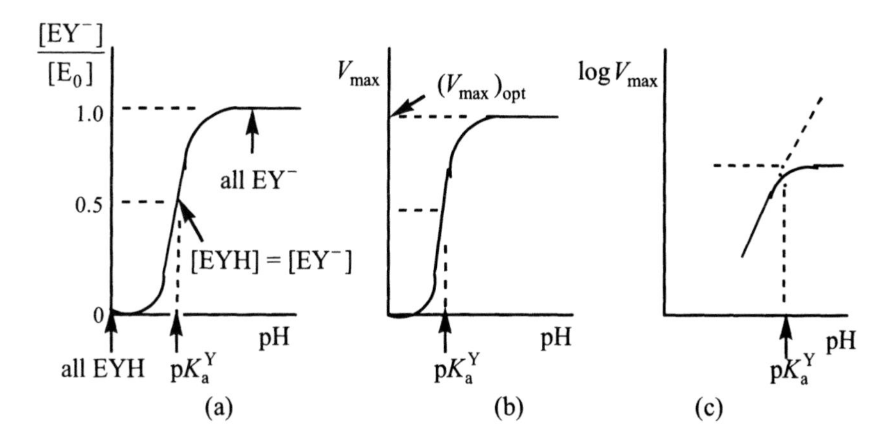

Fig. 10.1 – Graphs showing the effect of dissociation of the enzyme form E.YH, according to the Henderson-Hasselbalch equation: (a) a graph of $[E.Y^-]/[E_0]$ against pH; (b) a graph of V_{max} against pH, at constant $[E_0]$, where catalytic activity depends on the presence of $E.Y^-$; and (c) the corresponding graph of $\log_{10}V_{max}$ against pH.

Hence a graph of V_{max} against pH at constant $[E_0]$ (Fig. 10.1b) will be identical in shape to that shown in Fig. 10.1a. The pK_a of the ionizable group (the amino acid side chain) can be obtained from the point of inflection of the curve (where $V_{max} = \frac{1}{2}(V_{max})_{opt}$), thus making it possible to identify the amino acid residue in question. In plotting such graphs, it is important to ensure that the enzyme is fully saturated with substrate at each pH investigated, for the concentration required to achieve this may also vary with pH (see later).

The pK_a value may also be obtained by an alternative treatment of these data. The dissociation constant, K_a^Y, of E.YH is given by:

$$K_a^Y = \frac{[H^+][E.Y^-]}{[E.YH]}$$

$$\text{Also,} \quad [E_0] = [E.Y^-] + [E.YH]$$

$$\text{Hence,} \quad [E_0] = [E.Y^-] + \frac{[H^+][E.Y^-]}{K_a^Y} = [E.Y^-]\left(1 + \frac{[H^+]}{K_a^Y}\right)$$

$$\therefore \frac{[E.Y^-]}{[E_0]} = \frac{1}{\left(1 + \frac{[H^+]}{K_a^Y}\right)}$$

Substituting for $[E.Y^-]/[E_0]$ in the expression for V_{max} obtained above (10.12):

$$V_{max} = \frac{(V_{max})_{opt}}{\left(1 + \frac{[H^+]}{K_a^Y}\right)} \tag{10.13}$$

Taking logarithms,

$$\log_{10}V_{max} = \log_{10}(V_{max})_{opt} - \log_{10}\left(1+\frac{[H^+]}{K_a^Y}\right) \quad (10.14)$$

In situations where $[H^+] \gg K_a^Y$:

$$\left(1+\frac{[H^+]}{K_a^Y}\right) \approx \frac{[H^+]}{K_a^Y}$$

and the equation reduces to:

$$\log_{10}V_{max} = \log_{10}(V_{max})_{opt} - \log_{10}[H^+] + \log_{10}K_a^Y$$

$$= \log_{10}(V_{max})_{opt} + pH - pK_a^Y \quad (10.15)$$

Hence a plot of $\log_{10}V_{max}$ against pH, where $[H^+] \gg K_a^Y$, will be linear and of slope = + 1.

When $[H^+] \ll K_a^Y$, the expression for $\log_{10}V_{max}$ reduces to:

$$\log_{10}V_{max} = \log_{10}(V_{max})_{opt} - \log_{10}1$$

$$= \log_{10}(V_{max})_{opt}$$

Hence a plot of $\log_{10}V_{max}$ against pH, where $[H^+] \ll K_a^Y$, will be horizontal.

By combining the equations obtained under the two extreme conditions, it can be seen that the two linear graphs, when extrapolated, intersect where pH = pK_a^Y (Fig. 10.1c).

Let us now consider the situation of an enzyme whose catalytic activity depends on a certain amino acid side chain (–ZH) being able to act as a proton donor, according to the general equation: $E.ZH \rightleftharpoons H^+ + E.Z^-$. From the Henderson-Hasselbalch equation, the relationship between the proportion of enzyme in the form $E.Z^-$ and the pH will be exactly as shown for $E.Y^-$ in Fig. 10.1a. However, in this instance, E.ZH and not $E.Z^-$ is required for catalytic activity. Since $[E_0] = [E.ZH] + [E.Z^-]$, it readily follows that the relationship between the proportion of enzyme in the form E.ZH and the pH will be as in Fig. 10.2a, and a graph of V_{max} against pH, at constant $[E_0]$, for such enzyme will be as in Fig. 10.2b. The pK_a of the ionizable group is the pH where $[E.ZH] = [E.Z^-]$, so is the same regardless of whether [E.ZH] or $[E.Z^-]$ is plotted against pH and is given by the point of inflection of either graph. Therefore it is again possible to identify the amino acid residue in question from a graph of V_{max} against pH, or alternatively of $\log_{10}V_{max}$ against pH (Fig. 10.2c).

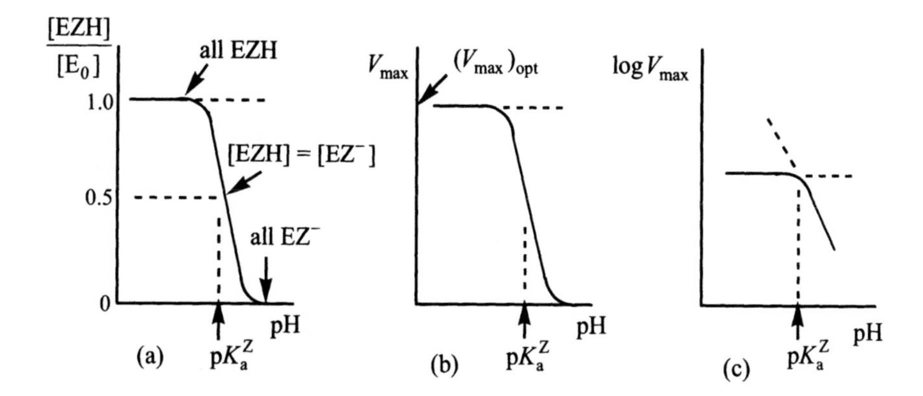

Fig. 10.2 – Graphs showing the effects of dissociation of E.ZH according to the Henderson-Hasselbalch equation: (a) a graph of [E.ZH]/[E_0] against pH; (b) a graph of V_{max} against pH at constant [E_0], where catalytic activity depends on the presence of E.ZH; and (c) the corresponding graph of $\log_{10}V_{max}$ against pH.

Finally, let us consider an enzyme whose catalytic activity depends on one amino acid side chain ($-Y^-$) being able to act as a proton acceptor and another (–ZH) being able to act as a proton donor. In this situation, a graph of V_{max} against pH at constant [E_0] will be a combination of Figs. 10.1b and 10.2b; the two points of inflection give the pK_a values of the two amino acid side chains involved. If these pK_a values are far apart (Fig. 10.3a), their determination from a plot of V_{max} against pH is relatively straightforward, for the points of inflection will occur where $V_{max} = \frac{1}{2}(V_{max})_{opt}$ at a particular enzyme concentration. If, however, the two pK_a values are less than 2 pH units apart (Fig. 10.3b), then the theoretical $(V_{max})_{opt}$ will never be achieved and the pK_a values cannot be accurately determined from such a graph or one of $\log_{10}V_{max}$ against pH. Nevertheless, reasonable estimates can be made, and more accurate values obtained from further mathematical treatment of the results, to give some indication as to the probable amino acid residues involved. It will be noted that this bell-shaped curve of V_{max} against pH is obtained for many enzyme-catalysed reactions.

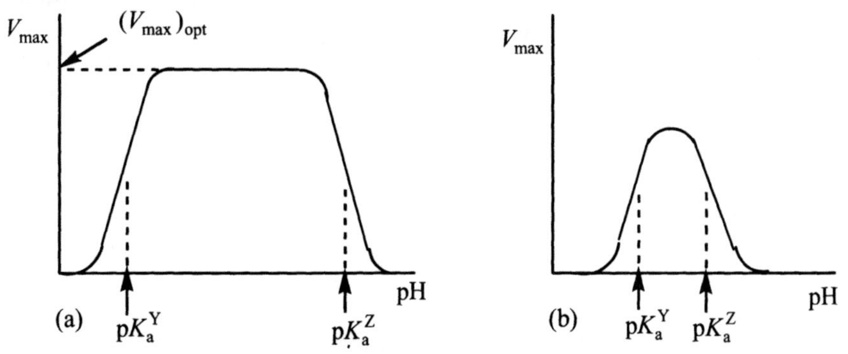

Fig. 10.3 – Graphs of V_{max} against pH, at constant [E_0], where catalytic activity depends on the simultaneous presence of $E.Y^-$ and E.ZH: (a) where pK^Y_a and pK^Z_a are more than 2 units apart; and (b) where pK^Y_a and pK^Z_a are less than 2 units apart.

So, in general, a plot of V_{max} against pH at fixed [E_0] for a particular enzyme-catalysed reaction may enable one or two pK_a values to be obtained for processes which are essential for catalytic activity. A comparison of these values with known pK_a values for amino acid side chains (Fig. 3.10) may identify the amino acid residues involved.

However, it is necessary at this point to register a few words of caution. A straightforward interpretation of the results is only valid if the assumptions discussed above are true of a particular enzyme-catalysed reaction and no other pH-dependent factors are involved. In practice, for example, the degree of ionization of the substrate may vary with pH. Also, V_{max} (or, more accurately, k_{cat}) may be a composite function of several rate constants, not all of which may be affected by pH to the same degree. Thirdly, the micro-environment in which a particular amino acid residue exists as part of an enzyme may affect the pK_a value of its ionizable side chain by as much as 3 pH units from that found in free solution.

For these and other reasons, results of pH studies have to be treated with caution, unless confirmed by other methods. Nevertheless, with this proviso, such studies may be useful since they are easy to perform and can give the first indication of which amino acid residues should be considered as essential components of a particular active site.

So, for example, the hydrolysis of chitin (poly-*N*-acetylglucosamine) by lysozyme was found to exhibit a bell-shaped curve of activity against pH, the pH optimum being about 5 and the two component pK_a values approximately 4 and 6. The lower of these suggested the presence of an essential carboxyl group, which was confirmed by other studies and shown to belong to aspartate-52. Less predictably, the other essential ionizable amino-acid side chain has also been shown to be a carboxyl group, this time belonging to glutamate-35. In contrast to aspartate-52, glutamate-35 is located in a non-polar region and is thus protonated at pH 5.

In general, the variation of V_{max} with changing pH will indicate the essential ionizing groups of the **enzyme-substrate complex**, i.e. those involved in the catalytic process. It is also possible that some ionizing groups might be involved in substrate-binding: these can be identified from the pK_a values obtained by plotting $1/K_m$ against pH, provided the substrate is non-ionizable. However, only where the Michaelis-Menten equilibrium-assumption (section 7.1.1) is valid will the results be independent of ionizations involved in the catalytic process.

Further information may be obtained by plotting V_{max}/K_m against pH, at constant [E_0]. This will indicate the essential ionizing groups of the free enzyme, i.e. those involved in both the binding and catalytic processes. Ionizable groups which are not involved in either of these processes will not usually contribute to changes in K_m or V_{max}.

Pancreatic **ribonuclease** was shown by Brian Rabin and colleagues (1961) to have two active site ionizable groups, their pK_a values being 5.2 and 6.8 in the free enzyme, but perturbed by the binding of substrate to 6.3 and 8.1 in the enzyme-substrate complex. From this and other data, it was concluded that both of the ionizable groups were histidine side chains (histidine-12 and histidine-119), one required to act as a proton donor and the other as a proton acceptor.

Myron Bender (1966) and others demonstrated that two distinct ionizable groups influence the reaction catalysed by **chymotrypsin**. One of these caused an increase in K_m at high pH, without a corresponding change in V_{max}. This was attributed to an N-terminal isoleucine residue (isoleucine-16, see section 5.1) whose α-amino group must be in a protonated form to maintain the enzyme in a structure capable of binding substrate.

The other ionization, which affected V_{max}, showed a pK_a of about 6.8. Aspartate-102 was the residue implicated, the anomalous pK_a value being due to its hydrophobic environment. A plot of V_{max}/K_m against pH, at constant [E_0] was consistent with the above, showing a bell-shaped curve with a maximum at about pH 8, and pK_a values of approximately 6.8 and 9.0.

10.2 THE INVESTIGATION OF THE THREE-DIMENSIONAL STRUCTURES OF ACTIVE SITES

From the above discussion, it must be clear that the active site of an enzyme is more than the sum of its parts. In order to understand how an enzyme works, it is necessary to know not only the identities of the amino acid residues which make up the substrate-binding sites and catalytic sites, but also the environment in which each site exists and the arrangement in space of the sites with respect to each other.

The technique which has given far more information than any other about the three-dimensional structure of enzymes is **X-ray crystallography** (see section 2.5). Of course it is the structure in solution which determines the properties of enzymes, but all available evidence suggests that structures determined in crystals can exist in solution; some crystals of enzymes, e.g. ribonuclease, even retain catalytic activity.

If this technique is applied to investigate the binding of the substrate to the active site of an enzyme, a major difficulty is encountered: X-ray diffraction analysis requires the structure under investigation to be static, but enzyme-substrate complexes break down rapidly to give reaction products. Hence, the structure of an enzyme-substrate complex is usually inferred from that obtained when the substrate is replaced by a slow-reacting analogue or an inhibitor (see section 10.1.2).

Enzyme crystals often consist of about 50% water, so it may be possible to diffuse these ligands (substrate analogue or inhibitor) into an existing crystal. Alternatively the enzyme-ligand complex may be crystallized from solution. For example, David Blow and his colleagues (1967) prepared **tosyl-chymotrypsin** (see section 10.1.3) by diffusing tosyl-fluoride into a crystal of α-chymotrypsin. They then used X-ray crystallography to elucidate its structure, the findings supporting much of the data discussed in section 10.1. The tosyl group was found to be bound to serine-195, which was in close proximity to histidine-57 and methionine-192 on the surface of the molecule (Fig. 10.4). The imidazole ring of histidine-57 was shown to be linked by a hydrogen bond to the side chain carboxyl group of aspartate-102, itself located in the hydrophobic interior of the molecule. Also, the protonated α-amino group of the N-terminal isoleucine residue (isoleucine-16) was found to form an electrostatic linkage with the side chain of aspartate-194.

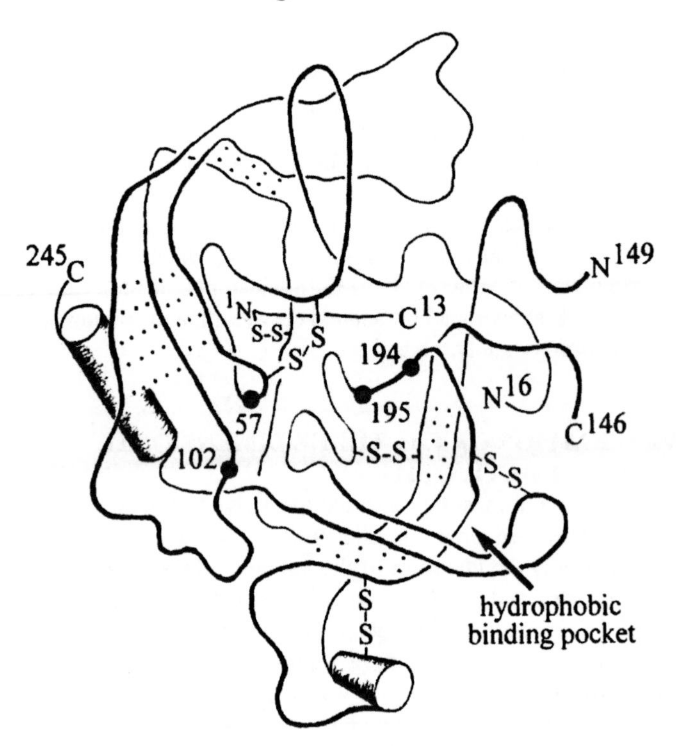

Fig. 10.4 – A simplified representation of the three-dimensional structure of α-chymotrypsin, as revealed by the X-ray diffraction studies of Blow and colleagues (1967). (Conventions as for Fig. 2.10.)

Over a period from 1969, Margaret Adams, Michael Rossmann and colleagues obtained crystal structures for dogfish muscle **lactate dehydrogenase** in the free LDH, LDH-NADH and LDH-NADH-pyruvate forms, and showed that a loop of the polypeptide chain in the region of residues 100-110 closed over the active site after the substrate was bound (see Fig. 12.7). The pyruvate could be seen to be very close to the key residues histidine-195, arginine-171, aspartate-168 and arginine-109.

John Holbrook and colleagues, in the 1970s and 1980s, confirmed and extended these conclusions by chemical modification and site-directed mutagenesis studies. The involvement of histidine-195 in the catalytic process was shown by treatment with diethylpyrocarbonate, which reacted more strongly with this histidine residue than any other, and in doing so inactivated the enzyme. Replacement of aspartate-168 with asparagine reduced the affinity for pyruvate, suggesting a link between aspartate and histidine residues, as in the serine proteases. The substitution of isoleuceine-250 (another residue found at the active site) by asparagine also reduced the affinity for pyruvate, suggesting a significant role for the hydrophobic side chain. Replacement of arginine-l09 by glutamine had little effect on binding, but lowered the rate of the subsequent reaction considerably. The insertion of a fluorescent tryptophan residue in place of glycine-106 enabled the movement of the loop region to be studied by stopped-flow techniques, and it was found that loop closure over the active site was the rate-limiting step for the overall process.

Further examples of X-ray crystallography in the elucidation of enzyme structure and function are discussed in Chapter 11.

Spectrometry in its various forms (see section 2.6) can give further information about enzyme-substrate interactions. Studies by **NMR,** for example, have provided information about interactions between particular atoms in an enzyme-substrate complex, and if carried out at various pH values, can indicate the pK_a values of individual groups (see section 11.3.2).

Various techniques have indicated that segments of a protein may be flexible and free to move within a crystal. Flexible, or disordered, segments are 'invisible' to X-ray diffraction analysis, since only regular arrangements of atoms give rise to well defined patterns. A disordered region often starts at a glycine residue, the absence of a side chain allowing maximum flexibility, and usually contains no aromatic residues.

Robert Huber and his colleagues (1977) showed that **trypsinogen** contains several disordered regions, but the active enzyme, **trypsin,** does not. The inactivity of trypsinogen may be due to the inability of the binding sites in the disordered regions to form bonds of sufficient strength with the substrate. The electrostatic linkage occurring in chymotrypsin between isoleucine-16 and aspartate-194 is also found in trypsin, but not in trypsinogen or chymotrypsinogen, and it appears to act as a clamp, holding the enzyme in the rigid structure required for activity.

In other enzyme systems, e.g. **acetyl-CoA carboxylase** (section 11.5.8), flexibility is essential for activity, since it appears that an intermediate may be transferred from one site to another on a flexible protein arm. In some enzymes, e.g. **lactate dehydrogenase** (see above) and **carboxypeptidase A** (section 11.4.4), important conformational changes have been shown to take place associated with the binding of a substrate. In general, proteins are likely to be more flexible in solution than when crystalline.

SUMMARY OF CHAPTER 10

A major problem in the identification of the amino acid residues involved in substrate-binding and catalysis is posed by the instability of the enzyme-substrate complex: substrate and product molecules will easily dissociate from the enzyme during investigation, making their positions of attachment difficult to determine. This problem may be overcome by the use of chemical procedures to prevent the breakdown of the enzyme-substrate complex, or by the use of substrate analogues, including irreversible inhibitors, which bind readily to the enzyme but do not easily dissociate from it.

If the chemical modification of a particular amino acid side chain results in a loss of catalytic activity, then it may be assumed that the residue in question is located at the active site of the enzyme, provided the modification can be prevented by the presence of excess substrate or competitive inhibitor. The introduction of site-directed mutagenesis in the 1980s, allowing the controlled substitution of amino acid residues, led to significant advances in knowledge about active site structure and function. The identity of amino acid residues involved in binding or catalysis may also be indicated by the activity profile of an enzyme at different pH values.

However, the techniques which have so far given most information about the structure of enzymes and the nature of enzyme-substrate interactions are X-ray crystallography and NMR spectrometry.

FURTHER READING

As for Chapters 2, 4 and 5, plus:
Blow, D. M. (1997), The tortuous story of Asp...His...Ser: structural analysis of α-chymotrypsin, *Trends in Biochemical Sciences,* **22,** 405-408.
Dixon, H. B. F. (1992), How groups of proteins titrate - a new approach, *Essays in Biochemistry,* **27**, 161-176, Portland Press.
Fersht, A. (1999), *Structure and Mechanism in Protein Science,* Freeman.
Smith, M., *Synthetic DNA and biology*, Text of 1993 Nobel lecture available at: http://nobelprize/org/nobel_prizes/chemistry/laureates/1993/smith-lecture.pdf.

PROBLEM

10.1 The hydrolysis of an ester substrate by papain was investigated at fixed temperature and enzyme concentration, but at a series of pH values. The following results were obtained:

	Initial velocity (μmol l^{-1}min^{-1}) at $[S_0]$ (mmol l^{-1})				
pH	$[S_0]$ = 2.0	2.5	3.33	5.0	10.0
2.0	1	1	1	2	2
3.0	11	12	14	16	20
4.0	83	93	105	122	143
5.0	156	172	196	222	263
6.0	157	173	197	223	264
7.0	156	173	196	223	263
8.0	139	156	179	208	250
9.0	43	52	66	89	141
10.0	5	7	9	13	25

What can you deduce about the possible involvement of ionizable amino acid side chains?

11

The chemical nature of enzyme catalysis

11.1 AN INTRODUCTION TO REACTION MECHANISMS IN ORGANIC CHEMISTRY

A covalent chemical bond involves the sharing of a pair of electrons between the atoms (X:Y). If the bond breaks by **homolytic fission**, each atom separates as a highly reactive free radical, possessing an unpaired electron (X· + ·Y). Such reactions are very uncommon in aqueous solutions, where most bonds are broken by **heterolytic fission**, leaving one of the atoms with both electrons. If both electrons are retained by a carbon atom, a **carbanion** is produced:

$$R_3C{:}X \rightleftharpoons \underset{\text{carbanion}}{R_3C{:}^-} + X^+ \quad (11.1)$$

If neither electron is retained by the carbon atom, a **carbonium ion** is produced:

$$R_3C{:}X \rightleftharpoons \underset{\text{carbonium ion}}{R_3C^+} + {:}X^- \quad (11.2)$$

Carbanions and carbonium ions may be stabilized by **delocalization** of the charge over other atoms in the molecule. For example:

$$CH_2 = CH - \overset{+}{C}H_2 \leftrightarrow \overset{+}{C}H_2 - CH{=}CH_2 \quad (11.3)$$

The curved arrow indicates the flow of a pair of electrons which would take place if one of these **canonical structures** of the carbonium ion was converted to the other. The actual structure is somewhere in between the two canonical structures written down: it should not be imagined that the two forms are being rapidly interconverted. **Conjugated** double bond systems (i.e. where double and single bonds occur alternately) are of particular importance for stabilization, for a charge can be delocalized over the entire system.

It should be noted that a benzene ring consists of conjugated double bonds, and hence of alternative canonical structures, which is why it is often represented by a circle within a hexagon, as may be seen in subsequent illustrations.

The formation of carbanions and carbonium ions is likely to occur as part of a complete reaction sequence. Organic compounds can participate in four main types of reaction: displacement (or substitution) reactions; addition reactions; elimination reactions; and rearrangements.

Substitution reactions may be **nucleophilic,** when the group attacking the carbon atom is an electron donor called a nucleophile since it is attracted to nuclei; alternatively they may be **electrophilic,** when the attacking group is an electron acceptor, or electrophile. Electrophilic substitution reactions often involve the displacement of hydrogen. For example:

$$C_6H_5\text{-}H + NO_2^+ \longrightarrow C_6H_5\text{-}NO_2 + H^+ \quad (11.4)$$

benzene nitrobenzene

In **nucleophilic substitution,** an atom other than hydrogen is usually displaced. Such reactions may have a **unimolecular** or a **bimolecular** mechanism.

In **unimolecular nucleophilic substitution (S_N1)** reactions, the rate-limiting step is the ionization of a single molecule to form a carbonium ion which then reacts with a nucleophile, as in the following example (with bonds going away from the reader being indicated by a dotted line – see Fig 2.2):

$$R'R''R'''C\text{—}Y \xrightarrow{-Y^-} \left[R'R''R'''C \right]^+ \xrightarrow{+X^-} R'R''R'''C\text{—}X \text{ or } X\text{—}CR'R''R''' \quad (11.5)$$

The carbonium ion is planar and, if the resulting product has four different groups attached to that particular carbon atom (as above), a racemic mixture of optical isomers will be obtained. The overall rate of the reaction is determined by the rate of ionization, and thus on the basicity of the leaving group (Y^-): weak bases are good leaving groups and depart readily; strong bases are poor leaving groups. The strength of the attacking nucleophile does not usually affect the rate of the reaction.

In **bimolecular nucleophilic substitution (S_N2)** reactions, the attacking nucleophile adds to the carbon atom at a point diametrically opposite the leaving group, which it displaces in one rapid step:

$$X^- + R'R''R'''C\text{—}Y \rightarrow \left[X\text{----}CR'R''R'''\text{----}Y \right]^- \rightarrow X\text{—}CR'R''R''' + Y^- \quad (11.6)$$

transition-state

It can be seen that the reaction involves an inversion of configuration at the carbon atom. The rate of the reaction depends, among other factors, on the relative strengths of X^- and Y^- as nucleophiles.

It is unnecessary here to go into details about other types of reaction. **Addition reactions** involve the addition of an electrophile or a nucleophile to form a relatively stable compound without any group being displaced, while **elimination reactions** involve the removal of a group, usually in a strongly basic solution, without it being replaced by another group. **Rearrangements** are reactions where bonds are formed and broken within a single molecule.

Information about the mechanism of a particular reaction can be obtained from the **stereochemistry** of the products, as discussed above for the S_N1 and S_N2 reactions. Another useful technique is the investigation of the effect of **isotopic substitution** on the reaction rate. For example, it is usually found that a C–D bond is broken more slowly than a C–H bond, so the use of deuterium-labelled reactants can give information as to which C–H bonds are normally broken as part of the rate-limiting step of the overall process.

11.2 MECHANISMS OF CATALYSIS

11.2.1 Acid-base catalysis

Acids can catalyse reactions by temporarily donating a proton; bases can do the same by temporarily accepting a proton. In section 6.2.3 we saw how a base could catalyse the hydrolysis of an ester by stabilizing the transition-state. Bases may also increase reaction rates by increasing the nucleophilic character of the attacking groups. Thus, hydroxide ions displace halides from alkyl halides more rapidly than do neutral water molecules. Similarly, acids facilitate the removal of leaving groups, where these are strong bases. For example, the following reaction would normally proceed extremely slowly:

$$\underset{\underset{X^-}{\uparrow}}{R_3C}\text{—O—R}' \longrightarrow R_3CX + {}^-OR' \qquad (11.7)$$

In the presence of an acid, however, conditions are more favourable:

$$R_3C\text{—O—R}' + H^+ \longrightarrow \underset{\underset{X^-}{\uparrow}}{R_3C}\text{—}\underset{\underset{H}{|}}{\overset{+}{O}}\text{—R} \longrightarrow R_3CX + HOR' \qquad (11.8)$$

Similarly, in the case of ester hydrolysis, protonation of the substrate leads to a transition-state which avoids the unstable two-charge arrangement occurring under neutral conditions (section 6.2.2), thus increasing the rate of reaction:

$$HO^{+}{=}C(R')(OR'') + H_2O \rightleftharpoons H{-}O{-}C(R')(OR''){-}O^{+}H{-}H \rightleftharpoons \text{products} \quad (11.9)$$

In general, acid or base catalysis may be shown to be operating by determining the rate constants of a reaction at different concentrations of acid or base.

11.2.2 Electrostatic catalysis

A transition-state may be stabilized by electrostatic interaction between its charged groups and charged groups on a catalyst. Thus, the positive charge on a carbonium ion can be stabilized by interaction with a negatively charged carboxylate ion; similarly, the negative charge on an oxyanion can be stabilized by a positively charged metal ion. For example, the hydrolysis of the methyl ester of glycine to produce glycine and methanol may be catalysed by cupric ions, the mechanism probably involving the following steps:

$$Cu^{2+}\cdots H_2N{-}H_2C{-}C(=O\cdots Cu^{2+}){-}OCH_3 + :OH_2 \rightleftharpoons Cu^{2+}\cdots H_2N{-}H_2C{-}C(O^-\cdots Cu^{2+})(OCH_3)({}^{+}OH_2) \rightarrow H_2N.CH_2COOH + CH_3OH \quad (11.10)$$

11.2.3 Covalent catalysis

In contrast to acid-base and electrostatic catalysis, where the transition-state is merely modified, covalent catalysis introduces a different reaction mechanism and is sometimes termed **alternative pathway catalysis**. In the case of nucleophilic catalysis, the catalyst is more nucleophilic than the normal attacking groups and so rapidly forms an intermediate which itself rapidly breaks down to give the products. For example, Myron Bender and co-workers (1957) showed that a variety of tertiary amines catalyse the hydrolysis of esters, as follows:

$$R_3N + R'C(=O){-}OR'' \rightleftharpoons R_3\overset{+}{N}{-}C(O^-)(R'){-}OR'' \rightleftharpoons R_3\overset{+}{N}{-}C(=O)R' + {}^{-}OR'' \xrightarrow{H_2O} R'C(OH){=}O + R_3N \quad (11.11)$$

In contrast, electrophilic catalysts act by withdrawing electrons from the reaction centre of an intermediate and may be termed **electron sinks**. This type of catalysis is found in the reactions of the coenzymes **thiamine pyrophosphate** (section 11.5.6) and **pyridoxal phosphate** (section 11.5.7). Sometimes, metal ions may also be regarded as acting in this way, as in the example in section 11.2.2, where Cu^{2+} facilitates the withdrawal of electrons from the reacting carbon atom.

11.2.4 Enzyme catalysis

All of the above mechanisms of catalysis are seen in enzyme-catalysed reactions. However, enzymes, because of their great size and range of properties, are able to impose their presence on a reaction to a far greater extent than most catalysts.

All catalysts act by reducing the energy of activation, or more correctly the free energy of activation ($\Delta G^{\neq}$) of the reaction being catalysed (see sections 6.2.3 and 6.6). As we saw in section 6.1.2, free energy is made up of an entropic and an enthalpic component, since $\Delta G = \Delta H - T\Delta S$. Hence $\Delta G^{\neq} = \Delta H^{\neq} - T\Delta S^{\neq}$, where $\Delta H^{\neq}$ is the enthalpy of activation and $\Delta S^{\neq}$ is the entropy of activation, and so anything which decreases $\Delta H^{\neq}$ or which makes $\Delta S^{\neq}$ more positive will lower the value of $\Delta G^{\neq}$ and help the reaction proceed.

There is a great loss of entropy when reactants leave their random existence in free solution to become bound in the transition-state, so $-T\Delta S^{\neq}$ will generally have a large positive value, possibly making up about half of the total free energy of activation where there is more than one reactant. In enzyme-catalysed reactions, there is inevitably a similar loss of entropy at some stage, but it occurs largely in the binding steps when enzyme-substrate complexes are formed, and not to the same degree in the actual conversion of substrates to products. The binding of substrate molecules in close proximity to each other on the enzyme surface effectively increases their concentrations and reduces the entropy loss for the subsequent formation of a transition-state. This has been called the **proximation effect**, the **approximation effect** and/or the **propinquity effect** by Thomas Bruice, William Jencks and others. The enzyme may also ensure that the reacting groups of the bound substrates approach each other with their electronic orbitals correctly orientated, thus ensuring that the reaction takes place under optimal conditions; Daniel Koshland termed this property of enzymes **orbital steering**.

The contribution of the catalytic sites on the enzyme to the reaction will generally be to the enthalpic factor by stabilizing the transition-state. It appears that the stability of the enzyme-bound transition-state may be the most important single factor in determining whether the reaction proceeds (section 4.6). It is possible that several different catalytic processes may be involved in the same enzyme-catalysed reaction, as may be seen in the examples discussed in sections 11.3-11.5. Another way in which the enthalpy factor may be reduced occurs if the positions of the binding sites on the enzyme do not correspond exactly to those on a substrate. As the substrate binds, some of its bonds may be distorted and weakened, resembling those found in the transition-state. This is the Haldane and Pauling concept of **strain**, discussed in section 4.5. Alternatively, the enzyme-substrate complex may be under stress from internal interactions which are absent from the enzyme-bound transition-state complex. Again, this would facilitate the forward reaction (section 4.5). A distortion of the protein to fit the substrate (the Koshland induced-fit hypothesis, section 4.4) would have no effect on enthalpy, but it would not preclude stress factors from playing a part.

Some reactions proceed better in organic rather than aqueous solutions. Most enzymes contain regions of non-polar character (section 2.5.2) and these may be of importance in the catalysis of such reactions.

So, enzymes create a suitable environment for the reaction to proceed. They reduce the total free energy of activation required and also spread out the free energy requirement over several stages. Nevertheless, some activation energy is still required for each stage of the reaction sequence. This cannot come from the substrate, because the translational energy that these possess in free solution is lost the moment they become bound to the enzyme. It seems likely that the required energy is obtained from collisions between solute molecules and the enzyme-substrate complex, but the mechanism for this is far from clear.

11.3 MECHANISMS OF REACTIONS CATALYSED BY ENZYMES WITHOUT COFACTORS

11.3.1 Introduction

Enzymes which operate without cofactors tend to be relatively small and have relatively straightforward reaction mechanisms. For these reasons, such enzymes were among the first to be investigated in detail. Some examples are given below. As before, and in subsequent sections, mention is made of historical findings which played a significant part in the development of current knowledge.

11.3.2 Chymotrypsin

Chymotrypsin is formed by the cleavage of several peptide bonds in the inactive monomeric protein, chymotrypsinogen, which is synthesized and secreted by mammalian pancreas. The active enzyme thus produced consists of three non-identical polypeptide chains (section 5.1.2).

Chymotrypsin catalyses the cleavage of peptide bonds at the carboxyl side of aromatic amino acid (phenylalanine, tyrosine or tryptophan) residues. It will also hydrolyse a variety of amides and esters, and these artificial substrates have been used to investigate the enzyme in great detail. In section 7.2.1 we discussed kinetic evidence showing that the chymotrypsin-catalysed hydrolysis of an ester proceeded via the formation of an acyl-enzyme. This is also true for the hydrolysis of amides, and we can write the reaction in general terms as follows:

$$\underset{\text{ester or amide}}{E + R.CO.Y} \rightarrow E.R.CO.Y \xrightarrow{\;-YH\;} \underset{\text{acyl enzyme}}{E.CO.R} \xrightarrow{H_2O} E + RCOOH \quad (11.12)$$

In sections 10.1.2, 10.1.3 and 10.1.6, we discussed evidence that the substrate could bind to serine-195; and also that histidine-57 was involved in the reaction mechanism. X-ray diffraction studies (see section 10.2 and Fig. 10.4) revealed the presence of a hydrophobic binding-pocket for aromatic side chains at the active site. Such studies also showed that aspartate-102, buried in the interior of the molecule, could be closely linked to the action of histidine-57 and serine-195, possibly setting up what was termed a charge relay system, with aspartate-102 removing a proton from histidine-57, thus making it easier for the latter to remove a proton from serine-195 during the course of the reaction.

Detailed ^{1}H, ^{13}C and ^{15}N NMR studies showed that histidine-57 has a relatively normal pK_a value near 7, at least in the free enzyme, casting doubt on the charge relay theory. However, investigations using site-directed mutagenesis (see section 10.1.5) have shown that the three amino acids, if not setting up a charge relay system as such, nevertheless act in a concerted fashion, perhaps explaining why several enzymes have a 'catalytic triad' of this type at the active site.

An outline reaction mechanism for the hydrolysis of an ester or amide by chymotrypsin (or other serine protease) is as follows. Histidine-57 acts as a base catalyst to enable the oxygen of serine-195 make a nucleophilic attack on the carboxyl group of the enzyme-bound substrate. An unstable tetrahedryl intermediate is formed whose negatively-charged oxygen atom may be stabilized by hydrogen bonding with the backbone -NH of glycine-193. Similarly, the positive charge on the imidazole ring of histidine-57 is stabilized by electrostatic interaction with aspartate-102. The imidazole group of histidine-57 then acts as an acid catalyst to facilitate the liberation of the first product (YH), leaving behind the acyl enzyme (note the covalent bond linking the acyl group to serine-195):

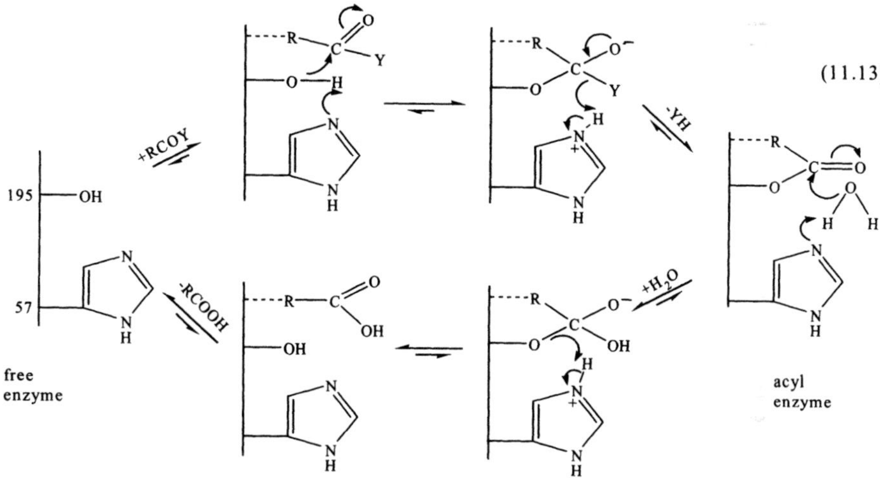

(11.13)

The second stage of the reaction, like the first, is initiated by a nucleophilic attack, this time by water. The liberation of the product (RCOOH) is again assisted by acid catalysis.

The same mechanism is thought to occur for peptide substrates, R′.NH.C(R″).CO.NH.R‴. If the substrate is again written in the form R.CO.Y, to be compatible with the illustration above, R– represents R′.NH.C(R″)–, where R″ is the side chain which fits into the binding pocket (the –NH forming a hydrogen bond with the backbone carboxyl group of serine-214); and –Y represents –NH.R‴.

11.3.3 Ribonuclease

Bovine pancreatic ribonuclease A has been studied in great detail by enzymologists and was the first enzyme to have its complete amino acid sequence determined. This was achieved by William Stein and Stanford Moore in 1963.

For their contributions to an understanding of the links between structure and catalytic activity (see also section 10.1.3), Stein and Moore were subsequently awarded a Nobel Prize.

Ribonuclease catalyses the cleavage of the phosphodiester backbone of ribonucleic acids by a reaction involving transfer of a phosphate group from the 5′-position of one nucleotide to the 3′-position of the next nucleotide in the chain (see section 3.1.1). In sections 10.1.3 and 10.1.6 we discussed evidence that histidine-12 and histidine-119 both contribute to the catalytic process, one possibly acting as a proton donor and the other as a proton acceptor; similar investigations have suggested that lysine-41 is also involved. The three-dimensional structure of the enzyme, as revealed by the X-ray diffraction studies of Gopinath Kartha and colleagues (1967), shows these three amino acids are sited close together near a substrate-binding cleft in the molecule (Fig. 11.1). Other noteworthy features of the structure are the presence of three α-helices and three β-hairpins (see section 2.5.2).

In 1962, Brian Rabin and co-workers suggested the following mechanism for the reaction:

(11.14)

. Histidine-12 removes a proton from the substrate, producing a nucleophilic oxygen atom which attacks the phosphate. Histidine-119 donates a proton to the complex, and the first product, R′OH, is liberated. The rest of the reaction is the reverse of the first part, with water replacing the R′OH.

This mechanism is consistent with the evidence now available. It is likely that lysine-41 helps to stabilize the complex, its positively charged amino group forming an electrostatic interaction with the negative charge of the phosphate.

11.3.4 Lysozyme

Lysozyme catalyses the hydrolysis of the polysaccharide consisting of alternating *N*-acetylglucosamine (NAG) and *N*-acetylmuramic acid (NAM) units, a component of peptidoglycan, which gives shape and rigidity to bacterial cell walls.

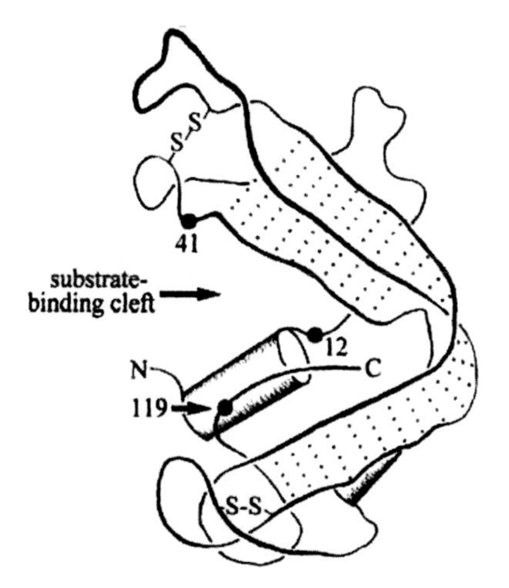

Fig. 11.1 – A simplified representation of the three-dimensional structure of ribonuclease A, as revealed by the X-ray diffraction studies of Kartha and colleagues (1967). (Conventions as for Fig. 2.10.)

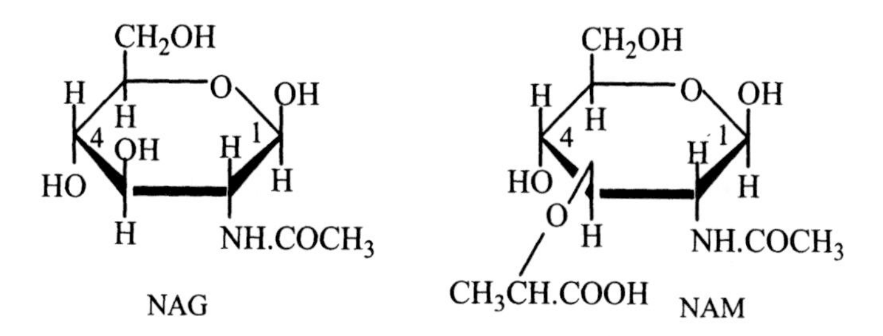

The lysozyme which has been most studied is that from hen egg-white, the first enzyme to have its complete three-dimensional structure determined by the use of X-ray crystallography. These investigations (by Colin Blake, David Phillips and colleagues, 1965), and subsequent ones, revealed that the enzyme has a large cleft which can accommodate six of the substrate's amino-sugar units (see Fig. 2.10). This may be represented diagrammatically as follows:

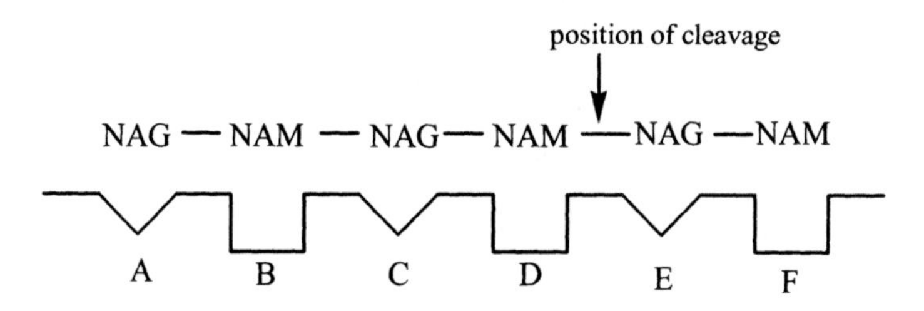

Each of the six amino-sugars is bound to the enzyme by hydrogen bonds, e.g. those in sites A and B to aspartate-l0l; that in site B to tryptophan-62 and tryptophan-63; and that in site F to asparagine-37. NAM, because of its large lactyl side chain, cannot bind to sites A, C or E, where there is restricted space.

The hydrolysis takes place between the units in sites D and E, and therefore must always involve the β-1,4 glycosidic link between Cl (the reducing end) of NAM and C4 of NAG. The amino acids near this bond are glutamate-35 and aspartate-52. Evidence that these are involved in the reaction has been discussed in sections 10.1.3 and 10.1.6.

In 1967, David Phillips and Charles Vernon both suggested the following mechanism for the reaction (many bonds and atoms not involved in the reaction being omitted from the diagram for the sake of clarity):

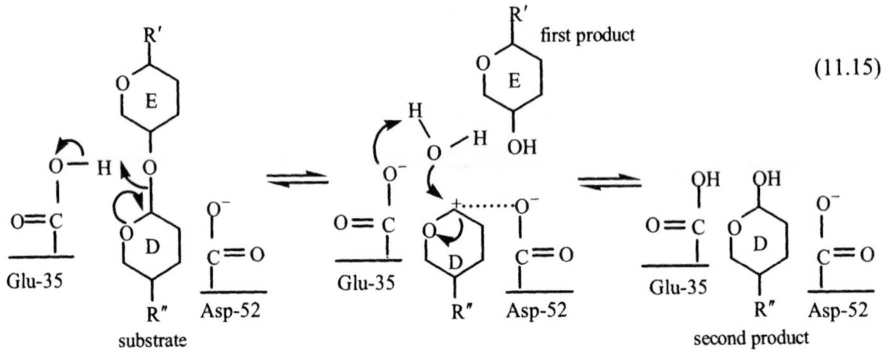

(11.15)

The substrate binds to the enzyme and hexose ring D is distorted to the less-stable half-chair conformation, which facilitates the formation of a carbonium ion also having the half-chair conformation.

chair conformation half-chair conformation

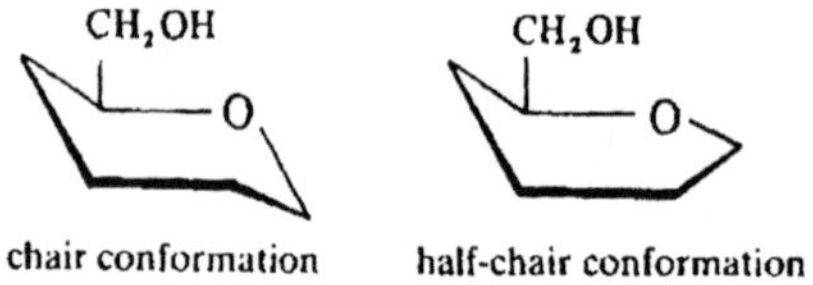

Glutamate-35, which is in a non-polar environment, and thus protonated at pH 5, can act as a general acid catalyst and donate its proton to the oxygen of the glycosidic bond, causing the bond to break. Aspartate-52 is in a more polar environment than glutamate-35 and is negatively charged at pH 5. Therefore, according to this mechanism, it can stabilize the carbonium ion by electrostatic interaction. The first product leaves, and the reaction is completed by a nucleophilic attack on the carbonium ion by water.

However, Stephen Withers and colleagues (2001), by a variety of techniques, including the replacement of the catalytic carboxyl group by an amide and the use of electrospray ionization mass spectrometry, demonstrated that this scheme (generally called the "Phillips mechanism") was only partially correct.

Although it seems that ring D of the substrate is indeed distorted on binding, the reaction proceeds via the covalent attachment of this ring (following its split from ring E) to aspartate-52, which thus has an important role as a nucleophile, rather than simply stabilizing a charged transition-state:

R″–(D)–O–C(=O)–| Asp-52

The reaction is presumably completed by the hydrolysis of the linkage between ring D and aspartate-52. This mechanism, involving covalent catalysis, is also common to other β-glycosidases.

11.3.5 Triose phosphate isomerase

Muscle triose phosphate isomerise (TIM, or TPI) is a dimeric enzyme which catalyses the interconversion of glyceraldehyde-3-phosphate (G-3-P) and dihydroxyacetone-phosphate (DHAP). X-ray crystallography has revealed that each sub-unit has an inner cylinder of eight strands of parallel β-pleated sheet linked by predominantly helical regions which form an outer cylinder. This arrangement, also found in some other enzymes, is termed an **α/β barrel**, or **TIM barrel**. Each of the two identical binding sites of TIM contains lysine-13, histidine-95, glutamate-165 and glycine-232 from one sub-unit and phenylalanine-74 from the other (Fig. 11.2).

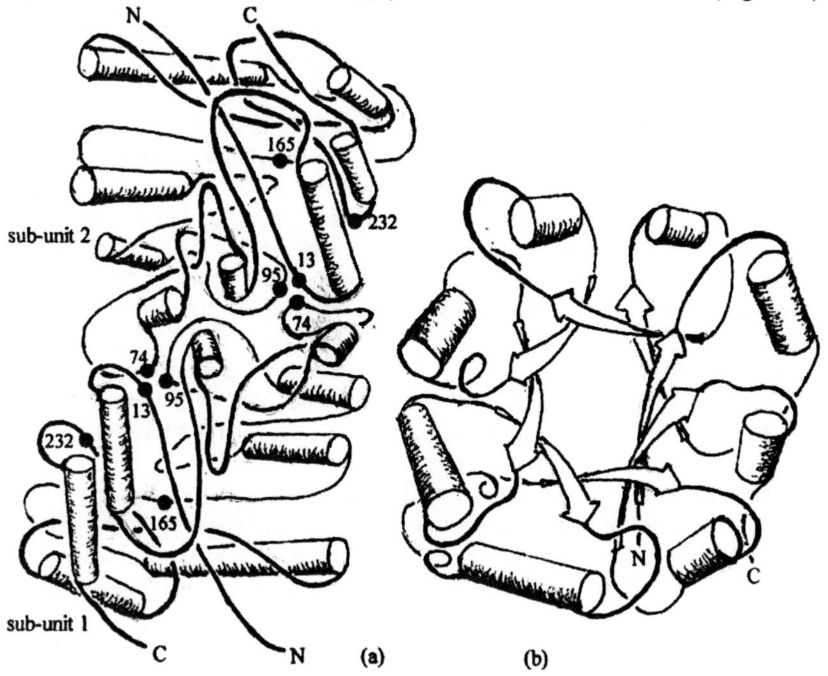

Fig. 11.2 – (a) A simplified representation of the three-dimensional structure of chicken muscle triose phosphate isomerise, as revealed by the X-ray diffraction studies of Phillips and colleagues (1975) (conventions as for Fig. 2.10); (b) the α / β barrel structure of each of the two sub-units (conventions as for Fig. 2.8)

The reaction mechanism apparently involves glutamate-165 acting as a base catalyst and histidine-95 as an acid one:

G-3-P ⇌ ⇌ DHAP (11.16)

His-95 His-95 His-95

Glu-165 Glu-165 Glu-165

11.4 METAL-ACTIVATED ENZYMES AND METALLOENZYMES

11.4.1 Introduction

More than a quarter of all known enzymes require the presence of metal atoms for full catalytic activity. Metal atoms usually exist as cations and often have more than one oxidation state, as with ferrous (Fe^{2+}) and ferric (Fe^{3+}) iron. We have noted that this positive charge can stabilize transition-states by electrostatic interactions, giving one mechanism for catalysis by metals (section 11.2.2). However, irrespective of the oxidation state and charge carried, a metal ion can bind a particular number of groups (**ligands**) by accepting free electron pairs to form co-ordinate bonds in specific orientations (see section 2.3.2).

Therefore, metal ions can be involved in enzyme catalysis in a variety of ways: they may accept or donate electrons to activate electrophiles or nucleophiles, even in neutral solution; they themselves may act as electrophiles; they may mask nucleophiles to prevent unwanted side reactions; they may bring together enzyme and substrate by means of co-ordinate bonds, possibly causing strain to the substrate in the process; they may hold reacting groups in the required three-dimensional orientation; or they may simply stabilize a catalytically-active conformation of the enzyme.

With **metalloenzymes,** the metal is tightly bound and retained by the enzyme on purification. With **metal-activated** enzymes, the binding is less tight and the purified enzymes may have to be activated by the addition of metal ions. There is no clear-cut division between the two groups.

Albert Mildvan (1970) pointed out that ternary complexes formed between an enzyme (E), metal ion (M) and substrate (S) may be enzyme bridge complexes (M–E–S), substrate bridge complexes (E–S–M) or metal bridge complexes (E–M–S). Metalloenzymes cannot form substrate bridge complexes, since the purified enzyme exists as E–M.

The involvement of metal ions in enzymes may be investigated by NMR, ESR and proton relaxation rate (PRR) enhancement techniques, as well as forming part of more general investigations of enzyme structure and function. Examples are given below.

11.4.2 Activation by alkali metal cations (Na^+ and K^+)

Alkali metal cations bind only weakly to form complexes with enzymes, but K^+, the most abundant intracellular cation, is known to activate a great many enzymes, particularly those catalysing phosphoryl transfer or elimination reactions.

It appears that the role of K^+ is largely to bind to negatively-charged groups on an inactive form of the enzyme and cause a change in conformation to a more active form. However, in some cases, it may also aid substrate binding. For example, muscle **pyruvate kinase,** a tetrameric enzyme which catalyses the reaction:

$$\underset{\text{phosphoenolpyruvate (PEP)}}{CH_2 = \underset{\underset{OPO_3^{2-}}{|}}{C}.CO_2^- } + H^+ + ADP \rightleftharpoons \underset{\text{pyruvate}}{CH_3\underset{\underset{O}{\|}}{C}.CO_2^-} + ATP \qquad (11.17)$$

has a requirement for alkali metal cations and for Mn^{2+} (or Mg^{2+}), all of which bind in the region of the active site. Various studies indicated that the carboxyl group of PEP binds to the enzyme-bound K^+, whereupon a conformational change takes place which facilitates the progress of the reaction via an E–Mn^{2+}–PEP complex. Note also that each sub-unit has an α/β barrel domain, like TIM (section 11.3.5)

11.4.3 Activation by alkaline earth metal cations (Ca^{2+} and Mg^{2+})

Oxygen atoms are often involved in the bonds of both alkali metal and alkaline earth metal cations, bonds of the latter being relatively stronger. The divalent cations, Ca^{2+} and Mg^{2+}, can form six co-ordinate bonds to produce octahedral complexes.

Mg^{2+} is accumulated by cells in exchange for transport of Ca^{2+} in the opposite direction. As might be expected, therefore, enzymes requiring Ca^{2+} for activation are mainly *extracellular* ones, e.g. the salivary and pancreatic α-amylases. The Ca^{2+} appears to play a role in maintaining the structure required for catalytic activity. A variety of *intracellular* enzymes require Mg^{2+} for activity and, in most cases, this requirement can be replaced *in vitro* by one for Mn^{2+}. Unlike Mg^{2+}, Mn^{2+} is paramagnetic, which enables the system to be more easily investigated. (Note that Mn^{2+} may, in its own right, be a component of a metalloenzyme, e.g. arginase, where it stabilizes a reactive hydroxide ion, ensuring that an activated nucleophile is available for catalysis. In the rat liver enzyme, Mn^{2+} is bound to histidine-101, aspartate-124, and aspartate-232 in each of the three sub-units, and has a catalytic interaction with aspartate-128.)

It has been shown that all possible types of ternary bridge complexes involving divalent cations can exist. Most **kinases** form E–S–M complexes, where S is the reacting nucleotide. Let us consider, as an example, the reaction catalysed by muscle **creatine kinase**: creatine + ATP $\rightleftharpoons$ ADP + phosphocreatine + H^+.

The true substrate is in fact Mg^{2+}ATP and the reaction proceeds via the formation of the complex creatine–E–ATP–Mg^{2+}. Mildred Cohn and colleagues (1971) showed that the divalent cation binds to the α- and β-phosphates of the nucleotide, but not to the terminal (γ) phosphate, which is transferred to creatine. The cation helps in the orientation of the complex, and may also assist in the breaking of the pyrophosphate bond by withdrawing electrons from the β-phosphate. The overall reaction has a random-order rapid-equilibrium kinetic mechanism, and dead-end complexes may be formed (see section 9.2.3). Identification of some of these, e.g. E–ADP–Mn^{2+}, helped in the elucidation of the mechanism of the reaction.

In contrast, the reaction catalysed by **pyruvate kinase** (see section 11.4.2) involves a cyclic metal bridge complex:

$$\mathrm{E}\begin{cases}\mathrm{Mg^{2+}} \\ \mathrm{ATP} \\ \mathrm{pyruvate}\end{cases}$$

Metal bridge complexes are found with many enzymes which use pyruvate or phosphoenolpyruvate as substrate. Another example is **enolase**, which catalyses the reaction:

$$\underset{\text{phosphoenolpyruvate}}{CH_2{=}\underset{\displaystyle OPO_3^{2-}}{\underset{|}{C}}.CO_2^-} + H_2O \rightleftharpoons \underset{\text{2-phosphoglycerate}}{HOCH_2.\underset{\displaystyle OPO_3^{2-}}{\underset{|}{C}H}.CO_2^-} \qquad (11.18)$$

Enolase is a dimeric enzyme, two Mg^{2+} ions being required to stabilize the active dimer; a further two Mg^{2+} ions are required if each of the two active sites binds a substrate. Thomas Nowak, Albert Mildvan and colleagues (1973) demonstrated that the enzyme-bound cation probably binds to water, forming a co-ordinated hydroxyl group which can attack the phosphoenolpyruvate:

$$\begin{array}{l} \qquad\qquad H_2C{=}\underset{\displaystyle OPO_3^{2-}}{\underset{|}{C}}{-}CO_2^- \\ H\diagdown \\ \quad O{-}H \\ \quad\ \vdots \\ E{-}Mg^{2+} \end{array} \qquad (11.19)$$

A few enzymes, e.g. *E. coli* **glutamate-ammonia ligase**, have a mechanism involving an enzyme bridge complex. Here the divalent cation presumably has a purely structural role.

11.4.4 Activation by transition metal cations (Cu, Zn, Mo, Fe and Co cations)

Transition metal ions bind to enzymes much more strongly than the metal ions discussed above, and usually form metalloenzymes; this makes their involvement in enzyme-catalysis relatively easy to investigate. They are found in only trace amounts in living organisms, for larger amounts can be toxic.

The trace metals Mo and Fe are found in **nitric-oxide reductase**, the nitrogen-activating complex of nitrogen-fixing bacteria, and Fe is a component of haemoglobin, the oxygen-carrying haemoprotein of the erythrocytes of vertebrates. Another trace metal, Co, is found in vitamin B_{12} (see section 11.5.10). We will now consider, in a little more detail, an example of a Cu- and a Zn-metalloenzyme.

Superoxide dismutase is a copper-metalloenzyme with an extensive "Greek key" β-pleated sheet region (see section 2.5.2). It catalyses the removal of the highly-reactive and destructive O_2^- (possessing a free radical) produced from oxygen by a variety of reactions and by irradiation). Irwin Fridovich and colleagues (1969) demonstrated that the dismutase reaction proceeds as follows:

$$2O_2^- + 2H^+ \rightleftharpoons H_2O_2 + O_2 \qquad (11.20)$$

Bovine erythrocyte superoxide dismutase is a dimeric protein containing two Cu^{2+} ions and two Zn^{2+} ions. The Zn^{2+} ions appear to have a structural rather than a catalytic role, while the Cu^{2+} ions are involved in the reaction sequence:

$$E-Cu^{2+} + O_2^- \longrightarrow E-Cu^+ + O_2 \qquad (11.21)$$

$$E-Cu^+ + O_2^- \xrightarrow{+2H^+} E-Cu^{2+} + H_2O_2 \qquad (11.22)$$

In contrast to the above, Zn^{2+} has a catalytic role in the reaction catalysed by **carboxypeptidase** A, where the C-terminal amino acid of a polypeptide is removed, provided it has a non-polar side chain. Carboxypeptidase A from bovine pancreas is a monomeric enzyme which contains one atom of zinc, with helical regions arranged around a core consisting of an 8-stranded mixed β-pleated sheet with a right-handed twist. X-ray crystallography studies (by William Lipscomb and colleagues, 1967) and the determination of the complete amino acid sequence (by Ralph Bradshaw and co-workers, 1969) revealed that the active site contains the co-ordinated Zn^{2+} ion bound to histidine-69, glutamate-72, histidine-196 and H_2O, as well as a groove to accommodate the polypeptide substrate and a hydrophobic pocket for binding the side chain of the C-terminal amino acid (Fig. 11.3); the terminal carboxyl group of the substrate forms an electrostatic interaction with arginine-145.

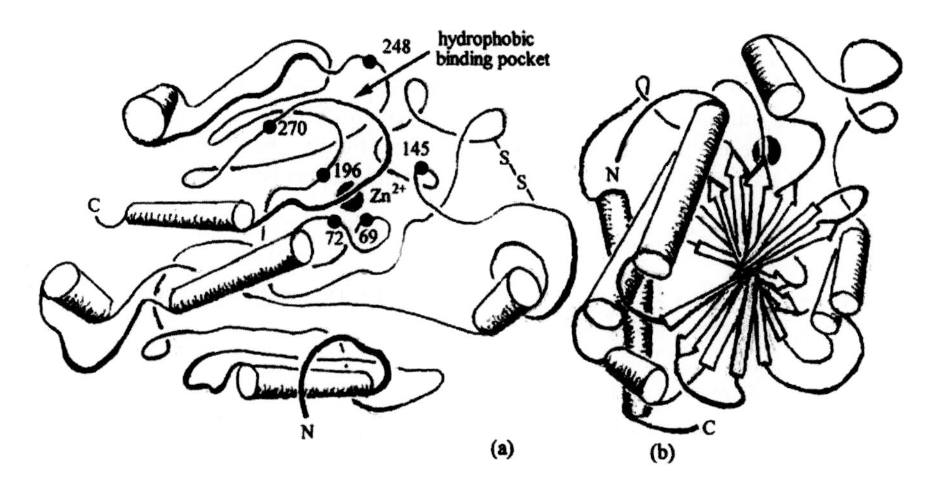

Fig. 11.3 – (a) A simplified representation of the structure of bovine carboxypeptidase A, as revealed by the X-ray diffraction studies of Lipscomb and colleagues (1967) (conventions as for Fig. 2.10); (b) a view from a different angle, illustrating the 8-stranded mixed (and twisted) β-pleated sheet at the core of the molecule (conventions as for Fig. 2.8).

During the reaction, the carbonyl oxygen of the peptide bond being hydrolysed replaces the water molecule bound to Zn^{2+}. The metal ion probably facilitates cleavage of the peptide bond by withdrawing electrons from this carbonyl group

Bert Vallee (1964) proposed a general mechanism for the reaction which involved acid and base groups on the enzyme. X-ray diffraction studies have shown that tyrosine-248 is located in such a position in the enzyme-substrate complex that it could donate a proton to the nitrogen of the peptide bond being hydrolysed, and there is also evidence that the carboxyl group of glutamate-270 acts as a general base catalyst to make the attacking water molecule nucleophilic. However, site-directed mutagenesis experiments by Stephen Gardell and colleagues (1985), in which tyrosine-248 was replaced by phenylalanine without affecting catalytic activity, have suggested that this tyrosine residue may simply have a binding function. Thus the reaction appears to involve:

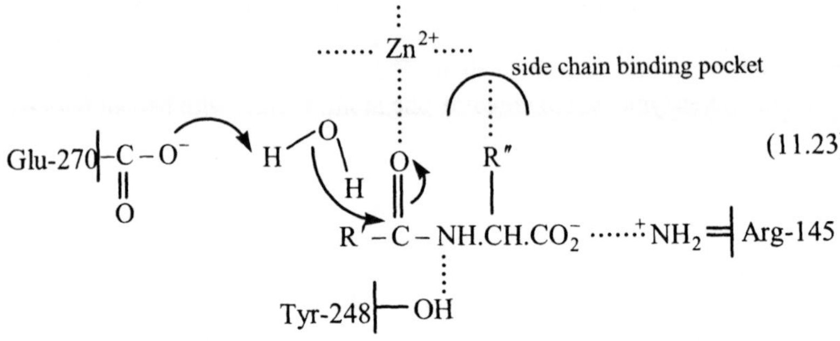

Zn^{2+}

side chain binding pocket

Glu-270 – C(=O) – O^- H–O–H O R″

$R' - C - NH.CH.CO_2^- \cdots {}^+NH_2 =$ Arg-145

Tyr-248 – OH

(11.23)

where R″ is the side chain of the C-terminal amino acid, and R′ represents the rest of the peptide molecule.

X-ray diffraction studies have also shown that substrate-binding results in arginine-145, glutamate-270 and especially tyrosine-248 moving to new positions close to the substrate, forcing water molecules out of the active site and thus creating a hydrophobic environment. The substrate is so tightly enclosed that it would not have been able to get into this position but for the mechanism of the conformational change.

11.5 THE INVOLVEMENT OF COENZYMES IN ENZYME-CATALYSED REACTIONS

11.5.1 Introduction

Coenzymes are organic compounds required by many enzymes for catalytic activity. They are often vitamins, or derivatives of vitamins. Sometimes they can act as catalysts in the absence of enzymes, but not so effectively as in conjunction with an enzyme.

As with metal-enzyme linkages, there is a range of bond strengths for coenzyme-enzyme links, the point of distinction between tightly-bound cofactor (prosthetic group) and loosely-bound cofactor being arbitrary. Coenzymes which are prosthetic groups form an integral part of the active site of an enzyme and undergo no net change as a result of acting as a catalyst. For this reason, some would exclude these from classification as coenzymes, simply calling them organic cofactors.

Conversely, loosely-bound coenzymes can be regarded as **co-substrates** since they often bind to the enzyme-protein together with the other substrates at the start of a reaction and are released in an altered form at the end of it. They are generally regarded as coenzymes since they usually bind to the enzyme before the other substrates are bound, since they participate in many reactions, and since they may be reconverted to their original form by many enzymes present within cells.

Some important coenzymes are discussed below.

11.5.2 Nicotinamide nucleotides (NAD^+ and $NADP^+$)

These are derived from the vitamin **niacin,** which is nicotinamide or nicotinic acid. The structure of nicotinamide adenine dinucleotide (NAD^+) in its oxidized and reduced forms is given below:

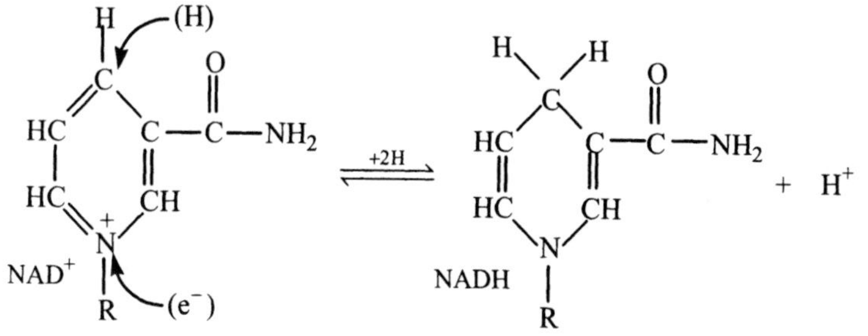

$$NAD^+ \; (\text{ring: HC, C, CH, N}^+\text{–R; C–C(=O)–NH}_2\text{; H} \leftarrow (H),\; N \leftarrow (e^-)) \underset{}{\overset{+2H}{\rightleftharpoons}} NADH \; (\text{ring: } CH_2, \text{HC, C, CH, N–R; C–C(=O)–NH}_2) + H^+ \qquad (11.24)$$

where –R represents D-ribose-phosphate-phosphate-D-ribose-adenine.

It can be seen that the reduction of NAD^+ to NADH requires two reducing equivalents per molecule, i.e. one electron (e^-) and one hydrogen atom ($H = H^+ + e^-$), which together may be regarded as a hydride ion (H^-), to add to the pyridine ring of nicotinamide. The pyridine ring is conjugated, so the positive charge may be delocalized, making several points vulnerable to nucleophilic attack. However, it seems that the reaction mechanism usually involves:

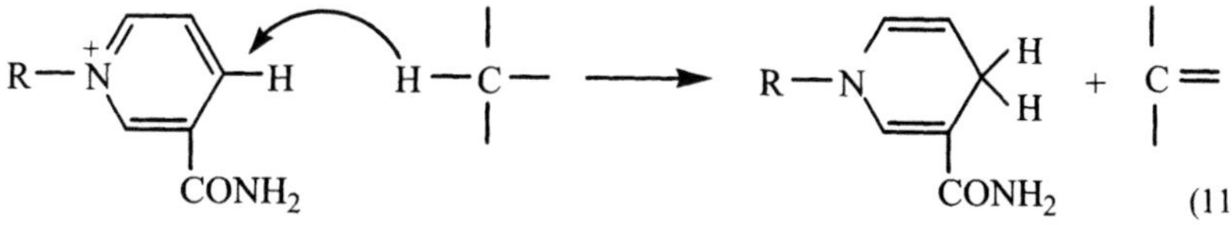

$$R{-}N^+(\text{pyridine, CONH}_2)\text{–H} + H{-}\overset{|}{\underset{|}{C}}{-} \longrightarrow R{-}N(\text{dihydropyridine, CONH}_2)(H)(H) + \overset{|}{\underset{|}{C}}{=} \qquad (11.25)$$

The transfer of the hydride ion shows stereochemical specificity: with some enzymes, the transferred hydride ion becomes the H_R atom of nicotinamide (that in the diagram coming out of the plane of the paper towards the reader), whereas with others it becomes the H_S atom (that going away from the reader).

Nicotinamide adenine dinucleotide phosphate ($NADP^+$) is identical to NAD^+, except that the 2′-position of the D-ribose unit attached to adenine is phosphorylated. This does not affect the oxidation/reduction properties, but results in NAD^+ and $NADP^+$ acting as coenzymes for different enzymes. Enzymes utilizing NAD^+ usually have a catabolic function, the NADH produced being an energy source for the cell. Anabolic enzymes, in contrast, frequently involve NADPH as coenzyme.

The names and abbreviations given above are those currently (and for the past 40 years) recommended by the International Union of Biochemistry and Molecular Biology. NAD^+ has also been known as diphosphopyridine nucleotide (DPN^+) or Coenzyme I; $NADP^+$ as triphosphopyridine nucleotide (TPN^+) or Coenzyme II.

Needless to say, NAD^+ and $NADP^+$ act as coenzymes for oxidation/reduction reactions. They are loosely-bound, and leave the enzyme in a changed form at the end of the reaction.

The kinetic mechanism of horse liver **alcohol dehydrogenase (ADH)**, an enzyme which involves NAD^+ as coenzyme in the catalytic oxidation of primary or secondary alcohols, was discussed in section 9.4. It was noted that NAD^+ is the first substrate to be bound and NADH the last product to leave, the dissociation of NADH from the enzyme being the rate-limiting step of the overall reaction, which is one reason why NAD^+ is regarded as a coenzyme rather than simply as a substrate. Horse liver ADH is a dimer, each sub-unit containing one binding site for NAD^+ and two sites for Zn^{2+}, although only one of these zinc ions is directly involved in catalysis. As with many other dehydrogenases, it has been found that the coenzyme-binding domain of ADH molecule is part of a 6-stranded parallel twisted β-pleated sheet structure termed a **Rossmann fold** (named after Michael Rossmann; and depicted in Fig 12.7), similar to the mixed 8-stranded structure found in carboxypeptidase A (see section 11.4.4).

From results of X-ray diffraction studies, Carl-Ivar Brändén and colleagues (1975) deduced that the ternary complex is of the form:

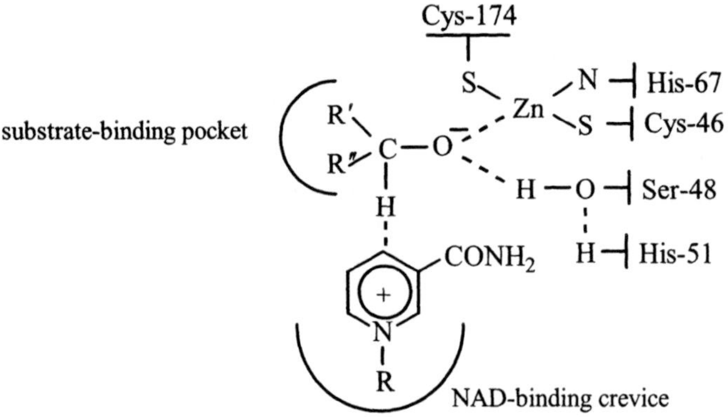

Cys-174
substrate-binding pocket
R′ R″ C—O⁻ ··· Zn (S—Cys-174, N—His-67, S—Cys-46)
H—O—Ser-48
H—His-51
H
CONH₂
N⁺
R
NAD-binding crevice

and the reaction mechanism involves:

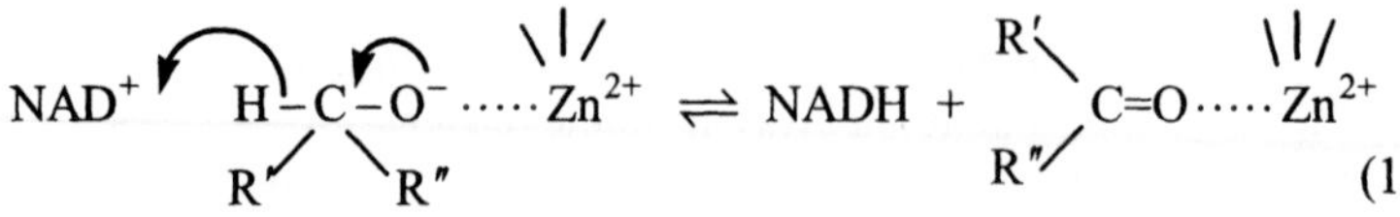

$$NAD^+ \; H{-}\underset{R' \quad R''}{C}{-}O^- \cdots Zn^{2+} \rightleftharpoons NADH + \underset{R''}{\overset{R'}{}}C{=}O \cdots Zn^{2+} \qquad (11.26)$$

The initial proton loss from the substrate in the hydrophobic environment of the active site to the surface of the enzyme appears to involve histidine-51, the 2′-hydroxyl group of the coenzyme and serine-48, possibly operating in a coordinated fashion.

The zinc ion then acts as an electrostatic catalyst to stabilize the negatively-charged transition-state, while a hydride ion is transferred to NAD^+, becoming the H_R atom of NADH. For the reverse reaction, the zinc ion acts as an electrophilic catalyst to enhance the polarization of the carbonyl group of the substrate and facilitate hydride transfer from the reduced coenzyme. As demonstrated by later studies, the catalytic process is facilitated by conformational changes in the enzyme as the reaction proceeds.

Dogfish muscle **lactate dehydrogenase (LDH)**, which catalyses the interconversion of lactate and pyruvate (see sections 1.3.4 and 5.2.2), is another NAD-utilizing dehydrogenase whose mechanism has been investigated in detail (see section 10.2 and Fig. 12.7). In this case, no metal ions are involved. LDH is a tetrameric enzyme, each sub-unit having a binding-site for NAD^+. As with ADH, the coenzyme binds to part of a "Rossmann fold". The overall reaction has a compulsory-order mechanism, the coenzyme binding first and bringing about a conformational change in the enzyme which enables the substrate to bind close to the nicotinamide ring.. Substrate-binding, in turn, causes a peptide loop of the enzyme to close over the active site. From the studies of Margaret Adams and others in the early 1970s, it was concluded that the reaction mechanism involves:

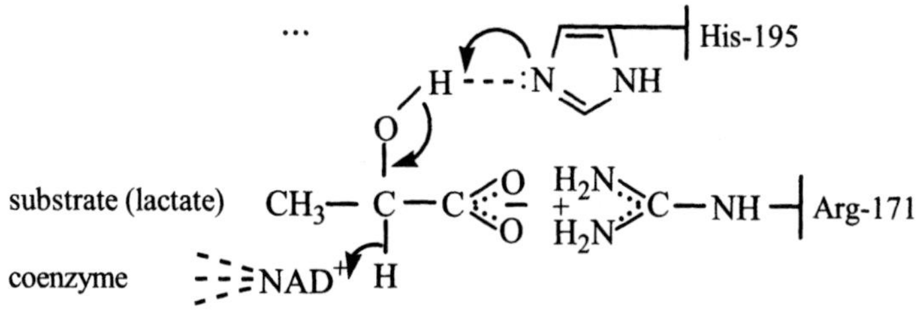

His-195

H - - - :N NH

O

substrate (lactate) CH_3— C —C(O)(O) H_2N + H_2N C—NH— Arg-171

coenzyme ---NAD^+ H

(11.27)

Arginine-171 binds the carboxyl group of the substrate by a two-point electrostatic interaction, which helps to orientate it. Histidine-195 then acts as an acid catalyst, removing a proton from lactate during its oxidation. This leaves histidine-195 in a positively-charged form which, together with arginine-109, can stabilize the transition-state by electrostatic catalysis. The positive charge on histidine-195 is in turn stabilized by electrostatic interaction with aspartate-168. The reaction is completed by the transfer of a hydride ion to NAD^+, in the same orientation as with ADH. The reactive nicotinamide ring of the coenzyme is held in a hydrophobic environment, one component of which, Ile-250, lies close to Arg-171 in the complex. It seems that this arrangement strengthens the interaction between Arg-171 and the substrate, and also increases the stability of NADH relative to NAD^+.

The adenine ring similarly binds in a hydrophobic environment, but can form hydrogen bonds with Tyr-85 and Asp-53. The latter also links by hydrogen bonding to the adenine ribose. These and other interactions have only a binding function.

11.5.3 Flavin nucleotides (FMN and FAD)

Flavin nucleotides are derived from **riboflavin, vitamin B_2**. Like the nicotinamide nucleotides, they function in oxidation/reduction reactions, the reducing equivalents being carried by the fused three-ringed system of flavin as shown below:

$$\text{FMN or FAD} \xrightleftharpoons{+2H} \text{FMNH}_2 \text{ or FADH}_2 \qquad (11.28)$$

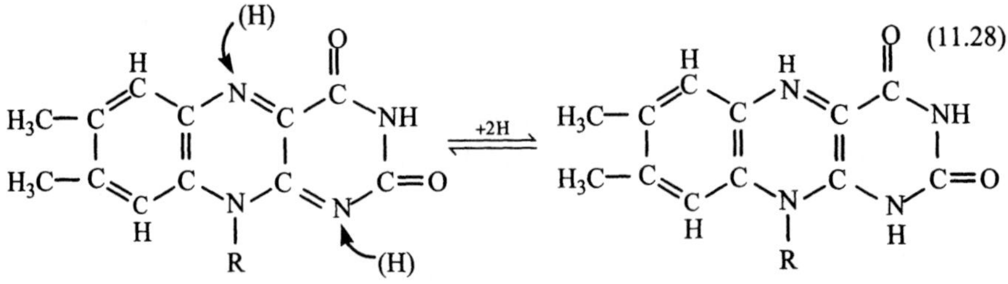

FMN or FAD FMNH$_2$ or FADH$_2$

For flavin mononucleotide (FMN), –R = –ribitol-phosphate; for flavin adenine dinucleotide (FAD), –R = –ribitol-phosphate-phosphate-D-ribose-adenine.

In contrast to NAD$^+$ and NADP$^+$, FMN and FAD are prosthetic groups and cannot generally be separated from the protein without denaturing it, the protein-flavin nucleotide complex being termed a **flavoprotein**. Hence, some enzymologists do not accept the classification of FMN and FAD as coenzymes (see section 11.5.1).

Because the flavin nucleotide does not have an independent existence, reactions catalysed by flavoproteins usually involve the transfer of reducing equivalents from a donor via the flavin to some specific external acceptor. For example, **glucose oxidase** (see section 4.1), which catalyses the reaction

$$\text{D-glucose} + O_2 \rightleftharpoons \text{D-glucono-}\delta\text{-lactone} + H_2O_2 \qquad (11.29)$$

utilizes FAD as prosthetic group and O_2 as hydrogen acceptor. Harold Bright and Quentin Gibson (1967) showed that this is a two-stage reaction:

$$\text{E–FAD} + \text{D-glucose} \rightleftharpoons \text{E–FADH}_2 + \text{D-glucono-}\delta\text{-lactone} \qquad (11.30)$$

$$\text{E–FADH}_2 + O_2 \rightleftharpoons \text{E–FAD} + H_2O_2 \qquad (11.31)$$

With some flavoproteins, the reduction of the flavin has been shown to be a two-step process, involving an unstable free radical **semiquinone** as intermediate:

$$\text{FMN or FAD} \xrightleftharpoons{+H} \text{flavosemiquinine} \xrightleftharpoons{+H} \text{FMNH}_2 \text{ or FADH}_2 \qquad (11.32)$$

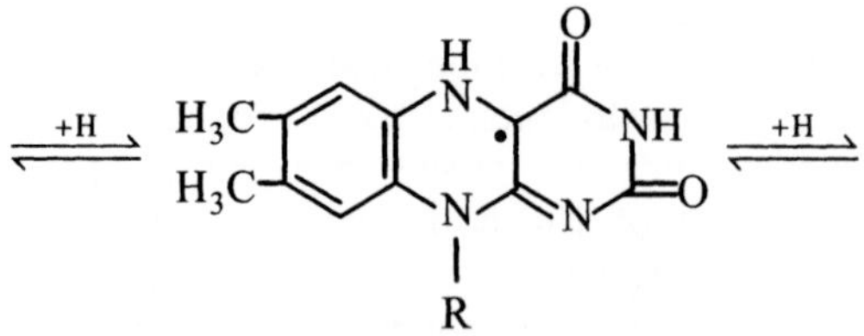

flavosemiquinine
(FMNH• or FADH•)

The unpaired electron, which can be delocalized about the ring system, has been revealed by ESR studies. Many flavoproteins are also metalloproteins and one of the roles of the metal ion could be to stabilize this semiquinone.

However, it seems likely that not all reactions involving flavin nucleotides as coenzymes proceed via an identical mechanism: the reaction catalysed by **NADH dehydrogenase** has been shown to involve semiquinone formation, but that catalysed by **glucose oxidase** (discussed above) apparently does not. Similarly, the reoxidation of E–$FADH_2$ can proceed by a variety of mechanisms: where molecular oxygen is the acceptor, the products may be H_2O_2 (with oxidases), H_2O and hydroxylated products (with hydroxylases), or the superoxide anion (O_2^-) and flavin semiquinone.

Flavin-dependent disulphide oxidoreductase enzymes, of which **glutathione reductase** is an example, use FAD to shuttle reducing equivalents from NADPH (or sometimes NADH) to a cysteine residue, which generally involves the cleavage of a disulphide bridge. Usually, two electrons are transferred simultaneously to FAD, so it never passes through the semiquinone form. Finally, the substrate is reduced by the cysteine.

X-ray diffraction analysis of human erythrocyte **glutathione reductase**, a large dimer consisting of identical 478-residue sub-units, was carried out by Georg Schulz, Heiner Schirmer and colleagues during the 1980s. These studies revealed that each sub-unit consists of NADPH, interface and FAD domains, the last-mentioned incorporating an N-terminal domain. The two sub-units link along the interface domains, with a catalytic site, consisting of residues from both sub-units, at each end. In each sub-unit, NADPH and FAD bind to their respective domains, which results in the catalytically-active components of each coenzyme being appropriately positioned. Each catalytic site consists of active residues cysteine-58, cysteine-63 and tyrosine-197 from the sub-unit providing the NADPH and FAD, together with glutamate-472 and histidine-467 from the other sub-unit.

In the absence of NADPH, the side chain of tyrosine-197 prevents solvent gaining access to the NADPH-binding site and to the flavin ring of FAD. When NADPH binds, it does so in such a position that its active pyridine ring lies parallel, and in close proximity, to the flavin ring of FAD. Equally close to the flavin ring, but on the opposite side to NADPH, lies the sulphur atom of cysteine-63.

A hydride ion is transferred from NADPH to FAD (the H_S atom of NADPH being the one involved), resulting in the formation of the $FADH^-$ anion. The electrons are then immediately passed on from $FADH^-$ to cysteine-63, causing the cleavage of its disulphide link with cysteine-58. Cysteine-63 has a transient covalent linkage with the flavin ring, but this changes to a non-covalent linkage as the amino acid accepts a hydrogen atom to finish in the –SH form, while cysteine-58 picks up a proton from solution to finish in the same form. Overall, the formation of the reduced cysteine residues may be written:

$$FADH^- \quad \underset{\text{Cys-63}}{\overset{S}{CH_2}}\!-\!\underset{\text{Cys-58}}{\overset{S}{CH_2}} \;(\to H^+) \xrightarrow{+\,H^+} \underset{\text{Cys-63}}{\overset{FAD\cdots SH}{CH_2}} + \underset{\text{Cys-58}}{\overset{SH}{CH_2}} \qquad (11.33)$$

$NADP^+$ may then be released from its binding-site to leave the enzyme in its reduced form. Again, the side chain of tyrosine-197 blocks access to the vacant coenzyme-binding site, thus preventing re-oxidation by the solvent before the enzyme has had chance to act upon the substrate, glutathione disulphide, and convert it to glutathione. The product, glutathione, is γ-glutamylcysteinylglycine, while the substrate consists of two glutathione molecules whose cysteine residues are linked by a disulphide bridge.

When glutathione disulphide binds to the reduced form of the enzyme, cysteine-58, made more nucleophilic by the concerted action of histidine-467 and glutamate-472 (operating rather like histidine-57 and aspartate-102 in chymotrypsin) initiates the attack on the substrate which results in the breaking of the disulphide bond. One half of the substrate becomes attached to cysteine-58, while the other half picks up a proton from histidine-467 and is released as the first of the two product molecules:

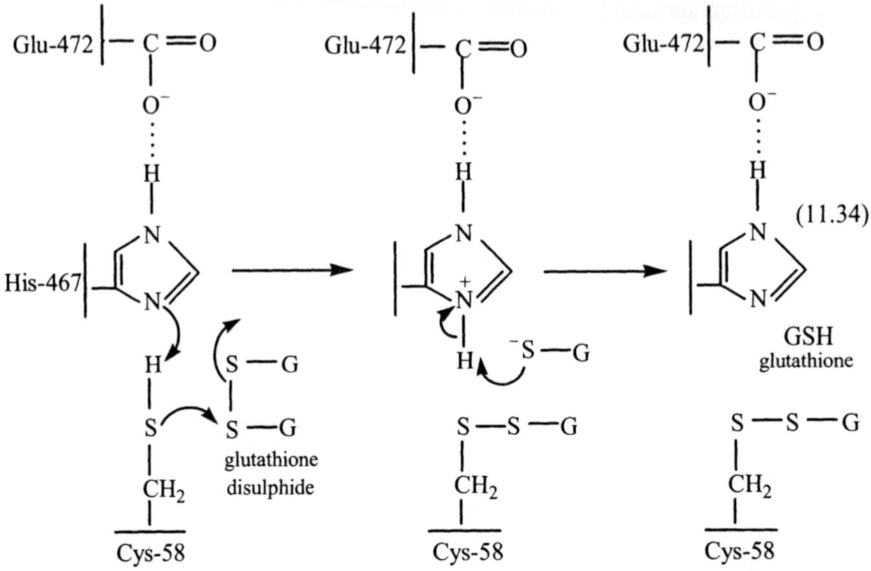

(11.34)

It can be seen that histidine-467 acts as a base catalyst to assist the splitting of the disulphide bridge of the substrate, and then as an acid catalyst to promote the formation of glutathione, the positive charge carried by the imidazole ring during part of the sequence being stabilized by glutamate-472.

The reaction is completed by nucleophilic attack by the sulphur atom of cysteine-63 on that of cysteine-58, which allows release of the second molecule of product, leaving the enzyme in its original oxidized form.

11.5.4 Adenosine phosphates (ATP, ADP and AMP)

The nucleoside phosphates ATP, ADP and AMP are involved in phosphate transfer reactions.

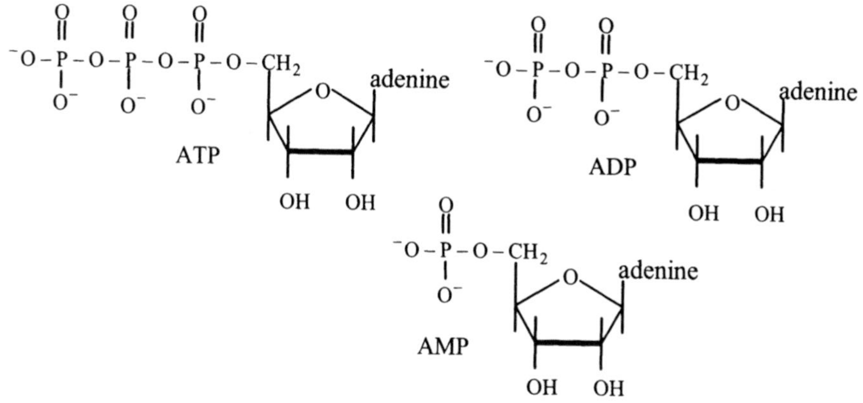

ATP

ADP

AMP

ATP and ADP may be interconverted by the reaction: ATP + $H_2O \rightleftharpoons$ ADP + P_i.

This tends to go strongly in the forward direction as written ($\Delta G^{\ominus\prime} = -31.0$ kJ mol^{-1}) because the four negative charges which are in close proximity on ATP make it an unstable molecule (although the product, ADP, itself has three negative charges close together), and because the reverse reaction requires negatively-charged ADP to react with negatively-charged P_i. The reaction can be coupled to others, so that phosphate may be transferred between ATP and other organic compounds without ever being present as free P_i (section 6.1.5).

The importance of ATP in energy metabolism is that, by comparison to other organic phosphates, it is only moderately unstable. Hence it may be synthesized by the transfer of phosphate to ADP from a more unstable organic phosphate (e.g. phosphoenolpyruvate) by substrate-level phosphorylation, or by oxidative phosphorylation. However, ATP is sufficiently unstable to be able to force the transfer of the phosphate to a whole variety of other compounds, thus driving such processes as biosynthesis, active transport and muscular contraction.

In the cell, adenosine phosphates are stabilized by binding to Mg^{2+} ions, and their metabolism is strictly mediated by enzymes (see section 11.4.3). In some instances, two phosphate groups, rather than one, may be removed from ATP to liberate inorganic pyrophosphate: ATP + $H_2O \rightleftharpoons$ AMP + $(PP)_i$

Adenosine phosphates, like the nicotinamide nucleotides, are loosely bound by enzymes and may be regarded both as coenzymes and as co-substrates/co-products of the reactions in which they participate.

11.5.5 Coenzyme A (CoA.SH)

Coenzyme A has the structure:

$HS.CH_2CH_2NH$-pantothenic acid-$OPO_3^-.PO_3^-$-ribose-3-P-adenine

With carboxylic acids it can form thioesters:

$$RCO_2H + HS.CoA \rightleftharpoons \underset{\text{acyl-CoA}}{R.COSCoA} + H_2O \tag{11.35}$$

These thioesters are of great importance in biochemical metabolism since they can be attacked by electrophiles (including other acyl–CoA molecules and CO_2) to form addition compounds, and by nucleophiles (including water) to displace the –SCoA group:

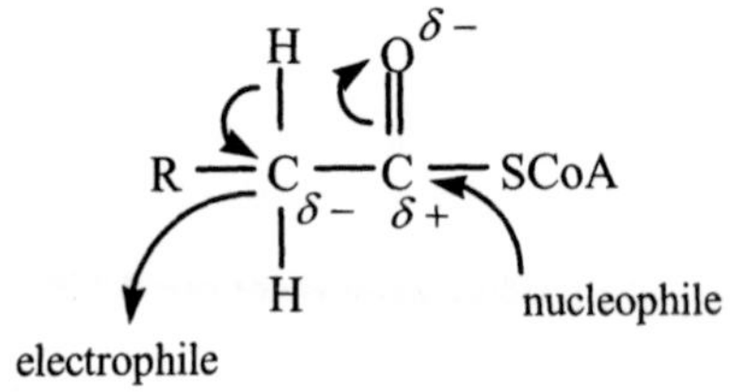

electrophile

nucleophile

(11.36)

Some examples are given in sections 11.5.6, 11.5.8 and 11.5.10.

11.5.6 Thiamine pyrophosphate (TPP)

Thiamine pyrophosphate (also called thiamine diphosphate) is derived from **vitamin B_1 (thiamine)** and has the structure:

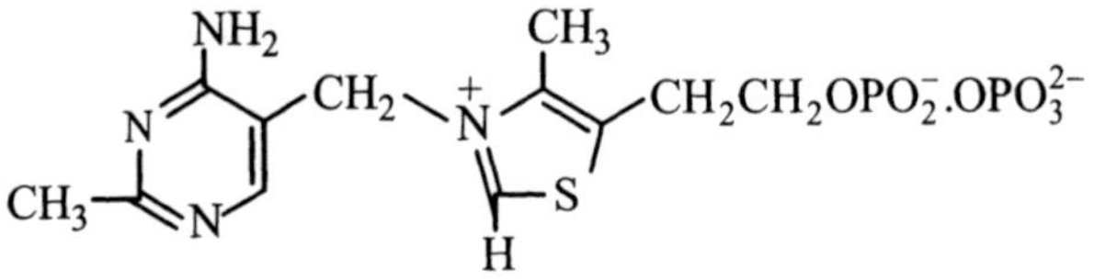

$CH_2CH_2OPO_2^-.OPO_3^{2-}$

The **thiazole ring** can lose a proton to produce a negatively-charged carbon atom:

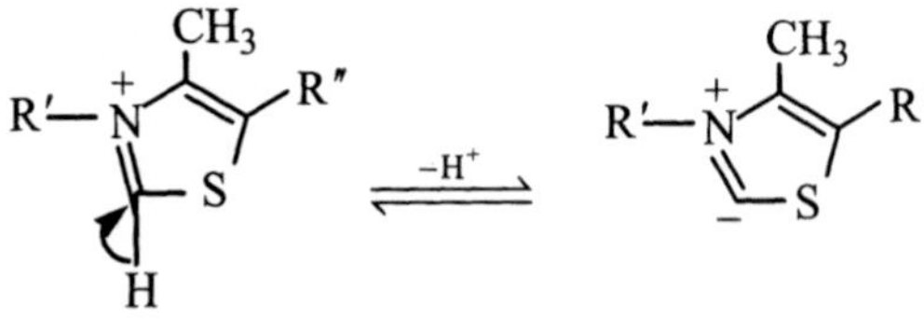

$-H^+$

(11.37)

This is a potent nucleophile and can participate in covalent catalysis, particularly with α-keto (oxo) acid decarboxylase, α-keto acid oxidase, transketolase and phosphoketolase enzymes.

For example, **pyruvate decarboxylase,** found in yeast and some other micro-organisms, utilizes TPP to catalyse the production of acetaldehyde from pyruvate. Ronald Breslow (1957) proposed the following reaction mechanism:

(11.38)

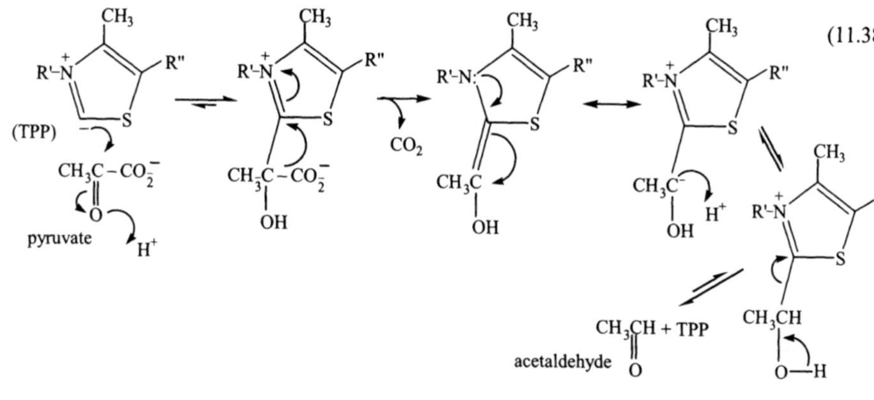

The actual decarboxylation step is facilitated by electrophilic catalysis as the thiazole ring withdraws electrons. The reaction will proceed in the absence of enzyme, but the acetaldehyde formed tends to react with the TPP–$C^{-}(CH_3)OH$ complex to produce acetoin as the final product. It is likely that the enzyme stabilizes the TPP–acetaldehyde complex and prevents this condensation from occurring.

The **pyruvate dehydrogenase multienzyme complex** (see section 5.2.5) also catalyses the decarboxylation of pyruvate, but it utilizes a second coenzyme, **lipoic acid,** to introduce an oxidation step and a third coenzyme, **coenzyme A (CoA.SH)**, to react with the acetyl-lipoamide complex, giving acetyl-CoA as the final product. Initially, the TPP–$C^{-}(CH_3)OH$ complex is formed as above, and then the reaction is thought to proceed as follows:

(11.39)

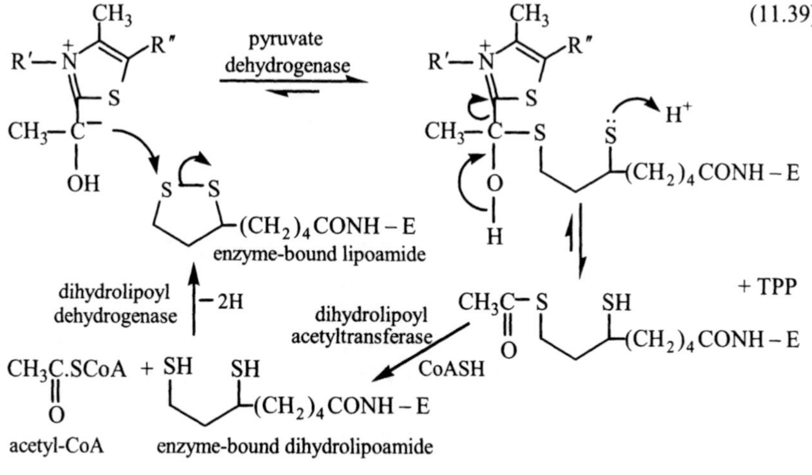

Dihydrolipoyl dehydrogenase has a very similar active site to glutathione reductase (section 11.5.3). Both are flavoproteins, and the two enzymes operate in much the same way, although normally in opposite directions. Whereas glutathione reductase utilizes NADPH to cleave a disulphide bond, dihydrolipoyl dehydrogenase uses NAD^+ to create one.

11.5.7 Pyridoxal phosphate

Pyridoxal 5′-phosphate is derived from pyridoxine or pyridoxamine, **vitamin B_6**. It has the structure:

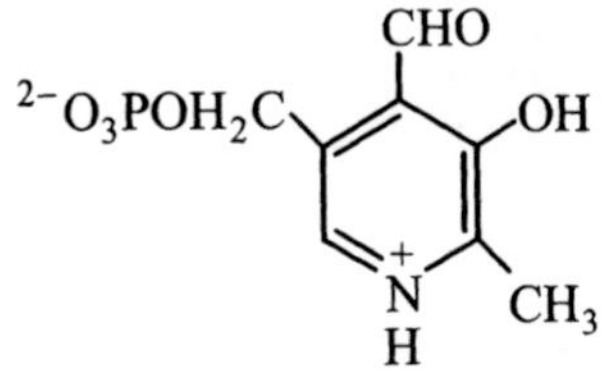

CHO
$^{2-}O_3POH_2C$ OH
$\overset{+}{N}$ CH_3
H

The coenzyme is important in amino acid metabolism, being involved in aminotransferase (transaminase), decarboxylase and racemase reactions. There is much evidence to suggest that in each case a **Schiff base** linkage (–N=CH–) is formed, involving the aldehyde group of the coenzyme (as suggested by Alexander Braunstein and Esmond Snell, 1953). The phenolic group of the coenzyme is also important, as it may help to stabilize the Schiff base intermediate:

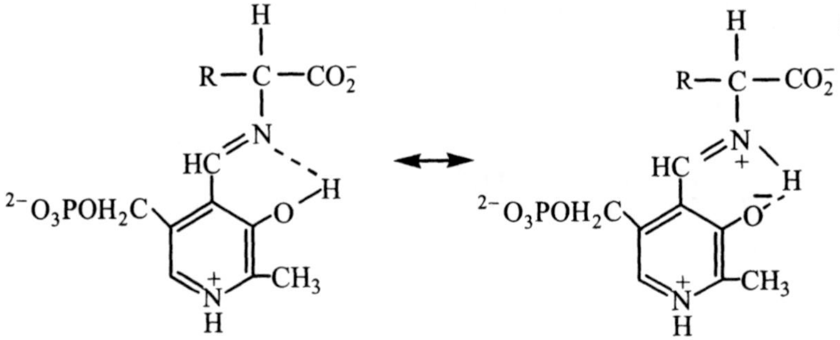

(11.40)

H
R—C—CO_2^-
HC=N
$^{2-}O_3POH_2C$ O H
CH_3
$\overset{+}{N}$
H

⟷

H
R—C—CO_2^-
HC=$\overset{+}{N}$ H
$^{2-}O_3POH_2C$ O^-
CH_3
$\overset{+}{N}$
H

A conjugated double-bond system links the pyridine ring with the substrate, and the positively-charged nitrogen atoms tend to withdraw electrons from the α-carbon of the amino acid, weakening its bonding to R, H and CO_2^-. Any of these three bonds may thus be cleaved as a result of this electrophilic catalysis to form an anion which is stabilized by the conjugated system.

The simplest mechanism to consider is that of the **amino acid racemase** enzymes, where a racemic mixture of D- and L-stereoisomers is produced from a single stereoisomeric form. The coenzyme binds to the enzyme by a Schiff base linkage to a side chain amino group of a lysine residue, and also by an electrostatic linkage between the coenzyme-phosphate and a positively-charged group on the enzyme. When the amino acid substrate binds to the enzyme, the carbon of the Schiff base undergoes electrophilic attack by the α-amino group of the substrate and a new Schiff base linkage is formed, this time between co enzyme and substrate. The weak C–H bond in the complex is then broken to release a proton:

(11.41)

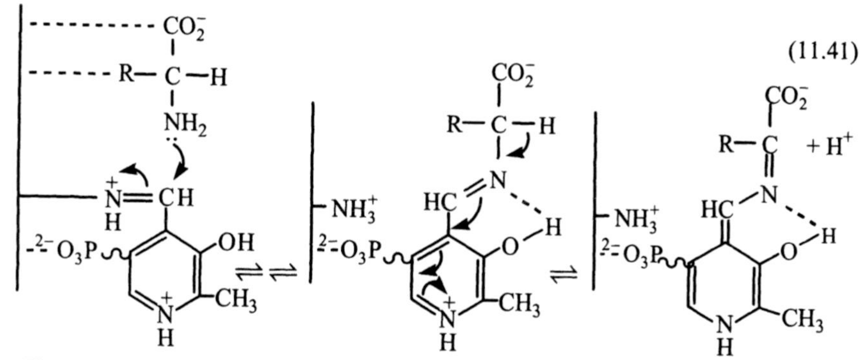

The whole process is reversible, so a proton could add on to the complex and re-form the free amino acid. However, the initial loss of the proton also results in the loss of asymmetry of the α-carbon atom, so the amino acid produced is not necessarily the same stereoisomer as that present at the start. Hence the overall outcome is racemization.

Aminotransferase (transaminase) reactions proceed as above, with loss of a proton from the Schiff base. However, the resulting complex is then attacked in a different place by a proton and undergoes hydrolysis to form pyridoxamine phosphate, with the liberation of an oxo acid:

(11.42)

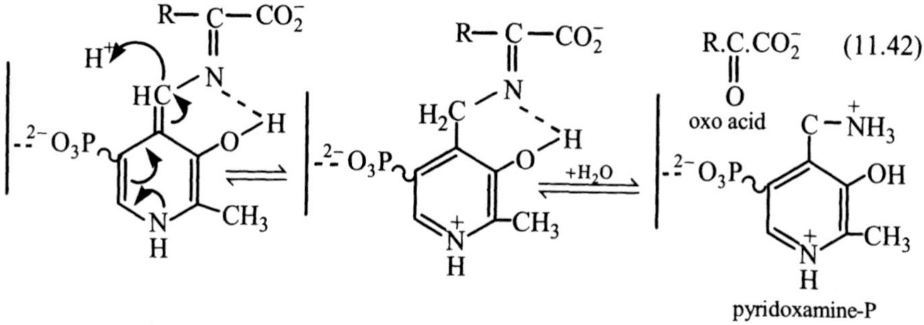

oxo acid

pyridoxamine-P

The reverse sequence may then take place, initiated by the attack on pyridoxamine phosphate by a different oxo acid, and so producing a different amino acid from that initially present. A mechanism of this type was established for **aspartate transaminase** by Terry Jenkins and co-workers (1966). The two-stage reaction sequence is:

L-glutamate + E-pyridoxal-P ⇌ 2-oxoglutarate + E-pyridoxamine-P (11.43)

oxaloacetate + E-pyridoxamine-P ⇌ L-aspartate + E-pyridoxal-P (11.44)

Kinetic studies have indicated a ping-pong bi-bi mechanism, which is consistent with the above chemical mechanism.

Decarboxylase reactions involving amino acid substrates commence in much the same way, but the Schiff base complex loses the α-carboxyl group rather than the hydrogen atom; the resulting complex is then hydrolysed to reform the coenzyme and give an amine (RCH_2NH_2) as product.

11.5.8 Biotin

Biotin, like lipoic acid, is always found firmly bound to a side chain amino group of one of the lysine residues of a protein. Protein-bound biotin can act as a carrier of CO_2, which is important in carboxylation reactions, e.g. that catalysed by acetyl-CoA carboxylase.

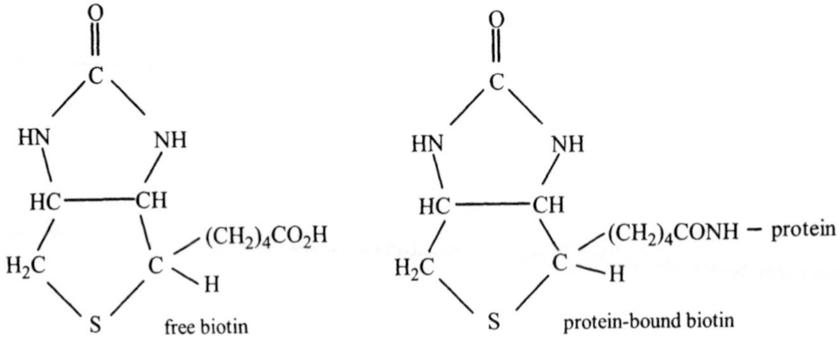

free biotin

protein-bound biotin

Acetyl-CoA carboxylase from *E. coli* was shown, by Daniel Lane and co-workers (1971) and Roy Vagelos and colleagues (1973), to dissociate into three distinct sub-units: one is a biotin carboxyl carrier protein (BCCP); another is biotin carboxylase, which catalyses the reaction:

$$\text{ATP} + \text{HCO}_3^- + \text{BCCP} \rightleftharpoons \text{BCCP}-\text{CO}_2^- + \text{ADP} + \text{P}_\text{i} \qquad (11.45)$$

and the third is carboxyltransferase, which mediates the reaction:

$$\text{BCCP}-\text{CO}_2^- + \text{acetyl-CoA} \rightleftharpoons \text{malonyl-CoA} + \text{BCCP} \qquad (11.46)$$

The BCCP appears to act as a flexible arm, transporting the CO_2^- from the active site of biotin carboxylase to that of carboxyltransferase, where it is presented to the acetyl-CoA. The chemical mechanism is likely to involve:

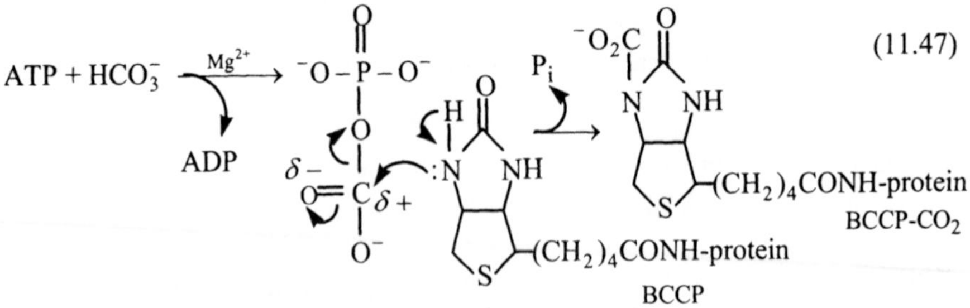

(11.47)

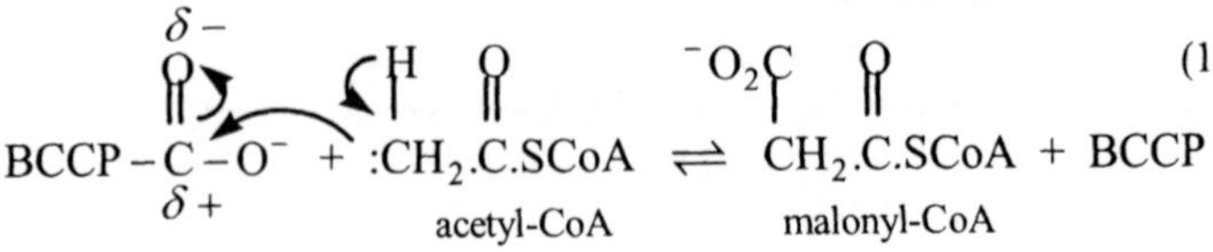

$\text{ATP} + \text{HCO}_3^- \xrightarrow{\text{Mg}^{2+}}$ ADP; $^-$O–P–O$^-$; P_i; (CH$_2$)$_4$CONH-protein BCCP; (CH$_2$)$_4$CONH-protein BCCP-CO$_2$

(11.48)

$$\text{BCCP}-\overset{\delta-}{\text{C}}(=\text{O})-\text{O}^- + :\text{CH}_2.\text{C}(=\text{O}).\text{SCoA} \rightleftharpoons {}^-\text{O}_2\text{C}-\text{CH}_2.\text{C}(=\text{O}).\text{SCoA} + \text{BCCP}$$

acetyl-CoA malonyl-CoA

11.5.9 Tetrahydrofolate

Tetrahydrofolate (THF, H_4F or FH_4) is derived from the B vitamin, folic acid, and has the structure:

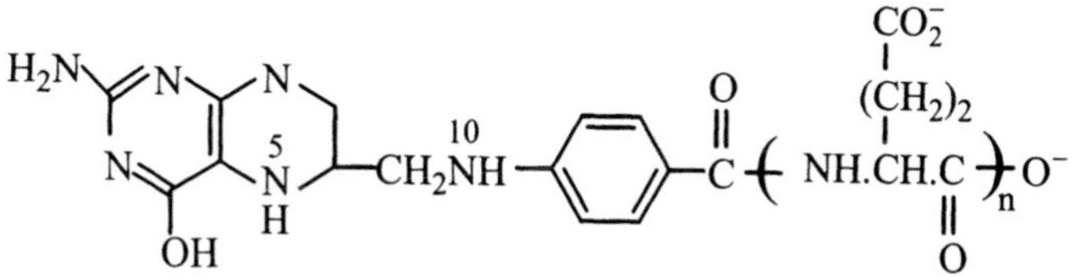

(n = 1 – 7)

Atoms N5 and N10 can carry **one-carbon units** for transfer to a suitable acceptor. **Formate-tetrahydrofolate ligase** (previously called formyltetrahydrofolate synthetase) catalyses the addition of formate to THF by the reaction:

$$\text{ATP} + \text{formate} + \text{THF} \xrightleftharpoons{Mg^{2+}} \text{10-formyl-THF} + \text{ADP} + P_i \quad (11.49)$$

The mechanism for this is likely to resemble that for the carboxylation of biotin (section 11.5.8), except that it involves addition of a formyl group and not a carboxyl group. The 10-formyl-THF synthesized can then react further to produce a variety of one-carbon groups carried by the coenzyme:

(11.50)

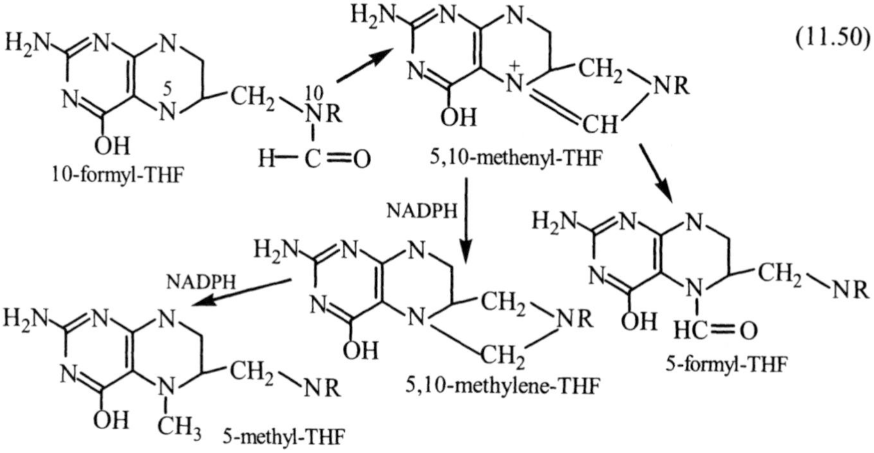

10-formyl-THF → 5,10-methenyl-THF → (NADPH) 5,10-methylene-THF → (NADPH) 5-methyl-THF; 5,10-methenyl-THF → 5-formyl-THF

These, and other, one-carbon units may then be transferred to an acceptor. Consider, for example, the reaction catalysed by **glycine hydroxymethyltransferase**:

$$\underset{\text{glycine}}{H_2N.CH_2CO_2H} + \text{5,10-methylene-THF} \xrightleftharpoons{+H_2O} \underset{\text{L-serine}}{H_2NCH(CH_2OH).CO_2H} + \text{THF} \quad (11.51)$$

Pyridoxal phosphate is also involved and, as is usual with reactions involving this coenzyme (see section 11.5.7), the substrate (glycine) binds by a Schiff base linkage and then goes on to lose a proton from the α-carbon atom. The methylene-THF complex then presents the one-carbon unit in a suitable orientation and the reaction is completed.

11.5.10 Coenzyme B_{12}

Dorothy Hodgkin (1956), in a study that was subsequently rewarded with a Nobel Prize, showed that **cobalamin (vitamin B_{12})** has a central core of a cobalt ion surrounded by a **corrin ring** (Fig. 11.4): the metal ion is linked to the four nitrogen atoms of the ring by one covalent and three co-ordinate bonds, but, because of resonance, the four bonds are almost equivalent. One of the pyrrole components of the corrin ring is joined to dimethylbenzimidazole, which co-ordinates Co from 'below' the ring; the sixth co-ordination site is believed to be occupied *in vivo* by OH^-. In this naturally occurring compound, **hydroxycobalamin (B_{12a})**, the metal ion has an oxidation state of +3, but possesses only a single positive charge because of its interactions with various groups; this is indicated as Co (III)$^+$.

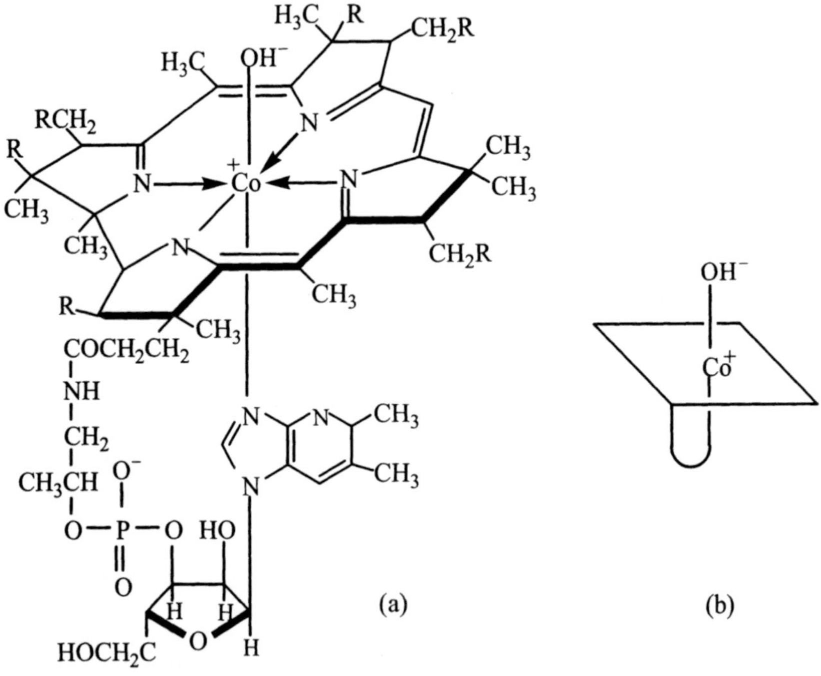

(a) (b)

Fig. 11.4 – Vitamin B_{12} (hydroxycobalamin): (a) complete structure ($-R = -CH_2CONH_2$); (b) simplified representation

The coenzyme forms are obtained after activation of the vitamin by NADH-linked reducing systems:

$$\underset{B_{12a}}{Co(III)^+} \xrightarrow[-OH]{e} \underset{B_{12r}}{\dot{C}o(II)} \xrightarrow{e} \underset{B_{12s}}{\ddot{C}o(I)} \qquad (11.52)$$

The Co(I) in B_{12s} is strongly nucleophilic and can be alkylated, a bond of largely covalent character being formed. Reaction with ATP, mediated by an adenosyltransferase, results in the formation of **5-deoxyadenosylcoenzyme B_{12}** (Fig. 11.5a), also called **5′-deoxyadenosylcobalamin.** Alternatively, methylation may take place, with the formation of **methylcobalamin** (Fig. 11.5b).

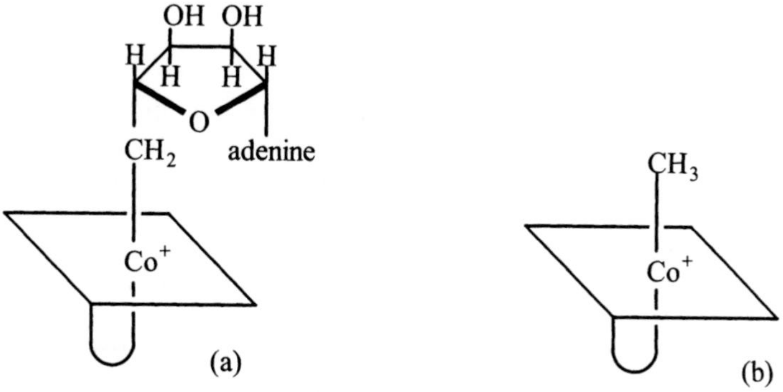

Fig. 11.5 – Simplified representation of the B_{12} coenzymes: (a) 5′-deoxyadenosylcobalamin; (b) methylcobalamin (for detailed structure see Fig. 11.4).

5′-Deoxyadenosylcobalamin acts as a coenzyme for a variety of transfer reactions, usually those of an intramolecular kind, in animals and bacteria. An example is the reaction catalysed by **methylmalonyl-CoA mutase**:

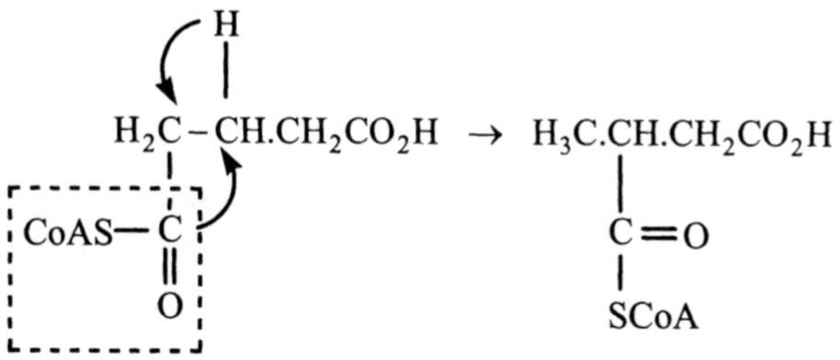

$$H_2C(C(=O)SCoA)-CH(H).CH_2CO_2H \rightarrow H_3C.CH(C(=O)SCoA).CH_2CO_2H \qquad (11.53)$$

There is evidence, as with all reactions of this type, that the hydrogen atom is transferred via the coenzyme.

Methylcobalamin acts as a coenzyme in several methyl-transfer reactions, e.g. in the conversion of homocysteine to methionine in some bacteria and mammals:

$$\text{N5-methyl-THF} + \text{HS.CH}_2\text{CH}_2.\underset{\text{homocysteine}}{\overset{^+\text{NH}_3}{\overset{|}{\text{C}}}\text{HCO}_2^-} \rightarrow \text{THF} + \text{H}_3\text{C.S.CH}_2\text{CH}_2\underset{\text{methionine}}{\overset{^+\text{NH}_3}{\overset{|}{\text{C}}}\text{HCO}_2^-} \quad (11.54)$$

The methyl group is transferred via the cobalt ion of B_{12}. In other bacteria, as well as in fungi and higher plants, this reaction does not involve B_{12}.

The detailed mechanisms of action of 5′-deoxyadenosyl- and methyl-cobalamin are not known for certain. They may well involve further changes in the oxidation state of cobalt.

SUMMARY OF CHAPTER 11

Enzyme-catalysis can include, within a single reaction mechanism, acid-base, electrostatic and covalent catalysis as well as proximity effects, orbital steering and stress/strain factors. In this way the total free energy of activation required for the reaction to proceed is reduced and also spread out over several stages. The free energy requirement for each stage may be provided in some way from the energy made available by collisions between the enzyme-substrate complex and solute molecules.

Some enzymes can catalyse reactions without requiring cofactors. Other enzymes require the presence of metal ions for full catalytic activity: these ions can play a structural role or act as catalysts in a variety of ways. Coenzymes are also required by many enzymes, in some cases in addition to metal ions. Coenzymes usually play a catalytic role.

FURTHER READING

As for Chapter 10, plus:

Allen, K. N. and Mariano, D. D. (2004), Phosphoryl group transfer: evolution of a catalytic scaffold, *Trends in Biochemical Sciences*, **29**, 495-503.

Ballou, D. (ed.) (1999), *Metalloproteins: Essays in Biochemistry*, **34**, Portland Press.

Banerjee, R. and Ragsdale, S. W. (2003), The many faces of vitamin B_{12} – catalysis by cobalamin-dependent enzymes, *Annual review of Biochemistry*, **72**, 209-247.

Bugg, T. (2004), *Introduction to Enzyme and Coenzyme Chemistry*, Blackwell.

Christianson, D. W. and Cox, J. D. (1999), Catalysis by metal-activated hydroxide in zinc and manganese metalloenzymes, *Annual Review of Biochemistry,* **68**, 33-57.

Dodson, G. and Wlodawer, A. (1998), Catalytic triads and their relatives, *Trends in Biochemical Sciences,* **23**, 347-352.

Eliot, A. C. and Kirsch, J. F. (2004), Pyridoxal phosphate enzymes, *Annual Review of Biochemistry*, **73**, 383-415.

Fraajie, M. W. and Mattevi, A. (2000), Flavoenzymes - diverse catalysts with recurrent features, *Trends in Biochemical Sciences,* **25**, 126-132.

Fridovich, I. (1995), Superoxide radical and superoxide dismutases, *Annual Review of Biochemistry,* **64**, 97-112.

Gerlt, J. A. and Babbitt, P. C. (2001), Divergent evolution of enzymatic function, *Annual Review of Biochemistry*, **70**, 209-246.

Gutteridge, A. and Thornton, J. M. (2005), Understanding nature's catalytic toolkit, *Trends in Biochemical Sciences*, **30**, 622-629.

Hammes-Schiffer, S. and Benkovic, S. J. (2006), Relating protein motion to catalysis, *Annual Review of Biochemistry*, **75**, 519-541.

Jencks, W. P. (1997), From chemistry to biochemistry to catalysis to movement, *Annual Review of Biochemistry,* **66**, 1-18.

Keeler, J. and Wothers, P. (2003), *Why Chemical Reactions Happen*, Oxford University Press.

Kraut, D. A., Carroll, K. S. and Herschlag, D. (2003), Challenges in enzyme mechanisms and energetics, *Annual Review of Biochemistry*, **72**, 517-571.

Loferer, H. and Hennecke, H. (1994), Protein disulphide oxidoreductases in bacteria, *Trends in Biochemical Sciences,* **19**, 169-172.

Mattevi, A. (2006), To be or not to be an oxidase: challenging the oxygen reactivity of flavoenzymes, *Trends in Biochemical Sciences*, **31**, 276-283.

McMurry, J. (2007), *Organic Chemistry: A Biological Approach*, Brooks/Cole (Chapter 5)

Methods in Enzymology, **279-282** (1997): Vitamins and coenzymes, Academic Press.

Miller, B. G. and Wolfenden, R. (2002), Catalytic proficiency: the unusual case of OMP decarboxylase, *Annual Review of Biochemistry*, **71**, 847-885.

Paetzel, M. and Dalbey, R E. (1997), Catalytic hydroxylamine dyads within serine proteases, *Trends in Biochemical Sciences,* **22**, 28-31.

Poorafshar, M (2000), *Chymotrypsin/Trypsin-Related Serine Proteases: A Structural, Functional and Evolutionary Analysis*, Uppsala Universitet.

Price, N. C. and Stevens, L. (1999), *Fundamentals of Enzymology,* 3rd edn., Oxford University Press (Chapters 5, 7).

Rees, D. C. (2002), Great metalloclusters in enzymology, *Annual Review of Biochemistry*, **71**, 221-246.

Sutton, R., Rockett, B. and Swindells, O. (2000), *Chemistry for the Life Sciences,* Taylor and Francis.

Todd, A. E., Orengo, C. A. and Thornton, J. M. (2002), Plasticity of enzyme active sites, *Trends in Biochemical Sciences*, **27**, 419-425.

Vocadlo, D. J., Davies, G. J. *et al* (2001), Catalysis by hen egg-white lysozyme proceeds via a covalent intermediate, *Nature*, **412**, 835-838.

12

The Binding of Ligands to Proteins

12.1 INTRODUCTION

In this chapter, we will discuss the binding of **ligands** to monomeric and oligomeric proteins. Anything which binds to an enzyme or other protein is a ligand, regardless of whether or not it is a substrate and undergoes a subsequent reaction. Here, in general, we will be considering binding processes where no subsequent reaction is taking place, e.g. the binding to a protein of a non-substrate, or of a substrate for a two-substrate reaction in the absence of the second substrate. However, we will briefly consider what effects the binding characteristics might have on the kinetics of any subsequent reaction. We will also take into consideration the possibility of interaction between binding sites, particularly in the case of oligomeric proteins where there are several identical binding sites for the same ligand (i.e. one on each identical sub-unit).

12.2 THE BINDING OF A LIGAND TO A PROTEIN HAVING A SINGLE LIGAND-BINDING SITE

Consider the binding of a ligand (S) to a protein (E), in the simplest possible system: $E + S \rightleftharpoons ES$.

The **binding constant K_b** is defined by the relationship:

$$K_b = [ES]/([E][S]) \qquad \text{(note that } K_b = 1/K_s\text{)} \tag{12.1}$$

The **fractional saturation (Y)** of the protein is given by:

$$Y = \frac{[ES]}{[E_0]} = \frac{[ES]}{[E]+[ES]} = \frac{K_b[E][S]}{[E]+K_b[E][S]} = \frac{K_b[S]}{1+K_b[S]} \tag{12.2}$$

From this, it can be seen that a plot of Y against [S] at constant $[E_0]$ will be hyperbolic (Fig. 12.1).

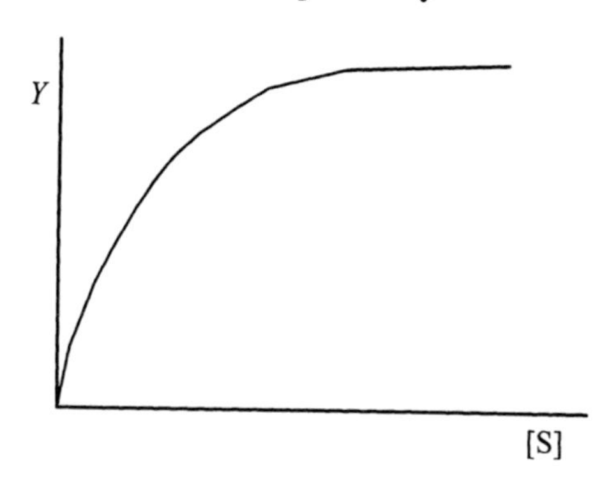

Fig. 12.1 – Graph of fractional saturation (Y) against ligand concentration ([S], at fixed concentration of a protein having a single binding-site for S.

Let us now consider the situation where the binding of S to E is the first step in a process whereby a product P is formed. If the reaction proceeds under steady-state conditions, where $[S_0] \gg [E_0]$ and $[S] \approx [S_0]$, then [ES] does not vary with time and, in the most straightforward system, v_0 is proportional to [ES]. Under these conditions,

$$\frac{v_0}{V_{max}} = \frac{[ES]}{[E_0]} = Y \qquad (12.3)$$

so a graph of v_0 against $[S_0]$ will be the same shape as that of Y against [S], i.e. hyperbolic. This hyperbolic relationship between v_0 and $[S_0]$ under steady-state conditions is, of course, predicted by the Michaelis-Menten equation (see sections 7.1.1 and 7.1.2).

If, on the other hand, the reaction proceeds in a way which is *not* consistent with all of the assumptions made in the derivation of the Michaelis-Menten equation, then the kinetic characteristics of the reaction will not usually run parallel to the binding characteristics.

12.3 COOPERATIVITY

If more than one ligand-binding site is present on a protein, there is a possibility of interaction between the binding sites during the binding process. This is termed **cooperativity**.

Positive cooperativity is said to occur when the binding of one molecule of a substrate of ligand *increases* the affinity of the protein for other molecules of the same or different substrate or ligand

Negative cooperativity occurs when the binding of one molecule of a substrate of ligand *decreases* the affinity of the protein for other molecules of the same or different substrate or ligand.

Homotropic cooperativity occurs when the binding of one molecule of a substrate or ligand affects the binding to the protein of subsequent molecules of the *same* substrate or ligand (i.e. the binding of one molecule of A affects the binding of further molecules of A).

Heterotropic cooperativity occurs when the binding of one molecule of a substrate or ligand affects the binding to the protein of molecules of a *different* substrate or ligand (i.e. the binding of one molecule of A affects the binding of B).

Cooperative effects may be positive and homotropic, positive and heterotropic, negative and homotropic, or negative and heterotropic. **Allosteric inhibition** (section 8.2.7) is an example of negative heterotropic cooperativity and **allosteric activation** an example of positive heterotropic cooperativity.

12.4 POSITIVE HOMOTROPIC COOPERATIVITY AND THE HILL EQUATION

Let us consider the simplest case of positive homotropic cooperativity in a dimeric protein. There are two identical ligand-binding sites, and when the ligand binds to one, it increases the affinity of the protein for the ligand at the other site, so the reaction sequence is:

$$M_2 + S \xrightarrow{\text{slow}} M_2S$$

$$M_2S + S \xrightarrow{\text{rapid}} M_2S_2$$

(where M is the monomeric sub-unit, termed a **protomer,** and M_2 is the dimeric protein).

If the increase in affinity is sufficiently large, M_2S will react with S almost immediately it is formed. Under these conditions, $[M_2S_2] \gg [M_2S]$ and

$$Y = \frac{[M_2S_2]}{[(M_2)_0]} \qquad (12.4)$$

where $(M_2)_0$ is the total concentration of dimer present. Also, a graph of Y against [S] will be sigmoidal (S-shaped) rather than hyperbolic (Fig. 12.2).

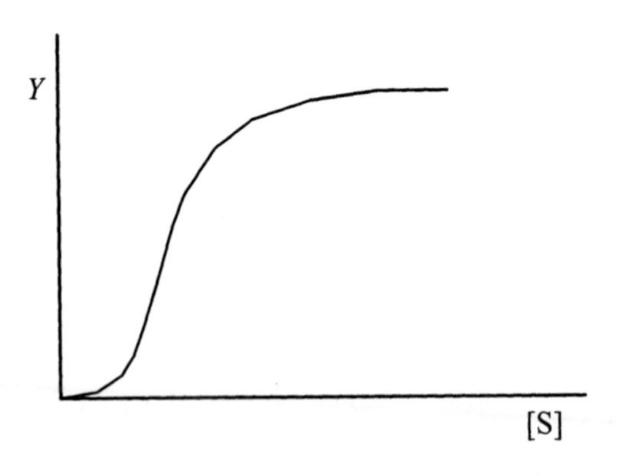

Fig. 12.2 – Graph of Y against [S], at fixed protein concentration, where the binding shows positive homotropic cooperativity.

For **complete cooperativity**, where each protein molecule must be either free of ligand or completely saturated, the reaction may be written

$$M_2 + 2S \rightleftharpoons M_2S_2$$

The binding constant of this reaction is given by the expression

$$K_b = \frac{[M_2S_2]}{[M_2][S]^2} \tag{12.5}$$

from which

$$Y = \frac{K_b[S]^2}{1+K_b[S]^2} \tag{12.6}$$

Alternatively, taking logs,

$$\log K_b + 2\log[S] = \log\left(\frac{[M_2S_2]}{[M_2]}\right) = \log\left(\frac{[M_2S_2]}{[(M_2)_0]-[M_2S_2]}\right) \tag{12.7}$$

In the general case of complete positive homotropic cooperativity of a protein with n identical binding sites, this becomes

$$\log K_b + n\log[S] = \log\left(\frac{[M_nS_n]}{[(M_n)_0]-[M_nS_n]}\right) = \log\left(\frac{Y}{1-Y}\right) \tag{12.8}$$

This is called the **Hill equation**, after its deriver, Archibald Hill. If it is obeyed, a graph of log $(Y/(1-Y))$ against log[S] will be linear with slope = n and intercept = log K_b. Such a graph is called a **Hill plot**, and its experimentally-determined slope is known as the **Hill coefficient** and given the general symbol h (Fig. 12.3).

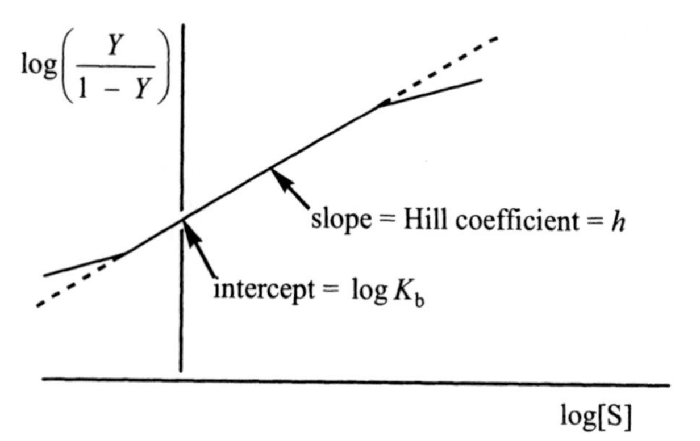

Fig. 12.3 – The Hill plot of log $(Y/(1-Y))$ against log [S], at fixed protein concentration, where the binding shows positive homotropic cooperativity.

At values of Y below 0.1 and above 0.9, the slopes of Hill plots tend to a value of 1, indicating an absence of cooperativity. This is because at very low ligand concentrations there is not enough ligand present to fill more than one site on most protein molecules, regardless of affinity; similarly, at high ligand concentrations, there are extremely few protein molecules present with more than one binding site remaining to be filled.

The Hill coefficient is therefore taken to be the slope of the linear, central portion of the graph, where the cooperative effect is expressed to its greatest extent (Fig. 12.3). For systems where cooperativity is complete, the Hill coefficient (h) is equal to the number of binding sites (n). Proteins which exhibit only a partial degree of positive cooperativity may still give a Hill plot with a linear central section, but in such cases h will be less than n, and the linear section is likely to be shorter than that for a system where cooperativity is more nearly complete.

In the case where S is a substrate and the reaction proceeds to yield products in such a way that the Michaelis-Menten equilibrium assumption is valid, then initial velocity is proportional to the concentration of enzyme-bound substrate, i.e. $v_0 \propto$ [MS], and

$$\frac{v_0}{V_{max}} = \frac{[\mathrm{MS}]}{[\mathrm{M}_0]} = Y \qquad (12.9)$$

(where [MS] is the number of substrate-bound sub-units present per unit volume, and $[\mathrm{M}_0]$ the total number of sub-units per unit volume, i.e. $[\mathrm{M}_0] = [\mathrm{M}] + [\mathrm{MS}]$.)

Under these conditions,

$$\frac{Y}{1-Y} = \frac{v_0}{V_{max} - v_0} \qquad (12.10)$$

So, a Hill plot of log $((v_0/(V_{max} - v_0))$ against $\log[\mathrm{S}_0]$ may be substituted for the one shown in Fig. 12.3. The slopes of the two graphs will have the same value and meaning. Note that, although the relationship $Y = v_0/V_{max}$ may be assumed valid for systems involving monomeric enzymes under general steady-state conditions, the same is not true for the more complicated systems involving oligomeric enzymes. In the latter case, $Y = v_0/V_{max}$ only if the binding process is at or very near equilibrium.

One of the main problems in constructing a Hill plot from kinetic data is to obtain an accurate estimate of V_{max}. This is particularly true for cooperative systems, since the primary plots (sections 7.1.4 and 7.1.5) are not linear. Nevertheless, an estimate of V_{max} can be obtained from an Eadie-Hofstee or other plot, enabling a Hill plot to be constructed and a Hill coefficient (h) determined. The primary plot can then be redrawn, substituting $[\mathrm{S}]^h$ for [S], which should give more linear results and a more accurate estimate of V_{max}. If this differs markedly from the initial estimate of V_{max}, the Hill plot should then be redrawn, incorporating the new (and better) estimate of V_{max}.

12.5 THE ADAIR EQUATION FOR THE BINDING OF A LIGAND TO A PROTEIN HAVING TWO BINDING SITES FOR THAT LIGAND

12.5.1 General considerations

Let us now investigate the binding of a ligand to a protein having a number of identical binding sites for that ligand, making no assumptions at all about cooperativity. The **intrinsic** (or **microscopic) binding constant (K_b)** for each site is defined as the binding constant which would be measured if all the other sites on the protein were absent. Since all the sites are identical in the example we are considering, each will have the same K_b. However the **actual**, or **apparent, binding constant** for each step of the reaction will not be the same. In the case of a dimeric protein (M_2) having two identical binding sites for a ligand (S), the two steps in the binding process are:

$$M_2 + S \rightleftharpoons M_2S \quad \text{apparent binding constant} = K_{b1}$$
$$M_2S + S \rightleftharpoons M_2S_2 \quad \text{apparent binding constant} = K_{b2}$$

Note that K_{b1} and K_{b2} depend solely on the position in the reaction sequence and do not refer to any particular binding site.

Fractional saturation Y is the number of protomers per unit volume which are bound to ligand divided by the total number of protomers per unit volume.

$$\therefore Y = \frac{[MS]}{[M_0]} = \frac{[MS]}{[MS]+[M]} \tag{12.11}$$

However, there are no isolated protomers present: they are part of the dimeric protein. Hence it is necessary to express Y in terms of the various protein-ligand complexes which are actually present.

The species M_2 consists of two protomers, both unbound;

the species M_2S consists of one bound and one unbound protomer; and

the species M_2S_2 consists of two protomers, both bound.

Therefore, the total concentration of ligand-bound protomers present ([MS]) is given by $[MS] = [M_2S] + 2[M_2S_2]$. Similarly, the total concentration of unbound protomers present ([M]) is given by $[M] = 2[M_2] + [M_2S]$.

Also, $[MS] + [M] = [M_2S] + 2[M_2S_2] + 2[M_2] + [M_2S]$

$$= 2([M_2] + [M_2S] + [M_2S_2])$$

$$\therefore Y = \frac{[MS]}{[MS]+[M]} = \frac{[M_2S]+2[M_2S_2]}{2([M_2]+[M_2S]+[M_2S_2])} \tag{12.12}$$

By definition, $K_{b1} = [M_2S]/([M_2][S])$, so $[M_2S] = K_{b1}\,[M_2][S]$.

Similarly, $K_{b2} = [M_2S_2]/([M_2S][S])$, so $[M_2S_2] = K_{b2}\,[M_2S][S] = K_{b1}K_{b2}\,[M_2][S]^2$.

Substituting for $[M_2S]$ and $[M_2S_2]$ in the expression for Y obtained above (12.12):

$$Y = \frac{K_{b1}[M_2][S] + 2K_{b1}K_{b2}[M_2][S]^2}{2([M_2] + K_{b1}[M_2][S] + K_{b1}K_{b2}[M_2][S]^2)}$$

$$= \frac{K_{b1}[S] + 2K_{b1}K_{b2}[S]^2}{2(1 + K_{b1}[S] + K_{b1}K_{b2}[S]^2)} \qquad (12.13)$$

This is the **Adair equation** (see section 12.9) for the binding of a ligand to a dimeric protein.

12.5.2 Where there is no interaction between the binding sites

Let us now look at the relationship between the intrinsic and apparent constants where there is no interaction between the binding sites. We will compare the reaction for the dimer with that for the hypothetical isolated protomer under identical conditions of molar concentration, assuming that each binding site behaves in an identical manner, regardless of its surroundings.

The **first step** in the reaction involving the dimer is

$$M_2 + S \rightleftharpoons M_2S \text{ (binding constant } K_{b1})$$

whereas the reaction for the protomer is

$$M + S \rightleftharpoons MS \text{ (binding constant } K_b).$$

In diagrammatic form, these reactions can be written:

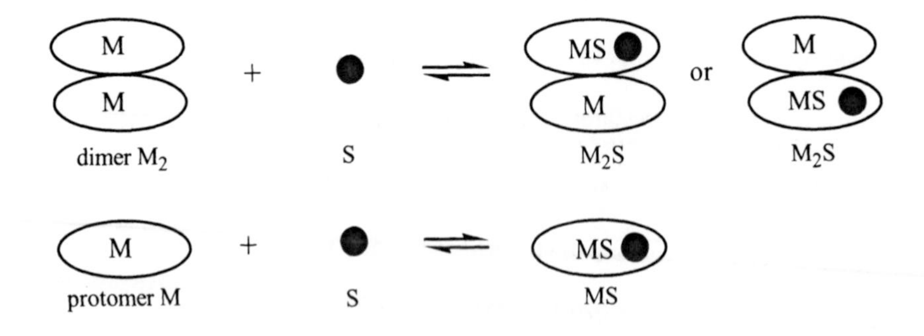

In the forward direction, the dimer has two free binding sites whereas the isolated protomer has only one. Therefore, the ligand is two times more likely to bind to a molecule of the dimer than to a molecule of the isolated protomer. In the reverse direction, in both cases, there is only one site from which S can dissociate, i.e. that to which it is attached. Hence there is no difference between the rates of dissociation of the dimer and the isolated protomer. Taking the forward and back reactions together, we see that $K_{b1} = 2K_b$.

The **second step** in the reaction involving the dimer is:

$$M_2S + S \rightleftharpoons M_2S_2 \text{ (binding constant } K_{b2})$$

In diagrammatic form, this is:

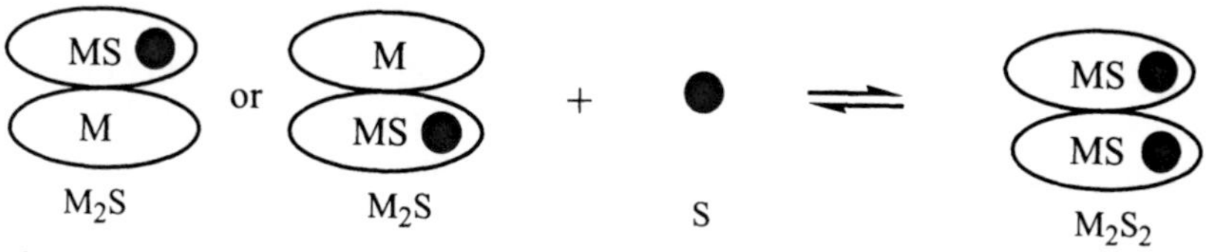

MS / M or M / MS + S ⇌ MS / MS

M_2S M_2S S M_2S_2

while for the isolated protomer we again have

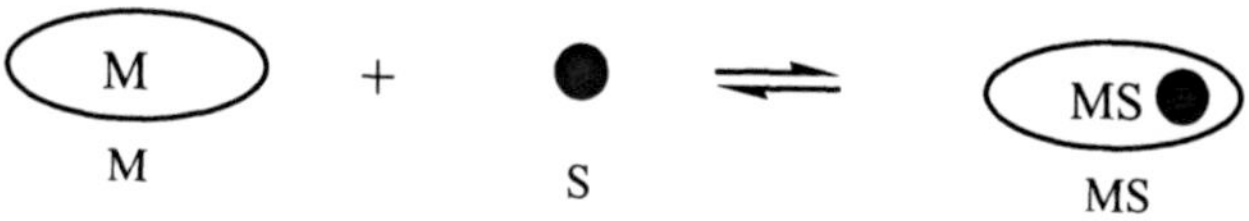

M + S ⇌ MS

M S MS

In the forward direction, both the dimer and the hypothetical isolated protomer have one free binding site and so the ligand is equally likely to bind to either. In the reverse direction, there are two sites in the dimer from which S can dissociate, but only one on the isolated protomer. Hence a molecule of ligand is twice as likely to dissociate from a molecule of dimer M_2S_2 than from a molecule of protomer MS. Therefore, for the overall reaction, $K_{b2} = \frac{1}{2}K_b$.

If we substitute these relationships in the general equation for Y (section 12.5.1),

$$Y = \frac{2K_b[S] + 2.2K_b.\frac{1}{2}K_b[S]^2}{2(1 + 2K_b[S] + K_b^2[S]^2)}$$

$$= \frac{K_b[S] + K_b^2[S]^2}{(1 + 2K_b[S] + K_b^2[S]^2)} = \frac{K_b[S](1 + K_b[S])}{(1 + K_b[S])^2}$$

$$= \frac{K_b[S]}{1 + K_b[S]}$$

This is identical to the expression obtained for a protein with a single ligand-binding site, which gives a hyperbolic plot of Y against [S] (equation 12.2 in section 12.2). In general, for the binding of a ligand (S) to a protein having several identical binding sites for the ligand, a hyperbolic plot of Y against [S] will be obtained provided there is no interaction between the binding sites. If this binding is the first step in a process by which S is converted to a product in such a way that the equilibrium assumption is valid and v_0 is directly proportional to [MS], then a plot of v_0 against $[S_0]$ will also be hyperbolic. This conclusion has already been stated (in section 7.1.3). Here we have seen the justification for that statement.

One further relationship can be obtained for the reaction involving the binding of a ligand to a dimeric protein with no interaction between the binding sites. From the above discussion, $K_{b1} = 2K_b$ and $K_{b2} = \frac{1}{2}K_b$. Hence $K_{b1} = 4K_{b2}$.

12.5.3 Where there is positive homotropic cooperativity

If the binding of the first molecule of the ligand increases the affinity of the protein for the ligand, the second step of the binding process will be faster than it is in the situation where there is no interaction between the binding sites, i.e. where $K_{b1} = 4K_{b2}$.

Hence, for positive homotropic cooperativity, $K_{b1} < 4K_{b2}$. According to the Adair equation, this relationship results in a sigmoidal plot of Y against [S] being obtained (see Fig. 12.4a); the sigmoidal character of the curve is more marked the greater the degree of cooperativity.

When cooperativity is complete:

$$Y = \frac{K_b[S]^2}{1 + K_b[S]^2}$$

where K_b in this case is the binding constant for the overall process $M_2 + S \rightleftharpoons M_2S_2$ (see section 12.4).

12.5.4 Where there is negative homotropic cooperativity

Negative cooperativity results in the second step of the binding process being slower than it would be if there were no interaction between the binding sites. Hence, for negative homotropic cooperativity, $K_{b1} > 4K_{b2}$. In this case, a plot of Y against [S] is neither sigmoidal nor a true rectangular hyperbola (see Fig. 12.4a).

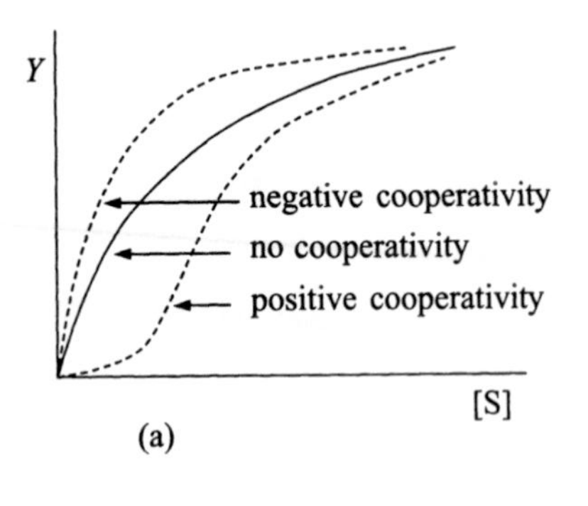

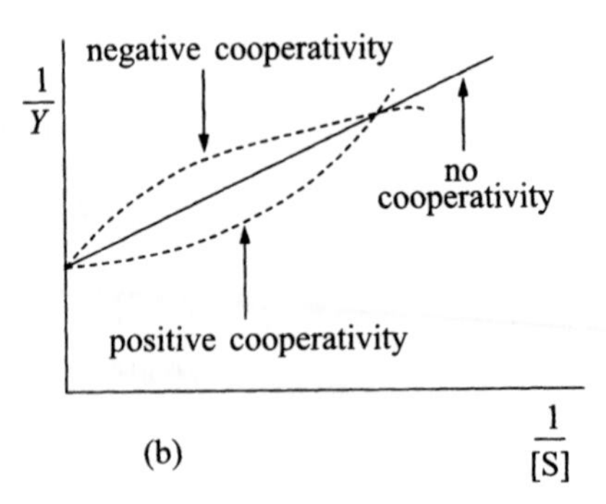

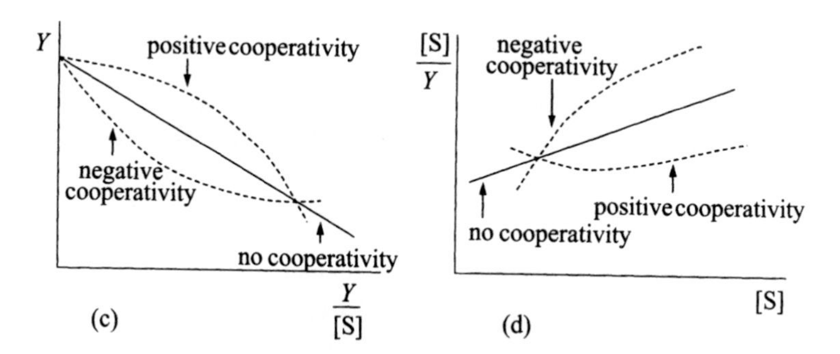

Fig. 12.4 – Plots of: (a) Y against [S]; (b) $1/Y$ against 1/[S]; (c) Y against Y/[S]; and (d) [S]/Y against [S]; all at constant [E_0], showing the effects of positive and negative homotropic cooperativity.

12.6 THE ADAIR EQUATION FOR THE BINDING OF A LIGAND TO A PROTEIN HAVING THREE BINDING SITES FOR THAT LIGAND

For a trimeric protein (M_3) having three identical binding sites for a ligand (S), there are three steps in the binding process:

$$M_3 + S \rightleftharpoons M_3S \quad \text{(apparent binding constant } K_{b1})$$
$$M_3S + S \rightleftharpoons M_3S_2 \quad \text{(apparent binding constant } K_{b2})$$
$$M_3S_2 + S \rightleftharpoons M_3S_3 \quad \text{(apparent binding constant } K_{b3}).$$

Using reasoning exactly as for the dimeric protein in section 12.5,

$$Y = \frac{K_{b1}[S] + 2K_{b1}K_{b2}[S]^2 + 3K_{b1}K_{b2}K_{b3}[S]^3}{3(1 + K_{b1}[S] + K_{b1}K_{b2}[S]^2 + K_{b1}K_{b2}K_{b3}[S]^3)} \qquad (12.14)$$

This is the **Adair equation** for a trimeric protein.

If there is no interaction between the binding sites,

$$K_{b1} = 3K_b\,; \quad K_{b2} = K_b\,; \quad \text{and} \quad K_{b3} = \tfrac{1}{3}K_b$$

Hence $K_{b1} = 3K_{b2}$; $K_{b2} = 3K_{b3}$; and the Adair equation reduces, as before, to:

$$Y = \frac{K_b[S]}{1 + K_b[S]}$$

If there is positive homotropic cooperativity, $K_{b1} < 3K_{b2}$ and $K_{b2} < 3K_{b3}$

and, if cooperativity is complete, the Hill coefficient (h) = 3.

If there is negative homotropic cooperativity, $K_{b1} > 3K_{b2}$ and $K_{b2} > 3K_{b3}$.

12.7 THE ADAIR EQUATION FOR THE BINDING OF A LIGAND TO A PROTEIN HAVING FOUR BINDING SITES FOR THAT LIGAND

A tetrameric protein (M_4) having four identical binding sites for a ligand (S) will have four steps in the binding process, with apparent binding constants K_{b1} , K_{b2} , K_{b3} and K_{b4} .

The Adair equation for a tetrameric protein is found to be:

$$Y = \frac{K_{b1}[S] + 2K_{b1}K_{b2}[S]^2 + 3K_{b1}K_{b2}K_{b3}[S]^3 + 4K_{b1}K_{b2}K_{b3}K_{b4}[S]^4}{4(1 + K_{b1}[S] + K_{b1}K_{b2}[S]^2 + K_{b1}K_{b2}K_{b3}[S]^3 + K_{b1}K_{b2}K_{b3}K_{b4}[S]^4)} \quad (12.15)$$

If there is no interaction between the binding sites,

$$K_{b1} = 4K_b \;;\; K_{b2} = \tfrac{3}{2}K_b \;;\; K_{b3} = \tfrac{2}{3}K_b \;;\text{ and } K_{b4} = \tfrac{1}{4}K_b$$

Under these conditions, $K_{b1} = \tfrac{8}{3}K_{b2}$; $K_{b2} = \tfrac{9}{4}K_{b3}$; $K_{b3} = \tfrac{8}{3}K_{b4}$; and the Adair equation again reduces to

$$Y = \frac{K_b[S]}{1 + K_b[S]}$$

If there is positive homotropic cooperativity,

$$K_{b1} < \tfrac{8}{3}K_{b2} \;;\; K_{b2} < \tfrac{9}{4}K_{b3} \;;\; K_{b3} < \tfrac{8}{3}K_{b4}$$

and if the cooperativity is complete, the Hill coefficient (h) = 4.

If there is negative homotropic cooperativity,

$$K_{b1} > \tfrac{8}{3}K_{b2} \;;\; K_{b2} > \tfrac{9}{4}K_{b3} \;;\text{ and } K_{b3} > \tfrac{8}{3}K_{b4} .$$

12.8 INVESTIGATION OF COOPERATIVE EFFECTS

12.8.1 Measurement of the relationship between *Y* and [S]

If there is some measurable difference between a ligand in its free and protein-bound forms, or between the free protein and the protein~ligand complex, then the relationship between fractional saturation (Y) and the free ligand concentration ([S]) is relatively easy to determine. For example, as mentioned in section 9.4.2, there is a difference in absorbance at 350 nm between free NADH and NADH bound to alcohol dehydrogenase; hence it is possible to investigate the binding of NADH to this enzyme at different NADH concentrations in the absence of all other substrates. Other methods for the investigation of ligand-binding to protein include the observation of changes in the fluorescence or NMR spectra, or the measurement by ion-selective electrodes of the loss of free ligand as binding takes place.

In general, for an oligomeric protein (E , or M_n) having n identical and non-interacting binding-sites for a ligand (S),

$$Y = \frac{K_b[S]}{1 + K_b[S]}$$

This is valid where M_n and S are at or near equilibrium, regardless of whether or not a product is being formed, since, in either case, [S] and [MS] may be assumed constant (section 12.5.2). If possible, it is best to investigate under conditions where equilibrium can be ensured, e.g. to determine the binding characteristics for one substrate of a multi-substrate reaction in the absence of the other substrates. This minimizes the assumptions being made, and excludes possible heterotropic effects.

It will be apparent that the relationship between Y and [S] in the absence of cooperativity is the equation of a rectangular hyperbola, like the Michaelis-Menten equation derived in section 7.1. As with the Michaelis-Menten equation, it is possible to manipulate the binding equation to obtain linear relationships between variables: if the equation is obeyed, linear plots are obtained of $1/Y$ against $1/[S]$, Y against $Y/[S]$ and $[S]/Y$ against [S] (exactly analogous to the Lineweaver-Burk, Eadie-Hofstee and Hanes plots of section 7.1). These are shown in Fig. 12.4.

Where positive homotropic cooperativity occurs, a sigmoidal plot of Y against [S] is obtained; the other plots are non-linear, as shown in Fig. 12.4. In general, it is considered that departures from linearity are more obvious on Eadie-Hofstee and Hanes-type plots than on those of the Lineweaver-Burk type.

Where negative homotropic cooperativity occurs, the plot of Y against [S] is neither sigmoidal nor a rectangular hyperbola, although it could easily be mistaken for the latter. For this reason, it is essential to investigate the other relationships, the plots for negative cooperativity being non-linear and of the opposite curvature to those for positive cooperativity (Fig. 12.4).

12.8.2 Measurement of the relationship between v_0 and $[S_0]$

If S is a substrate, and reacts to form products in such a way that the binding process remains at or near equilibrium, then [MS] is constant, v_0 is proportional to [MS] and $Y = v_0/V_{max}$. Under these conditions, and provided $[S_0] \gg [E_0]$, kinetic data may be used to plot the graphs shown in Fig. 12.4, with v_0 replacing Y and $[S_0]$ replacing [S]. The conclusions would be unchanged.

This gives a more versatile way of investigating cooperative effects, for only a limited number of binding processes can be monitored directly by the use of spectroscopy or ion-selective electrodes. However, more assumptions are involved, and complexities in the kinetic mechanism could give misleading results (see section 13.5).

12.8.3 The Scatchard plot and equilibrium dialysis techniques

For systems where a single ligand (S) binds to an oligomeric protein (E, or M_n) having n identical and non-interacting binding sites for that ligand,

$$Y = \frac{[MS]}{[MS]+[M]} = \frac{K_b[S]}{1 + K_b[S]} \quad \text{(see section 12.5)}$$

$$\therefore [\text{MS}] + [\text{MS}]K_b[\text{S}] = K_b[\text{S}][\text{MS}] + K_b[\text{S}][\text{M}]$$

$$\therefore K_b = \frac{[\text{MS}]}{[\text{M}][\text{S}]} = \frac{[\text{MS}]}{([\text{M}_0]-[\text{MS}])[\text{S}]}$$

$$\therefore \frac{[\text{MS}]}{[\text{S}]} = K_b[\text{M}_0] - K_b[\text{MS}] \tag{12.16}$$

George Scatchard (1949) pointed out that, under these conditions, a graph of [MS]/[S] against [MS] will be linear, with characteristics as shown in Fig. 12.5. Cooperativity will lead to non-linearity. Note that, since $Y = [\text{MS}]/[\text{M}_0]$, this is basically a plot of the Eadie-Hofstee type (Fig. 12.4c), with the axes reversed.

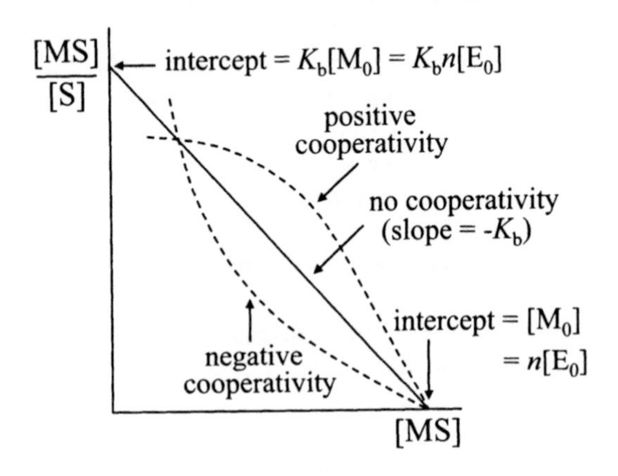

Fig. 12.5 – Scatchard plot of [MS]/[S] against [MS], at fixed $[\text{E}_0]$, showing the effects of positive and negative cooperativity.

This Scatchard plot may be used to determine the presence or type of cooperativity, and also the number of binding sites, from the results of **equilibrium dialysis** studies. A solution of protein of known concentration ($[\text{E}_0] = [(\text{M}_n)_0]$) is dialysed against a solution of ligand of known concentration ($[\text{S}_0]$) and allowed to come to equilibrium. (Note that this limits the use of such investigations to systems where binding is not a prelude to product formation, and to systems where both protein and ligand are stable for several hours.) The ligand will be able to pass freely through the dialysis membrane, but the protein will be trapped within its compartment (e.g. dialysis bag). The concentration of free ligand outside the protein compartment can be easily determined at any time, and at equilibrium it should be equal to the free ligand concentration within the protein compartment (= [S]) (Fig. 12.6). Radioactive-labelled ligands are often used for equilibrium dialysis experiments, since they result in greater sensitivity being obtained. If the volume of liquid within the protein compartment is negligible compared to the total volume of liquid present, then $[\text{S}] = [\text{S}_0] - [\text{MS}]$, from which [MS] may be calculated

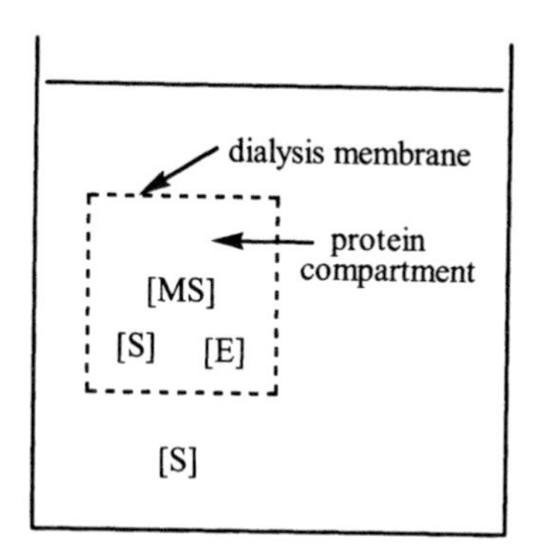

Fig. 12.6 – Diagrammatic representation of an equilibrium dialysis experiment, showing the concentrations present in the two compartments at equilibrium.

Alternatively, and without making this assumption, the total ligand concentration within the protein compartment (= [MS] + [S]) can be determined, and [MS] calculated as the difference between this and the total ligand concentration outside the protein compartment (= [S]). A Scatchard plot can then be drawn.

The binding of NAD^+/NADH to **lactate dehydrogenase** is one of the processes that has been investigated by such techniques, no interaction between the NAD binding-sites on the four sub-units being indicated. For general information, the points of interaction between the lactate dehydrogenase sub-units (revealed by X-ray crystallography), together with other important features on each sub-unit (see section 11.5.2), are shown in Fig. 12.7.

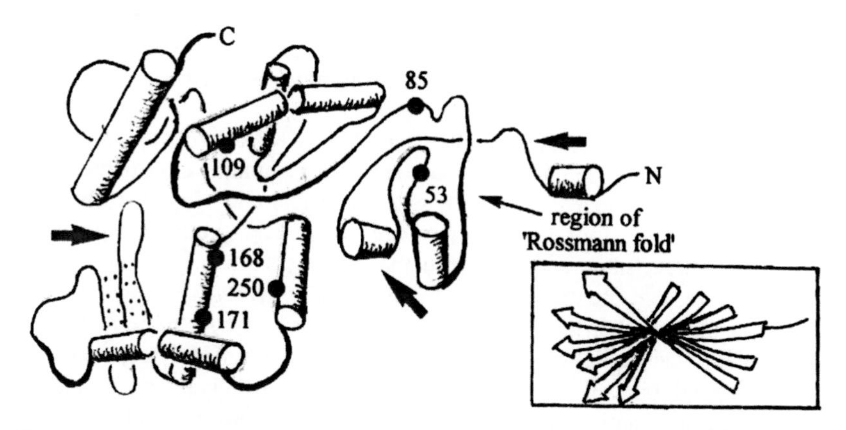

Fig. 12.7 – A simplified representation of the three-dimensional structure of one of the four identical sub-units of dogfish muscle lactate dehydrogenase, as revealed by the X-ray diffraction studies of Adams, Rossmann and colleagues (1972). (Conventions as for Fig. 2.10.) Note that the N-terminal domain, which includes the large NAD binding site (the nicotinamide end of the coenzyme interacting with residue 250 and the adenine end with residues 53 and 85), incorporates an extensive twisted β-pleated sheet structure known as a Rossmann fold (see insert, conventions as for Fig. 2.8). Note also that the loop which comes towards the reader left of residue 85 closes over the nicotinamide ring of NAD and the substrate after binding of the latter (to Arg-171). Areas of contact with the three other sub-units are indicated by thick arrows.

Similar results to those obtained in equilibrium dialysis experiments may be obtained by the use of ultracentrifugation or size-exclusion techniques, both of which involve moving an initially ligand-free protein through a solution of ligand and observing the changes which take place as it binds ligand.

12.9 THE BINDING OF OXYGEN TO HAEMOGLOBIN

The stimulus for much of the work described in Chapter 12 was experimental evidence regarding the binding of oxygen to haemoglobin. In 1904, Christian Bohr (father of the physicist, Niels Bohr) and co-workers showed that if the fractional saturation of haemoglobin with oxygen was plotted against the partial pressure of oxygen gas (equivalent to the concentration), a curve was obtained which was clearly sigmoidal.

Archibald Hill (1909) explained this on the basis of interaction between binding sites causing positive cooperativity. At that time it was known that each haem (iron protoporphyrin) group bound one oxygen molecule, and Hill correctly suggested that each haemoglobin sub-unit contained one haem group, but it was not known how many sub-units made up the oligomeric protein. Hill assumed that cooperativity was complete, so if there were n sub-units in the haemoglobin molecule, the overall reaction was $Hb + nO_2 \rightleftharpoons Hb\,(O_2)_n$. On this basis he derived what became known as the Hill equation (section 12.4) and found the Hill coefficient (h) to be about 2.8.

It was subsequently shown that there were four binding sites to each haemoglobin molecule, so cooperativity was far from complete. Gilbert Adair (1925) then developed the theory of ligand binding to protein which was described in general terms in section 12.5. He saw that oxygen molecules could bind to a haemoglobin molecule in four separate steps, each with a different apparent binding constant, and derived the Adair equation for a tetrameric protein. He also showed what the relationship between the apparent binding constants must be to explain positive cooperativity.

Results from X-ray diffraction studies, reported by Max Perutz and co-workers in 1960, showed that the four binding sites are in very similar environments, so the assumption that they behave identically is a reasonable one. However, these studies also showed that the four haem groups are completely spatially separate in the molecule, so direct interaction between the binding sites is impossible. It seems likely, therefore, that the mechanism of cooperativity involves interactions between sub-units at places other than the binding-sites (as we saw in the previous section, the sub-units of lactate dehydrogenase come into contact at widely-separated points).

With haemoglobin, all four C-terminal amino acid residues, and possibly some others, form electrostatic linkages with groups on other sub-units in the oxygen-free molecule (deoxyhaemoglobin), but not in the fully-oxygenated molecule (oxyhaemoglobin). Conformational changes also take place as the oxygen binds to the haemoglobin molecule, the binding site on each sub-unit being a Fe(II) atom attached to a histidine residue and to the four pyrrole groups of a protoporphyrin ring. In the unbound form, the Fe atom is too large to fit into the hole in the centre of the porphyrin ring, so lies about 0.75 Å out of .the plane of this ring. When oxygen fills the vacant sixth coordination position of the Fe atom it decreases the atomic radius, enabling the metal atom to move into the plane of the porphyrin ring.

This it proceeds to do, pulling the histidine residue after it and so altering the tertiary structure of the sub-unit. The tyrosine adjacent to the C-terminus is forced out of a pocket between two helical regions, where in deoxyhaemoglobin it plays a role in stabilizing the tertiary structure, and with it moves the C-terminal amino acid. As a result, the electrostatic linkages with other sub-units are broken and a less constrained (or more relaxed) conformational state is assumed.

Although it is still not entirely clear how this facilitates oxygen-binding to other sub-units, one relevant factor is that the breaking of some electrostatic interactions between sub-units when the first molecule of oxygen binds means that there are fewer such interactions remaining to be broken when subsequent molecules bind, so these processes are energetically more favourable than the first.

SUMMARY OF CHAPTER 12

If there are several ligand-binding sites on a protein, it is possible that there could be interaction between them: the binding of one ligand might increase or decrease the affinity of another site on the protein for the same or a different ligand. Such interaction between binding sites is called a cooperative effect: positive cooperative effects increase affinity, while negative effects decrease it; homotropic effects concern identical ligands, whereas heterotropic effects concern different ligands.

If the ligand is a substrate and goes on to give a product in such a way that the Michaelis-Menten equilibrium assumption is valid, then initial velocity is proportional to the concentration of enzyme-bound substrate and cooperative effects are reflected in the kinetics of the overall reaction. In the presence of cooperativity, Michaelis-Menten plots will not be rectangular hyperbolae, and other primary plots, e.g. those of Lineweaver-Burk and Eadie-Hofstee, will not be linear.

From initial studies on the binding of oxygen to haemoglobin, Hill derived an equation relating fractional saturation to ligand concentration. This is strictly valid only where positive homotropic cooperativity is total. Adair formulated an equation which is of more general application. It is valid for any oligomeric protein which has several identical binding sites for a particular ligand, since it makes no assumptions about cooperativity.

Cooperative effects can be investigated by the use of spectroscopy (to determine fractional saturation), by equilibrium dialysis experiments in association with the Scatchard plot, or by kinetic studies under steady-state conditions.

FURTHER READING

Bisswanger, H. (2004), *Practical Enzymology*, Wiley–VCH (Chapter 4).

Clarke, A. R., Atkinson, T. and Holbrook, 1. J. (1989), From analysis to synthesis: new ligand binding sites on the lactate dehydrogenase framework, *Trends in Biochemical Sciences,* **14,** 101-105, 145-148.

Kurtz, D. M. (1999), Oxygen-carrying proteins - three solutions to a common problem, *Essays in Biochemistry,* **34**, 85-100, Portland Press.

Nelson, D. L. and Cox, M. M. (2004), *Lehninger* Principles *of Biochemistry,* 4th edn., Worth (Chapter 6).

Voet, D. and Voet, J. G. (2004), *Biochemistry,* 3rd edn., Wiley (Chapter 15)

PROBLEMS

12.1 A single-substrate enzyme-catalysed reaction was investigated at fixed total enzyme concentration and the following results were obtained:

$[S_0]$ (mmol l^{-1}):	1.0	1.67	2.0	2.5	3.33	5.0	10.0
v_0 (μmol min^{-1}):	1.10	1.43	1.54	1.75	2.00	2.56	4.00

Draw Michaelis-Menten, Lineweaver-Burk, Eadie-Hofstee and Hanes plots of these data. Assuming the reaction was proceeding under steady-state conditions in each case, what type of cooperative effect is indicated?

12.2 The following results were obtained during an investigation of the binding of a ligand to a protein at fixed total protein concentration:

[ligand] (mmol l^{-1})	1.0	1.67	2.0	2.5	3.33	5.0	10.0
Fractional saturation:	0.06	0.14	0.19	0.24	0.35	0.53	0.80

What can you conclude about the binding of the ligand? Draw a Hill plot from these data and determine the Hill coefficient.

12.3 An enzyme was dialysed against one of its substrates at a series of different initial substrate concentrations. The system was allowed to come to equilibrium in each case and the total concentration of substrate inside and outside the dialysis bag was measured. The following results were obtained at equilibrium:

Total enzyme concentration (mmol l^{-1})		Total substrate concentration (mmol l^{-1})	
inside dialysis bag	outside bag	inside dialysis bag	outside bag
2.0	0	2.40	0.80
2.0	0	3.33	1.28
2.0	0	5.25	2.34
2.0	0	8.55	4.55
2.0	0	11.60	6.78
2.0	0	17.90	12.10
2.0	0	34.50	27.60

What can you deduce from these data about the binding of the substrate?

13

Sigmoidal Kinetics and Allosteric Enzymes

13.1 INTRODUCTION

In Chapter 12 we discussed how interaction between the ligand-binding sites of oligomeric proteins could give rise to cooperative binding, which would be reflected in departures from linearity of Lineweaver-Burk and similar plots if the ligand was a substrate. This is an important consideration, for many enzymes are oligomeric proteins made up of several identical sub-units or protomers. As we shall see later (section 13.5), similar departures from linearity may be seen in the absence of cooperative binding if the kinetic mechanism of the reaction is not straightforward. However, first we must consider in a little more detail how the cooperative binding of ligand to protein may occur. How do the binding sites interact?

It appears that with most proteins, as with haemoglobin (section 12.9), binding sites are clearly separated and so cannot interact directly. Hence it seems that the mechanism of cooperative binding must involve more general interactions between sub-units and the occurrence of conformational changes. The simplest treatment considers that each protomer can exist in two conformational forms: the **T-form** is that which predominates in the unliganded protein, whereas the **R-form** predominates in the protein-ligand complexes. On the basis of the findings with haemoglobin, the T-form may be taken to represent a tensed (or constrained) sub-unit, and the R-form a more relaxed one, but this is not necessarily always the case.

From this starting point, Jacques Monod, Jeffries Wyman and Jean-Pierre Changeux (1965), and Daniel Koshland, George Némethy and David Filmer (1966), put forward models to account for cooperative binding. These models do not give a detailed chemical explanation for cooperativity, but they provide a framework within which the factors involved may be discussed.

13.2 THE MONOD-WYMAN-CHANGEUX (MWC) MODEL

13.2.1 The MWC equation

The MWC model is sometimes referred to as the **symmetrical model**.

. This is because it is based on the assumption that, in a particular protein molecule, all of the protomers must be in the same conformational state: all must be in the R-form or all in the T-form, no hybrids being found because of supposed unfavourable interactions between sub-units in different conformational states.

The two conformational forms of the protein are in equilibrium in the absence of ligand, and the equilibrium is disturbed by the binding of the ligand. This alone can be the explanation for cooperative effects.

Let us consider a dimeric protein having two identical binding sites for a substrate or ligand (S). In the absence of ligand, there will be equilibrium between the two conformational forms of the dimer ($R_2 \rightleftharpoons T_2$), the equilibrium constant being termed the **allosteric constant** and given the symbol L. The hybrid RT is held to be unstable and ignored.

The ligand can bind to either of the sites on the R_2 molecule, each having an intrinsic dissociation constant K_R. In the simplest form of the hypothesis, it is assumed that S does not bind to T to any appreciable extent. Therefore, the only processes which need to be considered (apart from any subsequent reaction to form products) are:

$$R_2 \rightleftharpoons T_2 \qquad \text{(equilibrium constant } L)$$
$$R_2 + S \rightleftharpoons R_2S \qquad \text{(intrinsic dissociation constant } K_R)$$
$$R_2S + S \rightleftharpoons R_2S_2 \qquad \text{(intrinsic dissociation constant } K_R)$$

In diagrammatic form this may be written:

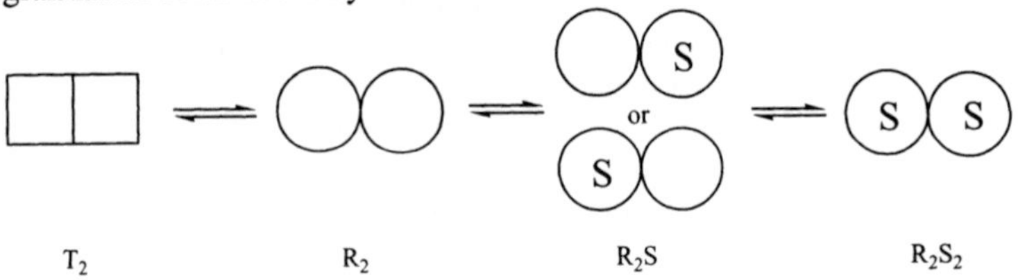

S or S S S

T_2 R_2 R_2S R_2S_2

Let us apply exactly the same logic to this sequence of reactions as we applied in section 12.5. Again, we are considering the binding of a ligand to a dimeric protein, but on this occasion we have the extra complication of two conformational forms. We will assume that the binding of one molecule of S to R_2 does not alter the affinity of the other binding site for S.

The concentration of bound sub-units present = $[R_2S] + 2[R_2S_2]$. The total concentration of sub-units present = $2[R_2] + 2[R_2S] + 2[R_2S_2] + 2[T_2]$.

$$\therefore \text{Fractional saturation } Y = \frac{[R_2S]+2[R_2S_2]}{2([R_2]+[R_2S]+[R_2S_2]+[T_2])}$$

$$= \frac{[R_2S]+2[R_2S_2]}{2([R_2]+[R_2S]+[R_2S_2]+L[R_2]\}} \qquad (13.1)$$

For the **first step** in the binding process, $R_2 + S \rightleftharpoons R_2S$, the apparent binding constant $K_{b1} = [R_2S]/[R_2][S]$. Therefore $[R_2S] = K_{b1}[R_2][S]$.

Since there are two unbound sites which may be filled in the forward reaction but only one bound ligand to dissociate in the reverse reaction, $K_{b1} = 2 \times$ intrinsic binding constant = $2/K_R$. Hence, substituting for K_{b1} in the expression for $[R_2S]$ above, $[R_2S] = (2/K_R)[R_2][S]$.

For the **second step** of the binding process, $R_2S + S \rightleftharpoons R_2S_2$, the apparent binding constant $K_{b2} = [R_2S_2]/([R_2S][S])$. Therefore, $[R_2S_2] = K_{b2}[R_2S][S] = K_{b1}K_{b2}[R_2][S]^2$

Since there is only one unbound site which may be filled in the forward reaction but two bound ligand molecules to dissociate in the reverse reaction, $K_{b2} = \frac{1}{2} \times$ intrinsic binding constant = $1/(2K_R)$.

Hence, substituting for K_{b1} and K_{b2} in the expression for $[R_2S_2]$ above,

$$[R_2S_2] = \frac{2}{K_R} \times \frac{1}{2K_R}[R_2][S]^2 = \frac{[R_2][S]^2}{(K_R)^2}$$

Now, substituting for R_2S and R_2S_2 in the expression for Y above (13.1),

$$Y = \frac{\left(\dfrac{2[R_2][S]}{K_R} + \dfrac{2[R_2][S]^2}{(K_R)^2}\right)}{2([R_2] + \dfrac{2[R_2][S]}{K_R} + \dfrac{[R_2][S]^2}{(K_R)^2} + L[R_2]}$$

$$= \frac{\dfrac{[S]}{K_R}\left(1 + \dfrac{[S]}{K_R}\right)}{L + \left(1 + \dfrac{[S]}{K_R}\right)^2} \tag{13.2}$$

This is the **Monod-Wyman-Changeux equation** for a dimeric protein. It may similarly be shown that for a protein consisting of n protomers, each with a binding site for the substrate or ligand (S), the MWC equation is

$$Y = \frac{\dfrac{[S]}{K_R}\left(1 + \dfrac{[S]}{K_R}\right)^{n-1}}{L + \left(1 + \dfrac{[S]}{K_R}\right)^n} \tag{13.3}$$

According to this equation, the greater the value of L, the more sigmoidal a plot of Y against [S]. If $L = 0$, a hyperbolic curve is obtained. A hyperbolic curve is also obtained, as would be expected, for a monomeric protein, i.e. where $n = 1$, and for the situation where the substrate can bind equally well to the R and the T conformational forms.

13.2.2 How the MWC model accounts for cooperative effects

The MWC equation is consistent with a sigmoidal binding curve, even though its derivation assumes that the binding of one molecule of ligand does not affect the affinity for the ligand of other binding sites on the molecule. The explanation for the cooperative effects lies in the R_n/T_n equilibrium.

When L is large, this equilibrium is in favour of the T_n form in the absence of ligand. If ligand is introduced, but at very low concentrations, there will not be enough present to react significantly with the small amounts of R_n present, so very little formation of R_nS, R_nS_2 and the other liganded species of protein will take place. At higher ligand concentrations, however, there will be enough ligand present to force formation of significant amounts of R_nS, R_nS_2 etc.. Thus, some free R_n will be removed from the system, thereby disturbing the R_n/T_n equilibrium and causing more R_n to be formed from T_n. This freshly-formed R_n can also react with ligand, resulting in yet more formation of R_nS, R_nS_2 and the other liganded forms. Hence the T_n species can be regarded as a reservoir of R_n which only becomes available when the ligand concentration is high enough to cause the formation of appreciable amounts of protein-ligand complex. There will be a surge in the binding curve in the region of the critical ligand concentration.

At still higher ligand concentrations, more of the reservoir of protein will be utilized, and this process will continue until a ligand concentration is reached which is high enough to force conversion of all T_n to R_n. At this point the protein will be fully saturated with ligand.

Thus, the overall binding curve will be **sigmoidal,** a characteristic of positive homotropic cooperativity. It will be apparent from the above that the MWC model *cannot* explain negative homotropic cooperativity.

13.2.3 The MWC model and allosteric regulation

One of the main reasons for the introduction of the MWC model was an attempt to explain the phenomena of allosteric inhibition and activation. Edwin (H. E.) Umbarger (1956) first found that isoleucine could inhibit **threonine dehydratase**, an enzyme involved in its biosynthesis in bacteria; other similar examples of **end-product inhibition**, and also of **allosteric activation**, were soon reported. In 1963, Monod, Changeux and Francois Jacob put forward the allosteric theory of regulation. They pointed out that these naturally-occurring metabolic **regulators** (also called **effectors** and **modifiers)** generally do not resemble the substrate in structure, so are likely to bind to the enzyme at a separate site and affect the binding of the substrate by heterotropic cooperativity. The word **allosteric** was originally used to stress the difference in shape between regulator and substrate (*allo* meaning *other*). Since then it has been used loosely to describe any kind of cooperative effect, homotropic as well as heterotropic.

According to the MWC model, allosteric inhibitors bind to the T-form of the enzyme, stabilizing it and thus increasing the value of L. Allosteric activators have the opposite effect, binding to and stabilizing the R-form and decreasing L. In either case, the binding of the modifier to one of the forms of the enzyme will disturb the R/T equilibrium and therefore show some degree of sigmoidal character if investigated in the absence of substrate.

However, the more important consideration is how such binding affects subsequent substrate-binding. Enzymes subject to allosteric control may fit into either of two categories: they may be **K-series** or **V-series** enzymes.

K-series enzymes are those where the presence of the modifier changes the binding characteristics of the enzyme for the substrate but does not affect the V_{max} of the reaction. The term K_m has no real meaning for an allosteric enzyme, particularly if the binding rather than the kinetic properties are being considered: a more appropriate term is $S_{0.5}$, which is the ligand concentration required to produce 50% saturation of the protein. For a K-series enzyme, $(S_{0.5})_{substrate}$, i.e. the substrate concentration required to half-saturate the enzyme, varies with the concentration of modifier. The MWC hypothesis is that the substrates of such enzymes bind preferentially to the R-form, giving a sigmoidal binding curve as discussed in section 13.2.1. The subsequent reaction is straightforward, so the shape of the Michaelis-Menten plot is determined simply by that of the binding curve. Allosteric inhibitors, by increasing the value of L, increase the sigmoidal nature of the binding curve for substrate. Thus they decrease the fractional saturation of an enzyme with its substrate at low and moderate substrate concentrations, decreasing the value of v_0 under these conditions (Fig. 13.1). Allosteric activators, on the other hand, tend to increase the hyperbolic nature of the substrate binding curve. In each case, the degree of allosteric effect depends on the concentration of modifier, but the value of V_{max} is not affected.

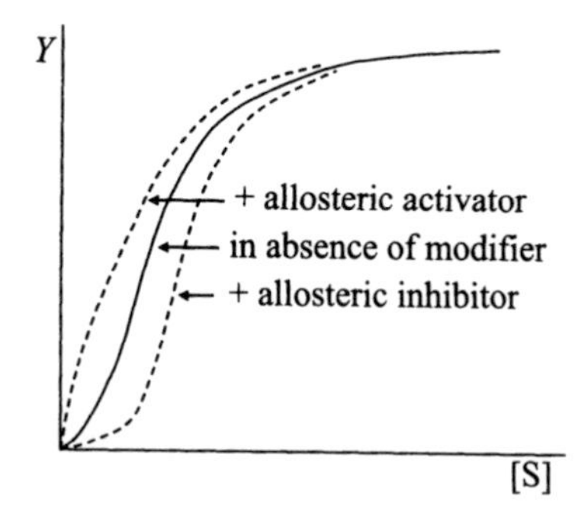

Fig. 13.1 – Effects of allosteric activators and inhibitors on the binding of a substrate to a K-series enzyme, at fixed concentrations of modifier and enzyme.

V-series enzymes are those where the presence of a modifier results in a change in the V_{max} but not in the value of the apparent K_m (or $S_{0.5}$) for the substrate. The binding curve (and Michaelis-Menten plot) for the substrate at constant modifier concentration is a rectangular hyperbola, but the binding curve for the modifier itself is sigmoidal. This can be explained, according to the MWC model, if the substrate can bind equally-well to the R- and T-forms of the enzyme, but the reaction catalysed by the R-form is faster than that catalysed by the T-form. V-series enzymes are much less common than K-series enzymes, but Keith Tipton and colleagues (1974) showed that possible examples include **fructose-l,6-bisphosphatase,** of which AMP is an allosteric inhibitor, and **pyruvate carboxylase,** activated by acetyl-CoA.

Enzymes are also likely to exist in which the R- and T-forms have different affinities for the substrate and also catalyse the reaction at different rates. In this case, allosteric modifiers would affect both the V_{max} and apparent K_m values.

13.2.4 The MWC model and the Hill equation

For the MWC model where the substrate binds only to the R-form of the enzyme, the fractional saturation, as we saw in section 13.2.1, is given by the expression

$$Y = \frac{\dfrac{[S]}{K_R}\left(1+\dfrac{[S]}{K_R}\right)^{n-1}}{L+\left(1+\dfrac{[S]}{K_R}\right)^{n}}$$

If L is very large, most of the enzyme will usually be in a form (T) which will not bind S, keeping the free substrate concentration [S] relatively high. Also, if it is the R-form that binds S, K_R will be relatively low. Hence $[S]/K_R$ will tend to be large, so

$$\left(1+\frac{[S]}{K_R}\right) \approx \frac{[S]}{K_R}$$

Under these conditions,

$$Y = \frac{\left(\dfrac{[S]}{K_R}\right)^{n}}{L+\left(\dfrac{[S]}{K_R}\right)^{n}} = \frac{\dfrac{[S]^n}{K_R^n L}}{\left(1+\dfrac{[S]^n}{K_R^n L}\right)}$$

However, K_R, L and n are all constant, so $1/(K_R{}^nL)$ is a constant ($= K'$).

$$\therefore Y = \frac{K'[S]^n}{1+K'[S]^n} \qquad (13.4)$$

This is a form of the Hill equation (see section 12.4) and implies that, if L is sufficiently large, the only enzyme species present are T_n, R_n and R_nS_n. Note that, since R is assumed to have a high affinity for S, the overall reaction will approximate to $T_n + nS \rightleftharpoons R_nS_n$, thus possessing the 'all-or-nothing' features characteristic of reactions exhibiting Hill-type kinetics.

Hence, if a Hill plot of log $(Y/(1 - Y))$ against log[S], or of log $(v_0/(V_{max} - v_0))$ against log$[S_0]$, is drawn from experimental data and the Hill coefficient (h) is found to be equal to the number of binding sites (n), as determined by an independent experiment, then this series of assumptions must be valid for the system under investigation.

A value of $h = n$ will therefore imply that the MWC model is operating in this instance, that S does not bind to the T-form of the enzyme and that L is very large. No other model has been proposed which is consistent with the Hill equation.

In the simple system discussed in section 12.4, a value of $h < n$ was taken to imply that cooperativity was not complete. In the slightly more complicated system being considered here, a value of $h < n$ would indicate that one (or more) of the assumptions made above was not valid for the enzyme under study. It would not in itself exclude the possibility that the MWC was operating because, for example, the value of L might not be large enough to enable a Hill-type equation to be obtained. Since allosteric inhibitors are assumed to increase the value of L, determination of the Hill coefficient in the presence of an allosteric inhibitor is likely to give the best indication as to whether or not the MWC model is operating. For example, Eduardo Scarano and co-workers (1967) showed that, for the reaction catalysed by donkey spleen **deoxycytidine monophosphate deaminase**,

$$\mathrm{dCMP} + H_2O \rightleftharpoons \mathrm{dUMP} + NH_3 \qquad (13.5)$$

the Hill coefficient in the presence of the allosteric inhibitor dTTP is 4. From other evidence, it was known that there are four binding sites for dCMP, so it was concluded that the MWC model operates for this reaction.

According to this model, the limiting value of h is n, this being obtained when the substrate binds only to the R-form of the enzyme and where L is very large. A value of $h > n$ should never be obtained.

13.3 THE KOSHLAND-NÉMETHY-FILMER (KNF) MODEL

13.3.1 The KNF model for a dimeric protein

The KNF model differs from the MWC one in that it does not exclude hybrids between the two conformational forms of the protein. Therefore, for a dimeric protein where each protomer can exist in R- and T-forms, the species R_2, T_2, R_2S, R_2S_2, R.TS, RS.TS, T_2S and T_2S_2 can all exist. However, in order to explain cooperative effects, some restrictions have to be made.

In the KNF linear sequential model, the only protein species present to any appreciable extent at (or near) equilibrium are T_2, T.RS and R_2S_2. The reaction sequence may therefore be written:

$$T_2 + S \rightleftharpoons \mathrm{T.RS} \text{ (apparent binding constant } K_{b1})$$
$$\mathrm{T.RS} + S \rightleftharpoons R_2S_2 \text{ (apparent binding constant } K_{b2})$$

There is no fundamental difference between this reaction sequence and that used in section 12.5 to derive the Adair equation for a dimeric protein existing in one conformational form. Hence, here too, Y is given by equation 12.13:

$$Y = \frac{K_{b1}[S] + 2K_{b1}K_{b2}[S]^2}{2(1 + K_{b1}[S] + K_{b1}K_{b2}[S]^2)}$$

If $K_{b1} = 4K_{b2}$, there is no cooperativity.
If $K_{b1} < 4K_{b2}$, there is positive homotropic cooperativity.
If $K_{b1} > 4K_{b2}$, there is negative homotropic cooperativity.

The KNF linear sequential model was developed from the induced-fit theory of Koshland (see section 4.4) and implies that the substrate or ligand induces a conformational change to take place (T $\rightarrow$ R) as it binds to the T-form of the protein: $T_2 + S \rightarrow T.TS \rightarrow T.RS$. However the same results could be obtained by an alternative pathway, in which there is an R/T equilibrium which strongly favours the T-form, but where S can only bind to the R-form and so disturbs the equilibrium: $T_2 + S \rightleftharpoons T.R + S \rightleftharpoons T.RS$. In both cases there are negligible amounts of T.TS, T.R and similar species present at (or near) equilibrium. The KNF linear sequential model may therefore be analysed in terms of either of these alternative pathways, and the one chosen was that where the substrate can only bind to the R-form. The following constants are introduced:

$\boldsymbol{K_t}$, an **equilibrium constant** for the conformational change $T \rightleftharpoons R$, so that $K_t = [R]/[T]$.

$\boldsymbol{K_b}$, a **binding constant** for the reaction $R + S \rightleftharpoons RS$, so that $K_b = [RS]/([R][S])$.

$\boldsymbol{K_{RT}}$ **,** $\boldsymbol{K_{RR}}$ **and** $\boldsymbol{K_{RT}}$**, interaction constants** indicating the relative stabilities of the various conformational forms of the oligomeric protein, such that:

$$K_{RT} = [RT]/([R][T]); \quad K_{RR} = [RR]/([R][R]); \quad K_{TT} = [TT]/([T][T]).$$

Since we are only interested in comparing the stabilities of these species, K_{TT} is arbitrarily given the value of 1. On this basis if, for example, K_{RT} has a value greater than 1, then RT will be more stable than TT, which will facilitate binding of S; on the other hand, if $K_{RT} < 1$, RT will be less stable than TT and binding of S will be difficult.

Let us now analyse the step $T_2 + S \rightleftharpoons T.RS$ in terms of these constants:

$$K_{b1} = 2K_t K_b \frac{K_{RT}}{K_{TT}}$$

(Note that the factor 2 is introduced because there are two equally-possible binding-sites in the forward direction.)

Similarly, for the step $T.RS + S \rightleftharpoons RS.RS$:

$$K_{b2} = \tfrac{1}{2} K_t K_b \frac{K_{RR}}{K_{RT}}$$

If $K_{TT} \approx K_{TT} \approx K_{RR}$, then $K_{b1} = 4K_{b2}$ and there is no cooperativity, all interactions between the protomers being identical.

If $K_{TT} \approx K_{RT} \ll K_{TT}$, then $K_{b1} < K_{b2}$ and positive homotropic cooperativity results. The binding of S to one protomer traps it in the R-form. This results in the other protomer staying mainly in the ligand-binding R-form, since R–R interactions are more favourable than R–T interactions.

If $K_{TT} \approx K_{RT} \gg K_{TT}$, then $K_{b1} > 4K_{b2}$ and negative homotropic cooperativity results. The binding of S to one protomer again traps it in the R-form. In this instance, this results in the other protomer staying mainly in the T-form which cannot bind ligands, since R–T interactions are more favourable than R–R interactions.

Note that if $K_{TT} > K_{RR} \gg K_{RT}$ we have the conditions assumed for the MWC model, interactions between RT hybrids being very unfavourable. Since $K_{RT} \ll K_{RR}$ we can confirm that positive homotropic cooperativity, but not negative homotropic cooperativity, is possible for the MWC model.

As pointed out by Manfred Eigen (1967), the KNF linear sequential model and the MWC model are special cases of a general model in which all combinations are possible (Fig. 13.2). The MWC model may be termed the **concerted** form of the general model.

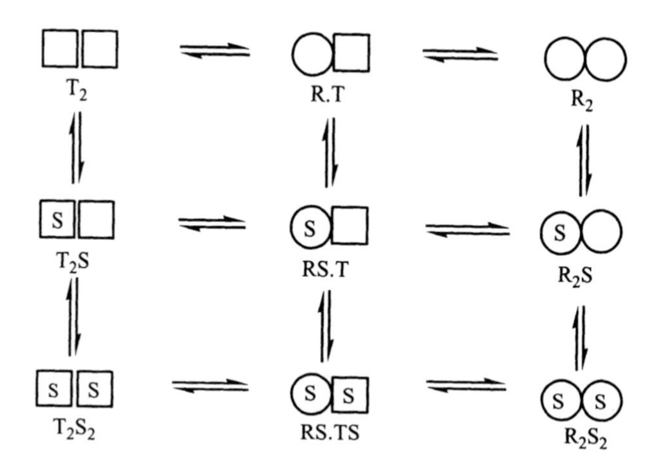

Fig. 13.2 – General scheme for the binding of a ligand (s) to a dimeric protein where each protomer can exist in two conformational forms. In the KNF linear sequential model, the only protein species present at (or near) equilibrium are T_2, RS.T and R_2S_2; in the simplest form the MWC (concerted) model, the only protein species present at (or near) equilibrium are T_2, R_2, R_2S and R_2S_2.

13.3.2 The KNF model for any oligomeric enzyme

A similar treatment to that discussed in section 13.3.1 can be applied to any oligomeric protein with a number of identical binding sites for a ligand, using the appropriate form of the Adair equation. The only extra complication is in deciding which of the protomers can interact, and thus which interaction constants have to be considered. In the case of a tetramer, for example, it is possible that each of the protomers can interact with the other three; this arrangement is called a **tetrahedral model** (Fig. 13.3). Alternatively, each protomer may only be able to interact with two other protomers, forming a **square model**.

A further possibility is a **linear model,** where two of the protomers can interact with one other protomer and two can interact with two other protomers. These models refer only to possible interactions and do not necessarily describe the arrangements of the sub-units in space. Also, with any oligomeric protein, there exists the possibility that the concerted form of the general model, corresponding to the MWC model, may operate.

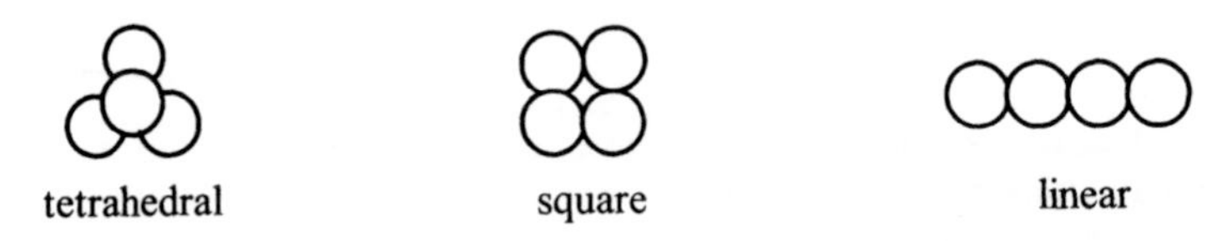

Fig. 13.3 – Model of possible interactions between sub-units in a tetrameric protein. Interactions take place only between sub-units visualized as being in direct contact

13.3.3 The KNF model and allosteric regulation

In contrast to the MWC model, where the explanation for allosteric regulation is relatively straightforward, the KNF model allows for the possibility that allosteric modifiers may act in a variety of ways. For example, the modifier could bind to the same form of the enzyme as the substrate and cause either the same or a different conformational change to take place. Also, the binding of the modifier might or might not prevent the subsequent binding of the substrate to the same sub-unit. The overall effect will depend on factors of this type and also on the various interaction constants involved.

13.4 DIFFERENTIATION BETWEEN MODELS FOR COOPERATIVE BINDING IN PROTEINS

There are a variety of ways of investigating whether the cooperative binding of a ligand to a protein results from a mechanism based on the concerted (MWC) model or on some other model. As we discussed earlier (section 13.2.4), if positive homotropic cooperativity is observed and the **Hill coefficient** is found to be equal to the number of binding sites, then it is likely that the MWC model is operating. On the other hand, if **negative homotropic cooperativity** is found, then the MWC model is excluded. For example, Koshland and colleagues (1968) reported that the binding of NAD^+ to rabbit muscle **glyceraldehyde-3-phosphate dehydrogenase** shows negative homotropic cooperativity, so the MWC model cannot be operating in this instance. The reaction catalysed by this enzyme is: D-glyceraldehyde-3-P + NAD^+ + $P_i \rightleftharpoons$ 3-phospho-D-glyceroyl-P + NADH + H^+.

Relaxation studies (see section 7.2.2) are ideally suited for the investigation of processes involving conformational changes, since these are likely to be extremely rapid and difficult to follow by any other technique. Kasper Kirschner, Manfred Eigen and co-workers (1966) used the **temperature-jump method** to investigate the binding of NAD^+ to yeast glyceraldehyde-3-phosphate dehydrogenase, which has four binding sites for this coenzyme. Rate constants for three different processes could be identified from the results, the slowest process being independent of NAD^+ concentration.

It was concluded that this supported the MWC model, the two fastest processes being the binding of NAD^+ to the R- and T-forms of the enzymes, the slowest being the $R_4 \rightleftharpoons T_4$ transformation. Four processes dependent on NAD^+ concentration would have been expected if the KNF model was operating. Thus it would appear that the binding of NAD^+ to yeast glyceraldehyde-3-phosphate dehydrogenase proceeds in a different way from the binding of NAD^+ to the same enzyme from rabbit muscle. However, it will be realized that if two processes have very similar rate constants, it is likely that they would appear to be a single process in relaxation studies. Hence, in general, findings from such studies must be supported by independent evidence before a firm conclusion can be reached.

The binding-curves predicted by the MWC and KNF models are not exactly the same, and on this basis computers may be able to help determine the most probable model in a particular instance if supplied with suitable experimental data. Needless to say, the degree of accuracy and reliability required of such data is extremely high.

With some proteins, it may be possible to investigate the fractional saturation of each conformational form at different ligand concentrations by the use of optical rotation or spectroscopic techniques. Again this may help to distinguish between possible binding models.

13.5 SIGMOIDAL KINETICS IN THE ABSENCE OF COOPERATIVE BINDING

13.5.1 Ligand-binding evidence versus kinetic evidence

Kinetic studies are often performed to investigate possible cooperative binding of a substrate to an enzyme, since they are generally easier to carry out than direct binding studies. Cooperative binding-effects are reflected in non-hyperbolic Michaelis-Menten plots and departures from linearity of Lineweaver-Burk and similar plots derived from the Michaelis-Menten equation. However, such kinetic findings cannot be said to prove the existence of cooperative binding unless there is corroborative evidence. The Michaelis-Menten plot only follows exactly the characteristics of the binding plot if the reaction is straightforward and proceeds at too slow a rate to significantly affect the equilibria of the binding processes (see section 12.4). Hence, if direct binding studies show that substrate-binding is *not* a cooperative process, but corresponding kinetic studies show non-hyperbolic Michaelis-Menten plots and departures from linearity in other primary plots, then it must be concluded that the kinetic mechanism of the reaction is not consistent with all of the Michaelis-Menten assumptions (section 7.1.1). Some situations where this might be found are discussed below.

It should also be mentioned that, if an enzyme preparation contains a mixture of isoenzymes having different K_m values, then both binding and kinetic plots may show irregularities which are not due to cooperative binding.

13.5.2 The Ferdinand mechanism

William Ferdinand (1966) showed that a random-order ternary-complex mechanism for a two-substrate enzyme-catalysed reaction can lead to sigmoidal kinetics being observed in the absence of cooperative binding:

$$\begin{array}{ccccc} & \overset{+AX}{\nearrow\!\!\swarrow} & E.AX & \overset{+B}{\searrow\!\!\nwarrow} & \\ E & & & & E.AX.B \longrightarrow \longrightarrow \longrightarrow \text{products} \\ & \underset{+B}{\searrow\!\!\nwarrow} & E.B & \underset{+AX}{\nearrow\!\!\swarrow} & \end{array} \qquad (13.6)$$

It is assumed that one of the pathways, let us say the one via E.AX, is kinetically preferred and will proceed faster than the other, even though both pathways are possible. Also, the affinity of E for B is less than that of E.AX for B. We will investigate this system at constant $[E_0]$ and $[B_0]$ but variable $[AX_0]$.

At low $[AX_0]$, E will react mainly with B and so the reaction will proceed via the slower pathway $E \rightleftharpoons E.B \rightleftharpoons E.AX.B \rightarrow$products. At higher $[AX_0]$ there will be a switch-over to the faster pathway $E \rightleftharpoons E.AX \rightleftharpoons E.AX.B \rightarrow$products. E will be depleted as a result of its rapid reaction with AX, so the EB which is formed will tend to dissociate back to E + B rather than to proceed to E.AX.B. This will provide yet more E to react with AX and ensure maximum utilization of the faster pathway.

Therefore, a graph of v_0 against $[AX_0]$ at constant $[B_0]$ will be sigmoidal, even where there is no possibility of cooperative binding. The surge in the curve is explained by the switch-over from the slow to the rapid pathway.

Roy Jensen and William Trentini (1970) showed that the reaction catalysed by **3-deoxy-7-phosphoheptulonate synthase** from *Rhodomicrobium vannielli* may be of this type. The enzyme catalyses the reaction, phosphoenolpyruvate + erythrose-4-phosphate $\rightleftharpoons$ 3-deoxy-D-arabino-heptulosonate-7-phosphate + P_i, and the preferred pathway is that where phosphoenolpyruvate binds first.

13.5.3 The Rabin and mnemonical mechanisms

Brian Rabin (1967) showed that even a single-substrate reaction catalysed by an enzyme with a single binding-site for the substrate can exhibit sigmoidal kinetics, provided the enzyme can exist in more than one conformational form. Consider the sequence for the forward reaction in 13.7(a) below.

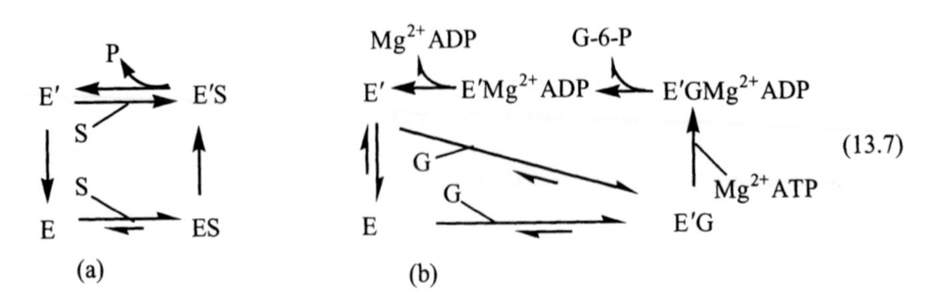

E is assumed to be thermodynamically more stable than the other conformational form of the enzyme, E′. The rate-limiting step of the whole sequence is ES→E′S, E′→E also being slow. The formation of E′S from E′ + S will be appreciable provided free E′ is present.

At low substrate concentrations, the overall rate of the reaction ES→E′S→E′ + P will be very slow compared to the rate of the reaction E′→E. Therefore, the amount of free E′ present will be low, and the supplementary pathway E′ + S→ E′S will not be used. At higher substrate concentrations, the rate of the reaction ES→E′S→E′ + P will be increased, so E′ will be formed faster than it can be converted back to E. Therefore, appreciable amounts of E′ will be present. This will result in the formation of more E′S via the supplementary route E′ + S→E′S and so there will be a further increase in the rate of product formation.

Hence, as soon as the substrate concentration is high enough to produce E′ appreciably faster than it can be converted back to E, the supplementary pathway E′ + S→E′S comes into operation and the overall rate of reaction escalates. A plot of v_0 against $[S_0]$ at fixed total enzyme concentration will therefore be sigmoidal. Allosteric modifiers could act on such a system by increasing or decreasing the rates of the isomerization steps E′→E and/or ES→E′S.

A variation of the Rabin mechanism has been proposed for some two-substrate reactions, e.g. that catalysed by rat liver ‘glucokinase’ (see section 4.1). This is the **mnemonical mechanism** shown in 13.7(b) above. The enzyme is monomeric, so cooperativity is out of the question, but the reaction exhibits sigmoidal kinetics with respect to variable glucose (G) concentration at high concentrations of Mg^{2+}ATP, possibly because the latter prevents the E/E′/E′G system coming to equilibrium.

SUMMARY OF CHAPTER 13

Cooperative binding in oligomeric enzymes can be explained by the Monod-Wyman-Changeux (MWC) or the Koshland-Némethy-Filmer (KNF) hypotheses; these are seen as special cases of a more general hypothesis. Only the MWC model can explain binding characteristics consistent with the Hill equation, but this model cannot explain negative homotropic cooperativity.

Allosteric inhibitors usually increase, and allosteric activators usually decrease, the degree of positive homotropic cooperativity in the binding of a substrate. The MWC model explains this on the basis of allosteric inhibitors stabilizing a conformational form of the enzyme which does not bind to the substrate, and allosteric activators stabilizing one which does. In the KNF model, allosteric modifiers may act in a variety of ways.

Sigmoidal kinetics can be seen in the absence of cooperative binding and may be a consequence of the kinetic mechanism of the reaction.

FURTHER READING

Bisswanger, H. (2002), *Enzyme Kinetics: Principles and Methods*, Wiley–VCH.

Cornish-Bowden, A. (1995), *Fundamentals of Enzyme Kinetics,* 2nd edn., Portland Press (Chapter 9).

Fersht, A. (1999), *Structure and Mechanism in Protein Science,* Freeman (Chapter 10).

Marangoni, A. G. (2002), *Enzyme Kinetics: A Modern Approach*, Wiley (Chapter 8).
Nelson, D. L. and Cox, M. M. (2004), *Lehninger Principles of Biochemistry*, 4th edn., Worth (Chapter 6).
Price, N. C. and Stevens, L. (1999), *Fundamentals of Enzymology*, 3rd edn., Oxford University Press (Chapter 6).
Schulz, A. R. (1994), Cambridge University Press (Chapters 11-16).

PROBLEMS

13.1 An enzyme-catalysed pseudo-single-substrate reaction was investigated in the absence and presence of a fixed concentration (1 mmol l^{-1}) of an allosteric inhibitor, I. The same concentration of enzyme was present in every case. The following results were obtained:

Initial substrate conc^n (mmol l^{-1})	Initial velocity of reaction (μmol product formed min^{-1}mg $protein^{-1}$)	
	In absence of I	In presence of I
0.65	0.56	0.43
0.77	0.78	0.50
0.89	1.28	0.57
1.00	1.74	0.73
1.13	2.25	1.08
1.35	3.06	1.74
1.62	3.68	2.64
1.90	4.03	3.24
2.46	4.26	3.95
3.09	4.31	4.26

The enzyme was found to dissociate into a number of identical sub-units and ultracentrifuge studies were performed on enzyme and sub-units. The sedimentation coefficient (s) of the sub-unit was 4.6×10^{-13}s and of the enzyme 1.6×10^{-12}s. The corresponding diffusion coefficients (D) were 5.96×10^{-7} cm^2s^{-1} and 5.31×10^{-7} cm^2s^{-1} (all values corrected for water at 20°C). The partial specific volume ($\bar{v}$) in each case is 0.736, and the density of water (ρ) at 20°C is 0.998. What can you conclude from the data? (Note that according to Svedberg's equation, molecular weight = $RTs/(D(1 - \bar{v}\rho))$, where $R = 8.314 \times 10^7$ erg $K^{-1}mol^{-1}$ and T = temperature (K).)

13.2 An enzyme-catalysed reaction of the form, $NADH + B \rightleftharpoons BH_2 + NAD^+$, was investigated as follows.

(a) The enzyme at a concentration of 2.5 mmol l^{-1} in a dialysis bag was dialysed against a series of different initial concentrations of NADH in a suitable buffer. No B, BH_2 or NAD^+ was present.

The volume of liquid within the bag was very small compared to the total volume of liquid present, and remained constant throughout each experiment. Each system was allowed to come to equilibrium, then a sample of the solution surrounding the bag was removed, diluted 1 in 100 in buffer, and the absorbance at 340 nm determined in 1 cm cells against a distilled water blank. The following results were obtained:

[NADH$_0$] (mmol l^{-1})	4	5	6	7	8	10	12
Absorbance	0.044	0.059	0.081	0.103	0.131	0.193	0.274

[NADH$_0$] (mmol l^{-1})	14	16	18	20
Absorbance	0.367	0.473	0.582	0.694

(Molar absorption coefficient of NADH at 340 nm = 6.22×10^3.)

(b) The rate of the forward reaction (as written above) at different initial concentrations of NADH and NAD$^+$ was investigated, and the results given below. The enzyme concentration was the same in each case. The initial concentration of B was always 3.0 mmol l^{-1}, and the initial concentration of BH$_2$ was always zero.

[NADH$_0$] (mmol l^{-1})	[NAD$^+_0$] (mmol l^{-1})	Absorbance (340 nm) at time t (min) = 0.5	1.0	1.5	2.0	2.5	3.0
1.25	0	0.470	0.455	0.440	0.427	0.413	0.400
1.25	1	0.472	0.460	0.447	0.434	0.423	0.410
1.25	2	0.475	0.463	0.452	0.440	0.430	0.423
1.5	0	0.467	0.447	0.430	0.410	0.395	0.378
1.5	1	0.468	0.453	0.437	0.420	0.405	0.390
1.5	2	0.470	0.458	0.445	0.435	0.418	0.410
2.0	0	0.460	0.437	0.413	0.390	0.370	0.350
2.0	1	0.463	0.443	0.422	0.400	0.380	0.365
2.0	2	0.467	0.449	0.430	0.412	0.398	0.383
2.5	0	0.457	0.430	0.403	0.377	0.353	0.330
2.5	1	0.460	0.437	0.413	0.388	0.365	0.346
2.5	2	0.464	0.442	0.420	0.400	0.380	0.362
3.33	0	0.454	0.423	0.393	0.362	0.333	0.307
3.33	1	0.457	0.430	0.400	0.373	0.348	0.325
3.33	2	0.460	0.437	0.412	0.386	0.365	0.345
5.0	0	0.450	0.413	0.378	0.342	0.310	0.280
5.0	1	0.452	0.420	0.385	0.353	0.320	0.302
5.0	2	0.455	0.425	0.395	0.364	0.338	0.314
10.0	0	0.442	0.400	0.360	0.317	0.278	0.242
10.0	1	0.445	0.405	0.364	0.325	0.290	0.260
10.0	2	0.445	0.407	0.370	0.332	0.303	0.277

(c) Equilibrium isotope exchange studies showed an increased rate of transfer of label from NADH to NAD^+ with increased concentration of B (maintaining a constant $[BH_2]$:[B] ratio) at all concentrations of B tested.

What can you conclude about the enzyme mechanism from these data?

13.3 A single-substrate enzyme-catalysed reaction gave the following results at fixed enzyme concentration in the presence or absence of a fixed concentration (10 mmol l^{-1}) of an inhibitor A.

	Product concentration (μmol l^{-1}) at time t (min) =					
$[S_0]$ (mmol l^{-1})	0.5	1.0	1.5	2.0	2.5	3.0
0.5 uninhibited	4.0	7.5	11.0	15.0	19.0	22.0
0.5 inhibited	0.15	0.3	0.45	0.6	0.75	0.9
1.0 uninhibited	9.5	19.0	28.5	38.0	47.5	57.0
1.0 inhibited	1.3	2.5	3.7	5.0	6.3	7.5
2.0 uninhibited	35.0	69.0	104	138	172	206
2.0 inhibited	10.0	20.0	30.0	40.0	50.0	60.0
5.0 uninhibited	165	330	495	660	825	990
5.0 inhibited	119	238	357	476	595	714
10.0 uninhibited	339	677	1020	1350	1690	2030
10.0 inhibited	357	714	1070	1430	1790	2140
20.0 uninhibited	438	876	1310	1750	2190	2630
20.0 inhibited	476	952	1430	1900	2380	2860
50.0 uninhibited	480	959	1440	1920	2400	2880
50.0 inhibited	499	997	1500	1990	2490	2990
100.0 uninhibited	490	980	1470	1960	2450	2940
100.0 inhibited	500	1000	1500	2000	2500	3000

SEC experiments gave a single band for the enzyme corresponding to a molecular weight of about 211 000. SDS-electrophoresis experiments also gave a single band for the enzyme, corresponding to a molecular weight of 70 500.

What can be concluded from these results about the mechanism of the reaction and the nature of the inhibition by A?

14

The Significance of Sigmoidal Behaviour

14.1 THE PHYSIOLOGICAL IMPORTANCE OF COOPERATIVE OXYGEN-BINDING BY HAEMOGLOBIN

Since many proteins show evidence of cooperative ligand-binding, it is reasonable to ask if there is a physiological advantage in them doing so. In the case of haemoglobin, the advantages of its cooperative oxygen-binding are easy to see.

Haemoglobin is a tetrameric protein which can bind four molecules of oxygen in a sigmoidal manner (section 12.9). It is found in the blood of vertebrates, where it transports oxygen from the alveoli of the lungs to the capillaries of muscle and other tissue. There the oxygen is released to diffuse into the cells. These cells possess no haemoglobin, but they do contain its monomeric relative, **myoglobin,** which has only one binding site for oxygen and so must bind in a hyperbolic fashion. The myoglobin can store the oxygen and, when required, facilitate its transport to **cytochrome oxidase** in the inner mitochondrial membrane of the cell, where the gas completes its physiological role and accepts electrons which have passed down the respiratory pathway.

If we consider the binding-curves for haemoglobin and myoglobin (Fig. 14.1) we see that the partial pressure of oxygen in the capillaries of the lung is sufficient to cause saturation of both haemoglobin and myoglobin. The oxygen tension in the tissues is much lower, but it is still sufficient for the almost complete saturation of myoglobin. However, haemoglobin, because of its different binding characteristics, is only about 40% saturated at this lower pO_2.

Thus we see that the sigmoidal oxygen-binding of haemoglobin provides a mechanism by which 60 % of the oxygen taken up in the lungs can be released in the capillaries of the tissues. If myoglobin was to act as an oxygen carrier in blood, a much greater differential in oxygen tension between lungs and tissues would be required before the required amount of oxygen could be released.

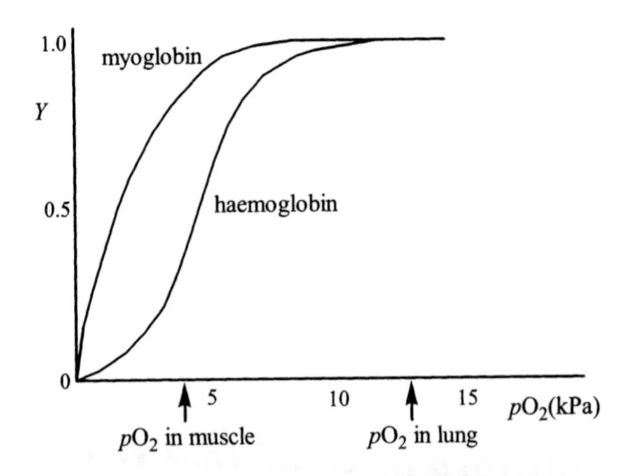

Fig. 14.1 – Graphs of fractional saturation (Y) against pO_2 for the binding of oxygen to myoglobin and haemoglobin under physiological conditions.

In general, fractional saturation is far more sensitive to changes in ligand concentration if the binding mechanism is sigmoidal rather than hyperbolic. According to the simple binding equation for a non-cooperative system (equation 12.2), an 81-fold increase in ligand concentration is required to change the fractional saturation from 0.1 to 0.9. In contrast, according to the MWC equation for the situation where a high degree of positive homotropic cooperativity (Hill-type) is present for a tetrameric enzyme (equation 13.4), the identical change in fractional saturation can be brought about by a three-fold increase in ligand concentration (also see Fig. 14.2).

In the case of haemoglobin, the degree of positive homotropic cooperativity, and hence of sigmoidal binding, is influenced by heterotropic factors. In isolation, its $(S_{0.5})_{oxygen}$ is about 0.1 kPa, which is much the same as that of myoglobin. However, the presence of **2,3-bisphosphoglycerate (BPG)** in red blood cells increases the $(S_{0.5})_{oxygen}$ value of haemoglobin *in vivo* to about 3.5 kPa (which is the value shown in Fig. 14.1). BPG fits into the central cavity of deoxyhaemoglobin and forms, through its phosphate groups, electrostatic interactions with three positively-charged groups in each of the β-chains. The central cavity of oxyhaemoglobin is too small to contain BPG, so the binding of oxygen results in the ejection of the modifier. BPG may be considered to have some of the characteristics of an MWC allosteric inhibitor, since it stabilizes the conformational form of the protein which has least affinity for oxygen.

The sigmoidal nature of oxygen-binding is also made more marked by increasing the concentration of H^+ and CO_2, and the binding of these in turn is affected by the partial pressure of oxygen. Three protons are taken up by haemoglobin as oxygen is released, because two terminal amino groups and one histidine residue are in more negatively-charged environments in deoxyhaemoglobin and can more readily accept a proton. Carbon dioxide may bind to any of the four terminal amino groups of haemoglobin to form a carbamate group: $RNH_2 + CO_2 \rightleftharpoons RNHCO_2^- + H^+$. Again, this takes place more readily when the haemoglobin is in the deoxy form.

Muscle produces a great deal of CO_2 and H^+ as metabolic end-products, and the presence of these in the muscle capillaries facilitates the release of oxygen from haemoglobin. This in turn makes the binding of protons and CO_2 to the haemoglobin more favourable, enabling them to be carried through the venous system back to the alveolar capillaries of the lungs. There, the high oxygen tension results in the binding of oxygen to haemoglobin and the concomitant release of protons and CO_2. This enables excess CO_2 to be removed from the body in expired air. The interrelation between the binding to haemoglobin of oxygen, H^+ and CO_2 is termed the **Bohr effect**.

14.2 ALLOSTERIC ENZYMES AND METABOLIC REGULATION

14.2.1 Introduction

Before we go on to discuss the possible roles of allosteric enzymes in metabolic regulation, we must first consider the environment in which this regulation takes place. Previously, we have usually restricted our discussions about the properties of enzymes to those observed under simple *in vitro* conditions, e.g. at fixed concentration of enzyme in a dilute solution in the absence of product and of other enzymes. In contrast, one of the essential features of living cells is that conditions such as concentrations of substrate, product and enzyme are, to a greater or lesser degree, constantly changing. Also, concentrations of enzymes *in vivo* are usually considerably greater than those used *in vitro* for steady-state investigations. Furthermore, the reaction catalysed by one enzyme is linked to reactions catalysed by others, forming a **metabolic pathway** in a highly organized environment.

Hence it cannot be blindly assumed that findings made *in vitro* are applicable to the situation *in vivo*. For example, although it has also been shown *in vitro* that the H_4 isoenzyme of LDH is inhibited by high concentrations of pyruvate (pyruvate and NAD^+ forming a dead-end ternary complex with the enzyme), it is not certain that pyruvate concentrations *in vivo* could ever reach high enough levels for this to be of significance in metabolic regulation; also, the characteristics of the LDH isoenzymes may vary according to whether they are freely soluble or membrane-bound, as some may be *in vivo*. For these and other reasons (see section 5.2.2), extensive *in vitro* investigations have so far failed to establish beyond doubt the physiological roles of the isoenzymes of LDH.

The total concentration of each enzyme present in a cell is determined by the rate of its **synthesis** and the rate of its **breakdown**. The former is influenced by such factors as **induction** and **repression** (mainly in prokaryotic cells) and by the presence of **hormones** (possible agents of transcriptional control in eukaryotic cells), thus ensuring that each cell synthesizes only the enzymes required at that time (see section 3.1.5). Similarly, the breakdown of enzymes (catalysed by proteolytic enzymes) is subject to some degree of control. In general, enzymes are broken down more rapidly when they represent the only source of energy available to the cell (e.g. in starvation) or when some change in function is taking place which requires the synthesis of different proteins (e.g. in germinating seeds). In addition to this, large enzymes and those involved in metabolic control tend to be broken down more rapidly than others.

On the other hand, many enzymes are more resistant to breakdown if their substrates are present in high concentrations.

Control mechanisms which affect the rates of synthesis or degradation can only serve as **coarse** (or **long-term) agents of metabolic regulation**, because at least several minutes are likely to elapse before they can bring about a significant change in the total concentration of the enzyme in question. Also, mechanisms of this type, and those hormonal mechanisms which control metabolism by regulating the entry of substrates into cells, are likely to affect a group of cells rather than a single one. Hence, in order to meet the needs of the moment in each individual cell, **fine** (or **acute) mechanisms of metabolic regulation** exist in which the **activity** of an enzyme, rather than its total **concentration**, is controlled. It is this subject of metabolic regulation by the control of enzyme activity that particularly concerns us in the present chapter.

14.2.2 Characteristics of steady-state metabolic pathways

Metabolic pathways often contain branch points, at which metabolites may enter or leave by alternative routes, as in the following example:

$$\begin{array}{ccccccc} P \rightleftharpoons\ \rightleftharpoons & Q & \rightleftharpoons\ \rightleftharpoons & R & \rightleftharpoons\ \rightleftharpoons & S & \rightleftharpoons\ \rightleftharpoons T \\ & \upharpoonleft\downharpoonright & & \upharpoonleft\downharpoonright & & \upharpoonleft\downharpoonright & \\ & \upharpoonleft\downharpoonright & & \upharpoonleft\downharpoonright & & \upharpoonleft\downharpoonright & \\ & X & & Y & & Z & \end{array} \qquad (14.1)$$

The sequences between successive branch points may be regarded as separate units, for each usually contains a regulated step. Let us consider one such unbranched sequence of steps: $A \overset{E_1}{\rightleftharpoons} B \overset{E_2}{\rightleftharpoons} C \overset{E_3}{\rightleftharpoons} D \overset{E_4}{\rightleftharpoons} E \overset{E_5}{\rightleftharpoons} F$.

Such a system is not likely to be found at equilibrium *in vivo,* because it would then be unable to provide any free energy for the organism (section 6.1.5). If, instead, it is assumed to be at steady-state, then there will be a net flux, J, through the system in one particular direction (let us say A to F) and the concentrations of all intermediates will be constant. This implies that A is being fed into the system (by metabolic routes or by transport from outside the cell or cellular compartment) at a constant rate and F is being dissipated (by further metabolic routes or by transport) at the same rate. This rate must also be the net flux through the system (J), i.e. the difference between the rates of the forward and back reactions for each step and for the overall process.

Although this steady-state assumption must obviously be a simplification of the situation *in vivo,* it is nevertheless a reasonable one: concentrations of metabolic intermediates are usually maintained within quite a narrow range within the living cell. In general, the level of each intermediate is found at a concentration slightly below the $(S_{0.5})$ value for the enzyme which utilizes it as substrate, i.e. the next one in the sequence. Therefore, in the example being considered, the concentration of B is usually slightly below the $(S_{0.5})_B$ value of the enzyme E_2.

This general non-saturation of enzymes is an important condition for the setting up and maintenance of a steady-state. If, for example, enzyme E_2 was usually found to be very nearly saturated with B, this would mean that there was just enough E_2 present to handle the B being produced from A at its normal rate, but without there being any margin of safety. If the rate of production of B from A should increase, the reaction B→C could not be speeded up by any significant factor in response to this and so the concentration of B would rise, possibly catastrophically. Although metabolic regulation might limit the rate of production of B from A, it is clearly essential that sufficient E_2 is present to cope with the maximum rate at which B is likely to be produced, and that is what is found. Also, since all reactions are reversible to some degree, the maintenance of a steady-state in a particular direction requires that the concentrations of substrate and product for each enzyme bear such a relationship to the characteristics of the enzyme that the forward reaction is favoured, e.g. if $[B] > (S_{0.5})_B$ for enzyme E_2, then $[C] < (S_{0.5})_C$ for the same enzyme.

Since the system is at steady-state, none of the individual steps can be at equilibrium. An indication of the disequilibrium of each step can be obtained from the **mass action ratio (Γ)** as compared to the **apparent equilibrium constant (K'_{eq})**. For the step $B \rightleftharpoons C$, $\Gamma = [C]/[B]$. The **disequilibrium ratio (ρ)** is defined such that $\rho = \Gamma/K'_{eq}$ so, for the step $B \rightleftharpoons C$, $\rho = [C]/([B]K'_{eq})$. If the step is near equilibrium, $\Gamma \approx K'_{eq}$ and $\rho \approx 1$; if the step is some way from equilibrium, $\Gamma \ll K'_{eq}$ and $\rho \ll 1$.

By extending the treatment of simple steady-state kinetics (section 7.1.2) to a system in which the product concentration cannot be ignored, it may be shown that, for the enzyme-catalysed reaction $B \rightleftharpoons C$,

$$v_f = \frac{V^B_{max} K^C_m [B]}{K^B_m K^C_m + K^C_m [B] + K^B_m [C]} \quad \text{and} \quad v_b = \frac{V^C_{max} K^B_m [C]}{K^B_m K^C_m + K^C_m [B] + K^B_m [C]}$$

where v_f is the rate of the forward reaction and v_b the rate of the back reaction.

$$\therefore \frac{v_b}{v_f} = \frac{V^C_{max} K^B_m [C]}{V^B_{max} K^C_m [B]}$$

But, from the Haldane relationship (equation 7.17 in section 7.1.7),

$$K'_{eq} = \frac{V^B_{max} K^C_m}{V^C_{max} K^B_m}$$

$$\therefore \frac{v_b}{v_f} = \frac{[C]}{[B]K'_{eq}} = \rho \qquad (14.2)$$

The disequilibrium ratio for the overall sequence A $\rightleftharpoons$ F is the product of the disequilibrium ratios of each of the individual steps, and must be less than 1 for a steady-state system in which there is net production of F from A. In fact, although in such a system $v_f - v_b$ must be the same for each step and must be positive, many steps in metabolic pathways are found to be quite near to equilibrium. This can be consistent with the steady-state assumption, provided both v_f and v_b are large.

14.2.3 Regulation of steady-state metabolic pathways by control of enzyme activity

Let us now consider how the steady-state discussed in section 14.2.2 will be affected by changing the activity of an enzyme. We will compare the results for an enzyme catalysing a step which is near equilibrium with those for one catalysing a step far from equilibrium. For simplicity, we will assume in each case that the characteristics of the enzymes are not altered, so that the change in activity is equivalent to a change in concentration of the enzyme.

First let us consider the situation where ρ for B $\rightleftharpoons$ C is 0.99, i.e. the reaction is almost at equilibrium. The overall flux, J, from B to C is given by

$$\begin{aligned} J &= v_f - v_b = v_f - 0.99v_f \\ &= 0.01v_f \end{aligned} \tag{14.3}$$

The steady-state expressions for v_f and v_b show that both of these terms are proportional to [E_0] (since V_{max}^{C} and V_{max}^{B} must both be proportional to [E_0]) and so the flux, J, must also be proportional to [E_0]

If the effective concentration of [E_0] is halved, then the immediate effect is that the values of J, v_f and v_b will all be halved. Since the rate of formation of B from A and the rate of conversion of C to D are not immediately affected, the concentration of C will start to fall, which will reduce the rate v_b still further without significantly altering v_f. When [C], and thus v_b, falls to such a level that the flux has been restored to its original value, then a new steady-state will have been set up. In fact, it may be shown that this is achieved when [C] has fallen by only 1%. At this new value of [C], the new rate for the forward reaction, v_f', is given by $v_f' = 0.5\,v_f$, and the new rate of the back reaction, v_b', by $v_b' = 0.5 \times 0.99\,v_b$. The new disequilibrium ratio, ρ', is given by $\rho' = 0.99\,\rho$, since [C] is reduced to 99% of its original value and [B] and K'_{eq} are unchanged.

According to the general relationship derived in section 14.2.2 (equation 14.2), $v_b' / v_f' = \rho'$

$$\begin{aligned} \therefore\ v_b' &= \rho' v_f' = 0.99\,\rho\, v_f' = 0.99 \times 0.99\ v_f' \\ &= 0.99 \times 0.99 \times 0.5\,v_f \\ &= 0.49\ v_f \end{aligned}$$

Therefore, under these new conditions, the flux, J', is given by

$$\begin{aligned} J' &= v_f' - v_b' = 0.50\ v_f - 0.49\ v_f \\ &= 0.01\ v_f \end{aligned} \tag{14.4}$$

This is the same as the value obtained in equation 14.3. Hence, the net flux through the system has not changed.

If, in contrast to the above, the value of ρ for $B \rightleftharpoons C$ is 1×10^{-4}, i.e. the reaction is far from equilibrium, the consequence is:

$$J = v_f - v_b = v_f - 10^{-4} v_f \\ \approx v_f \tag{14.5}$$

So, for such a reaction, v_f is approximately equal to the net flux through the system. If the effective concentration of $[E_0]$ is halved, the values of J, v_f and v_b will all be halved, as before. However, in this instance, compensation cannot come from a further decrease in v_b, for this is negligible to start with. This step will therefore be rate-limiting for the overall process, and a new steady-state will be set up with a lower net flux than before.

Thus it may be seen that an enzyme catalysing a step which is near equilibrium is not likely to be important in metabolic regulation, because a considerable change in enzyme activity may take place without affecting the flux through the system. In contrast, an enzyme catalysing a step which is far from equilibrium may well be important in metabolic regulation. Such steps are often rate-limiting, and changes in enzyme activity are accompanied by changes in the flux through the system.

The process of attempting to describe the changes in flux through metabolic systems in response to perturbations such as changes in activity of key enzymes is termed **metabolic control analysis**. Each enzyme in a metabolic pathway has a **flux control coefficient** ***C***, which provides a measure of the effect of a change in activity of that enzyme on the flux J through the pathway. By definition, for example, $C_1 = (\Delta J/J)/(\Delta e_1/e_1)$, where e_1 is the catalytic activity of enzyme E_1, and Δ indicates a change in value. The sum of the C values in a system = 1, with those of enzymes uninvolved in metabolic regulation making a negligible contribution.

Although the discussions above were all concerned with systems where changes in enzyme activity are not accompanied by changes in enzyme characteristics, the general conclusions are equally valid for systems where enzyme characteristics may be affected, as is the case with most allosteric enzymes. **Allosteric inhibitors** usually increase the sigmoidal nature of Michaelis-Menten plots, so the effect of the inhibition is most marked at low and moderate substrate concentrations, and possibly non-existent under conditions where the substrate is plentiful and need not be conserved. **Allosteric activators**, on the other hand, usually increase the hyperbolic character of Michaelis-Menten plots. Regardless of this, most allosteric modifiers act on enzymes catalysing steps far from equilibrium, i.e. where $\rho < 1 \times 10^{-2}$.

In pathways where the net flux may be in either direction *in vivo,* each regulatory step is usually associated with two enzymes. One catalyses the reaction in largely irreversible fashion in one direction while the other is responsible for the flux in the opposite direction. In general, for the reaction $B \rightleftharpoons C$, the isoenzyme with the lower $(S_{0.5})_B$ value will catalyse the reaction $B \rightarrow C$, while the isoenzyme with the lower $(S_{0.5})_C$ value will catalyse $C \rightarrow B$. An example of such a step occurs in glycolysis/ gluconeogenesis with the interconversion of fructose-6-phosphate and fructose-1,6-bisphosphate (see section 14.2.5).

In general, steps subject to allosteric inhibition occur early in metabolic sequences. For example, in the metabolic sequence whose main direction of flux is as follows:

$$\longrightarrow A \xrightarrow{E_1} B \longrightarrow C \longrightarrow D \longrightarrow E \xrightarrow{E_5} F \longrightarrow$$

(with downward arrows from A and from F)

the enzyme E_1, which catalyses the first step committed to the synthesis of F, might be inhibited by F or by one or more of the subsequent metabolites. This arrangement ensures that if F and its metabolic products are plentiful, more A is made available to be metabolized by alternative routes, and excessive build-up of intermediates such as B and C is prevented. In most cases, this **switching** between alternative pathways is most sensitive when controlled by inhibitors which affect $(S_{0.5})$ rather than V_{max} values, since the concentrations of most metabolites are found to be well below the levels required for enzyme saturation (section 14.2.2). Examples of allosteric inhibitors are discussed in sections 13.2.3, 14.2.5 and 14.2.6.

In contrast to the above, steps subject to allosteric activation often occur late in a metabolic sequence. Thus, enzyme E_5 in the sequence under consideration might be activated by A or by a metabolic precursor of A. This arrangement helps to prevent levels of metabolic intermediates falling excessively low. An example of allosteric activation is the effect of fructose-1,6-bisphosphate on pyruvate kinase, the final enzyme in the glycolytic pathway. Other examples are given in sections 14.2.5 and 14.2.6.

14.2.4 Allosteric enzymes and the amplification of metabolic regulation

Most modifiers of allosteric enzymes act to change the degree of sigmoidal character of a Michaelis-Menten plot, so have greatest effect on metabolic flux when the substrate concentration is low or moderate, and least effect when the substrate concentration is high (section 14.2.3). This applies regardless of whether the modifier primarily affects the substrate binding or the kinetic characteristics of the enzyme, although most known examples fit into the former category.

In many cases, the process by which the modifier itself binds to the enzyme has sigmoidal characteristics, and this makes a significant contribution to the regulatory effect. The degree of inhibition or activation produced must depend on the amount of modifier bound to an enzyme, so the sensitivity of a particular mechanism of metabolic control must depend on whether a given change in modifier concentration produces a large or a small change in the fractional saturation of the enzyme with modifier, all other factors being equal. If the modifier concentration *in vivo* is approximately equal to the $(S_{0.5})$ value for its binding to the enzyme, a not unreasonable assumption, then a system with a sigmoidal modifier-binding process will show far greater sensitivity to changes in modifier concentration than will one with a hyperbolic modifier-binding process (Fig. 14.2). Thus the sigmoidal binding of a modifier may be said to amplify the regulatory effect.

The sensitivity of fractional saturation to changes in ligand concentration for processes with different binding characteristics was discussed in general terms in section 14.1.

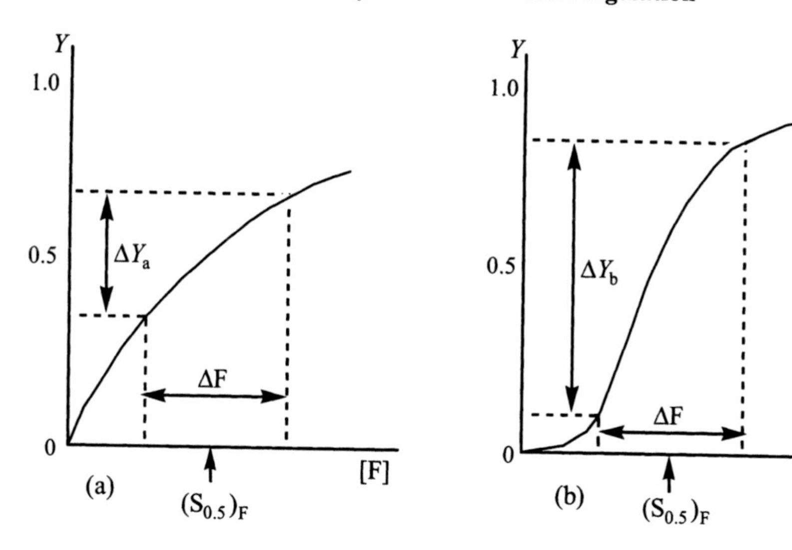

Fig. 14.2 – Graphs showing the change of fractional saturation of a protein with ligand, F, with changes of [F] at constant protein concentration for: (a) a system where the binding of F is hyperbolic; and (b) a system where the binding of F is sigmoidal. Consider, for example, the situation where this represents the binding of a feedback inhibitor, F, to the enzyme which catalyses the first step in the metabolic sequence A→B→C→D→E→F. If a given change in [F], ΔF, takes place near the $(S_{0.5})_F$ value in each case and produces a change in fractional saturation ΔY_a for the hyperbolic system and ΔY_b for the sigmoidal system, it can be seen that ΔY_b is much larger than ΔY_a, and will cause a greater increase in the sigmoidal character of the binding curve for the substrate, A, all other factors being equal. In other words, a regulatory system where the modifier binds by a sigmoidal process will show far greater sensitivity to changes in modifier concentration in this range than will a system where modifier-binding is hyperbolic.

14.2.5 Other mechanisms of metabolic regulation

Although in the present chapter we are concerned chiefly with the role of allosteric enzymes in metabolic regulation, it is necessary to mention briefly some other regulatory mechanisms. In some instances these may act in association with allosteric regulation, but not in every case.

One important factor in the regulation of metabolism is the availability of coenzymes such as NAD^+, NADH, ATP, ADP and AMP. Often these are co-substrates or co-products of the reaction being regulated (sections 11.5.2 and 11.5.4) but, in some instances, they may bind to allosteric sites (e.g. **phosphofructokinase** has two binding-sites for ATP: one forms part of the active site, whilst the other lies elsewhere on the molecule and has a purely regulatory function).

In general, the affinities of enzymes for coenzymes of this type are strong enough to ensure that the binding sites are more or less saturated at all times, i.e. the $(S_{0.5})$ values are considerably smaller than the physiological concentrations of the coenzymes (N.B. this is one reason why they are regarded as coenzymes and not simply as co-substrates or co-products). Also, these coenzymes exist in alternative forms (e.g. NAD^+ and NADH) which resemble each other sufficiently for both to be able to bind to the same site on the enzyme, so there will be competition between them for binding.

The total concentration of NAD^+ + NADH, and of ATP + ADP + AMP, within a cell is usually approximately constant, so the regulatory effects of these coenzymes depends on concentration *ratios* (e.g. of [NAD^+] to [NADH]) rather than on their individual concentrations. This applies regardless of whether the coenzyme functions as a co-substrate/co-product, or as an allosteric modifier. Note, however, that equilibrium as well as non-equilibrium reactions may be regulated by co-substrate/co-product concentration ratios.

Another important factor in the control of metabolism within the cell, and one which could not be deduced from simple *in vitro* studies, is the phenomenon of **cellular compartmentation.** The different organelles present in eukaryotic cells (Fig. 14.3) possess different enzymes and metabolic pathways from each other and from the cytosol. These organelles are enclosed by membranes which, to a greater or lesser extent, prevent the passage of molecules between organelle and cytosol, except where this passage is under the control of a specific carrier transport system. Therefore, the control of the transport of substrates, products and cofactors across membranes provides another mechanism for the regulation of metabolism.

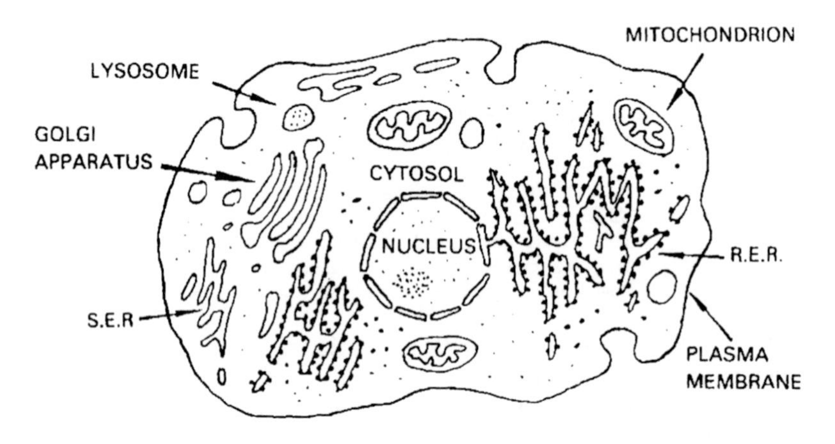

Fig. 14.3 – Diagrammatic representation of a 'typical' animal cell, showing the presence of the various membrane-enclosed organelles. The part of the cell between the nucleus and the plasma membrane is termed the cytoplasm. (R.E.R. = rough endoplasmic reticulum, i.e. that to which ribosomes are attached; S.E.R. = smooth endoplasmic reticulum.)

For example, the β-oxidation of fatty acids proceeds in mitochondria, but the fatty acids must first be activated to fatty acyl-CoA in the cytosol. The fatty acid moiety can only enter a mitochondrion if coupled to the cofactor carnitine, so any factor affecting the formation of fatty acyl-carnitine, and thus the transport of activated fatty acids into mitochondria, will affect the overall rate of fatty acid oxidation.

The inner mitochondrial membrane is impermeable to NAD^+ and NADH but, in heart, liver and kidney cells, reducing equivalents may enter a mitochondrion in the form of malate (Fig. 14.4). A carrier-protein in the inner mitochondrial membrane transports malate from the cytosol in exchange for 2-oxoglutarate. Adenine nucleotides may cross the mitochondrial membranes, but only in exchange for other adenine nucleotides (regardless of their phosphate content).

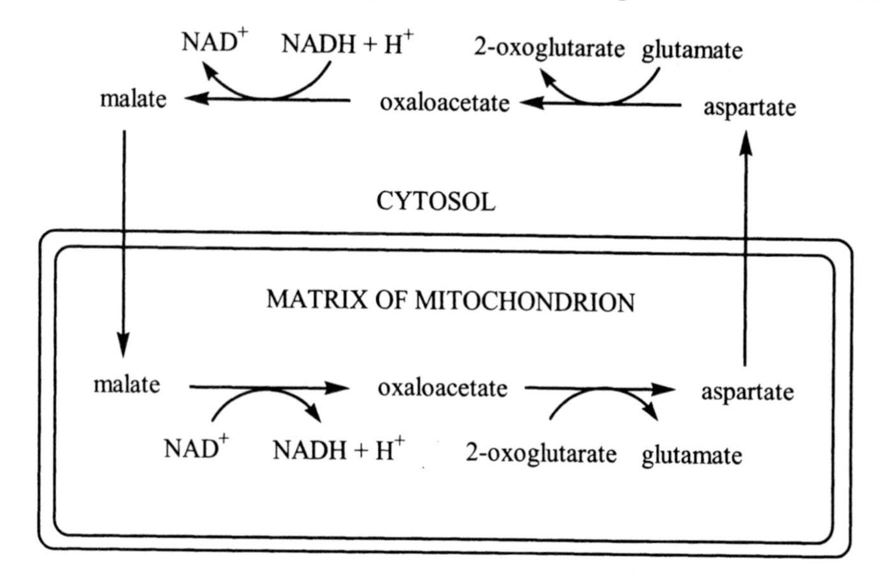

Fig. 14.4 – Transport of reducing equivalents as malate into a mitochondrion of a cell, as part of what is termed the malate-aspartate shuttle.

Thus there is clear evidence for compartmentation of nucleotide coenzymes in eukaryotic cells. There is also some evidence for this in prokaryotic cells, which is somewhat surprising, in view of the lack of membrane-enclosed organelles in these cells.

One reason for compartmentation within the cell is to provide a suitable environment for specialized enzymes. For example, the pH within lysosomes is low, to facilitate the operation of the hydrolytic enzymes found there, most of which have a pH optimum of around 4–5.

Another reason for compartmentation is to prevent unnecessary conflict between metabolic pathways which happen to have a common intermediate but which are otherwise completely independent. For example, carbamoyl phosphate may be metabolized via the urea cycle as part of a mechanism for removing excess nitrogen (especially toxic ammonia) from the cell or it may be used for the synthesis of pyrimidine nucleotides (see section 14.2.6). Both processes take place in the mammalian liver, but the former utilizes carbamoyl phosphate synthesized in mitochondria from ammonium ions and bicarbonate:

$$NH_4^+ + HCO_3^- + 2ATP \rightarrow H_2N.CO.OPO_3^{2-} + 2ADP + P_i \qquad (14.6)$$

whereas the latter proceeds from carbamoyl phosphate synthesized in the cytosol and uses glutamine rather than ammonia as the nitrogen source. *N*-acetyl glutamate activates the **carbamoyl phosphate synthase** enzyme present in mitochondria, but not that found in cytosol. Since there is normally little transport of carbamoyl phosphate across the mitochondrial membranes, this compartmentation enables the two different pathways to be regulated independently of each other.

Compartmentation may also take place in a chemical rather than a physical manner. A metabolite may exist in several isomeric forms, only one of them being able to act as substrate for a particular enzyme. Alternatively, a metabolite may be able to bind to a metal ion, affecting metabolite-enzyme affinity (e.g. **aconitate hydratase** will convert isocitrate to citrate, but will not use Mg-isocitrate as substrate). Enzyme activity may also, of course, depend on metal ion concentration (section 11.4). All of these factors may be able to play a role in metabolic regulation.

Enzymes which can exist in active and inactive forms usually have a very important regulatory function. In section 5.1 we saw that many hydrolytic enzymes are synthesized as inactive zymogens and activated by the removal of peptide fragments as required. This· is a crude mechanism, however, since the zymogens cannot be reformed from the active enzymes. Of more interest, from the point of view of fine control of metabolism, are systems where the enzyme is activated (or, in some cases, inactivated) by phosphorylation or adenylation: this is termed **covalent modification.**

Consider, for example, **glycogen phosphorylase** (also known simply as **phosphorylase**) of animal muscle cells, which catalyses, in a largely irreversible fashion, the reaction

$$\text{glycogen } (n \text{ residues}) + \text{P}_i \rightarrow \text{glucose-1-phosphate} + \text{glycogen } (n-1 \text{ residues}) \tag{14.7}$$

The enzyme is dimeric and exists in two forms. Phosphorylase *b* is strongly inhibited *in vivo* by glucose-6-phosphate and, in any case, is only active when AMP binds at an allosteric site, which rarely happens under physiological conditions because of competition by ATP for the same site. Phosphorylase *a*, which has a phosphorylated serine-14 residue on each of its two sub-units, is fully active under most physiological conditions. Detailed molecular structures were revealed in the late 1980s, by the X-ray diffraction studies of Louise Johnson, Robert Fletterick and others. The sub-units of phosphorylase *b* exist largely in the inactive T-state which, in contrast to the active R-state, has a loop (residues 282-286) obstructing a substrate-binding site. Phosphorylation of serine-14 causes a disordered N-terminal region (residues 5-16) to take up a helical structure which fits into a cavity at the sub-unit interface, linking to both sub-units through an extensive network of hydrogen bonds and van der Waals interactions. Thus, conversion of phosphorylase *b* to phosphorylase *a* brings the sub-units closer together, because of the introduction of interactions such as a hydrogen bond between the phosphorylated serine-14 in one sub-unit and arginine-43 in the other, together with similar arginine-10/leucine-115, aspartate-42/glutamine-72 and other links. These interactions preferentially stabilize the R-state of each sub-unit, and so lead to increased enzyme activity. Hence, the N-terminal region in its helical form may be regarded as an internal allosteric activator. In addition, conversion to phosphorylase *a* favours the binding of AMP, but not glucose-6-phosphate, which also favours the T to R transition.

The largely inactive phosphorylase *b* and the active phosphorylase *a* may be interconverted by a system involving two enzymes, **phosphorylase kinase** and **phosphorylase phosphatase**:

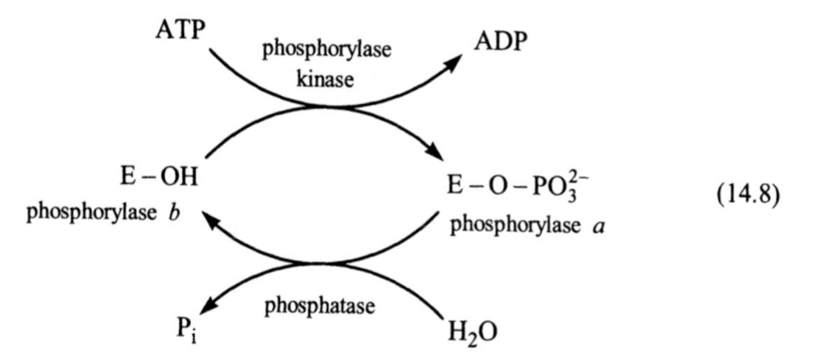

(14.8)

Muscle phosphorylase kinase has four different sub-units and the structure $(\alpha\beta\gamma\delta)_4$, the catalytic-sites being located on each γ sub-unit. The δ sub-unit is known as **calmodulin** (also found in other enzymes), which undergoes a conformational change in the presence of calcium ions and activates the γ sub-unit. For full activity, the α and β sub-units both require phosphorylating at serine residues. This is done by the action of another kinase **(protein kinase A),** which is activated by cAMP. Protein kinase has two catalytic sub-units, but these are inactive in the presence of two regulatory sub-units. However, cAMP, when present, forms complexes with the regulatory sub-units, causing them to dissociate away from the catalytic sub-units, which can then act on the phosphorylase kinase. Hence, an increased concentration of cAMP within a cell activates protein kinase, which in turn activates phosphorylase kinase. This can then convert glycogen phosphorylase *b* to phosphorylase *a,* facilitating the breakdown of glycogen under physiological conditions. **Cascade systems** of this type provide another mechanism for the **amplification** of metabolic regulation, since small changes in concentration of a key substance (in this case cAMP) can have a considerable effect on metabolism. Indeed, far greater amplification is possible from a cascade process than can be contributed by the sigmoidal binding of a modifier to a single allosteric enzyme.

Yet another amplification mechanism exists where there is an apparently **futile cycle** of intermediates. The best known example of this is the interconversion of fructose-6-phosphate (F-6-P) and fructose-l,6-bisphosphate (FBP), catalysed by the enzymes **phosphofructokinase (PFK)** and **fructose-l,6-bisphosphatase (FBPase)**:

ATP → ADP (PFK)

glucose-6-P ⇌ F-6-P ⇄ FBP ⇌ ⇌ ⇌ pyruvate (14.9)

P_i ← H_2O (FBPase)

PFK and FBPase both catalyse more-or-less irreversible reactions, and these reactions both appear to be taking place under most conditions, irrespective of the direction of flux through the system. Since the reaction catalysed by PFK is accompanied by the breakdown of ATP, this set-up seems to be wasteful of energy (hence the term ‘futile cycle ‘).

However, Eric Newsholme (1973) pointed out that this may be tolerated by the organism because the arrangement permits an extremely sensitive control mechanism to operate. A high [AMP]/[ATP] ratio results in inhibition of FBPase and activation of PFK: this favours glycolysis and the production of ATP at the expense of ADP and AMP (see sections 5.2.2. and 11.5.4). A low [AMP]/[ATP] ratio results in inhibition of PFK: this favours gluconeogenesis and the utilization of ATP. Thus it may be seen that ATP controls its own production and utilization.

In animals, another important regulatory mechanism in this enzyme system is the activation of PFK and inhibition of FBPase by fructose-2,6-bisphosphate, whose synthesis is linked to the cAMP concentration via a cascade system similar to that outlined above.

Some aspects of the characteristics of PFK have been explained by X-ray diffraction studies. The enzyme is a tetramer, and each of the four identical sub-units can exist in R and T forms, the transitions apparently occurring in symmetrical fashion as in the Monod-Wyman-Changeux model. Tilman Schirmer and Philip Evans (1990) showed that the four sub-units of bacterial PFK rotate with respect to each other during the T/R transition, which causes changes in the vicinity of binding sites for F-6-P, located at interfaces between sub-units. In particular, a reversal in the positions of the side chains of glutamate-161 and arginine-162 in one sub-unit greatly facilitates the binding of F-6-P to the opposing sub-unit in the R-form. Also, there are stabilizing hydrogen bonds between T-T and R-R sub-units, in the latter case involving intervening water molecules, whereas T-R hybrids might not be stabilized in this way. Thus the reasons for sigmoidal binding of F-6-P and the concerted T/R transitions may be understood. However, detailed explanations for the action of ATP and other allosteric effectors have yet to be obtained.

The amplification of regulation which may be achieved by the existence of a futile cycle, such as that involving PFK and FBPase, is far greater than that possible for a single enzyme, and can enable the direction of overall flux to be reversed (note that for a near-equilibrium reaction being catalysed by a single enzyme, an allosteric modifier will usually affect the forward and back reactions in a similar fashion, as discussed in section 14.2.3. In contrast, a modifier of the enzyme involved in a futile cycle may affect the forward and back reaction in different ways, i.e. may activate one and inhibit the other, thus providing a much more sensitive mechanism for metabolic regulation).

Note that, whereas glycogen breakdown is catalysed by glycogen phosphorylase, an enzyme activated by cAMP via a cascade system (as discussed above), glycogen synthesis is catalysed by a separate enzyme, glycogen synthase, which is *inactivated* by cAMP, a similar cascade system being involved. Also, the synthesis of cAMP in muscle cells is stimulated by the hormone epinephrine (adrenaline), via yet another cascade system. The binding of the hormone to a membrane-bound receptor causes GTP to replace GDP on the α-sub-unit of an associated G-protein, which can then activate adenylate cyclase to convert ATP to cAMP.

14.2.6 Some examples of allosteric enzymes involved in metabolic regulation

There is no typical mechanism for the fine control of metabolism. Already we have discussed examples of a variety of regulatory processes, most of them involving allosteric enzymes.

In the previous section (14.2.5) we considered how the involvement of PFK and FBPase in a futile cycle provides a very sensitive mechanism for metabolic regulation. We also saw how covalent modification of enzymes such as muscle glycogen phosphorylase can result in an amplification of regulation. Many other allosteric enzymes act without being involved in such systems.

Aspartate carbamoyltransferase (ACTase), also called aspartate transcarbamoylase (ATCase), from *E. coli* was the first enzyme in which the active and regulatory sites were shown to be clearly separated. Indeed, not only are they found in different parts of the molecule, they are located on different sub-units (Fig. 14.5). Although this has facilitated the investigation of the characteristics of the enzyme, it is an unusual finding. Most allosteric enzymes are oligomers consisting of *identical* sub-units. ACTase catalyses the reaction:

$$\underset{\text{aspartate}}{{}^{-}O_2C{-}CH_2{-}CH(NH_3^+){-}CO_2^-} + \underset{\text{carbamoyl phosphate}}{H_2N{-}C(=O){-}OPO_3^{2-}} \rightleftharpoons \underset{N\text{-carbamoyl aspartate}}{{}^{-}O_2C{-}CH_2{-}CH(CO_2^-){-}NH{-}C(=O){-}NH_2} + P_i \tag{14.10}$$

This is the first in a sequence of reactions leading to the biosynthesis of the pyrimidine nucleotides UMP, UDP, UTP and, finally, CTP. Two important control mechanisms regulate pyrimidine formation in *E. coli*: the synthesis of ACTase is repressed in the presence of uracil, and the metabolic end-product, CTP, controls its own synthesis by feed-back inhibition of ACTase. John Gerhart and Arthur Pardee (1962) showed that CTP increases the sigmoidal nature of a Michaelis-Menten plot for the reaction catalysed by ACTase, V_{max} being unchanged, whereas the activator ATP (a purine nucleotide) decreases the sigmoidal nature of the curve. From this, they and others developed the concept of allosteric regulation.

Gerhart, together with Howard Schachman, dissociated the enzyme by the action of *p*-mercuribenzoate and showed that one type of sub-unit (c) possesses all the catalytic activity, while another (r) contains the regulator sites. The reaction as catalysed in the absence of the regulatory sub-units shows hyperbolic characteristics and is not affected by ATP or CTP.

Sequence analysis by Klaus Weber (1968) and X-ray diffraction studies by William Lipscomb and co-workers (1968 and following years) showed that the holoenzyme structure is $c_3(r_2)_3c_3$, two catalytic trimers (each of M_r 100 000) being separated by three regulatory dimers (each of M_r 33 000). The structural integrity of the holoenzyme is stabilized by Zn^{2+} ions (Fig. 14.5).

Each regulatory dimer can bind two molecules of CTP but, although the two sub-units are apparently identical, the affinity for the first molecule of CTP is considerably greater than that for the second. The precise explanation for this negative homotropic cooperativity has yet to be found. Also, it is known that ATP and CTP compete for the same site, probably causing different conformational changes to take place on binding.

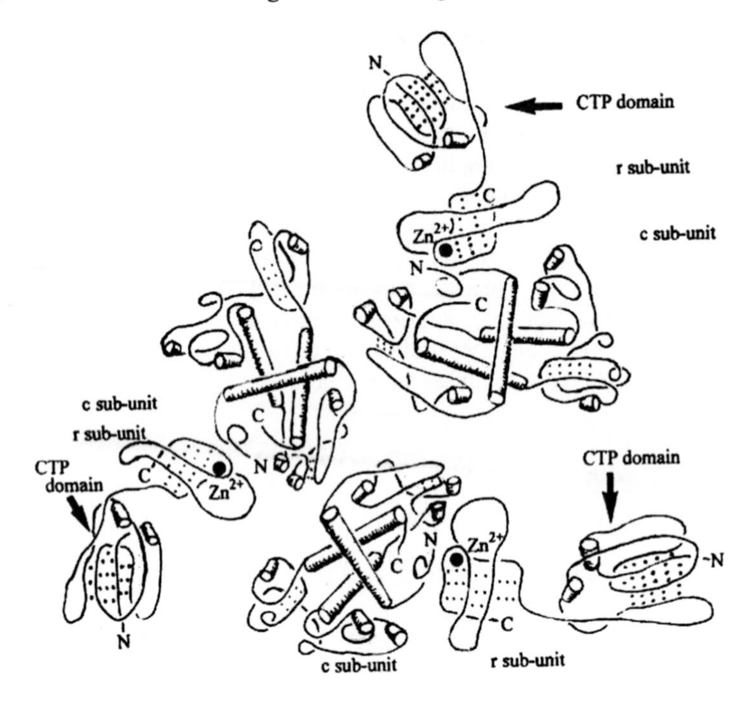

Fig. 14.5 – A simplified representation of the three-dimensional structure of half (c_3r_3) a molecule of *E. coli* ACTase as revealed by the X-ray diffraction studies of Lipscomb and colleagues (1978). The active sites are at or near the interface between adjacent c sub-units. The other c_3r_3 unit is behind the one shown, contact being via the r sub-units. (Conventions as for Fig. 2.10.)

Evidence from stopped-flow and relaxation measurements suggested that the homotropic interactions (between catalytic sites) involve a concerted mechanism, whereas the heterotropic interactions (between catalytic and regulatory sites) involve a mechanism which is sequential. Subsequently, in the 1980s, Evan Kantrowitz, William Lipscomb and colleagues used X-ray crystallography and site-directed mutagenesis to take this further. It emerged that each catalytic sub-unit has separate domains for binding the aspartate and carbamoyl phosphate substrates, which are further apart in the T-state than the R-state. Also, in the former state, a loop (residues 230-245) of each catalytic sub-unit forms linkages with a sub-unit in the catalytic trimer opposite. In particular, glutamate-239 in the loop interacts with both lysine-164 and tyrosine-165 in the other sub-unit, whose glutamate-239 in turn interacts with lysine-164 and tyrosine-165 in the first sub-unit. For the molecule as a whole, three such sets of interactions occur, holding the two catalytic trimers close to each other and hindering access to the active sites. None of these interactions are present in the R-state, allowing the trimers to move apart and facilitate substrate access. Furthermore, the aspartate and carbamoyl phosphate domains of each catalytic sub-unit are brought closer together because of the interaction between glutamate-50 and both arginine-167 and arginine-234, and between serine-171 and both glutamine-133 and histidine-134, helping to create an active site pocket. For these and other reasons, the R-state has much more catalytic activity than the T-state, and the change from T-state to R-state takes place in a concerted fashion.

The modifiers of most allosteric enzymes are relatively small molecules. However, **lactose synthase** (section 5.2.3) is an exception. Although the modifier, α-lactalbumin, may be regarded as a sub-unit of the enzyme, it binds only weakly to the other protein component (*N*-acetylactosamine synthase/galactosyltransferase), but in doing so it completely changes the specificity of the enzyme, enabling glucose to act as a substrate. John Morrison and Kurt Ebner (1971) concluded from their studies that the reaction proceeds by a compulsory-order mechanism, as follows:

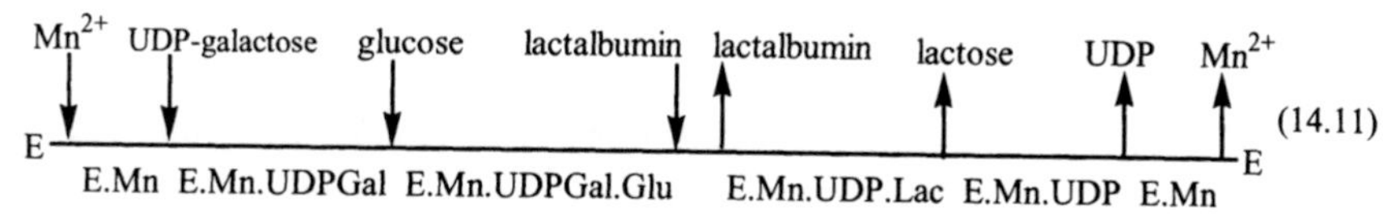

(14.11)

Boopathy Ramakrishnan and Pradman Qasba (2001-3) showed that the binding of UDP-galactose results in a conformational change to the active site of the enzyme. In the absence of α-lactalbumin, glucose can bind only transiently, but it forms hydrogen bonds with α-lactalbumin, when present, stimulating further adjustments to the active site which maximise interactions with glucose. The molecule is thus held in a suitable orientation long enough for lactose formation to take place.

The **pyruvate dehydrogenase multienzyme complex** of *E. coli* (see sections 5.2.5 and 11.5.6) is subject to competitive product inhibition by NADH and acetyl-CoA. Also, the E_1 component may be activated by AMP and inhibited by GTP. The regulation of the mammalian system has similar features, but is more complicated than in *E. coli*: covalent modification is involved, with **kinase** and **phosphatase** enzymes being present within the complex. Lester Reed (1974) showed that the kinase, bound to the acetyltransferase (E_2), inactivates the complex by catalysing the phosphorylation of a serine residue in the pyruvate dehydrogenase component (E_1). This kinase is itself activated by acetyl-CoA and NADH, and inhibited by ADP and pyruvate. The phosphatase, which catalyses the removal of this phosphate, is activated by Ca^{2+} and Mg^{2+}. Thus, increased ratios of [ATP]/[ADP], [NADH]/[NAD^+] and [acetyl-CoA]/[CoASH] within mitochondria, where the complex is located, will tend to reduce its overall activity. The situation in such a multienzyme complex, where a metabolite is passed directly from one enzyme to another, would appear to be ideal for sensitive metabolic regulation.

SUMMARY OF CHAPTER 14

The fractional saturation of a protein with a ligand is far more sensitive to changes in ligand concentration if the binding is sigmoidal than if it is hyperbolic. This is important in the physiological roles of haemoglobin and of allosteric enzymes.

The fine (or acute) control of metabolic regulation frequently involves the activation or inhibition of allosteric enzymes, particularly of those catalysing reactions which are far from equilibrium. Allosteric inhibitors usually act on enzymes occurring early in metabolic sequences, whereas allosteric activators may act on enzymes lying towards the end of metabolic pathways: this helps to maintain the concentrations of most metabolic intermediates within reasonable limits.

The effects of allosteric modifications are most marked at low or moderate substrate concentrations, when it is particularly important that the substrate is utilized (or conserved) according to the precise needs of the cell.

Sigmoidal binding of allosteric modifiers provides a mechanism for the amplification of regulation. Other amplification mechanisms include covalent modification and the establishment of 'futile' cycles.

Metabolic regulation *in vivo* may be extremely complex. Factors such as substrate or enzyme compartmentation and ratios of concentrations of pairs of coenzymes (e.g. [ATP]/[ADP] may be involved, together with allosteric effects.

FURTHER READING

Alberts, B., Johnson, A. *et al* (2002), *Molecular Biology of the Cell,* 4th edn., Garland (Chapters 3, 11-15).

Bender, D. A (1997), *Introduction to Nutrition and Metabolism,* 2nd edn., Taylor and Francis.

Carling, D. (2004), The AMP-activated protein kinase cascade: a unifying system for energy control, *Trends in Biochemical Sciences*, **29**, 18-24.

Danon, A. (2002), Redox reactions of regulatory proteins, *Trends in Biochemical Sciences*, **27**, 197-202.

Fell, D. (1997), *Understanding the Control of Metabolism*, Portland Press.

Gianchandani, E. P., Brautigan, D. L. and Papin, J. A. (2006), Systems analyses characterize integrated functions of biochemical networks, *Trends in Biochemical Sciences*, **31**, 284-291.

Hardie, D. G., Carling, D. and Carlson, M. (1998), The AMP-activated/SNFl protein kinase subfamily - metabolic sensors of the eukaryotic cell?, *Annual Review of Biochemistry,* **67**, 821-855.

Hudmon, N. and Schulman, H. (2002), Neuronal Ca^{2+}/calmodulin-dependent protein kinase II, *Annual Review of Biochemistry*, **71**, 473-510.

Kantrowitz, E. R. and Lipscomb, W. N. (1990), *E. coli* aspartate transaminase, *Trends in Biochemical Sciences,* **15**, 53-59.

Kemp, B. E., Mitchelhill, K. I., et *al* (1999), Dealing with energy demand: the AMP-activated protein kinase, *Trends in Biochemical Sciences,* **24**, 22-25.

Kim, J.-W. and Dang, C. V. (2005), Multifaceted roles of glycolytic enzymes, *Trends in Biochemical Sciences*, **30**, 142-150.

Marianayagam, N. J., Sunde, M. and Matthews, J. M. (2004), The power of two: protein dimerization in biology, *Trends in Biochemical Sciences*, **29**, 618-625.

Nelson, D. L. and Cox, M. M. (2004), *Lehninger Principles of Biochemistry,* 4th edn., Worth (Chapters 5, 14, 15, 23).

Papin, J. A., Price, N. D. *et al* (2003), Metabolic pathways in the post-genomic era, *Trends in Biochemical Sciences*, **28**, 250-258.

Qasba, P. K., Ramakrishnan, B. and Boeggeman, E. (2005), Substrate-induced conformational changes in glycosyltransferases, *Trends in Biochemical Sciences*, **30**, 53-62.

Salter, M. Knowles, R. G. and Pogson, C. I. (1994), Metabolic control, *Essays in Biochemistry,* **28**, 1-12, Portland Press.

Soderling, T. R. (1999), The Ca^{2+}-calmodulin-dependent protein kinase cascade, *Trends in Biochemical Sciences,* **24**, 232-236.

Voet, D. and Voet, J. G. (2004), *Biochemistry,* 3rd edn., Wiley (Chapters 10, 16-18, 21, 27).

Woodgett, J. (ed.) (2000), *Protein Kinase Functions,* Oxford University Press.

Part 3

Application of Enzymology

15

Investigation of Enzymes in Biological Preparations

15.1 CHOICE OF PREPARATION FOR THE INVESTIGATION OF ENZYME CHARACTERISTICS

It should be clear from previous chapters (particularly sections 7.1.3 and 14.2.1) that there is no ideal preparation in which to investigate the properties of an enzyme: different properties may have to be investigated in different preparations. **Highly purified preparations** (see section 16.2) are essential for the investigation of enzyme structures; they also allow enzyme characteristics (e.g. K_m and k_{cat}) to be determined and reaction mechanisms to be studied without complications arising from the presence of other enzymes and reactions. However, the removal of an enzyme from its natural environment may give a distorted impression of its characteristics *in vivo* and make the elucidation of its physiological role more difficult. In order to get the complete picture, therefore, it is necessary to investigate enzymes in a purified form and also in preparations where some cellular organization has been retained.

The investigation of enzymes in single-cell micro-organisms is relatively straightforward, but the extra organization of animals and plants presents further problems. **Whole animal** (or **plant**) **experiments** can give useful information about metabolic pathways (e.g. by studying the metabolic fate of ingested or injected radioactive isotopes), but reveal little about the function of individual enzymes.

Tissue slice techniques, as introduced by Otto Warburg (1923), enable *in vitro* studies to be carried out on cells which retain a considerable degree of their organization. A thin slice is cut from a tissue using a sharp implement such as a **microtome** (a device incorporating a specimen holder, a sharp blade and an advance mechanism, so that slices of reproducible thickness may be cut from the tissue). The slice is usually about 0.5 mm thick, because a thinner one would contain too large a portion of cells damaged during the slicing, and a thicker one would present problems relating to diffusion (see later).

After preparation, the sliced tissue is incubated in a suitable medium at controlled pH and temperature, and the uptake and release of substances of interest may be measured. Therefore, in contrast to *in vivo* studies, the metabolism being investigated is that of one particular type of tissue, e.g. liver. However, it is still difficult to relate the results to any particular enzyme. Also, in the absence of an intact vascular system, ingredients from the medium can only reach the innermost cells by diffusion through the others.

An alternative approach to the study of a tissue, and one which avoids causing damage to any cells, is the use of **perfusion techniques.** An intact tissue is removed from an animal and placed in a fluid environment which provides sufficient oxygen to keep the cells functioning while they are investigated. If the tissue is small and thin enough for each cell to be near the exterior, it may be incubated directly in the oxygenating medium; otherwise, it must be oxygenated by perfusion through blood vessels, when uptake and release of substances via these vascular routes may also be investigated. In either case, the main difference from the situation *in vivo* is that the isolated tissue is not linked to a nervous system.

Problems that result from different cells of a tissue not being equally accessible to the medium may be overcome by separating the cells from each other (i.e. by preparing a **tissue culture**). This is often done by treating the isolated tissue with collagenase or other proteolytic enzyme to break down the matrix and enable the individual cells to be dispersed. The conditions of proteolysis should be sufficiently mild so that each isolated cell retains most of its *in vivo* characteristics, but it is very difficult to prevent some damage occurring to surface proteins. Under favourable conditions, isolated cells in culture may grow and divide, exactly like micro-organisms, and thus increase the number of cells available for investigation. The main technical problem in long-term culture is the prevention of contamination, particularly by micro-organisms, but there is also a tendency for cells to lose their precise function (i.e. to de-differentiate) as time passes. So far, the most successful cultures of normal (as opposed to tumour) cells have come from skin biopsies, while cultures from liver samples have proved very difficult to grow.

All investigations of enzymes in intact cells must be complicated by transport through the **plasma membrane** (i.e. the membrane enclosing each cell). A simple way to overcome this problem is to prepare a **cell-free system.** This may be done by the **homogenization** of a tissue in a suitable isotonic medium (e.g. 0.25 M sucrose), the aim usually being to disrupt the cells and release the contents without damaging the sub-cellular organelles. Alternatively, a cell-free system may be obtained by **reconstitution** from previously isolated organelles and cytoplasm. Homogenates of soft tissues, such as liver, are often prepared by forcing the tissue and medium through the narrow gap (about 0.3 mm) between the walls of a glass tube and a close-fitting pestle. This pestle, which is made of Teflon or glass, is moved up and down the tube like a piston and also rotated, either by hand (as in the TenBroeck apparatus) or by an electric motor (as in the Potter-Elvehjam homogenizer). Plants, whose cell walls contain a considerable amount of structural material, and tough, fibrous animal tissues, such as heart, may be homogenized by the use of a Waring blender (which operates on the same principle as a domestic kitchen blender): the rapidly rotating blades successfully disrupt the cells, but unfortunately tend to disrupt the sub-cellular organelles as well.

Cell-free preparations are frequently produced from micro-organisms by subjecting them to high-frequency sound waves **(sonication)**; this procedure too causes considerable damage to cell membranes. It is important to maintain low temperatures throughout the production of *in vitro* preparations, for reasons discussed in section 15.2.1.

Cell-free preparations provide a convenient and reproducible means of investigating enzymes and metabolism in the presence of some important cellular components, and interesting deviations from findings in pure enzyme preparations are sometimes revealed. For example, different estimates for the K_m of liver 'glucokinase' have been obtained in homogenates and in pure solution. However, regardless of the method by which cell-free systems are prepared, a considerable degree of cellular organization is lost in the process. Hence the results obtained still cannot be taken to be an exact representation of what occurs in the intact organism. For the same reason, there should be caution in the interpretation of findings made with organelles which have been isolated from other cellular components by density-gradient centrifugation or some other method (see section 15.3.2).

In summary, therefore, it is only possible to investigate the characteristics of individual enzymes in preparations where cellular organization is totally or partially absent, so the characteristics observed are not necessarily identical to those existing *in vivo*. The differences between the preparation and the whole organism must always be borne in mind during the interpretation of results.

15.2 ENZYME ASSAY

15.2.1 Introduction

The purpose of enzyme assay is to determine how much of a given enzyme, of known characteristics, is present in a tissue homogenate, fluid or partially purified preparation. It is important that the specimens be treated carefully prior to assay, if the results are to have any meaning. The maintenance of cellular organization and integrity is a characteristic of life, and cellular destruction by natural processes (autolysis), accompanied by changes in enzymes and breakdown of cofactors, commences on the death of an organism or the isolation of a tissue. Autolysis is minimized if a tissue is kept cold, so specimens for enzyme assay are usually stored at temperatures below 4°C, both before and after the preparation of homogenates. Many enzymes are easily denatured at high or even moderate temperatures, so this is a further reason why all types of specimens or preparations should be kept cold prior to enzyme assay. (Note that these factors apply to the preparation of specimens for the investigation of enzyme characteristics as well as for enzyme assay.)

Since changes due to autolysis and denaturation must increase with time of storage, it will be apparent that the assays should be performed with a minimum of delay. If some delay is unavoidable, it may be necessary to store the specimens at very low temperatures (e.g. – 60°C) to prevent loss of enzyme activity. However, freezing of tissues might increase the disruption of cells, so there can be no general rule about storage. The stability of the enzyme in question at various temperatures, and the nature of the specimen or preparation (see section 17.3), all need to be taken into account.

15.2.2 Enzyme assay by kinetic determination of catalytic activity

By far the most convenient way to estimate the concentration of a particular enzyme in a preparation or fluid is to determine its catalytic activity, i.e. to find out how much substrate it is capable of converting to product in a given time under specified conditions. As discussed in section 7.1.2, the initial velocity of an enzyme-catalysed reaction taking place under conditions where the Briggs-Haldane steady-state assumptions are valid is given by $v_0 = k_2[E_0][S_0]/([S_0] + K_m)$, where k_2 is the rate constant relating to product formation (i.e. k_{cat}).

If the value of $[S_0]$ used in a particular assay is fixed, the only variables are v_0 and $[E_0]$, so v_0 = constant × $[E_0]$. In other words, for any system where the steady-state assumptions are valid (see section 7.1.3) and the initial substrate concentration is fixed, the initial velocity of an enzyme-catalysed reaction is directly proportional to the concentration of the enzyme, i.e. there is a linear relationship between enzyme concentration and catalytic activity.

Although in theory this relationship is true provided the substrate concentration is fixed, regardless of the actual value, in practice it is more reliably valid when the substrate concentration is high enough to be approximately saturating. There are two main reasons for this. Firstly, integration of the appropriate rate equation (see section 7.2.1) shows that the linear, steady-state phase of the reaction is more prolonged as the degree of enzyme saturation is increased (all other factors being equal). In simple terms, this is because the rate of utilization of substrate becomes less significant in relation to the total concentration of substrate present as $[S_0]$ is increased. For this reason more accurate estimates of v_0 can be obtained at high rather than at low $[S_0]$ values. Secondly, at high $[S_0]$, $([S_0] + K_m) \approx [S_0]$, so $v_0 \approx k_2[E_0]$ and v_0 does not vary with small changes in $[S_0]$. This means that reproducible results can be obtained from an assay system without it being necessary to fix the substrate concentration at an extremely precise value, provided $[S_0]$ is sufficient to almost saturate the enzyme. Hence the setting-up of a reliable assay system is more convenient the higher the chosen value of $[S_0]$.

According to the Michaelis-Menten equation, an enzyme is only completely saturated by substrate at infinite substrate concentration, so it is necessary to talk in terms of near-saturation rather than complete saturation. This equation also enables the v_0/V_{max} ratio to be calculated for each initial substrate concentration, provided the K_m value is known. So, for example, if $[S_0] = 0.5\ K_m$, $v_0 = 0.33V_{max}$; if $[S_0] = 5K_m$, $v_0 = 0.83V_{max}$; and if $[S_0] = 50K_m$, $v_0 = 0.98V_{max}$. For a simple system, the v_0/V_{max} ratio will also be the fractional saturation of the enzyme (see section 12.4).

It should be clearly understood that the v_0/V_{max} ratios and fractional saturation values are independent of $[E_0]$, provided $[S_0] \gg [E_0]$, as discussed in sections 7.1.1 and 7.1.2. However, the actual values of v_0 and V_{max} do vary with $[E_0]$, which is the whole point of enzyme assay by kinetic methods.

It might seem from the above discussion that a procedure being used for enzyme assay could always be improved by increasing the substrate concentration, without limit, but of course other factors have to be taken into consideration. These include the solubility of the substrate, the cost of the substrate, and the possibility of substrate inhibition (see section 8.2.6). Hence the actual substrate concentration chosen must be a compromise.

If a reaction involves more than one substrate, then the concentrations of each must be fixed, preferably at near-saturating levels.

The rate of an enzyme-catalysed reaction may also depend on the concentration of a cofactor (or cofactors). As with substrates, each cofactor must be present at a fixed concentration, and preferably in excess, if a reliable and reproducible system for enzyme assay is to be obtained.

Another prerequisite for a successful assay system is that the reaction being catalysed should be capable of being accurately monitored, i.e. there should be a change in optical, electrical or other properties as substrate is converted to product (see section 18.1). If this is not the case, it may be possible to follow the course of the reaction indirectly by coupling it to one where some such change does take place (section 15.2.3).

Regardless of this, the reaction should be carried out under fixed and suitable conditions of pH, ionic strength and temperature. Enzymes are often assayed at their optimal pH (section 3.2.2), but this is not essential. An enzyme might not operate at its exact optimal pH *in vivo,* so there is no fundamental reason for investigating its activity at this pH *in vitro;* also, the pH optimum may vary with temperature and other factors. Nevertheless, the pH chosen for an assay system must be sufficiently near the optimum for an appreciable rate of reaction to take place. It must also be one where the enzyme is relatively stable, for enzyme stability can vary with pH. With some reversible reactions, the pH may influence the forward and back reactions differently. For example, assays involving **lactate dehydrogenase** are best performed in the direction of lactate production at pH 7, but in the direction of pyruvate formation at pH 10.

The presence of salts may affect enzyme-catalysed reactions by shifting the equilibrium of any of the steps involved. They may also effectively reduce the concentration of a substrate by complexing with it. Hence enzyme assays must be performed under carefully controlled conditions of ionic strength and composition.

The choice of temperature is governed by two conflicting factors: reaction rate and enzyme stability. Specimens are usually stored at low temperatures prior to assay in order to prevent loss of enzyme activity (section 15.2.1). However, if they were assayed at these same temperatures, the rate of reaction would be extremely low (or even zero), so the sensitivity of the assay would be very poor. At higher temperatures, the reaction would proceed at a faster rate, giving a more sensitive assay, but the stability of the enzyme would decrease (see section 3.2.3). An important consideration is the time-scale of the assay procedure: it is essential that there is no significant change in the activity of the enzyme over the period in which v_0 is calculated, but immaterial what happens after that. Therefore, in general, it might be possible to use a particular temperature for a short assay procedure but not for a longer procedure involving the same enzyme (Fig. 15.1).

The most common temperatures at which enzyme assays are performed are 25°C (as recommended by IUB in 1961), 30°C (the 1964 IUB recommendation) or 37°C (as preferred by clinical chemists). Many enzymes are somewhat unstable at 37°C, but this temperature provides conditions which are the nearest approximation to those found in human beings and other mammals, and reaction rates are faster than at 25°C or 30°C.

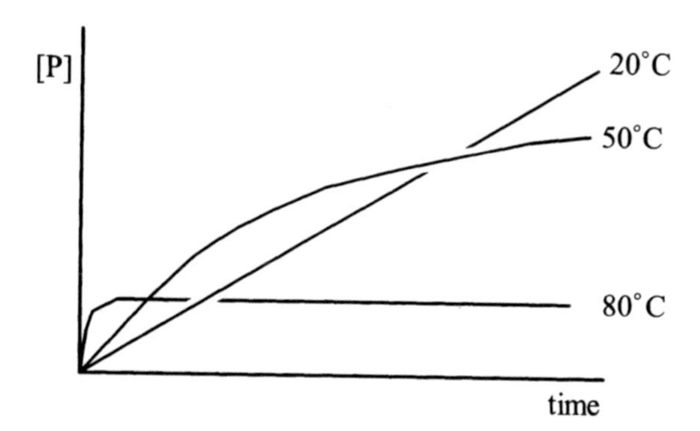

Fig. 15.1 – Plots of product concentration against time for an enzyme-catalysed reaction at 20°C, 50°C and 80°C, without pre-equilibration of the enzyme at the reaction temperature, all other conditions being identical.

The rates of many enzyme-catalysed reactions are approximately doubled if the temperature is increased by 10°C in this range. Hence, regardless of which temperature is chosen for an assay system, it is clear that it must be maintained to within about ±0.2°C if the results are to be reproducible.

Enzyme activity is measured as the amount of substrate lost (or product gained) per unit time, and it should also be related in some way to the amount of specimen used for assay. In 1961 the enzyme commission of the IUB defined an **Enzyme Unit (U)**, later to be known as an **International Unit (IU)** (although the original term continued to be widely used), as the amount of enzyme causing loss of 1 μmol substrate per minute under specified conditions. Later, in 1973, the Commission on Biochemical Nomenclature introduced the **katal (kat)** as the **Système International (SI)** unit of enzyme activity: this is defined as the amount of enzyme causing loss of 1 mol substrate per second under specified conditions. Both units (and all three terms) are in current usage.

If, for convenience, the rate of reaction is being determined by a single reading rather than by continuous monitoring, it is necessary to demonstrate by a preliminary experiment whether a graph of [P] against time is linear over the whole of the incubation period. If it is not, the results cannot be expressed either as International Units or katals, and must be given in terms of the total incubation time (e.g. μmol substrate converted per 20 minutes).

Obviously there must be a direct relationship between the number of units of activity found and the amount of specimen assayed. Therefore, in order for the results of an assay to have any meaning, they must be expressed in terms of the amount of sample assayed, and, where appropriate, extrapolated back to the original tissue. So, for example, results might be given as International Units per litre blood plasma, or as katals per gram plant seed, and so on. However, even these units may not be ideal. For example, liver is likely to contain irregular deposits of glycogen and other non-protein material, so any results expressed as, say, katals per gram liver could vary according to which part of the same liver was cut off, weighed, homogenized and assayed. To minimize problems of this type, units of enzyme activity may be related to the **total protein** content of the sample being assayed (see section 16.2.3), rather than to its weight or volume.

This is termed **specific activity**, and may be expressed as, for example, International Units per mg protein, or as katals per kg protein. In the example being considered, a liver homogenate would be prepared and an aliquot used for enzyme assay. Another aliquot of the same homogenate would be analysed for its total protein content (see section 3.2.1), and so the total amount of protein in the sample used for enzyme assay could be calculated and related to the observed activity (see problem 15.1).

Specific activity is quite different from **molar catalytic activity (or turnover number, k_{cat})**, the latter, by either name, being used to relate the observed activity of an enzyme to its known molar concentration, regardless of any other proteins which may be present. Molar catalytic activity may be expressed as, for example, katals per mole enzyme.

As discussed above, the conditions for the assay of an enzyme are chosen so that the concentrations of substrates, cofactors etc. are fixed and non-limiting. If this is successfully achieved, there will be a linear relationship between the initial velocity of the reaction and the concentration of enzyme in the incubation mixture. One way of testing if a system is satisfactory is to perform an assay and then repeat it using a different sample volume of the same enzyme preparation (and adjusting the volume of water or buffer added so that the total incubation volume is unchanged). The rates of reaction for the two assays should be different, but the calculated specific activity should be the same.

Enzyme assays based on the determination of catalytic activity can distinguish between **active** and **inactive** forms of enzymes, for they do not detect the presence of the latter. On the other hand, they cannot distinguish between **isoenzymes,** so the activity measured will be the sum of the contributions of all the active forms of the enzyme being assayed. Since these will not necessarily have the same molar activity, the relationship between activity and enzyme concentration could be less straight-forward than suggested above. Even if it is known that only one isoenzyme is present, it cannot be assumed that its molar activity in, say, a homogenate is the same as that obtained in a pure preparation.

Hence, kinetic assays are usually used to give an indication of the relative concentration of an enzyme in a preparation, without any attempt being made to interpret the results in terms of actual molar concentrations. The results are usually expressed simply in terms of units of activity.

15.2.3 Coupled kinetic assays

If easily measurable changes take place as a particular enzyme-catalysed reaction proceeds, this reaction can readily be used for the kinetic assay of the enzyme which catalyses it (section 15.2.2). Even if no such easily measurable change takes place, it may still be possible to develop a kinetic assay for the enzyme by coupling the reaction to one where suitable changes do occur. For example, in the reaction catalysed by alanine transaminase, L-alanine + 2-oxoglutarate $\rightleftharpoons$ L-glutamate + pyruvate, no reactant or product is coloured, nor does any one of them significantly absorb ultraviolet light.

However, the reaction may be monitored spectrophotometrically if it is coupled to a second reaction, in which the pyruvate formed in the first reaction acts as a substrate for lactate dehydrogenase (LDH) and is reduced to lactate by the action of the coenzyme (and co-substrate) NADH, as follows: pyruvate + NADH + H^+ $\rightleftharpoons$ lactate + NAD^+. NADH absorbs light at 340 nm, but NAD^+ does not (see Fig. 6.6), so the course of the reaction may be followed by monitoring at this wavelength. If conditions are chosen carefully, the rate of the second reaction will be an indicator of the rate of the first reaction.

Let us consider the general situation: $A \xrightarrow{E_1} B \xrightarrow{E_2} C$, where E_1 is the enzyme catalysing the primary reaction and E_2 the enzyme catalysing the indicator reaction. At zero time, the concentrations of B and C will be zero, and the concentration of A should be fixed and non-limiting (as discussed in section 15.2.2). If there are any second substrates or cofactors for E_1 or E_2 (e.g. NADH is a co-substrate in the example where E_2 is LDH), then these should also be present initially at fixed and non-limiting concentrations. As the reaction proceeds, the concentration of B (pyruvate in the example where E_2 is LDH) will start to rise from its initial value of zero. In general, the rate of change of [B] is given by $d[B]/dt = v_1 - v_2$, where v_1 is the velocity of the primary reaction and v_2 the velocity of the indicator reaction. As with the velocity of any single-enzyme system, v_1 should reach a constant value almost instantaneously and remain at this value over the period of interest (the next few minutes). In contrast, v_2 will have an initial value of zero (since [B] is initially zero) and will rise as [B] rises, according to the Michaelis-Menten equation (7.12)

$$v_2 = \frac{V_{max}^{B}[B]}{[B]+K_{m}^{B}}$$

where V_{max}^{B} and K_{m}^{B} relate to the reaction catalysed by the enzyme E_2. Therefore, over the period of interest

$$\frac{d[B]}{dt} = v_1 - \frac{V_{max}^{B}[B]}{[B]+K_{m}^{B}} \qquad (15.1)$$

For a satisfactory coupled assay system, the crucially important point is that [B] should reach a constant value very quickly. In other words, a steady-state should be set up for the overall system as quickly as possible and this should then remain in operation for the rest of the period of interest. Integration of the expression for $d[B]/dt$ shows that the time taken for this steady-state to be established (or at least very nearly established) is directly proportional to K_{m}^{B}/V_{max}^{B}. Now, K_{m}^{B} is a constant characteristic of the enzyme E_2, but V_{max}^{B} is directly proportional to the concentration of E_2.

Hence, the larger the concentration of E_2, the quicker is the establishment of a near steady-state for the overall system, with $v_1 = v_2$. If there is so little E_2 present that V^{B}_{max} is less than v_1, then a steady-state can never be set up, [B] will continue to rise and v_2 can never reach the value of v_1. It is therefore important that sufficient of the indicator enzyme is added to a coupled assay system to ensure that there is a minimum of delay before a steady-state is established, and that the rate of the indicator reaction at steady-state is a reasonable approximation to the rate of the primary reaction. As with a single-enzyme system (section 15.2.2), the validity of a coupled assay procedure can be checked by assaying different amounts of the same enzyme preparation: the specific activity should be found to be the same in each case.

One further factor which needs to be taken into account in the setting-up of a coupled assay procedure is the pH profiles of the various enzymes involved. Clearly it is essential that all the component enzymes are active at the pH used for assay, and the narrow ranges of activity of some enzymes pose problems in this respect. Consider, for example, the sequence:

$$\text{sucrose} \xrightarrow[\searrow \text{ fructose}]{\text{invertase}} \text{glucose} \xrightarrow[\text{O}_2 \curvearrowright \text{gluconolactone}]{\text{glucose oxidase}} \text{H}_2\text{O}_2 \xrightarrow[\text{chromogen} \curvearrowright \text{dye}]{\text{peroxidase}} \text{H}_2\text{O} + \tfrac{1}{2}\text{O}_2 \qquad (15.2)$$

A coupled assay procedure involving three enzymes, while more complicated than one involving two enzymes, may nevertheless be perfectly satisfactory provided the principles discussed above are adhered to. Hence, in theory, invertase could be assayed in presence of non-limiting amounts of sucrose, oxygen, glucose oxidase, peroxidase and a chromogen (e.g. guaiacum), by following the rate of appearance of the coloured dye. However, George Guilbault (1970) drew attention to a possible but unpractical system in which invertase was active around pH 5, glucose oxidase around pH 7-8 and peroxidase around pH 10, there being no pH value when all three enzymes were active enough to be coupled together. It should be pointed out, though, that today these enzymes are readily available in forms where the pH optima are much closer together than this. Of particular importance, fungal glucose oxidase and horseradish peroxidase are both sufficiently active at pH 5 to make practicable a coupled system for the assay of glucose oxidase (or, more commonly, of the substrate glucose, according to the principles discussed in section 17.2.2).

15.2.4 Radioimmunoassay (RIA) of enzymes

As a prerequisite for RIA, the substance to be assayed must be an antigen (Ag) for which a specific antibody (Ab) can be obtained: such an antibody will have a specific site for the antigen. It must also be possible to obtain a pure specimen of antigen labelled with a radioisotope (Ag*).

The sample to be assayed is mixed with antibody and with a small amount of the labelled antigen. The following reactions take place, and are allowed to come to equilibrium:

$$\mathrm{Ag} + \mathrm{Ab} \rightleftharpoons \mathrm{Ag.Ab} \quad (15.3)$$

$$\mathrm{Ag^*} + \mathrm{Ab} \rightleftharpoons \mathrm{Ag^*.Ab} \quad (15.4)$$

The antibody, with its bound antigen, is then separated from the free antigen and the distribution of radioactive isotope between the free and bound antigen fractions is investigated. It will be realized that the labelled antigen competes with the antigen in the sample for the available binding sites on the antibody, so the higher the concentration of antigen in the sample, the less radioactive antigen will be able to bind to the antibody and the greater will be the radioactive content of the free antigen fraction. In this way, the concentration of antigen in the sample can be estimated.

Antibodies for the RIA of enzymes may be prepared in the form of antisera by immunizing rabbits with the required enzyme. For example, the blood of a rabbit immunized in the footpad with human pancreatic α-amylase contains sufficient antibodies within a few weeks to be usable as an antiserum. However, even so, there remains the problem of obtaining pure specimens of radioactively-labelled enzyme (but see below).

A further problem is that, since the antigen is a protein like the antibody, it is more difficult to separate the free antigen from the antigen-antibody complex than would be the case if the antigen were small (when, for example, ion exchange resins might be employed). One way to overcome this problem is to use an antibody immobilized by attachment to an insoluble matrix. Another is to introduce further antigen-antibody interactions. For example, if an antiserum from rabbit is employed, the antigen-antibody complex might be precipitated with goat anti-rabbit γ-globulin.

The part of the enzyme which binds to the antibody is likely to be quite distinct from the active site, so RIA and catalytic assay of enzymes can give different information. Catalytically-inactive forms of enzymes (e.g. proenzymes) may be detected by RIA if they contain the structural part which is recognized by the antibody. On the other hand, each RIA procedure is likely to be specific for one particular isoenzyme (that used to produce the antibody). Hence it can be seen that catalytic assay and RIA procedures are not necessarily alternatives but can be used to complement each other. RIAs also have the advantage of being particularly sensitive (up to a thousand times more sensitive than catalytic assays). However, their use has so far been restricted because of the problems of obtaining the required materials.

Immunoradiometric assay (IRMA), a modified form of RIA, utilizes an immobilized antibody to which the sample antigen can bind, together with a separate, radio-labelled antibody which binds to a different site on the antigen. This technique can be applied to proteins, since these are large enough to have two separate binding-sites for antibody, and it will be noted that there is no requirement for a radio-labelled sample of antigen.

Antibodies produced as outlined above are always heterogeneous, i.e. they form a mixed population recognizing a number of sites (determinants) on the antigen. However, César Milstein and Georges Köhler showed that determinant-specific **monoclonal antibodies** may be produced by the following procedure: a mouse is immunized with an antigen and, several weeks later, cells from its spleen are removed and fused with mouse myeloma cells (cancer cells which can be cultured *in vitro)*; hybrid cells with a desired immunological specificity may then be selected and cultured to produce a large number of identical copies (clones). These monoclonal antibodies are analytical reagents of extremely high sensitivity and specificity, and may be used in RIA or EIA (see section 17.2.3) procedures. Milstein and Köhler were awarded the Nobel Prize for Medicine in 1984.

Antibodies (monoclonal or otherwise) may also be used to precipitate some isoenzymes from solution to allow others to be assayed by a kinetic procedure (see section 19.1.1).

15.3 INVESTIGATION OF SUB-CELLULAR COMPARTMENTATION OF ENZYMES

15.3.1 Enzyme histochemistry

The sub-cellular location of many enzymes may be revealed by microscopy, provided suitable fixation and staining procedures are followed.

Tissues are frozen to below –20°C and sliced using a **cryostat** (a refrigerated microtome). This procedure ensures that the tissue is rigid, essential for obtaining thin slices (about 10μm thick) of satisfactory quality, and also minimizes loss of enzyme activity. The slices are then usually fixed in **formaldehyde** (formalin), HCHO, to prevent any subsequent diffusion of proteins taking place. Formaldehyde brings about cross-linking between side chain amino groups of proteins:

$$\text{Protein-NH}_2 + \text{HCHO} + \text{H}_2\text{N-Protein} \rightarrow \text{Protein-NH-CH}_2\text{-NH-Protein} \quad (15.5)$$

Most enzymes retain at least some activity after treatment. However, some do not, and these have to be investigated without prior fixation. In either case, the tissue section is treated with a buffered solution of a staining mixture containing the substrate of the enzyme under investigation. The staining process requires the conversion of this substrate to the appropriate product, and so can only take place in the regions where the enzyme is present. If possible, the product is immobilized after formation to prevent it diffusing away from the location of the enzyme.

For example, the substrate solution for the widely used **lead-salt method** for **acid phosphatases** includes sodium β-glycerophosphate and lead ions in buffer at pH 5.5. Any acid phosphatase present in the tissue hydrolyses the β-glycerophosphate to produce glycerate, and the phosphate liberated in the process reacts immediately with the lead ions to form insoluble lead phosphate ($PbPO_4$). This may be converted to a more conspicuous product, also insoluble, by developing with a solution of ammonium sulphide:

$$\text{PbPO}_4 \xrightarrow{\text{ammonium sulphide}} \text{PbS}\,. \quad (15.6)$$

In this way, a black precipitate of lead sulphide may be visualized wherever acid phosphatase is located.

Many stains which show up clearly on **optical microscopy** are not suitable for use in **electron microscopy**, since their constituent atoms are no more opaque to an electron beam than are the atoms normally present in the tissue. Only atoms of large atomic weight, e.g. those of heavy metals, scatter an electron beam sufficiently to cause their locality to be significantly more opaque than the rest of the tissue. Thus the lead-salt method, discussed above, is one which can be applied to tissues being visualized by electron microscopy, since electron-opaque lead phosphate is precipitated in the regions where acid phosphatase is present. (Note that for electron microscopy, unlike optical microscopy, there is no advantage in developing to convert lead phosphate to lead sulphide, since both are equally electron-opaque.) However, ultra-thin sections (about 80 nm thick) are required for electron microscopy, so the sections prepared and stained as above must be dehydrated (replacing water with alcohol) and embedded in epoxy resin before being further sectioned on an **ultramicrotome**. Such investigations have revealed, for example, that acid phosphatase is characteristically associated with lysosomes.

Antibodies raised against an enzyme from one species can often show cross reactivity with the same enzyme from a different species. **Immunohistochemistry** utilises the specificity of antibodies to determine the cellular location of enzymes in fixed tissue slices or cultured whole cells. The initial primary antibody-antigen interaction (usually with a monoclonal antibody to reduce non-specific interactions) can be visualised indirectly by an additional incubation with a horseradish peroxidase- or alkaline phosphatase-labelled secondary antibody (usually a polyclonal antibody to enhance the detection) raised against the immunoglobulin species of the primary antibody. The addition of substrates for the enzyme conjugated to the secondary antibody will result in the deposition of coloured material at the site of the antibody-antigen interaction (Fig. 15.2a). Alternatively, the detection can be visualized directly with a fluorescence microscope or confocal microscope, if the primary antibody is labelled with a fluorophore, e.g. fluorescein isothiocyanate (FITC). Dual labelling of two different enzymes on the same tissue can be achieved using two antibodies with different fluorophores attached, e.g. FITC (excitation 490 nm and emission 525 nm) and Texas Red (excitation 596 nm and emission 625 nm) (see Fig 15.2b).

A simple method of locating the distribution of enzymes in tissue sections of animals or plants is **tissue printing**. The tissue to be analysed is cut with a sharp blade and lightly pressed onto a nitrocellulose membrane, which retains an imprint of the proteins in the tissue. The distribution of some enzymes, e.g. peroxidises, can be visualized directly on the membrane using reagents (e.g. chloro-naphthol and hydrogen peroxide in Tris buffer) which deposit a coloured product onto the membrane, reflecting the distribution of peroxidase in the tissue section. Alternatively, if an antibody is available for the enzyme of interest, its location can be visualized, after printing, by **immunoblotting**. The excess adsorption-sites on the membrane should be blocked with milk-protein (casein) or bovine serum albumin, before being probed with the primary-antibody and visualized with the enzyme-conjugated secondary-antibody (see above). The location of the enzyme is indicated by a comparison of the results with ones obtained with a general protein stain.

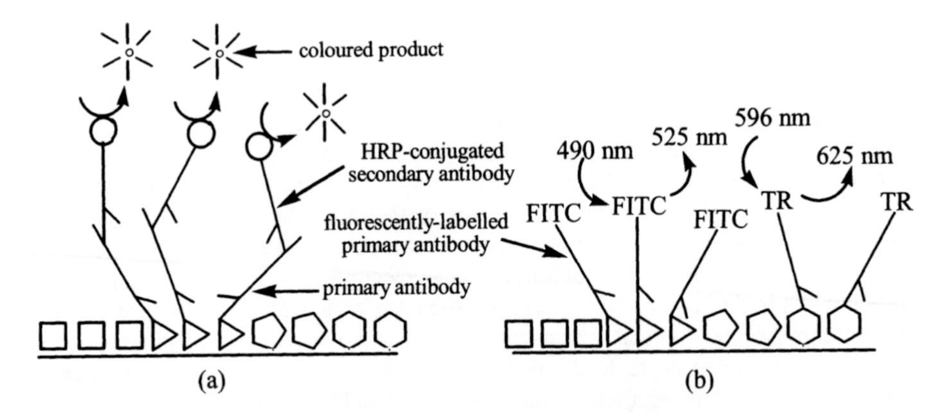

Fig. 15.2 – Diagrammatic representation of immunohistochemical processes: (a) involving a secondary antibody labelled with horseradish peroxidase (HRP); and (b) involving primary antibodies labelled with fluorescein isothiocyanate (FITC) and Texas Red (TR). See text for details. (Note that different proteins in the tissue section are indicated by squares, triangles, pentagons and hexagons).

15.3.2 The use of centrifugation

The sedimentation characteristics of the various sub-cellular organelles are different, so it is possible to separate them by centrifugation of a tissue homogenate, and then to investigate which enzymes are associated with each cell fraction.

Any particle suspended in a liquid medium and being spun in a centrifuge is acted upon by a centrifugal field (G) which is determined by the angular velocity (ω) of the rotor and the radial distance (r) of the particle from the axis of rotation, according to the relationship

$$G = \omega^2 r = (2\pi.\mathrm{RPM}/60)^2 r \tag{15.7}$$

where RPM is the number of revolutions per minute (rev min^{-1}) of the rotor, ω normally being measured in radians per second (Fig. 15.3).

It is convenient to express the centrifugal field as a multiple of the gravitational constant (g), this being termed the **relative centrifugal field (RCF)**.

$$\mathrm{RCF} = \frac{G}{g} = \left(\frac{2\pi\mathrm{RPM}}{60}\right)^2 \frac{r}{g} \tag{15.8}$$

But $g = 980$ cm s^{-1}, so, provided r is measured in centimetres,

$$\mathrm{RCF} = 1.11 \times 10^{-5}(\mathrm{RPM})^2 r \tag{15.9}$$

The sedimentation rate of the particle = $s\omega^2 r$, where s is the sedimentation coefficient, i.e. the rate per unit centrifugal field. Its units, if the other terms are measured in the units stated above, are seconds, but a more practical unit is the Svedberg (S), of numerical value 10^{-13} seconds.

The sedimentation coefficient depends on the viscosity and density of the suspending medium and on the shape, size and density of the particle. For a spherical particle,

$$s = \frac{2(\rho_p - \rho)r_p^2}{9\eta} \qquad (15.10)$$

where ρ_p = density of particle, ρ = density of medium, r_p = radius of particle and η = viscosity of medium.

The time (t) taken for the particle to sediment from the meniscus of the suspending medium to the bottom of the centrifuge tube is given by:

$$t = \frac{9}{2(\rho_p - \rho)r_p^2}\frac{1}{\omega^2}\log_e\left(\frac{r_b}{r_t}\right) = \frac{1}{s}\frac{1}{\omega^2}\log_e\left(\frac{r_b}{r_t}\right) \qquad (15.11)$$

where r_t = radial distance of liquid meniscus from the axis of rotation and r_b = radial distance of bottom of tube from axis of rotation (Fig. 15.3).

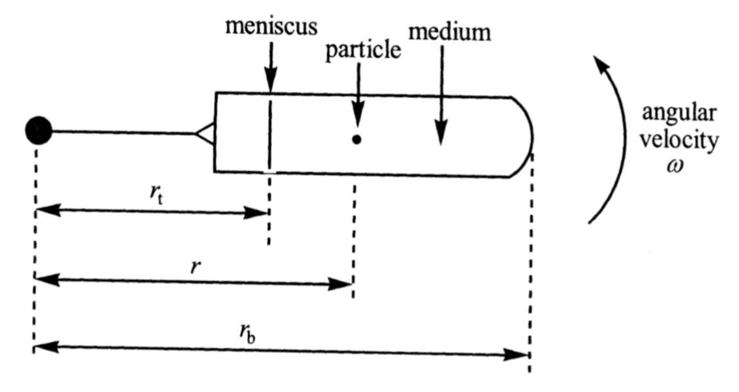

Fig. 15.3 – Diagrammatic representation of a typical centrifugation procedure, as viewed from above. See text for details.

Thus it can be seen that all particles of a given size and density will have reached the bottom of the tube by a time determined by the applied centrifugal field. Note, however, that the density of the suspending medium is another relevant factor and that only those particles which are more dense than the medium will travel towards the bottom of the tube on centrifugation.

The simplest and most widely used method of separating the various sub-cellular organelles from each other is **differential centrifugation**, which developed from the work of George Hogeboom and Walter Schneider (1948). A tissue homogenate is prepared in a medium of low density (e.g. 0.25 M sucrose) and centrifuged in a series of stages, the centrifugal field of each step being higher than for the previous one. At the end of each stage, the sedimented **pellet**, consisting of particles of similar sedimentation characteristics, is removed.

A simplified scheme for the fractionation of a rat liver homogenate is shown in Fig. 15.4. Similar schemes could be drawn up for the fractionation of homogenates of other animal and plant tissue. Although heavy particles sediment faster than lighter particles, there is originally a homogeneous distribution of particles, so some relatively light particles must be present at the bottom of the tube when centrifugation commences. Hence each pellet must be contaminated by lighter particles which happen to be at the bottom of the tube throughout centrifugation. This problem can be overcome to some extent by resuspending each pellet in fresh homogenizing medium and repeating the centrifugation. Another problem is that all sedimented particles tend to diffuse back up the tube towards less concentrated regions: this is of greatest significance when the applied centrifugal field is low. A third problem with differential centrifugation is that some particles, e.g. mitochondria and lysosomes, show similar sedimentation characteristics ($s \approx 2 \times 10^4$ S) because they are roughly the same size, even though they have quite different average densities.

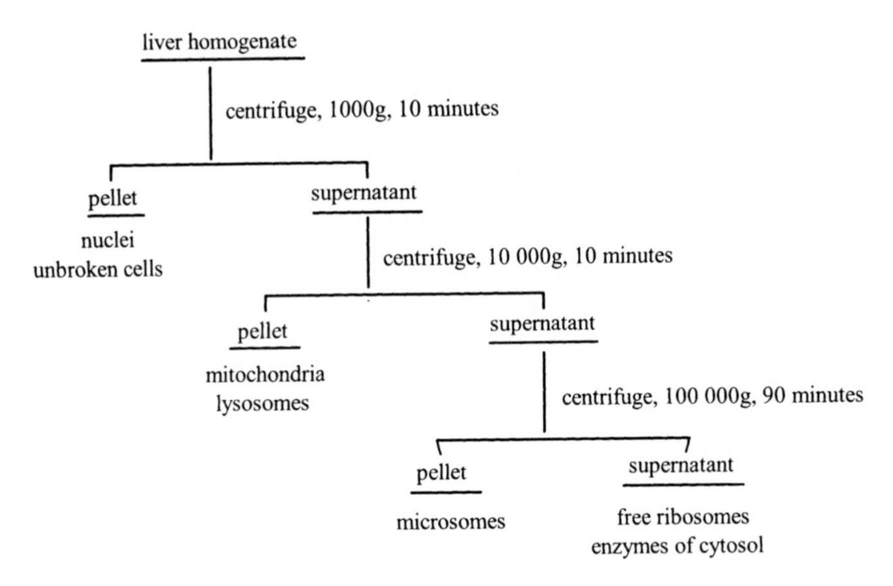

Fig. 15.4 – Simplified scheme for the fractionation by differential centrifugation of a 10% rat liver homogenate in 0.25 M sucrose at 0°C. Note that microsomes are not organelles but particles produced during homogenization, largely from endoplasmic reticulum, and constituting a specific sedimentation fraction.

Centrifugation techniques where the density of the suspending medium is not uniform throughout, i.e. **density-gradient centrifugation**, can help to overcome some or all of these problems. The simplest density gradients are step-wise (or discontinuous) ones, prepared by the careful layering of one solution on top of another in the centrifuge tube. For example, if a tissue homogenate or resuspended pellet in 0.25 M sucrose is layered on top of a pure solution of 0.5 M sucrose and centrifuged, the pellet formed will not be contaminated by lighter particles since no particles will be originally present at the bottom of the tube.

Even better separations of components can be achieved with continuous density gradients, prepared by the use of a **gradient-maker**. Sucrose or ficoll (a sucrose polymer) may be used as solute, the latter allowing the coverage of the same density range with smaller changes in osmolality.

In general, the aim of density-gradient centrifugation is to separate different groups of particles into zones along the density gradient, rather than to produce a pellet in the bottom of the tube. The **zonal rotor** (or **Anderson rotor)** is usually used in place of a simple centrifuge tube for these procedures. It is essentially a bowl with an assembly of vanes and ducts, designed so that a medium with a preformed density gradient may be introduced into it, after which a sample is applied to the top (least dense part) of the gradient. Centrifugation is then performed, and finally the medium is run out and collected in fractions, still largely maintaining the original density gradient and the separation of the zones achieved during centrifugation.

Two main types of density-gradient zonal centrifugation may be performed: **rate-zonal** and **isopycnic-zonal centrifugation**. With **rate-zonal centrifugation,** the density gradient is shallow and covers a range below the density of the lightest of the particles being separated. The sample is applied to the top of the gradient and centrifugation is carried out for long enough to allow separation of the different zones, without allowing sufficient time for the heaviest particles to sediment completely. The purpose of the gradient is to minimize convection currents and thus prevent mixing of the zones.

Isopycnic (equal density)-zonal centrifugation employs a much wider density gradient, and centrifugation is continued until each particle sediments to the level at which its density is equal to that of the medium. It is then buoyant and will sediment no more.

Regardless of the method of separation, each fraction may then be investigated further. Enzymes still trapped in membrane-bound organelles (e.g. those inside mitochondria) may be liberated by techniques harsher than those used for tissue homogenization (see section 16.1.2). Further centrifugation can be used to separate soluble enzymes liberated in this way from enzymes bound to the membrane of the organelle. Each fraction (or sub-fraction) may then be subjected to enzyme assay to find which enzymes are associated with it.

Two possible sources of error in the interpretation of such results should be pointed out: the presence of an enzyme in a certain fraction could be an artefact, due to the enzyme breaking away from a component of another fraction during homogenization or fractionation; also, an enzyme might be present in a fraction but have no catalytic activity because of loss of cofactors or structural organization during preparation of the fraction. Nevertheless, much useful information has been obtained from such investigations about the function of different sub-cellular compartments.

15.3.3 Some results of the investigation of enzyme compartmentation

The main enzymes of the cell **nucleus** are those involved in the replication and transcription of nucleic acids, including those which play a role in providing energy for these processes (e.g. the enzymes of glycolysis and the tricarboxylic acid cycle).

The **matrix of mitochondria** contains the enzymes of the tricarboxylic acid cycle, together with enzymes of fatty acid oxidation, protein synthesis and other processes. Reducing equivalents produced as the tricarboxylic acid cycle operates are passed down the respiratory pathway and used to synthesize ATP by oxidative phosphorylation; the components of the respiratory pathway and oxidative phosphorylation are found associated with the **inner mitochondrial membrane**.

The **thylakoid membranes** of **plant chloroplasts** contain assemblies which utilize light energy to generate ATP and NADPH. The chloroplast matrix (or stroma) contains enzymes which use this ATP and NADPH to convert CO_2 to hexose (the **dark reaction** of **photosynthesis**), together with enzymes of protein and nucleic acid synthesis.

The **matrix of lysosomes** contains mainly hydrolytic enzymes which have optimal activity at acidic pH values; many of these enzymes are glycoproteins. The single **lysosomal membrane** contains ATPase and NADH dehydrogenase.

Endoplasmic reticulum has been identified as the site of synthesis of phospholipids and of elongation of fatty acids. It is also important in many detoxification mechanisms.

The **cytosol** of cells contains the enzymes for the pathways of glycolysis and fatty acid biosynthesis. Also present are the enzymes of amino acid activation (for protein biosynthesis) and an ATP-dependent system for protein degradation involving the 76-residue peptide, ubiquitin (the discovery of which led to Aaron Ciechanover, Avram Hershko and Irwin Rose receiving a Nobel Prize in 2004).

The **plasma membrane** contains a Na^+, K^+ -dependent ATPase, whose function is to transport sodium ions out of the cell in exchange for potassium ions, and so maintain a high intracellular concentration of K^+ and a high extracellular concentration of Na^+. The red blood cell Na^+,K^+-pump has an $\alpha_2\beta_2$ structure, each α sub-unit traversing the membrane and possessing ATPase activity on the cytosol side. The β sub-units, with attached hydrophilic carbohydrate units, are located mainly towards the outer surface of the membrane, and their role is apparently to stop the molecule inverting through the hydrophobic membrane and so lose directional specificity. ATP and sodium ions bind to an α-sub-unit on the cytosol side, and hydrolysis of the ATP, attaching a phosphate group to an aspartyl residue of the enzyme, results in a conformational change which carries the sodium ions to the opposite side of the membrane. A reversal of the conformational change (following the hydrolytic removal of the phosphate from the enzyme) carries potassium ions in the opposite direction.

Also associated with the plasma membrane is adenylate cyclase, which is responsible for the production of cAMP from ATP. This enzyme is activated by the hormones adrenaline (in muscle) and glucagon (in liver). The mechanism involves a membrane-bound G-protein, and plays an important role in glycogen breakdown (see section 14.2.5) and other processes.

Enzyme-catalysed reactions which are associated with intracellular membranes in eukaryotic cells are often linked with the plasma membrane in prokaryotic cells, where there are no intracellular membranes: examples include the components of the respiratory pathway and the enzymes of phospholipid synthesis, discussed above.

SUMMARY OF CHAPTER 15

Whole-organism studies give useful information about metabolic pathways, but can reveal little about individual enzymes. In order to investigate an enzyme in detail, it is necessary to isolate a tissue and obtain a suitable preparation. This inevitably results in a loss of cellular organization, which should be borne in mind when the results are being interpreted. Tissue preparations are kept cold (below 4°C) whenever possible to minimize autolysis and loss of enzyme activity.

Enzyme assays may be performed on tissue homogenates or other preparations. The aim is to determine the concentration (actual or relative) of the enzyme, whose characteristics must already be known. In a catalytic assay procedure, all other factors which influence the rate of reaction are made fixed and non-limiting so that the initial velocity is proportional to the concentration of enzyme present. From the rate of reaction observed and the amount of sample assayed, the activity of the enzyme preparation can be calculated. If the procedure involves the coupling of the primary reaction to a second, indicator, reaction, the concentration of the indicator enzyme must be made high enough to be non-limiting. In a radioimmunoassay procedure, the basis of the assay is the binding of the enzyme to a specific antibody, catalytic activity not being involved at all. The two types of assay procedure are complementary.

The sub-cellular compartmentation of enzymes may be investigated by enzyme histochemistry (including immunohistochemistry) or by assay of the fractions obtained by differential or density-gradient centrifugation. In differential centrifugation, particles are separated mainly according to size, the largest and heaviest particles sedimenting fastest. In density-gradient centrifugation, the average density of particles is important, particularly in isopycnic-zonal centrifugation, where it forms the basis of separation.

FURTHER READING

Alberts, B., Johnson, A. *et al* (2002), *Molecular Biology of the Cell,* 4th edn., Garland (Chapters 11-12, 15).

Davis, W. C. (ed.) (1995), *Monoclonal Antibody Protocols,* Humana Press.

Goddard, J.-P. and Raymond, J.-L. (2004), Recent advances in enzyme assays, *Trends in Biotechnology*, **22**, 363-370.

Gul, S., Sreedharan, S. K. and Brocklehurst, K. (1998), *Enzyme Assays*, Wiley.

Hoogenboom, H. R. (1997), Designing and optimizing library selection strategies for generating high-affinity antibodies, *Trends in Biotechnology,* **15**, 62-70.

Kaplan, J. H. (2002), Biochemistry of Na^+,K^+-ATPase, *Annual Review of Biochemistry*, **71**, 511-535.

Kosloff, M. and Selinger, Z. (2001), Substrate-assisted catalysis: application to G-proteins, *Trends in Biochemical Sciences*, **26**, 161-166.

Marchese, A., Chen, G. *et al* (2003), The ins and outs of G-protein receptor trafficking, *Trends in Biochemical Sciences*, **28**, 369-376.

McCollough, K. C. and Spier, R. E. (1990), *Monoclonal Antibodies in Biotechnology,* Cambridge University Press.

Methods in Enzymology, **1** (1955): Preparation and assay of enzymes; **70** (1980), **92** (1983), **121** (1986), **150** (1987), **162, 163** (1988): Immunochemical techniques; Academic Press.
Nelson, D. L. and Cox, M. M. (2004), *Lehninger Principles of Biochemistry,* 4th edn., Worth (Chapters 11, 12).
Pickart, C. M. (2000), Ubiquitin in chains, *Trends in Biochemical Sciences*, **25**, 544-548.
Stein, K. E. (1997), Overcoming obstacles to monoclonal antibody product development and approval, *Trends in Biotechnology,* **15**, 88-90.
Voet, D. and Voet, J. G. (2004), *Biochemistry*, 3rd edn., Wiley (Chapters 6, 16, 19-20, 32).
Wilson, K. and Walker, J. (eds.) (2000), *Principles and Techniques of Practical Biochemistry,* 4th edn., Cambridge University Press (Chapters 1-7).

PROBLEM

15.1 Aliquots of an enzyme preparation were added to suitable incubation mixtures to give, in each case, a total volume of 5.0 cm^3, and assayed at 25°C. The following results were obtained:

Time (s)	Decrease in substrate concentration (mmol l^{-1})	
	0.2 cm^3 enzyme preparation	0.3 cm^3 enzyme preparation
30	0.17	0.25
60	0.33	0.49
90	0.49	0.74
120	0.64	0.96
150	0.80	1.20
180	0.93	1.38

The protein concentration of the preparation was found to be 720 mg l^{-1}. Calculate the specific activity of the enzyme preparation and comment on the validity of the results.

16

Extraction and Purification of Enzymes

16.1 EXTRACTION OF ENZYMES

16.1.1 Introduction

Before attempting to extract an enzyme from an organism, some preliminary considerations should be made.

For a commercial venture (see Chapters 17-20) the most important requirement would be to use a source which enabled large amounts of a suitable enzyme to be extracted by some convenient procedure. Any source which fulfilled this requirement could be chosen. On the other hand, a programme of scientific investigation might require the extraction of a given isoenzyme from a specified source, no alternatives being permissible. In the present chapter, therefore, we will discuss principles of extraction which are of general application.

First of all, it is essential to know the location of the enzyme within the cell, and whether the enzyme is present in free solution or bound to a membrane. This information may be obtained by the use of the techniques discussed in section 15.3. Only then is it possible to devise a suitable extraction procedure.

16.1.2 The extraction of soluble enzymes

Soluble cytoplasmic enzymes are the simplest to extract, for any disruption of the plasma membrane will enable the enzyme to pass into the surrounding medium. Soluble enzymes present in the organelles of eukaryotic cells are also easy to liberate, but the disruption of intracellular membranes often requires harsher conditions than disruption of the plasma membrane. To minimize the extraction of unwanted enzymes, cell fractionation (section 15.3.2) may be carried out prior to the disruption of organelles. The extraction procedures used depend on the type of organism acting as source.

Animal tissue is removed as soon after death as possible and kept cold to minimize autolysis. Obvious non-enzyme matter (e.g. fat deposits and connective tissue) may be discarded.

The remaining tissue is then treated to disrupt membranes using a technique appropriate to the location of the enzyme being extracted. Homogenization in an isotonic medium disrupts the plasma membrane but leaves most organelles intact (section 15.1). A Potter-Elvehjem or Douce homogenizer can be used for small amounts of animal tissue whereas a blender or Polytron/Ultraturax disruptor may be required for larger amounts of tissue.

The extraction of soluble enzymes from **higher plants** is complicated by the presence of the cell wall surrounding the plant cell, secondary metabolites (including tannins and phenols which can bind to proteins, causing precipitation) and phenol oxidases (which convert the phenols into reactive quinones which can bind to proteins). Freezing the tissue in liquid nitrogen and grinding to a powder with acid-washed sand in a mortar prior to the addition of extraction buffer can be used for small amounts of tissue, whereas homogenization in a blender can be used for larger amounts of tissue. The fibrous material can then be removed by filtration through several layers of muslin or cheesecloth. The extraction medium should be cold, set at a high pH, contain high millimolar concentrations of a reducing agent (e.g. 10mM 2-mercaptoethanol), include an anti-oxidant (e.g. ascorbic acid), insoluble polyvinylpolypyrrolidone (which provides an alternative substrate to proteins for reactive species) and a chelating agent such as diethyldithiocarbamate to reduce the deleterious effect of phenol oxidases.

Micro-organisms, in common with plants, have a cell wall and so tend to be more resistant to disruption than are animal cells. In general, methods of extraction fall into two main categories: mechanical disruption and the use of lytic enzymes. Cell pastes collected by centrifugation or filtration can be frozen in liquid nitrogen and ground in a mortar before the addition of extraction medium. Other mechanical procedures may involve grinding with carborundum or glass beads, applying pressure (e.g. with Hughes or French presses) or sonication. A popular alternative to mechanical procedures is the use of enzymes such as lysozyme and/or β-1,3-glucanases, particularly where these are available in an insolubilized form and can be recovered. This treatment makes the cell more susceptible to disruption by relatively mild techniques, e.g. osmotic shock or sonication. The enzymes can be also be used in combination with detergents to weaken the bacterial cell wall and to solubilize the membrane.

In general, the method of choice depends not only on the type of cell being disrupted and the sub-cellular location of the enzyme being extracted but also on the characteristics of this enzyme (e.g. its stability under the conditions of the extraction procedure). It must be admitted that the final choice is often made empirically, but a consideration of the principles discussed above (and in section 16.1.3) can help to limit the number of possibilities.

16.1.3 The extraction of membrane-bound enzymes

Enzymes which are bound to plasma or intracellular membranes cannot normally be extracted into the surrounding medium simply by the disruption procedures discussed in section 16.1.2. However, some enzymes are loosely bound and some tightly bound (many actually forming an integral part of the membrane), so there can be no general rule about their extraction.

Membranes, besides containing enzymes and other proteins, are made up largely of **amphipathic lipids,** i.e. lipids which contain hydrophilic and hydrophobic regions. The main classes of membrane lipids are phospholipids (fatty acid/phosphoric acid esters of glycerol and sphingosine) and glycolipids (fatty acid esters, mainly of sphingosine, attached to a sugar). Some membranes in eukaryotic cells also contain sterols (Fig. 16.1)

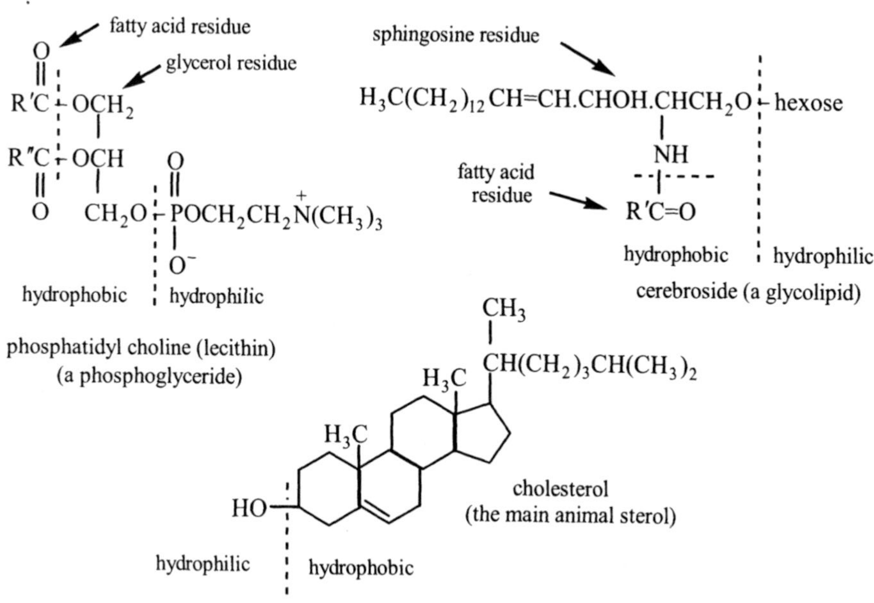

Fig. 16.1 – Some important examples of lipids found in membranes.

In each case the hydrophilic portion of the molecule is small compared with the hydrophobic part (note that in Fig. 16.1 the R-groups of the fatty acid residues are long hydrocarbon chains). An amphipathic lipid may therefore be represented as a molecule with a hydrophilic head and one or more hydrophobic tails, for example:

head (hydrophilic) tail (hydrophobic)

The model of membrane structure which best fits the available evidence (including that obtained by electron microscopy and NMR spectrometry) is that proposed by Jonathan Singer and Garth Nicolson (1972): this is termed the **fluid-mosaic** model (Fig. 16.2). According to this model, lipids and proteins can move laterally about the membrane, but find greater difficulty in moving in a transverse direction, from one membrane surface to the other. Proteins may be **peripheral**, when they are bound to the surface of the membrane, or they may be **integral**, when they are totally or partly embedded in the lipid layers.

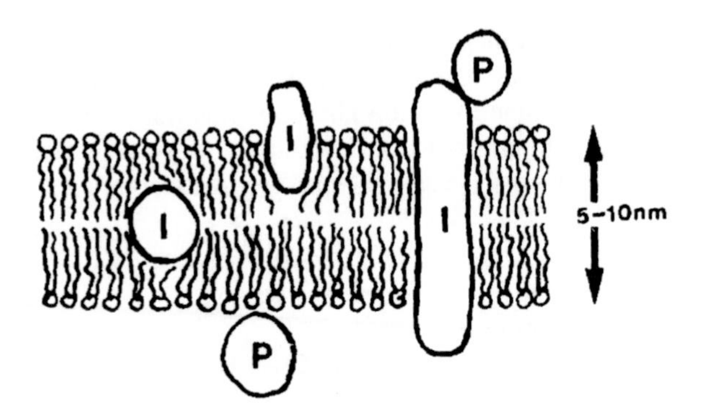

Fig. 16.2 – Representation of a typical membrane, according to the fluid-mosaic model of Singer and Nicolson (P = peripheral protein and I = integral protein).

Intrinsic integral membrane proteins traverse the phospholipid bilayer, having a domain containing predominantly hydrophobic amino acids (see Fig. 2.9) which spans the hydrophobic interior of the membrane. Extrinsic integral membrane proteins are located in either the cytoplasmic or extracellular leaflet of the phospholipd bilayer and have a hydrophobic domain or a fatty acid or farensyl/gerenyl-gerenyl isoprenoid anchor (added after translation). The hydrophobic domain of the integral protein is embedded in the membrane and the polar domains maintain contact with the hydrophilic cellular/extracellular environment. In addition, many proteins on the extracellular leaflet of the phospholipids bilayer can be glycosylated.

Peripheral proteins are apparently linked by electrostatic or hydrogen bonds to the polar heads of lipids or to integral proteins and can easily be dissociated from the membrane, e.g. by washing with a buffer of high ionic strength (1.0M NaCl) or by changing the pH of the buffer. Integral proteins, on the other hand, can only be extracted by breaking the hydrophobic interactions between lipid and protein (i.e. by dissociating a lipoprotein complex). This means either the use of a solvent or a detergent. Furthermore, for the extraction of an integral protein which is an enzyme, this must be done in such a way that enzyme activity is not lost during the process. A major problem is that an integral protein in its natural environment has some parts of its surface in contact with the hydrophobic membrane and other parts in contact with water, so structural changes are likely to take place no matter whether it is extracted into an organic or an aqueous medium. Again, the best method for each enzyme must be found largely by trial-and-error.

The most common method for the extraction of membrane proteins is the use of **detergents**, which disrupt membranes and extract lipoprotein fragments into aqueous media as components of micelles (aggregates of amphipathic molecules arranged with their hydrophobic parts in the interior of the micelle and their hydrophilic parts on the surface). These detergents include natural bile salts such as cholate; non-ionic synthetic detergents like Triton X-l00; and ionic synthetic detergents which may be zwitterionic, as in the case of CHAPS, anionic like sodium dodecyl sulphate (SDS) or cationic, e.g. CTAB.

In addition to the protein-detergent micelles, molecules of a detergent will themselves form micelles if present at a concentration above the critical micelle concentration (CMC), which is dependent on the temperature and, if the detergent is ionic, also on the ions present in solution. In general, the bile salts form the smallest micelles (M_r below 2000) and have the highest CMC (1-10 mM), while the non-ionic detergents form the largest micelles (M_r up to 100 000) and have the lowest CMC (in the order of 0.1 mM); detergents which form large micelles are difficult to remove from proteins. Knowledge of these characteristics may be helpful in the further purification of the extracted enzyme (see section 16.2.2). The choice of detergent will be influenced by whether or not the enzyme is required in an active form, and what further purification procedures are intended.

A detergent widely used in membrane studies is sodium dodecyl sulphate (SDS), which will sever most protein-lipid (and protein-protein) interactions and enable the M_r of a liberated protein to be determined (see section 16.3). However, for extraction of an active enzyme, non-ionic detergents are usually preferred. To maintain the solubility of the membrane protein during a purification process, low concentrations of detergent will be required in all buffers. This can present problems as some detergents can interfere with protein assays and may also absorb light in the UV at the same wavelength as proteins (the amino acids tyrosine and tryptophan present in a proteins structure absorb light at 280nm) making detection of protein-rich fractions from chromatographic runs difficult. The presence of a detergent requires careful thought about the choice of the chromatographic procedure. Size-exclusion chromatography (SEC) (section 16.2.2) is compatible with detergents and is a popular choice as the first chromatographic procedure in the purification of membrane proteins. To avoid detergents binding to an ion exchange (IEX) resin, neutrally-charged or oppositely-charged detergents are required. Many membrane proteins are also glycosylated, so the use of a detergent-tolerant lectin affinity resin provides a highly selective tool for the purification of membrane proteins.

Lipoprotein fragments may also be extracted by the use of **organic solvents**, and some free proteins may be liberated on subsequent dispersion in aqueous media. However, the use of solvents to extract proteins usually results in the denaturation of the protein and loss of biological activity, although solvent extraction can be useful to concentrate membrane proteins prior to an electrophoretic procedure e.g. isoelectric focusing or SDS-PAGE. Alternatively, the lipid-protein interactions may be broken by the action of **specific enzymes**: for example, phospholipase and triacylglycerol lipase have been used to extract alkaline phosphatase.

It will be realized that enzymes can exist in cells complexed to nucleic acids or carbohydrates, but, unless they are also linked to lipids in membranes, their extraction will be relatively straightforward

16.1.4 The nature of the extraction medium

Proteins exist within the cell at high concentrations in a reducing environment. Once cells have been disrupted, by whatever method, the environment is now oxidising and the contents are diluted with several volumes of aqueous medium. To minimise the possible damage to the enzyme, several factors have to be considered.

In general, a considerable amount of debris (mainly membrane fragments) is likely to be present, so it is essential that the enzyme of interest can be easily separated from this by being soluble in the extraction medium. As discussed in section 3.2.3, four main factors govern the solubility of proteins: **salt concentration, pH,** the **organic solvent content** and **temperature.** The temperature of the medium is usually kept below 4°C, despite the reductions in solubility that this entails, in order to minimize loss of activity of the enzyme.

Proteins are least soluble at their **isoelectric point**, so the extraction of an enzyme must be carried out at a pH value far from its isoelectric point, but nevertheless at a value where the enzyme is stable. The salt and organic contents of the medium are chosen, largely on an empirical basis, to ensure that the enzyme is soluble but many other proteins are not, thus achieving a preliminary purification.

For example, Edwin Cohn and co-workers (1951) showed that glutamate dehydrogenase, arginase, alkaline phosphatase and acid phosphatase from a bovine liver homogenate were soluble in 19% ethanol at pH 5.8 and ionic strength 0.02, but catalase, peroxidase and D-amino acid oxidase were not. However, these three enzymes were soluble at pH 5.8 in the absence of ethanol and at ionic strength 0.15. Also, various proteinases which were insoluble in both of these media were soluble if the pH of the second medium was increased to 7.4.

The solubility of an enzyme can sometimes be increased by the addition of its **substrate**. For example, alkaline phosphatase is more soluble in the presence of β-glycerophosphate than otherwise. Hence the extraction of an enzyme can be facilitated by adding its substrate to the medium.

Stabilizers, such as **dithiothreitol** (Cleland's reagent) or **2-mercaptoethanol (2-ME)** are sometimes added to extraction media to prevent oxidation of sulphydryl groups and thus loss of enzyme activity. Inhibitors of proteolytic enzymes, e.g. DFP (see section 8.3), PMSF (phenylmethylsulphonylfluoride), leupeptin, pepstatin, bestatin, E-64 or antipain, may also be added to prevent these attacking the enzyme of interest. Polyols (e.g. glycerol) or chelating agents (e.g. EDTA) also have a use in stabilizing solubilized enzymes

16.2 PURIFICATION OF ENZYMES

16.2.1 Preliminary purification procedures

As mentioned in section 16.1, a preliminary purification step is usually included in the procedure used to extract an enzyme. For example, some degree of cell fractionation is often carried out before extraction, so that only enzymes in the same sub-cellular compartment as the relevant enzyme are extracted. Also, the composition of the extraction medium is chosen so that the enzyme of interest is soluble but many others are not.

Extracted nucleic acids may cause problems by increasing the viscosity but they can be digested by the addition of nuclease enzymes or precipitated by the addition of basic substances (e.g. streptomycin, protamine, polyethyleneimine), or divalent metal ions (e.g. $MnCl_2$ or $MgCl_2$). Long fragments of nucleic acids can also be sheared by sonication. All precipitates and cell debris are then removed by centrifugation and discarded.

The next stage of purification is usually to precipitate the enzyme of interest from solution, thus concentrating the protein and separating it from mono- and oligo-saccharides, nucleotides, free amino acids etc. and from many other proteins which remain in solution. Proteins show minimal solubility at their **isoelectric point (pI)** and can be precipitated by adjusting the pH to the pI of the enzyme. However, the biological activity can be difficult to recover after isoelectric precipitation. This is also true for enzymes precipitated using organic solvents such as ice-cold acetone. The most enzyme-compatible method of precipitating proteins is by increasing the salt concentration: ammonium sulphate is often used for this purpose, because it is extremely soluble in water. The salt concentration is usually increased in stages, the aim being to precipitate other proteins (which are then discarded) before precipitating the fraction containing the relevant enzyme. This is then redissolved in buffer, and residual ammonium sulphate removed by dialysis or size-exclusion chromatography (see section 16.2.2).

Other methods suitable in laboratory-scale purification include ultrafiltration, polymer precipitation (e.g. polyethylene glycol, PEG) or the use of a chromatographic resin (e.g. ion exchange or hydrophobic interaction chromatography). In an industrial-scale purification, ultrafiltration, the use of stirred cells or expanded beds would be considered.

Although a considerable degree of purification can be achieved by these procedures, many other proteins will still be present because of overlap of solubility ranges. One possible way to remove some of these is to raise the temperature of the medium for a few minutes to a value where the enzymes being purified is known to be stable, but others might be denatured and precipitate from solution. Enzymes are often particularly stable in the presence of their substrates, so the relevant substrate can be added to increase the effectiveness of this procedure.

16.2.2 Further purification procedures

The crude extract, partially purified as described in section 16.2.1, may be treated in a variety of ways to increase the purification of the relevant enzyme. Chromatography based upon the partition (distribution) of material between two immiscible phases (a mobile phase and a stationary phase) represents the most popular collective technique for the purification of enzymes. There are a wide variety of protein-compatible resins available to which different functional groups can be attached to exploit the differences in protein structure.

There are a limited number of physical properties of proteins which can be utilised to enable the purification of an enzyme from a complex mixture. Enzymes have different biospecificity, surface charge, hydrophobicity, size, solubility and possibly post-translational modifications. A purification schedule should aim to exploit the different properties of proteins, avoiding techniques based upon similar principles.

Ion exchange chromatography (IEX) is a versatile technique with high binding-capacity for the purification of enzymes, relying on the attraction of oppositely-charged components.

At 1 pH unit above the pI of a protein, its surface will have an overall negative charge, so, at this pH, the protein will bind to a positively-charged resin, i.e. an anion exchange resin, such as one with a diethylaminoethyl (DEAE) functional group, $-OC_2H_5.^+NH(C_2H_5)_2$, attached to a support such as cellulose. At 1 pH unit below the pI of a protein, its surface will have an overall positive charge so, at this pH, the protein will bind to a negatively-charged resin, such as one with a carboxymethyl (CM) functional group, $-OCH_2CO_2^-$, attached to a suitable support. The binding can be reversed by lowering the pH in the case of anion exchange chromatography or by raising the pH in the case of anion exchange chromatography. Proteins will elute as the pH approaches their pI and the overall charge on the molecule approaches zero. However, many enzymes denature as they approach their isoelectric point, so bound proteins are usually eluted from IEX resins by raising the concentration of salt (0 - 2.0 M NaCl or KCl) in the eluting buffer. The ions in the salt compete with the bound protein for the charge on the resin, and proteins elute according to their overall surface charge, with the lowest surface-charge proteins eluting first. The salt elution of enzymes bound to an IEX resin results in good recovery of activity, with usually a 10-fold increase in the purity.

.Another chromatographic technique with high binding-capacity is **hydrophobic interaction chromatography (HIC)**. This technique depends on the fact that although amino acids with polar side chains predominate at the surface of enzymes (section 2.5.2), some non-polar ones will be present which can form hydrophobic interactions with column packings such as phenyl-Sepharose or octyl-Sepharose, particularly when the ionic strength is high (usually achieved with ammonium sulphate – see section 16.2.1). Elution can be brought about by decreasing the ionic strength or introducing an organic solvent. Recovery of enzyme activity from HIC is usually good. **Reversed-phase HPLC** also relies upon interaction between the hydrophobic areas of a protein and aliphatic ligands (e.g. C_{18}, C_{12}, C_8 or C_4) bonded to silica beads (3-10 mm diameter). The interaction between an enzyme and a reversed-phase resin is strong because there are more ligands bonded to the reversed-phase resin than to the HIC resin. However, the elution conditions of buffered aqueous and organic solvent mixtures rarely results in the recovery of biologically-active material.

In HIC, protein-binding to the ligand is determined by the concentration of **lyotropic** salts (e.g. ammonium sulphate). In **hydrophobic charge induction chromatography (HCIC)**, pH controls the binding of proteins to the hydrophobic ligand. The ligand is generally 4-mercapto-ethylpyridine (4-MEP) bonded to a modified cellulose resin. The resin was produced as an alternative to protein G for the purification of immunoglobulins and involves both 'pseudoaffinity' molecular recognition by interacting with the **Fc** region on the antibody and hydrophobic interactions at neutral pH. Elution is achieved by lowering the pH to around pH 5.0 and provides a means to enrich immunoglobulins essential free from contaminating albumin. This resin is one of a number (which include mercapto-benzimidazole sulphonic acid (MBI), hexylethylamino (HEA) and phenylpropylamino (PPA) resins produced by Pall; and Capto™ MMC produced by GE Healthcare) that operate in what is described as **mixed mode chromatography (MMC)**.

Protein interactions with MMC resins are not clear-cut and involve a mixture of different interactions. Potentially they provide the means to purify enzymes not easy resolved by other procedures.

As we noted in section 2.3.2, **covalent bonding** is characterized by an electron being shared with another atom, whereas, in **coordinate (dative) bonding**, both electrons are provided by one atom for another atom to share. The divalent metal ions Cu^{2+}, Hg^{2+}, Cd^{2+}, Zn^{2+}, Ni^{2+}, Co^{2+} or Mn^{2+} can accept pairs of electrons from electron-rich atoms to form coordinate bonds. At a physiological pH, some amino acids have functional groups with atoms that are rich in electrons including the imidazole group of histidine, the thiol group of cysteine and the indole group of tryptophan, which can form dative bonds with metal ions. In addition, the amino acids with carboxylic acid residues (glutamic and aspartic acid) and the phosphate group on **phosphorylated** amino acids (e.g. phosphoserine, phosphothreonine and phosphotyrosine) can form coordinate bonds with trivalent metal ions (e.g. Fe^{3+}, Al^{3+} and Ga^{3+}) at pH values around 3.0. **Immobilized metal ion affinity chromatography (IMAC)** resins typically have either iminodiacetate (IDA) or nitrilotriacetate (NTA) bound via a spacer-arm to Sepharose or agarose. These groups on the resin can bind di- or trivalent metal ions with coordinate bonds while leaving vacant orbitals which can then subsequently interact with the functional groups on amino acids already mentioned.

At pH 7.2 the primary interaction between Cu^{2+} bound to an IMAC resin and a protein will be with imidazole group of histidine residues. Elution of bound enzymes from the resin will depend upon the enzyme's stability and can be achieved by lowering the pH to below 5.0 or by displacement at pH 7.2 with increasing concentration of imidazole. The strength of the interaction between the bound protein and the bound metal ion will depend on the number and local environment of the histidine residues present. The purification of recombinant proteins (see below) can be facilitated by the addition of a **hexahistidine tag (6His)** on either of the termini of the protein. The presence of the 6His tag results in strong binding of the recombinant protein to a metal charged IMAC resin, allowing the unwanted cellular proteins to be eluted prior to the elution of the recombinant protein.

Hydroxyapatite (HA) is the crystalline form of calcium phosphate ($Ca_{10}(PO_4)_6(OH)_2$) prepared by mixing calcium chloride and sodium phosphate and can be encapsulated in agarose (HA Ultrogel: Pall) or coated to a solid ceramic support (CHT: Bio-Rad). The surface of hydroxyapatite is comprised of pairs of positively-charged calcium ions, triplet clusters of negatively-charged phosphates and hydroxyl groups. Proteins added to HA show a mixed interaction with the resin which is difficult to predict but depends upon the protein's amino acid content and the pH and ionic strength of the buffer. Positively-charged amino groups on the surface of the protein are attracted to the phosphate groups but also repelled by the positively-charged calcium ions. The carboxyl groups on a protein's surface are repelled by the phosphate groups and are attracted to the positively-charged calcium groups. The interaction between the carboxyl groups and the calcium is greater than mere electrostatic attraction and involves the formation of co-ordination complexes (see above). Phosphate groups on proteins will also co-ordinate with the calcium groups. Elution can be achieved by changes in the buffer's pH and ionic strength.

Enzymes are synthesised for a specific biological role and **affinity chromatography** exploits the biospecificity of a protein. Theoretically, affinity chromatography has the resolving power to isolate a pure enzyme from a complex mixture but the technique is rarely used in isolation, except in the purification of recombinant proteins (see section 21.1). The bio-specific nature of these affinity techniques means that a high degree of purification can be achieved. On the other hand, it also means that a polymer-ligand complex prepared for the purification of one enzyme is limited to that particular application. The column is packed with an inert matrix (e.g. agarose) to which ligands for the required enzyme have been attached. These immobilized ligands may be, for example, substrate analogues, and must be covalently linked to the matrix in such a way that they can still bind to their enzyme. If the ligand is small (e.g. a metabolite) a **spacer arm** (e.g. a hydrocarbon chain) can be included between matrix and ligand to avoid steric hindrance to the formation of an enzyme-ligand complex. If the ligand is large (e.g. a protein such as an antibody - see below) a spacer arm may not be required. The attachment of the spacer arms to —OH groups of the matrix is often achieved by the use of CNBr reagent (see section 20.2.1).The protein mixture is applied to the column, and the relevant enzyme is trapped by the immobilized ligands while all other proteins pass through and are discarded. The enzyme is then liberated from the column either by eluting with a **deforming buffer** at a pH which changes the characteristics of the enzyme and no longer allows it to bind to the immobilized ligand, or by the use of a **competitive counter-ligand**, which displaces the immobilized ligand on the enzyme. In both cases, the enzyme passes through the column and can be collected, now free of other proteins (Fig. 16.3). An example of an enzyme extracted by the use of such a technique is β-galactosidase with the substrate-analogue *p*-aminophenyl β-D-galactoside as ligand, the deforming buffer being 0.1 M borate at pH 10.

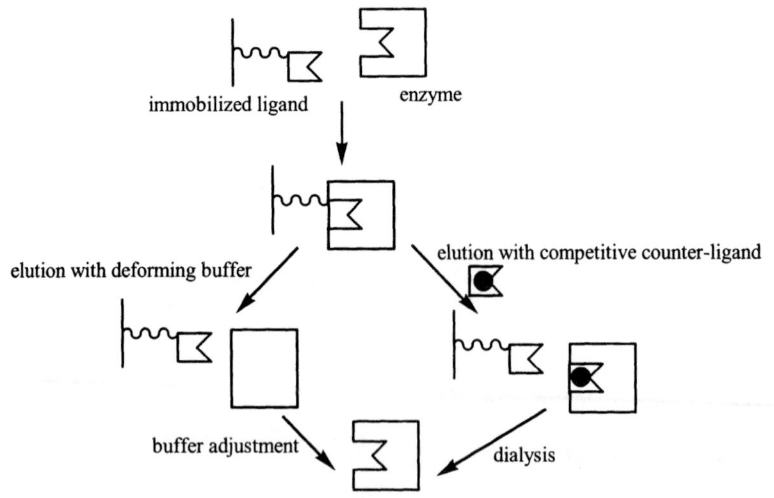

Fig. 16.3 – Diagrammatic representation of the processes involved in affinity chromatography; the treatment with deforming buffer and with competitive counter-ligand are alternatives.

The exploitation of an enzyme's biospecificity in theory allows for purification of that enzyme from a crude homogenate in a single chromatographic step. However, because of the cost of producing affinity ligands and the non-specific interactions of contaminating proteins present in crude extracts, affinity ligands are usually used after prior enrichment by precipitation and at least one chromatographic procedure (e.g. IEX, MMC or HIC).

If a target enzyme is present in a homogenate at low levels, it may not be feasible to process the large volumes of homogenate necessary to provide enough protein to produce crystals for structural studies (see section 2.5) or enough enzyme for industrial applications (see section 20.3).

However, if the gene for the enzyme is available, it is possible to incorporate the gene in an expression plasmid (see section 21.1) under the control of an inducible operon (e.g. isopropylthiogalactoside can induce the *lac* operon in *E. coli.*) and over-express the enzyme in an unrelated cell (e.g. *E. coli*, yeast, *Pichia* or *bacclovirus*), which can then be bulked up to produce enough protein. The fact that the protein can be over-expressed to represent a sizeable percentage of the total protein present significantly aids the purification process. However, because prokaryotic expression systems lack the necessary chaperones and enzymes responsible for eukaryotic post-translational modifications (e.g. glycosylation), the expressed eukaryotic protein may be incorrectly folded. These incorrectly folded proteins interact with each other, accumulating in the cytoplasm as opaque structures called inclusion bodies. To recover the expressed enzyme after cell lysis, these inclusion bodies have to be solubilized with detergents (e.g. SDS) and/or chaotropic agents (e.g. urea) and refolded by gradually removing the detergent by dialysis or size exclusion chromatography prior to purification.

To facilitate the rapid purification of recombinant proteins, the enzyme can be altered to aid the purification process and alterations at the N-terminus of a protein have been found to have a negligible effect on the function of a protein. The addition of poly-histidine (6His) results in stronger binding of the recombinant protein to divalent metal charged IMAC resins (see above) than host prokaryotic proteins, allowing an effective single-step purification of the recombinant proteins. The advantage of IMAC for the purification of recombinant proteins from prokaryotic hosts is that binding is still strong in the presence of a detergent. The addition of poly-arginine increases the net positive charge on the recombinant protein, resulting in stronger binding to cation exchange resins, but care must be taken in the use of the correct detergent to solubilize inclusion bodies. Alternatively, recombinant proteins can be fused to a protein (e.g. β-galactosidase, glutathione-S-transferase, IgG, or maltose-binding protein) which have an uncomplicated method of purification by affinity chromatography. After cell lysis, the fusion protein can be purified by affinity chromatography followed by treatment with a proteolytic enzyme to cleave a sequence (e.g. Factor Xa cleavage sequence IQGR) spanning the protein fusion. The choice of peptide cleavage sequence depends on whether it occurs in the sequence of the target enzyme. The fusion protein and target enzyme can then be separated by reapplication to the affinity resin used to isolate the fusion protein from the cell extract.

A closely related procedure is **immunoaffinity chromatography**, where the immobilized ligand is an antibody specific for the enzyme which is being purified. The use of an immobilized monoclonal antibody (see section 15.2.4) as ligand gives even greater specificity than use of a more conventional antibody.

Another form of affinity chromatography involves the use of reactive **triazine-based dyes**, e.g. Cibacron Blue F3G-A and dyes of the Procion series, as ligand. Under alkaline conditions the chlorotriazine dye is linked, usually directly, to a matrix such as agarose by a triazine bond. The reaction involves a hydroxyl group on the matrix and a chloride on the dye. Such immobilized dyes bind a wide range of NAD- and NADP-dependent dehydrogenases and other enzymes.

They are, of course, less specific than biological group-specific ligands, and their specificity is difficult to predict, but they have many advantages: they have a greater protein-binding capacity; they are extremely resistant to chemical and enzymic degradation; they can be used and re-used in a variety of applications; and the triazine bond is more stable than the isouronium linkage introduced by the CNBr activation of agarose. Elution of enzyme from the ligand is usually achieved by changing pH or increasing the salt concentration.

Affinity chromatography is a separation usually conducted in a column, relying upon an interaction between a binding site on a protein and a ligand bound to a resin. A variation upon column-based affinity chromatography is affinity precipitation. In **bis-affinity precipitation**, which involves 2 ligands (bis) joined by a spacer arm, the target protein must have a quaternary structure to form a precipitable complex and the length of the spacer arm used in the affinity ligand is important to prevent **intramolecular** and encourage **intermolecular** interactions. The precipitate can be separated by centrifugation and the complex competitively displaced with high concentrations of the substrate of the enzyme. The use of bis-affinity ligands was developed for the purification of lactate dehydrogenase (LDH) using a bis-NAD^+ ligand such as N(2),N(2)′-adipodihydrazido-bis-(N(6)-carbonylmethyl)-NAD). The NAD^+ molecules on the ends of the spacer arm can interact with two different LDH molecules and the subunits on each of these molecules can be linked to other affinity ligands forming a network which eventually becomes large enough to form a precipitate. Proteins without a quaternary structure can be isolated using a polymer (with unique solubility parameters) to which many ligands are attached. After incubation of the enzyme with the affinity polymer, the conditions are altered (e.g. a change in temperature or pH) to precipitate the polymer/enzyme complex from the other proteins present. The polymer can be collected, resolubilized and the target enzyme eluted by competitive displacement.

An enzyme's structure may be stabilized by the presence of covalent disulphide bridges (predominately occurring in enzymes exported from the cell) formed by the oxidation of two cysteine residues. In addition, there may be reduced cysteine residues present on the surface of proteins or at the active site of enzymes which do not form disulphide bridges. **Covalent chromatography** is a separation technique developed to isolate thiol-containing proteins and peptides. A **thiol** residue attached to a resin can react with a protein's surface and/or active site sulphydryl groups to form a covalent mixed disulphide. The covalently-bound protein can then be eluted by a buffer containing reducing agents such as L-cysteine, 2-ME or dithiothreitol.

Glycoproteins present in the extracellular leaflet of the plasma membrane in eukaryotic cells and those released from the cell into the extracellular matrix have oligosaccharides attached. The post-translational addition of oligosaccharides may occur during synthesis (**N-linked**: attached to arginine) or during maturation, as proteins progress through the endomembrane system (**O-linked**: attached to serine; and **GPI-linked**: attached to phosphatidylinositiol). The saccharides contained within these structures can interact with lectins (saccharide-binding proteins). Lectins have quaternary structure (usually tetrameric) and, if composed of different subunits, they can bind more than one type of saccharide. When attached to gels such as agarose, they can be used to isolate glycoproteins. For example, the lectin concanavalin A isolated from *Canavalia ensiformis* (jack bean) binds α-D-mannose with free hydroxyl groups at C3, C4 and C6. Lectin affinity chromatography can be used to isolate membrane glycoproteins using a lectin which can tolerate detergents e.g. one isolated from *Lens culinaris*

Size-exclusion (gel filtration, gel permeation or molecular-sieve) chromatography (SEC) is carried out with columns packed with swollen porous beads which separate the components of a sample on the basis of the molecular mass. Large molecules, because of their size and shape, cannot access the pores of the beads and are excluded (size exclusion) from the volume of buffer contained within the beads. They will only travel through the column in the volume of buffer surrounding the beads, passing quickly through the column. Smaller molecules will be able to access the same volume of liquid surrounding the beads as the larger molecules but in addition they can travel into the cortex of the porous beads, effectively moving through a larger volume of buffer. In SEC, molecules elute in decreasing molecular size. Cross-linked dextrans (Sephadex), cross-linked agarose (Sepharose) and cross-linked polyacrylamide (Biogel) are commonly used in SEC and can be purchased with different pore sizes to fractionate different mass ranges. For the successful separation of an enzyme of interest, the SEC gel must be chosen with a fractionation range which will allow the enzyme to enter the pores of the gel.

Separation in SEC is independent of the buffer, the only criterion for the buffer being that the proteins are stable in it and of sufficiently high ionic strength (e.g. 150 mM) to counteract the few charges which may be present on the gel. In the case of membrane proteins (see section 16.2.1) solubilized in detergents, SEC is widely used to separate protein-detergent micelles from excess detergent. This is particularly easy to achieve if detergent molecules are kept in the monomeric form, e.g. by ensuring that the detergent concentration is less than the CMC, or by adding a small amount of a bile salt. In addition, SEC can be a useful method in the refolding of proteins after denaturation, as the reduced diffusion in SEC promotes the correct refolding of proteins by suppressing non-specific protein-to-protein interactions. In addition, during SEC, the aggregates and solubilising buffers can be removed in a single experiment

Electrophoresis is mainly an analytical procedure, since it is ideally suited to the separation of small amounts of material, but it has also been used for the purification of proteins. The rate and direction of migration of a protein in an electric field depends upon a propelling force contributed by the field strength and the charge on the molecule, and a retarding force contributed by the size and shape of the molecule and the viscosity of the medium.

In **zone electrophoresis,** a mixture of proteins is introduced at a common point (the **origin)** and allowed to move in an electric field, each molecule travelling in the same zone as others of the same charge and size. To minimize diffusion, the whole process is usually carried out in a solid but porous **support medium** through which the buffer and proteins permeate, as in, for example, **polyacrylamide gel electrophoresis (PAGE).** Enzymes retain their native conformation in **non-denaturing PAGE**, when the experiment is conducted at a low temperature. At the end of the electrophoretic run, a chromogenic substrate for the target enzyme can be overlaid onto the gel to visualise its position or positions. Many dehydrogenases, phosphatases and proteolytic enzymes can be detected by using this technique.

The areas within the gel which show positive reactions for the target enzyme can be excised, homogenized in buffer and the enzyme allowed to diffuse into the buffer before discarding the acrylamide fragments. This results in a significant enrichment of the target enzyme. Electrophoresis can be viewed as incomplete electrolysis, because the experiment is stopped before the sample reaches the oppositely-charged electrode, and the samples within the gel are visualised by staining. Preparative zone electrophoresis allows the target enzyme to leave the end of the polyacrylamide, but it is prevented from contacting the electrode by dialysis tubing, and the sample is moved to a fraction collector.

Isoelectric focusing (IEF) (see section 16.2.3), which separates proteins according to their charge, is employed mainly for analytical purposes but can also be used in preparative work. Preparative IEF apparatus (Rotofor or Mini-Rotofor: Bio-Rad) can be used to concentrate and fractionate proteins by liquid-phase IEF. The proteins (μg-g amounts) are focused into chambers and collected under vacuum into fractions. By initially using a wide pH range (3-10) followed by a narrow pH range (1.0 pH scale spanning the pI of the target protein) highly purified fractions can be obtained which can then be subjected to additional chromatographic or electrophoretic techniques.

A major advance in protein purification over the past few years has been the emergence of protein-compatible robust resins (for SEC, IMAC, HIC, IEX and affinity), manufactured in smaller bead sizes (5-15 μm) which are able to withstand higher flow-rates. The smaller bead sizes increase the number of potential interactions with the functional groups but increase the back pressure. Computer-controlled equipment allows the production of accurate elevated flow-rates and gradients, which helps to improve the ability of the operator to determine a workable method in a short time.

When, after the use of one or more of these procedures, it is considered that the relevant enzyme has been separated completely from other proteins (see section 16.2.3), mineral salts and any small molecules present may be removed by **dialysis** or SEC, if this is desired.

The purified enzyme preparation is likely to be quite dilute, so it might require concentrating before being used for the purpose for which it was prepared. **Lyophilization** or precipitation using ammonium sulphate or a polymer (e. g. PEG) (see section 16 2.1) can be used, followed by the redissolving of the enzyme in a small volume of liquid.

Another suitable procedure is **ultrafiltration**. This involves forcing the solvent molecules through a membrane of chosen porosity (e.g. by the application of nitrogen gas pressure) and thus separating them from the protein molecules, these being too large to pass through the pores.

The ideal way to complete a purification procedure is to **crystallize** the enzyme, if this is possible. Sometimes it may be achieved by adding not quite enough ammonium sulphate to cause precipitation of the enzyme and leaving this solution in a coldroom for several days. In general, successful conditions have to be determined empirically by attempting crystallization at a variety of salt and organic solvent concentrations and pH values, and by showing much patience.

Membrane enzymes may be inactive after purification unless re-introduced into a phospholipid environment. For example, phospholipid may be added to purified protein-detergent micelles and concentrations adjusted so that the phospholipid replaces detergent in the micelles; released detergent, in the monomeric form, may then be removed by SEC, dialysis or ultrafiltration. An example of a membrane enzyme that has been characterized is Na^+-K^+ ATPase (see section 15.3.3).

16.2.3 Criteria of purity

At each stage of a purification procedure, an assay for the enzyme being purified should be performed on all fractions. This will show, regardless of preconceived ideas, precisely which fractions contain the relevant enzyme. It is also important that the **total protein** content of each fraction is determined. A variety of different methods are available for the assay of total protein (see section 3.2.1), each having positive and negative features. The simplest method is the measurement of absorbance (A) at 280nm, which covers the range 0.2 – 2.0 mg ml^{-1} (higher concentrations of protein are measurable, of course, but require dilution of the sample prior to analysis). In general, an A_{280nm} of 1.0 approximates to a concentration of 1.0 mg ml^{-1}. However, in crude extracts, other compounds (e.g. free amino acids and nucleic acids) also absorb light at 280nm. The formula, protein (mg ml^{-1}) = $1.55\ A_{280nm} - 0.76\ A_{260nm}$, corrects for the presence of nucleic acids.

The biuret method is also easy to carry out and less affected by the presence of contaminants, but it is relatively insensitive (range 1.0 – 5.0 mg ml^{-1}). The Lowry procedure is more sensitive (0.1 – 1.0 mg ml^{-1}) but is subject to interference by many compounds. The Coomassie Blue binding assay is quick and easy to use, as well as being sensitive (standard assay 0.1 – 1.0 mg ml^{-1}; micro-assay 0.05 – 0.1 mg ml^{-1}) but cannot routinely be used in the presence of a detergent. The bicinchoninic acid (BCA) is less straightforward to carry out but is equally sensitive (standard assay 0.1 – 1.0 mg ml^{-1}; micro-assay 0.05 – 0.1 mg ml^{-1}) and can be used in the presence of detergents. All protein assays have limitations, but in most cases the limitations can be overcome by the inclusion of appropriate buffer blanks. There are also a number of variations on the standard assay to overcome any interference by contaminating compounds.

In colorimetric protein assays, bovine serum albumin (BSA) is generally used as a standard protein to construct a calibration graph. The unknown samples are treated in the same manner as the standards and the colour generated is converted into a protein concentration by reference to the standard calibration curve.

It should be remembered that because the readings for the sample are in reference to a standard protein (BSA), this implies that the proteins in the sample have the same content of reactive amino acid residues (e.g. ones with amine or aromatic side-chains) as BSA. This will not be true for every protein, but the resulting small loss in accuracy is normally traded off for the convenience of the method used.

The determination of total protein serves two main purposes: it will indicate whether protein is being lost from the system as a whole as the purification proceeds, e.g. because of the presence of proteolytic activity; and it will enable the **specific activity** (see section 15.2.2) of the important enzyme in each fraction to be calculated, providing information about the progress of the purification.

The *total* activity of this enzyme will be unchanged as a result of a purification step, unless some has been lost during purification. Under no circumstances can it be increased. However, some of the contaminating proteins which were originally present in the same preparation as the enzyme should now be in different fractions. Hence, if the fractions containing the relevant enzyme are combined, the total activity of the enzyme in the combined fraction should be the same as that started with, whereas the total amount of protein in this fraction will be less than that originally present. The *specific* activity of the enzyme in the combined fraction should therefore be greater than that in the preparation before purification, and the increase in specific activity will be a measure of the purification achieved.

With each successive purification step, the specific activity of the fraction (or fractions) containing the enzyme should be greater than before until complete purification is achieved and the specific activity reaches a limiting value.

However, the finding of the same specific activity value before and after a purification step does not necessarily mean that the enzyme preparation is completely pure: it could simply mean that contaminating proteins have passed through the procedure in the same fraction as the enzyme. Similarly, crystallization cannot be taken as proof that only one protein is present, for many mixed protein crystals have been found. Hence other criteria of purity have to be considered.

If a particular enzyme is known to be a possible contaminant, then a logical step is to carry out an assay for that enzyme and demonstrate its absence in that way. However, that would not exclude the possibility that other contaminants were present.

Electrophoresis in various forms is very effective in separating mixtures of proteins (see section 16.2.2) so it is used extensively to investigate the purity of enzyme preparations. Of particular importance in this context is **denaturing PAGE**, in which the protein sample is denatured by heating in the presence of a detergent and a reducing agent (see section 16.3). When the denatured protein sample is applied to the polyacrylamide gel, the smaller polypeptides are subjected to less restriction and move further down the gel. There is a linear relation between mobility of the denatured polypeptide and its $\log_{10}M_r$. The buffer-system is **discontinuous**, i.e. the buffers in the electrodes are of a different composition and pH to the buffers in the large-pore **stacking (or spacer) gel** and in the **resolving (or running or separating) gel**, where the actual fractionation takers place (Fig. 16.4). This arrangement, which results in sample being concentrated in the stacking gel before reaching the resolving gel, is termed **discontinuous (disc) electrophoresis**.

The resolving gel has a higher percentage of acrylamide and consequently smaller pores than the stacking gel. Polyacrylamide gels can be cast with different percentages of acrylamide to alter the fractionating properties of the gel. A 10% acrylamide gel will fractionate proteins in the M_r range of 15 000 to 200 000, whereas a 5% acrylamide gel will fractionate proteins in the M_r range 60 000 to 350 000. Smaller pores prevent entry of large proteins onto the gel. This is called molecular sieving and it should be remembered that proteins with a mass larger than the upper limit will not be able to gain entry into the gel. On completion of the electrophoresis, and after staining for proteins, it is easy to see if more than one band is present, for each is extremely narrow and well-defined.

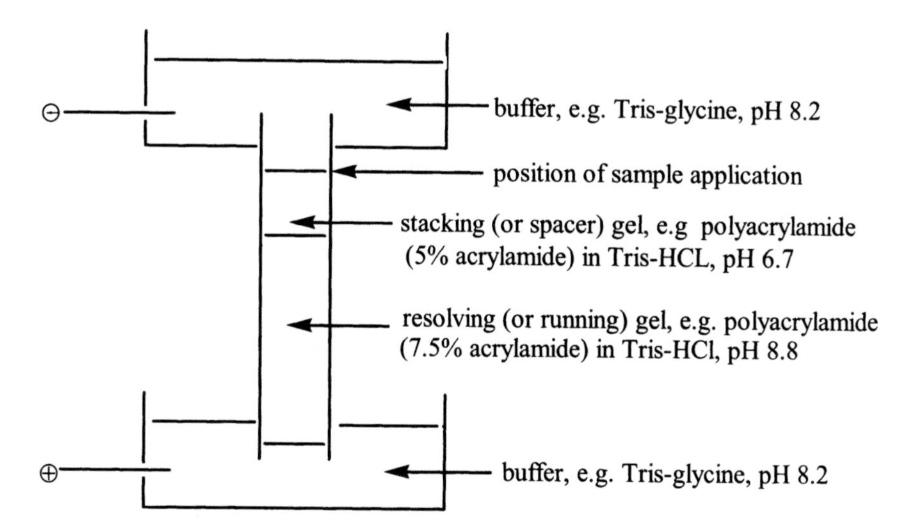

Fig. 16.4 – Diagrammatic representation of an apparatus set up for discontinuous (disc) electrophoresis of proteins.

Capillary electrophoresis (CE) can be used to determine the purity of an enzyme fraction using low sample volumes and fast separation times. Narrow bore glass capillaries (typically 50-100 cm with a 50 μm internal diameter) are used with high voltages (10-50 kV). The ends of the capillary are submerged into two separate buffer chambers containing either the anode or cathode. Depending on the instrument, the sample can be injected into one end of the capillary by syphoning, high pressure or high voltage. The capillary is then removed from the sample, replaced in the buffer chamber and the high voltage reapplied to commence the separation. Capillary electrophoresis of free solutions is the most widely practised technique, but the charged surfaces can adsorb proteins which may lead to band broadening or the total loss of the sample. Chemically-modified capillaries can be used, in which the silanol groups are capped with a neutral reagent. This eliminates the electroendosmotic flow, and the charged species will migrate towards the oppositely-charged electrode. The detector is usually placed at one end of the capillary and will only detect the charged species flowing towards that electrode. By controlling the pH of the buffer, proteins can be made to migrate to the chosen electrode.

Capillaries can be filled with a gel (either agarose or polyacrylamide) and separations undertaken which are similar to SDS-PAGE (see section 16.3) or IEF (see below). A single peak on an acrylamide-filled capillary in the presence of SDS and a single peak on a free solution run would be a good indication of the purity of a protein sample.

Another relevant technique is **isoelectric focusing (IEF),** which is an example of **moving boundary electrophoresis** rather than zone electrophoresis. In IEF, a pH gradient is generated within a polyacrylamide (4%) gel by the inclusion of **ampholytes** (synthetic and natural amino acids selected so that their individual isoelectric points cover the required pH range) in the monomer mixture prior to polymerisation. Alternatively, strips suitable for IEF can be purchased containing ampholytes covering both wide and narrow pH ranges. The pH gradient is set up within the polymerised gel by the application of an electrical field for a short period of time. The sample can then be applied (anywhere on the gel) and the electrical field is switched back on.

The proteins in the sample will migrate until they reach a pH in the gradient equivalent to their pI. At its pI, a protein will have no charge and will remain immobile within the gel. When the experiment has finished, the gel is washed with 10% (w/v) trichloroacetic acid to fix the proteins in the gel and to wash away the ampholytes. The gel can then be stained to visualise the samples. The zones containing each protein are very sharp as a result of this focusing, and proteins whose isoelectric points differ by as little as 0.02 pH units can be distinguished by this method, making the technique useful for the detection of isoenzymes or post-translational modifications such as phosphorylation

Analytical ultracentrifugation is widely used to investigate the purity of enzyme preparations. With high centrifugal fields (e.g. 500 000 g), diffusion effects are largely overcome (see section 15.3.2). Hence, as a randomly distributed group of identical protein molecules sediment in an **analytical cell** under the effect of such a field, a sharp boundary is formed between the portion of the medium which has been cleared of protein molecules and that which still contains them. This boundary refracts light, and its position can be determined through a quartz window in the cell. For example, it shows up as a peak if the refractive index gradient at each point in the cell is measured using the **Schlieren optical system**. The movement of this boundary as sedimentation proceeds can be constantly monitored, and automatically recorded, and the data used to calculate the characteristics of the molecule in question (see section 16.3). If the protein is pure, only one boundary should be observed: detection of extra boundaries suggests the presence of contaminants.

Analytical HPLC procedures may also be used to investigate the purity of a protein preparation (see section 16.2.1).

No single technique, however sensitive, can establish enzyme purity because of the possibility that a contaminating protein might behave in an identical fashion to the important protein under the conditions used. However, if several characteristics are investigated, and the conditions of investigation varied, the chances of the contaminant avoiding detection are reduced. Therefore, evidence of purity must be obtained by the use of several of the above procedures before it can be concluded that the sample is indeed pure.

16.3 DETERMINATION OF MOLECULAR WEIGHTS OF ENZYMES

When an enzyme has been purified, its molecular weight, more correctly termed relative molecular mass, M_r (see section 2.4.2) may be determined.

The most popular method of determining the M_r of an enzyme after purification is polyacrylamide electrophoresis (PAGE) in the presence of a detergent SDS (sodium dodecyl sulphate, $CH_3(CH_2)_{11}SO_3^-Na^+$) and a reducing agent (2-mercaptoethanol). SDS is an anionic detergent whose hydrocarbon chain can become linked to hydrophobic regions in the interior of globular proteins, leaving the ionized sulphonate ($-SO_3^-$) group jutting out into the surrounding medium, converting a protein's native globular or fibrous tertiary structure into a linear one.

SDS molecules bind to the protein's structure approximately every 2 amino acids, disrupting its natural shape and giving it a large negative charge, regardless of the net charge originally possessed by the protein. Large protein molecules will complex with more SDS than small ones, so all will have roughly the same charge/mass ratio, and all will have much the same shape.

Therefore, if molecules of different proteins are complexed with SDS and subjected to zone electrophoresis in an unrestrictive medium, all should travel together, regardless of their original characteristics. However, if electrophoresis is performed in a support medium with a restrictive pore size (e.g. in polyacrylamide gel), the complexes from each of the different proteins will be separated, entirely on the basis of size. Since the size of each protein-SDS complex must be dependent on the original size of the protein molecule, a system can be calibrated by the use of marker proteins of known molecular weight; hence the molecular weight of a test protein may be determined. One point that should be borne in mind, however, is that SDS will disrupt all weak polypeptide-polypeptide interactions, so, for an oligomeric protein, this technique will give the molecular weight of the sub-units rather than of the complete protein.

Size-exclusion chromatography (SEC), which effectively separates molecules on the basis of size (see section 16.2.2), provides one way of determining the M_r of an oligomeric protein. The elution volume (void volume V_0) of a large molecule which is totally excluded from the gel's pores (Blue dextran, M_r 2 000 000) and a small molecule which can occupy the total volume (V_t) of the pores (e.g. vitamin B_{12}, M_r 1355) is measured. The column can be calibrated by the application of a series of standard proteins of known M_r. The elution volumes (V_e) of these proteins can be measured and their K_{av} values (K_{av} being an approximation to the partition coefficient K_d) determined: $K_{av} = (V_e - V_0)/(V_t - V_0)$. The K_{av} of the standard proteins is plotted against the $\log_{10}M_r$ of the standard proteins. A target enzyme sample can be applied to the column, the K_{av} for the target protein measured and, using the calibration graph, an estimate of the M_r of the protein in its **native conformation** can be determined.

One source of error in this procedure is that it takes no account of molecular shape. In fact, a protein molecule which is somewhat elongated will pass down the column more slowly than a spherical one of the same molecular weight. To overcome this discrepancy, the experiment (including the standard proteins) can be conducted in the presence of urea, producing a random coil formation which will give a better estimate of the M_r of the protein.

Ultracentrifugation (see section 16.2.3) is also widely used for the determination of molecular weights of proteins. For a spherical protein of molecular weight M_r sedimenting through a medium in a centrifugal field, the centrifugal force per gram-mole is $M_r(1-\bar{v}\rho)\omega^2 r$, where $\bar{v}$ is the partial specific volume of the molecule (i.e. the increase in volume in cm^3 when 1 g solute is added to a large volume of solvent; the value for a protein is usually about 0.74) . The other terms are defined as in section 15.3.2. The centrifugal force is opposed by an equal frictional force. This frictional force per gram-mole is equal to the sedimentation rate multiplied by *RT/D,* where *R* is the gas constant (in ergs per mole per degree), *T* is the absolute temperature and *D* is the diffusion constant of the molecule.

$$\therefore M_r(1-\bar{v}\rho)\omega^2 r = \frac{RT}{D} \times \text{sedimentation rate} \qquad (16.1)$$

But sedimentation rate = $s.\omega^2 r$ (see section 15.3.2).

$$\therefore M_r(1-\bar{v}\rho)\omega^2 r = \frac{RT}{D} s\omega^2 r$$

$$\therefore M_r = \frac{RTs}{D(1-\bar{v}\rho)} \qquad (16.2)$$

This is known as the **Svedberg equation,** after Theodor Svedberg, the pioneer of ultracentrifugation.

As discussed in section 16.2.3 the rate of movement of the boundary separating the protein-free zone from the rest of the medium can be determined, and this is a measure of the sedimentation rate. Hence the sedimentation coefficient (*s*) can be calculated, since ω and *r* are easily determined. For most proteins, s lies in the range 1-200 S. The diffusion constant (*D*) can similarly be calculated from moving boundary determinations, this time in the absence of a centrifugal field.

Once these and the other relevant factors have been determined, the molecular weight of the protein can be calculated by the use of the Svedberg equation. However, as with SEC, there is a possible source of error in the assumption that all protein molecules are spherical.

SUMMARY OF CHAPTER 16

Soluble enzymes may be extracted from cells simply by the disruption of membranes, the actual technique employed depending on the nature of the cell, the intracellular location of the enzyme and the stability of the enzyme. In each case, the pH, salt and organic solvent concentration of the medium must be carefully chosen to ensure that the enzyme being extracted remains in solution while the cell debris is removed by centrifugation.

Special procedures have to be employed to extract enzymes which form an integral part of membranes (e.g. extraction with a detergent): otherwise they would be discarded with the cell debris.

Techniques for the purification of the extracted enzyme include IEX, HIC, MCAC, SEC, MMC, HCIC, affinity chromatography and electrophoresis. Enzyme assay should be performed after each purification step, and the specific activity of each fraction determined. As purification proceeds, the specific activity of the enzyme preparation should rise to a limiting value. When this has been achieved, the purity of the preparation should be checked by a variety of techniques, including ultracentrifugation, electrophoresis, isoelectric focusing and by high performance liquid chromatography.

After purification, the molecular weight of the enzyme may be determined by SDS-PAGE, SEC or ultracentrifugation.

FURTHER READING

Bonner, P. L. R. (2007), *Protein Purification (Basics)*, Taylor and Francis.

Cutler, P. (ed.) (2004), *Protein Purification Protocols*, 2nd edn., Humana Press.

Gupta, M. N. (2002), *Methods for Affinity-Based Separation of Enzymes and Proteins*, Birkhauser Verlag.

Hames, B. D. (1998), *Gel Electrophoresis of Proteins - A Practical Approach,* 3rd edn., IRL.

Irwin, J. A. and Tipton, K. F. (1995), Affinity precipitations - a novel approach to protein purification, *Essays in Biochemistry,* 29, 137-156, Portland Press.

Krebs, M. O., Noorwez, S. M. *et al* (2004), Quality control of integral membrane proteins, *Trends in Biochemical Sciences*, **25**, 648-655.

Levy, D. (1996), Membrane proteins which exhibit multiple topological orientations, *Essays in Biochemistry,* **31**, 49-60, Portland Press.

Methods in Enzymology, **22** (1971), **104** (1984), **182** (1990): Enzyme purification; **34** (1974), Affinity chromatography; **270**, **271** (1996): High resolution separation and analysis of biological macromolecules; Academic Press.

Op den Kamp, J. A. F. (ed.) (1994), *Biological Membranes - Structure, Biogenesis and Dynamics,* Springer- Verlag.

Patel, D. (2001), *Separating Cells*, Biosis.

Price, N. C. and Stevens, N. (1999), *Fundamentals of Enzymology,* 3rd edn., Oxford University Press (Chapters 2, 8).

Quinn, P. J. and Cherry, R. J. (1992), *Structural and Dynamic Properties of Lipids and Membranes,* Portland Press.

Raghupathi, R. N. and Diwan, A. M. (1994), A protocol for protein estimation that gives a nearly constant colour yield with simple proteins and nullifies the effects of four known interfering agents, *Analytical Biochemistry*, **219**, 356-359.

Reed, R., Holmes, D. *et al* (1998), *Practical Skills in Biomolecular Sciences,* Longman.

Roe, S. (ed.) (2001), *Protein Purification: Methods*, Oxford University Press.

Roe, S. (ed.) (2001), *Protein Purification: Applications*, Oxford University Press.

Rosenberg, I. M. (2005), *Protein Analysis and Purification*, 2nd edn., Birkhauser

Rothe, G. M. (1994), *Electrophoresis of Enzymes - Laboratory Methods,* Springer-Verlag.

Sachs, J. N. and Engelman, D. M. (2006), Introduction to the membrane protein reviews, *Annual Review of Biochemistry*, **75**, 707-712.
Sheehan, D. (2000), *Physical Biochemistry - Principles and Applications,* Wiley.
Shu-Sheng, Z. and Lundahl, P. (2000), A micro-Bradford membrane protein assay, *Analytical Biochemistry*, **284**, 162-164.
Voet, D. and Voet, J. G. (2004), *Biochemistry,* 3rd edn., Wiley (Chapter 6).
Wilson, K. and Walker, J. (ed.), (2000), *Principles and Techniques of Practical Biochemistry,* 5th edn., Cambridge University Press.

PROBLEM

16.1 An attempt was made to purify further a malate dehydrogenase preparation by precipitating protein with ammonium sulphate at 55% saturation. The precipitate was redissolved in water, and this solution was found to have a protein content of 1.48 g l^{-1}. A 1 in 500 dilution was made and aliquots of this were used for malate dehydrogenase assay, with excess malate and NAD^+, in a total incubation volume of 3.1 cm^3. When 10 μl diluted solution was used for assay, the initial velocity (measured as rate of increase of absorbance at 340 nm) was 0.11 units min^{-1} and when 15 μl were used the initial velocity was 0.165 units min^{-1}. The supernatant from the ammonium sulphate precipitation was found to have a protein content of 2.05 g l^{-1}. A 1 in 1000 dilution was made, and assayed for malate dehydrogenase as above. When 10 μl diluted solution was used for assay, the initial velocity was 0.08 units min^{-1} and when 15 μl was used, the initial velocity was 0.12 units min^{-1}. Calculate the specific activities of the two fractions, and comment on the usefulness of the purification step. (Molar absorption coefficient of NADH = 6.2×10^3 at 340 nm. Cells with a light path of 1 cm were used throughout.)

17

Enzymes as Analytical Reagents

17.1 THE VALUE OF ENZYMES AS ANALYTICAL REAGENTS

Every enzyme catalyses a reaction which is both substrate-specific and product-specific (section 4.1). Because of this, enzymes are extremely valuable as analytical reagents. In particular, they can be used for the estimation of specific substances, possibly present at very low concentrations, in the presence of other, chemically similar, substances. Ordinary chemical reagents might not be able to distinguish between several of the components of a sample, and costly separation procedures might be required before a satisfactory analysis can be carried out. Even then, several different products might be formed because of side-reactions, thus reducing the sensitivity of the method. In contrast, enzyme-based methods of analysis are both specific and sensitive, which means that little or no sample preparation is required in many cases.

Another characteristic which makes enzyme-catalysed reactions suitable for analytical applications is that they proceed under relatively mild conditions (e.g. near neutral pH and around room temperature). Hence they are usually simple to set up, and can be used for the analysis of substances which would be unstable under more extreme conditions.

On the other hand, enzymes themselves are quite unstable. For this reason, supposedly identical enzyme preparations may be found to have different activities, as may the same preparation on different occasions, particularly if correct storage and handling procedures (section 17.3) are not strictly adhered to. However, such variations in activity should not affect the results obtained by enzyme-based methods of analysis, provided the activity present in each case is a reasonable approximation to that specified, and provided appropriate calibration experiments are always carried out.

Another disadvantage in the use of enzymes is that they are often expensive and sometimes difficult to obtain. Hence it is important that they are not used in a wasteful fashion. One way of minimizing wastage is by the use of immobilized enzymes (section 20.2), which have become increasingly available.

These can be recovered intact at the end of a procedure and used again. Enzymes in this form are sometimes more stable than in free solution. Another method of reducing costs is to scale down the reagent volumes so that the experiment can be conducted using a microplate, which typically has 96 wells (each of volume 300 μl), although 384 and 1536 well microplates are also available (see section 18.3). The end result is measured (by absorbance, fluorescence or luminescence) using a microplate reader.

Enzyme-based analytical procedures may be designed to determine the concentrations of substrates, coenzymes, activators and inhibitors. All of these possible applications, together with enzyme assay, are included in the term enzymatic analysis and are discussed in the next section (section 17.2).

17.2 PRINCIPLES OF ENZYMATIC ANALYSIS

17.2.1 End-point methods

End-point methods of enzymatic analysis, also called **total change** or **equilibrium methods**, allow the reaction to go to equilibrium, whereupon the amount of product formed is determined. Ideally, the nature of the reaction and the conditions should be such that the position of equilibrium is strongly in favour of product formation, the reaction going close to **completion**, with almost all the substrate being converted to product. Since the concentrations of activators and inhibitors do not affect the situation at equilibrium (only its rate of attainment), they cannot be estimated by end-point methods, which are used only for the analysis of **substrates**.

Analysis should be performed under conditions where the concentration of the added enzyme is high, to ensure rapid progress towards equilibrium, while the concentration of the substrate under investigation should be low enough for the reaction to be first-order with respect to this substrate (see Fig. 6.5). The amount of sample used for analysis is therefore chosen accordingly, from a knowledge (or estimate) of the likely upper limit of concentration of the appropriate substrate in the sample. The concentrations of any extra substrates (including co-substrates such as NAD^+) should be in excess, so they do not limit the reaction nor change its first-order characteristics. No products are initially present, so the concentrations of all products (including co-products such as NADH) at equilibrium will be dependent solely on the initial concentration of the substrate being analysed. Therefore, if one of these product concentrations can be measured (e.g. NADH by absorbance at 340 nm) and suitable calibration experiments performed, then the initial concentration of substrate may be calculated. If it is known that the reaction has gone to completion, then the final product concentration will be a measure of the initial substrate concentration, and calibration may not be required.

For any first-order (or pseudo first-order) reaction $S \rightarrow P$, the rate of reaction at time t, $-d[S]/dt$, is given by $-d[S]/dt = k[S]$, where [S] is the concentration of S at time t, and k is a constant (see section 6.3.1). By integration,

$$t = \frac{1}{k}\log_e\left(\frac{[S_0]}{[S]}\right) = \frac{2.303}{k}\log_{10}\left(\frac{[S_0]}{[S]}\right) \qquad (17.1)$$

where $[S_0]$ is the concentration of S at zero time.

For an enzyme-catalysed reaction $S \rightarrow P$ which goes to completion in the absence of any appreciable back reaction, a steady-state may be assumed to exist throughout. Therefore the rate of reaction at time t, $-d[S]/dt$, is given by

$$-\frac{d[S]}{dt} = \frac{V_{max}[S]}{[S]+K_m} \quad \text{(see section 7.1.2)}$$

Note that, under these circumstances (i.e. with no appreciable back reaction), this applies to the rate at any time, not just to the initial velocity, provided $[S_0]$ is not substituted for [S]. At very low substrate concentrations $[S] \ll K_m$, and so $([S]+K_m) \approx K_m$. Therefore, under these conditions

$$-\frac{d[S]}{dt} = \frac{V_{max}}{K_m}[S] \qquad (17.2)$$

Hence, by comparison to the general rate equation for a first-order reaction, it can be seen that $k = V_{max} / K_m$. Therefore, for an enzyme-catalysed reaction progressing under first-order (or pseudo first-order) conditions in the absence of any appreciable back reaction,

$$t = 2.303\frac{K_m}{V_{max}}\log_{10}\left(\frac{[S_0]}{[S]}\right) \qquad (17.3)$$

Now, K_m is a characteristic of the enzyme and should be known. Also, V_{max} is proportional to the enzyme concentration, and may be calculated from the activity of the enzyme preparation and the amount of this used in the analytical procedure. Hence it is possible to determine the time for such an enzyme-catalysed reaction to reach 99% completion (i.e. by substituting $[S_0]/[S] = 100$, together with the calculated values for K_m and V_{max} in the expression for t derived above). If, for a given procedure, the time for the reaction to reach 99% completion seems inconveniently long, it can be reduced by increasing the value of V_{max}, i.e. by adding more enzyme.

However, it will be apparent that this calculation applies only to those enzyme-catalysed reactions which are essentially irreversible. Most, in fact, will reach a position of equilibrium some way short of completion, and although end-point methods may still be applied to such reactions, their sensitivity and reliability will be considerably improved if the reactions can be driven towards completion in some way. One way of doing this is by **trapping the product** and thus continually disturbing the equilibrium. For example, aldehydes and ketones may be trapped by the use of hydrazine. Another possible way of displacing the equilibrium in the direction of completion is by the use of analogues of second substrates or coenzymes. For example, acetyl pyridine adenine dinucleotide ($APAD^+$) has this effect if used in place of NAD^+ to oxidize substrates.

A third possibility is the use of **regenerating systems** to convert a co-product back to co-substrate, thus driving the main reaction towards completion. For example, pyruvate and LDH might be added to reform NAD^+ from NADH, as in the glutamate dehydrogenase (GDH) system for the estimation of glutamate:

$$\text{glutamate} + H_2O + NAD^+ \xrightarrow{\text{GDH}} NADH + H^+ + \text{2-oxoglutarate} + NH_3$$
$$\text{lactate} \xleftarrow{\text{LDH}} \text{pyruvate} \qquad (17.4)$$

(Note, however, that in this system, the final NADH concentration does not give a measure of the original glutamate concentration.)

In general, an enzyme-catalysed reaction being used for substrate analysis by an end-point method should have a product (or co-product) whose concentration can easily be determined. If this is not the case, the reaction may be **coupled** to another to enable the analysis to be performed, a product of the **primary** reaction acting as substrate for the **indicator** reaction. Consider, for example:

$$A \xrightarrow{E_1} B \xrightarrow{E_2} C$$

where A is the substrate whose concentration is being determined, E_1 is the enzyme of the primary reaction and E_2 the enzyme of the indicator reaction. The initial concentrations of B and C should be zero, and any additional substrates of E_1 or E_2 present in excess. For a satisfactory analytical procedure, both reactions should quickly reach equilibrium, preferably at or near completion. Since, under these conditions, the rate of the primary reaction is

$$\frac{V_{\max}^{A}}{K_{m}^{A}}[A]$$

and that of the indicator reaction is

$$\frac{V_{\max}^{B}}{K_{m}^{B}}[B]$$

where $V_{\max}^{A}$ and K_{m}^{A} relate to E_1 , and $V_{\max}^{B}$ and K_{m}^{B} to E_2 (see above), it is important that the enzyme concentrations are chosen such that

$$\frac{V_{\max}^{B}}{K_{m}^{B}} > \frac{V_{\max}^{A}}{K_{m}^{A}} \qquad (17.5)$$

This ensures that the B being formed from A is quickly converted to C. If the coupled procedure is satisfactory, the final concentration of C (or of a co-product of the indicator reaction) is a measure of the initial concentration of A.

The main advantage of end-point methods of analysis is that they do not require constant attention. A procedure may be set up and left to come to equilibrium, after which the final product concentration can be determined at any convenient moment. However, now that automatic monitoring and recording devices are widely available, this advantage over other methods of analysis is considerably less important than it once was.

17.2.2 Kinetic methods

At the present time, most procedures for enzymatic analysis involve steady-state kinetic methods: the initial velocity of the reaction is determined, at fixed temperature and pH, and used to calculate the concentration of the substance being investigated. Such procedures may be used for the analysis of enzymes (i.e. enzyme assay), substrates, coenzymes, activators and inhibitors.

As discussed in section 15.2.2, the initial velocity is directly proportional to **enzyme concentration** if all other relevant factors are made fixed and non-limiting. This is the general principle of enzymatic analysis by kinetic methods. It is arranged that all substances which influence the rate of reaction are present in fixed and non-limiting concentrations, with the exception of the one being analysed, whose concentration may thus be determined from the observed reaction rate.

The relationship between **substrate concentration** and initial velocity is apparent from equation 7.12, the Michaelis-Menten equation, and, since $V_{max} = k_2[E_0]$:

$$v_0 = \frac{V_{max}[S_0]}{[S_0]+K_m} = \frac{k_2[E_0][S_0]}{[S_0]+K_m} \quad (17.6)$$

Therefore, at constant $[E_0]$ in the absence of inhibitors and at fixed concentrations of any activators and additional substrates, there is a hyperbolic relationship between v_0 and $[S_0]$ (Fig. 7.1). This could be used to determine substrate concentration, provided suitable calibration experiments were performed.

However, the best analytical procedures are those where there is a linear relationship between the concentration of the substance being analysed (in this case $[S_0]$) and the experimentally determined property which is dependent on it (in this case v_0): this is because the procedure is then equally sensitive over the whole of the calibrated concentration range; also it enables errors in individual calibration values to be easily detected.

At low substrate concentrations, there is a linear relationship between v_0 and $[S_0]$, since if $[S_0] \ll K_m$, then $([S_0] + K_m) \approx K_m$ and $v_0 = k_2[E_0][S_0]/K_m$.

Hence, the calibrated values for substrate analysis should be restricted to the approximate range 0 to 0.1 K_m. The test sample should be diluted if necessary and the amount used for analysis chosen so that the concentration of the appropriate substrate *in the reaction mixture* lies in the required range. This is straightforward if the likely range in the sample is known; otherwise a series of different dilutions may have to be made to ensure that one gives readings in the calibrated range.

Another advantage of restricting the investigation of substrate concentrations to this range is that it avoids possible confusion resulting from substrate inhibition at high concentrations (section 8.2.6).

Coenzymes which act as co-substrates for enzyme-catalysed reactions may be analysed in exactly the same way as any other substrate.

Activators may also be determined in much the same fashion. These substances increase the catalytic performance of enzymes. Some enzymes can function to a limited extent in the absence of their activators, whereas others cannot. In either case there is usually a linear relationship between initial velocity and activator concentration over the lower part of the concentration range, at fixed and non-limiting concentrations of enzyme and all substrates and coenzymes. At higher activator concentrations, the enzyme becomes fully activated and initial velocity reaches a maximum value (Fig. 17.1).

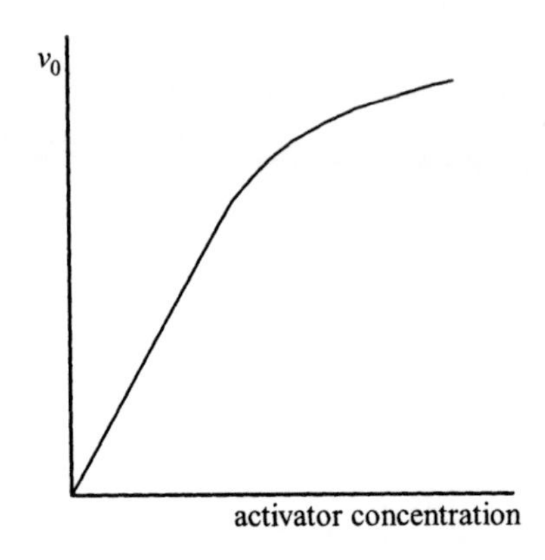

Fig. 17.1 – Typical plot of initial velocity against activator concentration at fixed and non-limiting concentrations of enzyme and all substrates and coenzymes, for a system which cannot function in the absence of activator.

A system may be calibrated, as in Fig. 17.1, and used to determine the activator concentration in a test solution. However, as before, it is best to work only in the lower, linear part of the concentration range and to dilute the sample as necessary to bring the test results into this range.

An example of the use of such a system is the determination of Mn^{2+} by means of its activation of **isocitrate dehydrogenase**: this provides an extremely sensitive method of analysis, enabling extremely low concentrations of Mn^{2+} ions to be accurately determined.

Enzyme **inhibitors** may also be determined by sensitive procedures of this kind. As discussed in Chapter 8, many different types of enzyme inhibitors exist. However, as far as enzymatic analysis of inhibitors is concerned, the type of inhibition is immaterial. A system is calibrated at fixed and non-limiting concentrations of enzyme and all substrates, coenzymes and activators, and a calibration curve plotted. This may take either of two forms: initial velocity may be plotted against inhibitor concentration (Fig. 17.2a); alternatively, percentage inhibition may be used instead of initial velocity (Fig. 17.2b).

$$\text{Percentage inhibition} = \frac{v_0\,(\text{uninhibited}) - v_0\,(\text{inhibited})}{v_0\,(\text{uninhibited})} \times 100 \qquad (17.7)$$

where all initial velocities are determined under identical conditions, apart from inhibitor concentration.

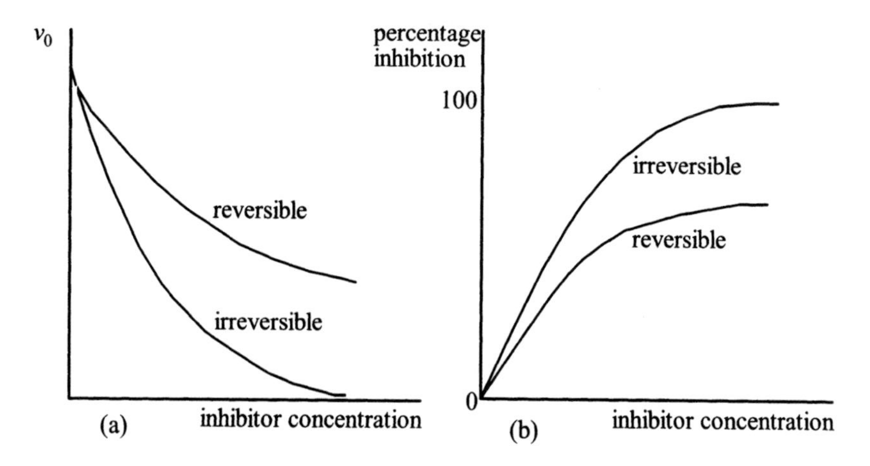

Fig. 17.2 – Plots of (a) initial velocity against inhibitor concentration, and (b) percentage inhibition against inhibitor concentration, for typical reversible and irreversible inhibitors, at fixed and non-limiting concentrations of enzyme and all substrates, coenzymes and activators.

Irreversible inhibitors, by definition, give 100% inhibition if they are present in excess, but reversible inhibitors do not give complete inhibition, no matter how high their concentrations. The inhibitor concentration giving 50% inhibition, under specified conditions, is sometimes termed the **I_{50} value.**

Once a system has been calibrated, it may be used to determine the concentration of an inhibitor in a test solution. Again, only the lower part of the concentration range, where there is a more-or-less linear relationship between initial velocity and inhibitor concentration, should be used. One such procedure, involving liver **carboxyl esterase**, enables small amounts of fluoride to be accurately determined in the presence of large amounts of phosphate, a factor which complicates many other procedures for the analysis of fluoride.

However, although enzymes are highly specific towards substrates, coenzymes etc., they may be affected by a large number of inhibitors. This should be borne in mind in all forms of enzymatic analysis, particularly where the test sample contains many different components. Many metal ions, for example, are enzyme inhibitors, so it is not possible to use enzymatic analysis to determine the concentration of one of these if others which inhibit the same enzyme are also present. In enzymatic analysis in general, it is important to know what substances may be present in significant amounts in the sample, in addition to that being analysed.

The **coupling** of reactions for the purpose of enzyme assay was discussed in section 15.2.3. Similar coupling procedures may be required in any type of enzymatic analysis where the rate of the main reaction under investigation cannot be determined directly by any convenient method. Consider the general situation:

$$A \xrightarrow{E_1} B \xrightarrow{E_2} C$$

where E_1 is the enzyme catalysing the **primary reaction** (i.e. the main reaction under investigation) and E_2 the enzyme catalysing the **indicator reaction**. The initial concentrations of B and C should always be zero, and the initial concentrations of any additional substrates of E_1 and E_2 should be fixed and non-limiting.

In the case of enzyme assay, the primary reaction is usually made to proceed under zero-order conditions which makes the design of the coupling procedure relatively straightforward (see section 15.2.3). However, substrate (or co-substrate) analysis is normally performed under conditions where the primary reaction is of first-order with respect to the substrate (or co-substrate) being analysed, which makes the relationship between the primary and indicator reactions less easy to establish. The initial velocity of the primary reaction at steady-state is given by the Michaelis-Menten equation:

$$v_0 = \frac{V_{max}^{A}[A_0]}{[A_0]+K_m^{A}}$$

which under first-order conditions (where $[A_0] \ll K_m^A$) simplifies to:

$$v_0 = (V_{max}^{A}/K_m^{A})[A_0] \tag{17.8}$$

To ensure that B does not accumulate and that the rate of the indicator reaction is a measure of the rate of the primary reaction over the period of interest, the concentration of E_2 (and hence V_{max}^B) is made sufficiently large so that

$$\frac{V_{max}^{B}}{K_m^{B}} \gg \frac{V_{max}^{A}}{K_m^{A}} \tag{17.9}$$

where V_{max}^A and K_m^A relate to E_1, and V_{max}^B and K_m^B to E_2. To give a reasonable margin of safety, the ratio

$$\frac{V_{max}^{B}}{K_m^{B}} : \frac{V_{max}^{A}}{K_m^{A}}$$

is usually made in the order of 100: 1 for substrate determinations and for most other forms of enzymatic analysis.

Another important requirement for coupled reactions is that the enzymes involved are active in the same pH range (see section 15.2.3).

The simplest, though most time-consuming way to perform a kinetic method of enzymatic analysis is to remove samples at timed intervals from the reaction mixture. The reaction in each sample tube is terminated immediately on removal, e.g. by rapid change of temperature or pH, or by the use of a protein precipitant such as perchloric acid, and then the product concentration in each is determined. From these data, the initial velocity can be calculated.

However, the use of such laborious **manual** procedures has now largely been discontinued. The methods used at the present time are mainly **instrumental** ones and, in particular, ones where the course of the reaction can be monitored and recorded automatically (see section 18.2). Since the initial velocity alone is required, many instruments can now determine this and display results within a minute or so of the start of the reaction.

17.2.3 Immunoassay methods

The determination of enzyme concentration by radioimmunoassay (RIA) was discussed in section 15.2.4.

Enzymes may also be involved in other immunoassay procedures, although it must be stressed that the role of the enzymes in these is a secondary one: they are used to replace radioisotopes as markers, since they are not such a hazard to health and can be detected by techniques which are more generally available. Any enzyme which has a sensitive and convenient assay procedure can be employed. There are currently two main types of these **enzyme-immunoassay (EIA) procedures: ELISA and EMIT**.

ELISA (enzyme-linked immunosorbent assay) resembles RIA in that it involves a heterogeneous system which must undergo separation as part of the analysis procedure. For example, a pure specimen of antigen may be labelled by attaching it to an enzyme (just as it is labelled with a radioisotope in RIA) in such a way that the activity of the enzyme is not affected. Then, specified amounts of antibody and enzyme-labelled antigen are mixed with the sample which contains an unknown amount of the antigen. The free antigen (Ag) and enzyme-labelled antigen (E.Ag) then compete for binding to the available antibody (Ab), according to the following reactions:

$$\text{Ag} + \text{Ab} \rightleftharpoons \text{Ag.Ab} \quad \text{(equation 15.3)}$$

$$\text{E.Ag} + \text{Ab} \rightleftharpoons \text{E.Ag.Ab} \qquad (17.10)$$

After equilibrium has been reached, the antibody-bound fractions are separated from the rest (the use of immobilized antibodies, i.e. antibodies attached to an insoluble matrix, enables this to be done conveniently) and the enzyme activity in one or both of the fractions is determined. From the distribution of the enzyme between the free and antibody-bound fractions, the concentration of antigen in the sample may be calculated. The more antigen in the sample, the greater will be the enzyme activity of the free antigen fraction after equilibration. Steroid hormones such as progesterone and cortisol have been determined by procedures of this type, and so have proteins such as insulin. Enzymes which have been used as labels include peroxidase, alkaline phosphatase and β-galactosidase.

An alternative form of ELISA involves attaching the enzyme to the *antibody* and the use of an immobilized antigen. Specified amounts of enzyme-labelled antibody and immobilized antigen are mixed with the sample. The antigen in the sample (Ag) then competes with the immobilized antigen (X.Ag) for the available enzyme-linked antibody (Ab.E) as follows:

$$\text{Ag} + \text{Ab.E} \rightleftharpoons \text{Ag.Ab.E} \qquad (17.11)$$

$$\text{X.Ag} + \text{Ab.E} \rightleftharpoons \text{X.Ag.Ab.E} \qquad (17.12)$$

After equilibration, the immobilized fraction is separated from the rest and the distribution of the enzyme is determined. This enables the antigen content of the sample to be calculated, there being an inverse relationship between the enzyme activity of the insoluble fraction and the amount of antigen in the sample. An assay for α-fetoprotein has been based on this principle.

EMIT (enzyme multiplied immunoassay technique) is a homogeneous procedure, since no separation of components is required. As with some types of ELISA, an enzyme is attached to a specimen of antigen to act as a label. However, in contrast to ELISA, the subsequent binding of the enzyme-labelled antigen to the antibody results in a significant change in activity of the enzyme. In most types of EMIT procedure, enzyme activity is lost completely on binding to the antibody, either as a result of steric hindrance or conformational changes. In such an assay, specified amounts of antibody and enzyme-labelled antigen are mixed with the sample. The antigen in the sample (Ag) then competes with the enzyme-labelled antigen (E.Ag) for the available antibody (Ab) according to the following reactions:

$$\text{Ag} + \text{Ab} \rightleftharpoons \text{Ag.Ab (equation 15.3)}$$

$$\text{E.Ag} + \text{Ab} \rightleftharpoons \text{E}'\text{.Ag.Ab} \qquad (17.13)$$

The enzyme-antigen-antibody complex formed (E′.Ag.Ab) has no catalytic activity, so the total activity present is contributed by E.Ag. Hence, the antigen content of the sample may be calculated from the total enzyme activity at equilibrium. The more antigen there is in the sample, the less E.Ag is able to bind to the antibody, so the greater will be the total enzyme activity.

EMIT procedures are more convenient than ELISA ones, since they do away with the separation step. However, the technical problems of obtaining enzyme-labelled antigen suitable for EMIT are considerable, particularly since the reasons for the loss of activity on binding to antibody are not fully understood and may vary from case to case. Because of this, only a limited number of EMIT procedures are currently available, but this number is increasing.

EMIT has so far been used principally for the determination of relatively small (i.e. non-protein) molecules, e.g. barbiturates. The enzymes which have been involved include lysozyme and malate dehydrogenase.

Automated instruments designed specifically for immunoassay procedures, using enzyme or other labels, have now been available for a number of years. Examples include the Abbott AxSYM and the PerkinElmer AutoDELFIA systems.

17.3 HANDLING ENZYMES AND COENZYMES

Enzymes easily become denatured and lose catalytic activity, so careful storage and handling is essential if this is to be prevented. The functional groups of some amino acids can be subjected to alteration on storage, which may alter the properties of the enzyme. The amide nitrogen in glutamine and asparagine can be lost at most pH values, converting the amino acids to glutamic and aspartic acid respectively, thus increasing the net negative charge on the protein.

The functional groups of cysteine, methionine and tryptophan are prone to oxidation and reducing sugars can be added to free amino groups at the N-terminus (or lysine residues). In addition, protein aggregation (giving rise to precipitates which are usually removed by centrifugation at 13 000g for 20 min) or proteolysis can occur at most temperatures.

The stability of proteins during storage is not easy to predict and has to be determined empirically, using different buffers at different pH values and at different temperatures. In general, temperatures higher than 40°C and extremes of pH (i.e. below pH 5 and above pH 9) should be avoided at all times. If the pH of an enzyme solution is being changed, continuous mixing helps to prevent denaturation taking place around the added drops of acid or alkali. However, too vigorous mixing causes the formation of froth, which can be counter-productive, since many enzymes are denatured at air-liquid interfaces.

The presence of various additives may improve storage stability in solution. These include low concentrations of cofactors, substrates, reversible inhibitors, activators, reducing agents, chelating reagents, proteolytic inhibitors and anti-microbial agents (e.g. 0.05% (w/v) sodium azide or 0.01% (w/v) thiomersal). Other additives may include non-reducing sugars such as sucrose, cyclodextrins, ethanediol and glycerol (which increases the viscosity of the medium).

Enzymes can be stored as an ammonium sulphate precipitates (covered in saturated ammonium sulphate at –25°C) for long periods with little loss of activity. At the required time, the precipitate is redissolved in buffer and the ammonium ions removed by dialysis or SEC (see chapter 16.2.2). Alternatively, lyophilization (freeze drying) can be used to remove water by sublimation from enzyme solutions, resulting in a powder. Volatile buffers (e.g. diethanolamine) can be used in the last chromatographic procedure prior to lyophilization to ensure that only the enzyme remains as a powder. The additives described above (perhaps in various combinations, determined empirically) can be used to improve the recovery of enzyme activity after lyophilization.

Some dry enzymes are quite stable at 0-4°C, whereas others need storing at lower temperatures (e.g. by use of liquid nitrogen at –196°C or a freezer at –80°C or –25°C). Enzyme preparations should also be stored at low temperatures, but not necessarily frozen, since freezing and thawing may cause loss of activity. This depends on the nature of the preparation. Crystalline enzymes suspended in ammonium sulphate solutions are best stored at 0-4°C, as are lyophilized enzyme preparations. However, the latter may be stored frozen if it is certain that they contain no moisture. In contrast, solutions of enzymes should always be stored frozen at temperatures as low as possible, but repeated freezing and thawing should be avoided. For this reason, such solutions are best divided into small portions for storage. Enzyme preparations are sometimes stored in aqueous solutions of glycerol, since this enables them to be kept at temperatures below 0°C without freezing taking place.

Most coenzymes must be stored as solids at 0-4°C, preferably in a desiccator, since many are subject to hydrolysis. Some coenzymes must also be protected from light or oxygen. NADH, for example, is destroyed by light to quite an appreciable extent, while also reacting with any atmospheric water to form inhibitors of dehydrogenases. In solution, NADH is particularly unstable in acid conditions, whereas NAD^+ is alkali-labile.

The sulphydryl group of coenzyme A can be oxidized by atmospheric oxygen, and its pyrophosphate bond is easily hydrolysed if moisture is present. Some of the more stable coenzymes may be stored in frozen solution, but, as with enzyme solutions, repeated freezing and thawing should be avoided.

All the water (preferably ultrapure, resistivity 18.2 MΩ cm at 25°C) and reagents used to make up the buffers should be of the highest grade available to minimise the presence of heavy metal ions and should be filtered through 0.2 μm membranes to remove bacterial contamination. The storage containers should have low protein-binding characteristics, and be pre-sterilized or autoclaved.

SUMMARY OF CHAPTER 17

Enzymes have great value as analytical reagents because of their specificity of action. Enzyme-based methods of analysis are very sensitive and are usually carried out under mild conditions. Such procedures may be used to determine the concentration of a substrate, coenzyme, activator or inhibitor of the enzyme in question. The general name for investigations of this type, including enzyme assay, is enzymatic analysis.

The principle of enzymatic analysis by kinetic methods is that the initial velocity of reaction is dependent solely on the concentration of the substance being analysed, regardless of whether it is an enzyme, substrate, coenzyme, activator or inhibitor, provided all other relevant factors are made fixed and non-limiting. End-point methods of analysis, on the other hand, involve measurement of product (or co-product) concentration when a reaction has gone to (or near to) completion and can only be used for the analysis of substrates. Both types of analysis may involve the coupling of the primary reaction to an indicator reaction.

Enzymes may also be used as an alternative to radioisotopes as markers in immunoassays. Such procedures, termed enzyme-immunoassays, have been used for the determination of a variety of proteins and hormones.

Enzymes easily lose their catalytic activity, so require careful handling. Exposure to high temperatures, extremes of pH and high concentrations of organic solvents should be avoided at all times. In common with coenzymes, they require storage at low temperatures.

All solutions used in enzymatic analysis procedures must be free from contamination by micro-organisms and metal ions.

FURTHER READING

Crowther, R. J. (2000), *The ELISA Guidebook,* Humana Press.

Gosling, J. P. (2000), *Immunoassay: A Practical Approach*, Oxford University Press.

Ryan, O., Smyth, M. R. and Fagain, C. O. (1994), Horseradish peroxidase - the analyst's friend, *Essays in Biochemistry,* **28**, 129-146, Portland Press.
Schoeff, L. E. and Williams, R. H. (eds.) (1993), *Principles of Laboratory Instruments,* Mosby (Chapter 19).
Wild, D. (ed.) (2005), *The Immunoassay Handbook,* 3rd edn., Elsevier.
Wilson, K. and Walker, J. (eds.) (2000), *Principles and Techniques of Practical Biochemistry,* 5th edn., Cambridge University Press (Chapters 4, 7).

PROBLEMS

17.1 An enzyme used in an end-point method of analysis has a K_m of 9×10^{-4} mol l^{-1} with respect to the appropriate substrate and its activity in the reaction mixture is 2.0 IU cm^{-3}. If the reaction is largely irreversible and is proceeding under first-order conditions, calculate the time required for it to reach 99% completion.

17.2 Sketch, from memory or deduction, the following plots for enzyme-catalysed reactions which obey the Michaelis-Menten equation. In each case, state clearly what restrictions must be imposed on other relevant factors to enable the plot you draw to be obtained.

(a) v_0 against $v_0/[S_0]$;
(b) v_0 against $[E_0]$;
(c) v_0 against coenzyme (e.g. NADH) concentration;
(d) [P] against t (under first-order and under zero-order conditions);
(e) v_0 against $[S_0]$ in the absence and presence of a competitive inhibitor;
(f) $1/v_0$ against $1/[S_0]$ showing the effects of substrate inhibition;
(g) v_0 against activator concentration, for a system which does not function in the absence of activator;
(h) v_0 against $[I_0]$ for an irreversible inhibitor.

18

Instrumental Techniques Available for Use in Enzymatic Analysis

18.1 PRINCIPLES OF THE AVAILABLE DETECTION TECHNIQUES

18.1.1 Introduction

A variety of detection procedures are used, or have been used in the past, for enzymatic analysis by either end-point or kinetic methods. They may also, of course, be used for the investigation of enzyme characteristics. The most important ones are discussed below.

In most cases, the instrument detecting the changes which take place as the reaction proceeds can be coupled to a recorder and/or computer, enabling the course of the reaction to be followed in detail without requiring constant human attention. Also, the enzyme involved may be in an immobilized form (see section 20.2) to allow it to be recovered at the end of the procedure. However, these refinements, whose use may depend on the resources available, do not affect the principles of the detection techniques.

18.1.2 Manometry

If any of the substrates or products of an enzyme-catalysed reaction is a gas, the course of the reaction may be followed by manometric techniques, as developed by Otto Warburg (1924). The basic system must be air-tight, and consists of a reaction vessel attached to a U-shaped tube. The sample and reagents are put into different compartments in the reaction vessel and mixed together at zero time; then, as the reaction proceeds, the uptake or evolution of gas is indicated by the movement of fluid in the manometer tube.

However, it will be realized that as the manometer fluid moves towards or away from the reaction vessel, the level of liquid in the two arms of the U-shaped tube will no longer be the same. This will introduce pressure changes which will affect the measurement of the progress of the reaction. Various features are therefore added to the basic system to ensure that all readings are made under identical conditions.

In **constant volume manometry**, the manometer is usually connected to a reservoir, so that fluid may be added to or removed from the manometer as required. The amount of fluid in the manometer is adjusted prior to taking a reading so that the level in the limb nearest the reaction vessel is at a fixed point. The difference between the fluid levels in the two limbs of the manometer gives an indication of the pressure in the reaction vessel, which changes as gas is taken up or evolved, and so may be used to follow the course of the reaction.

In **constant pressure manometry**, the volume of the reaction chamber is changed by measurable amounts (e.g. by connecting it to a micrometer, as in the **Gilson differential manometer**) sufficient to keep the manometer fluid level at a fixed point as the reaction proceeds. The amount of fluid in the manometer is not changed, so this procedure ensures that the levels in *both* arms remain fixed, and no pressure changes take place. The course of the reaction may thus be followed in terms of the volume changes required to maintain constant pressure.

Manometric methods may be used to investigate reactions where a gas is consumed, e.g. that catalysed by glucose oxidase, where molecular oxygen is a substrate. They may also be used when a gas is produced, e.g. to monitor CO_2 production in reactions catalysed by decarboxylases. Reactions where an acid is produced (including all reactions involving NAD^+ reduction to $NADH + H^+$) may be investigated by manometry by adding bicarbonate to the reaction mixture and measuring the evolution of CO_2,

However, such techniques usually involve a margin of error greater than 5%, so they are rarely used today for enzymatic analysis, since more accurate and reliable procedures are now available.

18.1.3 Spectrophotometry

If a substrate or a product absorbs light at a characteristic wavelength, then the reaction can be monitored by following the changes in absorbance at this wavelength. The absorbance, A, is given by $A = \log_{10}(I / I_0)$, where I_0 is the intensity of the incident light and I is the intensity of the light transmitted through the sample. The absorbance is related to the concentration of the light-absorbing substance in the sample by the **Beer-Lambert law**: $A = \varepsilon cl$, where ε is the molar absorption (or absorbance) coefficient, c is the molar concentration and l is the length of the light-path through the sample (usually measured in centimetres).

This is generally true for any wavelength of visible light (wavelength range 400-700 nm) or ultraviolet light (200-400 nm). The wavelength required can be selected from the spectrum obtained when light is passed through a prism or a diffraction grating. In simpler and cheaper instruments (i.e. **colorimeters**), a waveband, rather than a particular wavelength, is selected by passing the light through a coloured-glass filter: such instruments are less reliable than spectrophotometers and cannot be used in the ultraviolet region. For enzymatic analysis, the temperature of the reaction mixture should be maintained to within ±0.2°C, which necessitates the use of a jacketed cell-holder connected to a circulating water bath.

Procedures which involve the calculation of initial velocity should preferably be ones where an increase (rather than a decrease) of absorbance with time is observed. They will then be operating in the most sensitive part of the absorbance scale (the lower part) during the crucial early stages of the reaction.

Such procedures are best followed by instruments which give direct absorbance readings rather than by **null-point instruments** which require separate adjustments for every reading. **Direct-reading instruments** may be coupled to recorders so that the whole course of the reaction may be traced out. **Double-beam instruments,** which automatically compare the absorbance of the reaction mixture to that of a reagent blank at each point, are preferable in these circumstances, since changes of absorbance with time of the blank may be taken into account without having to re-set the instrument during the analysis. Computers may also be attached, to store and process the readings.

Spectrophotometric methods are particularly useful for monitoring reactions which involve NAD^+ or $NADP^+$ as coenzyme (see section 6.5.1), or ones which can be coupled to such reactions.

There are two main problems associated with spectrophotometric methods. Firstly they can only be used satisfactorily over the lower part of the absorbance range. At high absorbance values there is, in addition to reduced sensitivity, often some departure from the Beer-Lambert law because of interactions between solute molecules in concentrated solution. Secondly, errors may be introduced when there is a high background absorbance or when the sample contains material in suspension (e.g. blood cells or micro-organisms), thus scattering the incident light. For this reason, crude samples may often have to be partially purified (e.g. by dialysis, selective precipitation and/or centrifugation) before one of its components may be investigated by spectrophotometric procedures.

18.1.4 Spectrofluorimetry

Compounds are said to be **fluorescent** when they absorb light of one wavelength and then emit light of a longer wavelength. At low concentrations, the intensity of fluorescence (I_f) is related to the intensity of the incident light (I_0) of appropriate wavelength by the relationship: $I_f = I_0 \times 2.3\varepsilon clq$, where ε is the molar absorption coefficient, c the molar concentration, l the length of the light-path and q the quantum efficiency (i.e. the number of quanta fluoresced divided by the number of quanta absorbed). The fluorescence is usually measured at right angles to the incident beam, to avoid detection of any transmitted light. The wavelength of the fluorescence measured, and the wavelength of the incident beam, are usually selected by the use of diffraction gratings. In general, spectrofluorimetry is several times (theoretically up to a hundred times) more sensitive than spectrophotometry. However, fluorescence effects vary with temperature, so it is particularly important to maintain a fixed temperature throughout.

NADH and NADPH exhibit fluorescence, absorbing light at 340 nm and re-emitting it at about 460 nm. Therefore, any reactions utilizing these as coenzymes may be followed by the use of sensitive spectrofluorimetric procedures. Spectrofluorimetry is also widely used for the investigation of hydrolytic reactions, using synthetic non-fluorescent substrates which are esters of highly fluorescent alcohols or amines. For example, George Guilbault and David Kramer (1963) devised an assay for **triacylglycerol lipase** based on the following reaction:

$$\underset{\text{(non-fluorescent)}}{\text{dibutyryl fluorescein}} \xrightarrow{\text{lipase}} \underset{\text{(fluorescent)}}{\text{fluorescein}}$$

In a similar way, fluorescein diacetate may act as a substrate for esterases, and derivatives of highly-fluorescent 4-methylumbelliferone are often used in the assay of esterases, glycosidases, phosphatases and sulphatases.

Tyrosine and tryptophan fluoresce at 330-350 nm, and, since residues of these are usually present in enzymes, they would give a high background emission in the ultraviolet region. Hence, it is advisable to restrict the use of spectrofluorimetry in enzymatic analysis to the detection of substances which emit light in the visible region. Apart from this, the main problem associated with spectrofluorimetric techniques is **quenching**, where the extra energy possessed by the excited molecules after absorption of a photon of light is transferred to another molecule or group (e.g. to iodide or dichromate ions, if these are present) instead of being released as another photon. **Concentration quenching** may also occur, due to some net absorption of either the incident or the emitted light taking place.

18.1.5 Electrochemical methods

Many different types of electrochemical procedures have been used in enzymatic analysis. Often these involve **potentiometric techniques**, where an electrical potential is generated which is dependent on the concentration and properties of substances in the test solution; the change in potential as the reaction proceeds can be taken to be a measure of the rate of the reaction (see section 18.2.3). It should be noted that a potential, which is caused by a transfer of electrons between an electrode and a solution, cannot be measured in isolation: the half-cell consisting of the electrode dipped in the test solution must be connected to a reference half-cell, of fixed potential, to enable the potentials of the two half-cells to be compared.

Of particular importance are techniques involving **ion-selective electrodes**, where the potential generated can be related to the concentration of a specific substance. The best known ion-selective electrode is the **glass electrode** used in the **pH meter**. In this, an internal electrode (e.g. silver wire coated with silver chloride) is dipped in a solution containing HCl at known pH. This is separated from the test solution by a thin glass wall and, in a way not fully understood, a potential is generated which is dependent on the pH difference across the glass. The potential of the entire half-cell, which includes the contribution of the internal electrode as well as that resulting from the pH difference, may be measured by linking the internal electrode to an external reference electrode (e.g. a calomel electrode) and completing the circuit by means of a salt bridge (Fig. 18.1).

The complete cell consists of the following interacting compartments:

Ag	AgCl	HCl	glass	test-sample	saturated KCl acting as a salt bridge	external reference electrode

The potentials produced at each of these junctions will be constant for a particular instrument, the one exception being the important one between glass and the test-sample. Thus, a suitably calibrated instrument may be used to determine the pH of the sample.

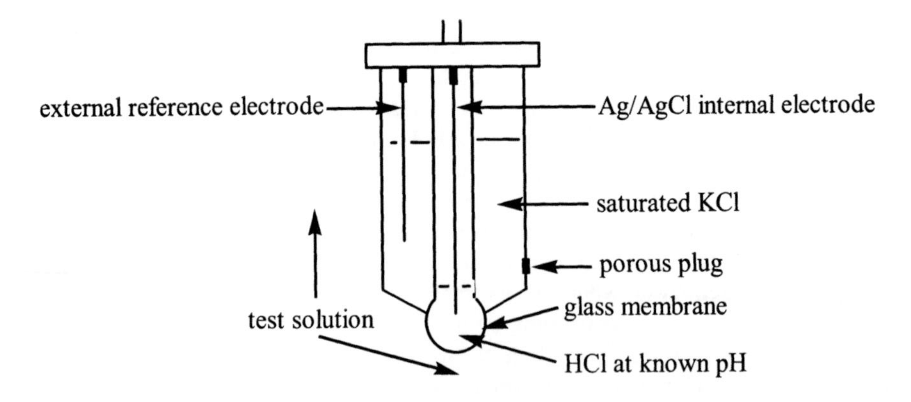

Fig. 18.1 – Diagrammatic representation of a combined glass electrode and external reference electrode, as used in the pH meter.

Many enzyme-catalysed reactions involve the production of an acid or base from a neutral compound (or vice versa), so the course of such reactions could be followed by measuring the change in pH as the reaction proceeds. However, this is rarely done because a change in pH would cause the activity of the enzyme to change. The best way of monitoring the course of such reaction is by the use of a **pH-stat**. This usually consists of a pH meter coupled to a motor-driven burette syringe and a circuit-breaker designed to switch off the burette motor when the pH of the reaction mixture is at the original pre-set value. As the reaction proceeds and the pH of the reaction mixture starts to change, the motor switches on and forces alkali (or acid) from the burette into the reaction vessel, with continuous stirring, to restore the pH to its original value, a recorder (or computer) automatically noting the amount added from the burette in relation to time. Thus the pH is maintained at an approximately constant value throughout, and the rate of addition of alkali (or acid) required to achieve this is a measure of the rate of reaction. Such techniques may be used, for example, to monitor the hydrolysis of esters.

Other types of electrochemical procedure include those based on **polarography** or **voltammetry**, where an increasing voltage is applied between two electrodes immersed in the test solution, whose composition determines the current which flows at each instant.

If a *fixed* voltage is applied between the electrodes, the current depends on the electrical conductance of the solution, which is the sum of the contribution of all the ions present (since these must be the carriers of the current through the solution). It is also the reciprocal of the electrical resistance between the electrodes. For those enzyme-catalysed reactions where there is a change in the number of charged species as substrates are converted to products, the course of the reaction may be followed by determining the change of electrical conductance with time: this is termed **conductometry.** Such procedures usually make use of alternating rather than direct current, to avoid electrolysis taking place, which would tend to reduce the applied voltage. Since the reaction mixture must be buffered, which will contribute a background conductance, a low concentration of a buffer of low intrinsic conductivity (e.g. Tris) is usually employed.

An example of a reaction which is ideally suited to enzymatic analysis by conductometric procedures is the urease-catalysed hydrolysis of urea: $O{=}C(NH_2)_2 + H_2O \rightleftharpoons 2NH_4^+ + CO_3^{2-}$. An uncharged molecule is converted into ions as the reaction proceeds, so the conductance must increase. This has been used for the analysis of the substrate, urea, and gives a linear response up to a concentration of about 22 μmol l^{-1}, above which there is interaction with buffer ions. Samples containing urea at higher concentrations than this should be diluted prior to analysis.

A specific electrode which is based on similar principles is the **oxygen electrode.** In the type developed by Leland Clark (1956), there is a central cathode of gold or platinum surrounded by an annular silver anode and separated from it by an epoxy resin casing (Fig. 18.2).

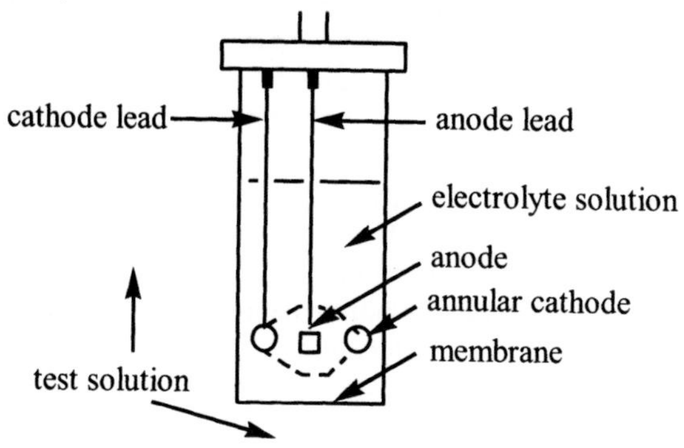

Fig. 18.2 – Diagrammatic representation of a typical oxygen electrode.

Oxygen dissolved in the reaction mixture (which should be in a thermostatted, air-tight container and stirred continuously) can freely diffuse to the electrodes through a membrane (e.g. Teflon or polypropylene) and an electrolyte solution (usually buffered KCl). A constant voltage of between 0.5 and 0.8V is applied across the electrodes, providing a supply of electrons at the cathode which can reduce oxygen atoms, $4e^- + O_2 + 2H^+ \rightarrow 2H_2O$, whereas at the anode there is a corresponding transfer of electrons *to* the electrode: $4Ag + 4Cl^- \rightarrow 4AgCl + 4e^-$.

Thus, a current flows between the electrodes which is proportional to the pO_2 of the sample. By monitoring how this current changes with time, the course of a reaction involving oxygen production or consumption (e.g. that catalysed by glucose oxidase) can be followed. This is a far more sensitive procedure than manometry (section 18.1.2), a possible alternative.

As a refinement, the oxygen electrode (or other specific electrode) may be coupled with an immobilized enzyme to form an **enzyme electrode** for the analysis of the appropriate substrate. For example, Stuart Updike and G. P. Hicks (1967) trapped glucose oxidase in polyacrylamide gel and layered it over the membrane of an oxygen electrode, thus constructing an instrument which could be used to analyse D-glucose by a sensitive, specific and convenient procedure. Such enzyme electrodes form a major category of what have been termed **biosensors**.

In general, electrochemical methods have an advantage over certain other procedures (e.g. spectrophotometry) in that contaminating substances can be present in suspension without affecting the analysis.

18.1.6 Enthalpimetry

Enzymatic analysis by enthalpimetry (microcalorimetry) is of potential importance because of the sensitivity, freedom from interference and almost universal possibilities of application of these techniques. In most reactions, heat (enthalpy) is gained or lost as the reaction proceeds (see section 6.1). In enthalpimetry, the reaction is carried out under controlled conditions in a calorimeter and the temperature change is monitored. The change is small (usually 10^{-2} to 10^{-4} °C) and well within the range of possible fluctuations in the ambient temperature. Hence extremely accurate thermostatting or excellent insulation is required, together with very sensitive temperature sensors. Suitably accurate temperature measurements may be obtained by the use of **thermistors** (temperature-sensitive resistors consisting of sintered mixtures of metal oxides in glass or plastic casings) or of **thermopiles** (thermocouples connected in series).

Secondary reactions with the buffer can be used to increase the signal and thus improve the sensitivity of the procedure. For example, a proton produced by the primary reaction can be trapped by buffer with evolution of heat. Tris is very useful in this respect, for its heat of protonation is large. David McGlothlin and Joseph Jordan (1975) used enthalpimetry to estimate blood glucose by means of a hexokinase-catalysed reaction in Tris buffer at pH 8:

$$\begin{aligned} &\text{glucose} + \text{ATP} \rightarrow \text{glucose-6-phosphate} + \text{ATP} + \text{H}^+ \quad &\Delta H = -28\ \text{kJ mol}^{-1} \\ &\text{Tris} + \text{H}^+ \rightarrow \text{Tris–H}^+ \quad &\Delta H = -47\ \text{kJ mol}^{-1} \end{aligned} \tag{18.1}$$

Note that the secondary reaction liberates more heat than the primary reaction.

However, such procedures have yet to be used routinely.

18.1.7 Radiochemical methods

The use of a radioactively-labelled substrate can be valuable in enzymatic analysis. The isotopes most commonly used for labelling purposes are ^{3}H (tritium), ^{14}C, ^{32}P, ^{35}S and ^{131}I. All of these isotopes emit β-radiation (electrons) as they decay.

After the enzyme-catalysed reaction has progressed for a specified period, it is terminated. The substrate is then separated from the product, usually by chromatography or electrophoresis, and the product concentration is determined indirectly by measuring the radioactivity of the product fraction.

High-energy β-emission, e.g. from ^{32}P or ^{131}I, may be detected and counted using a **Geiger-Müller counter**.

Low-energy β-emission, e.g. from ^{3}H, ^{14}C or ^{32}S, is usually monitored by **liquid scintillation counting**: some organic solvents fluoresce when bombarded with β-radiation, giving out light of very short wavelength; this is not suitable for efficient detection, but it may be converted into light of longer wavelength by the fluorescence of one or more **fluors**, or **scintillants** (e.g. 2,5-diphenyloxazole, PPO), and detected by a photomultiplier.

A typical example of enzymatic analysis by a radiochemical procedure is that involving the cholinesterase-catalysed hydrolysis of [^{14}C]-acetylcholine:

$$\underset{[^{14}\text{C}]\text{-acetylcholine}}{^{14}CH_3.CO_2CH_2CH_2N^+(CH_3)_3} \rightarrow \underset{[^{14}\text{C}]\text{-acetic acid}}{^{14}CH_3.CO_2H} + \underset{\text{choline}}{HOCH_2CH_2N^+(CH_3)_3} \qquad (18.2)$$

After incubation, the unreacted substrate and the choline may be removed by an ion exchange resin, and the radioactivity of the acetic acid fraction determined.

In general, radiochemical procedures are extremely sensitive (far more sensitive than photometric ones). They can be used in a straightforward fashion for enzyme assay, and, by means of isotope dilution techniques, also for other forms of enzymatic analysis. However, they have certain disadvantages when compared to other possible techniques. These include the health hazard inevitable in the handling of radioisotopes, and the separation steps which have to be performed. Also, it is impossible to use them to monitor continuously the course of a reaction, and any estimate of the initial velocity which is obtained can only be based on a limited number of points (possibly only *one*). A further possibility of error can result from a **quenching** of the radiation between source and detector. This is particularly likely in the case of tritium, whose emission is so weak, it can be absorbed by paper.

18.1.8 Dry-reagent techniques

For a number of years, **test strips** have been available for certain routine analytical procedures. The strip, now usually made of porous plastic, incorporates a pad at one end impregnated with the reagents, which may include enzymes. This end of the strip is dipped into the test solution, and, after a specified time, the colour produced in the pad is compared with those on the chart provided. A semi-quantitative assessment of the concentration of a reactant in the test solution is thus obtained. Such a procedure is widely used to give an indication of the D-glucose content of urine (see section 19.1.3).

Processes of this type have been developed to make them fully quantitative. Examples include the Kodak Ektachem system, widely-used during the 1990s, and the more recent Roche Reflotron systems.

The Kodak Ektachem procedure for the analysis of blood urea involves placing a drop of blood on a special slide. The sample then diffuses down through a layer of reflective titanium dioxide to a reagent layer impregnated with urease, which converts urea to ammonia. This in turn diffuses down through a semi-permeable membrane to a layer containing leuco-dye reagent, where a coloured product is formed. The intensity of the colour is measured by reflectance densitometry through the transparent plastic base of the slide, reflected light from a suitable source being detected by a photodiode. Similar slides for different tests can be used with the same instrumentation, including enzyme assay as well as the use of enzymes as reagents.

The advantage of such dry-reagent or non-liquid reagent systems is simplicity, for no reagents have to be prepared, and the materials are disposable. This makes them useful, for example, in diagnostic **point-of-care testing (POCT)** in hospital ward side-rooms or clinics.

18.2 AUTOMATION IN ENZYMATIC ANALYSIS

18.2.1 Introduction

As noted in section 18.1, most of the detectors used to monitor the course of enzyme-catalysed reactions can be coupled to automatic recorders. However, if each sample still has to be mixed with reagents and introduced to the detector by hand, the overall process can only be said to be **semi-automatic**. Here in section 18.2 we are concerned with **fully-automatic** procedures, where simultaneous or sequential analysis can be carried out on a batch of samples, without any manual steps being involved other than in sample preparation and in loading the batch on to the analyser, and possibly in the calculation of the results from recorder traces. Many analysers are now linked to computers, to enable the results to be calculated and assessed automatically. The choice of instrument and of its range of facilities is of course limited by financial considerations.

Most fully-automatic techniques used for enzymatic analysis give a result based on an estimate of the initial velocity of the reaction: in the case of relatively simple instruments, this may be derived from just one or two readings; more complex instruments monitor the reaction throughout and give the best statistically-derived estimate of initial velocity. Regardless of this, the analysis must always be carried out at constant pH and temperature.

When an instrument allows only a limited number of readings to be made for each analysis, it is advisable, before using it for a particular application, to perform preliminary experiments to ensure that the progress of the reaction is linear with respect to time over the period of interest. Under suitable conditions, most enzyme-catalysed reactions quickly reach a linear, steady-state phase which is then maintained for a few minutes before the rate of reaction starts to drop (section 7.2.1). However, some reactions (e.g. that catalysed by creatine kinase) have quite an appreciable lag-phase (i.e. one or more minutes) before the steady-state phase is reached. Even the others may, in practice, take seconds to reach the linear phase, because complete mixing of the sample and reagents cannot be achieved instantaneously. The duration of the linear-phase depends mainly on the relative concentrations of enzyme and substrate (see section 15.2.2). Hence, by preliminary investigations, it should be possible to choose analytical conditions which ensure that a **two-point procedure** gives a reasonable estimate of initial velocity provided the test substance falls within a specified concentration range (Fig. 18.3).

However, ideal conditions for a **one-point procedure**, which assumes linearity from zero time to the time of the reading, can never be achieved if there is any appreciable lag-phase. Although linearity over the period of interest should be aimed at in both one- and two-point procedures, reasonable results may nevertheless be obtained, even if there are slight departures from linearity, provided suitable standards of known composition are included among each batch of test samples. These standards may be solutions made up from solids, or they may be samples which have been analysed previously by a reliable method.

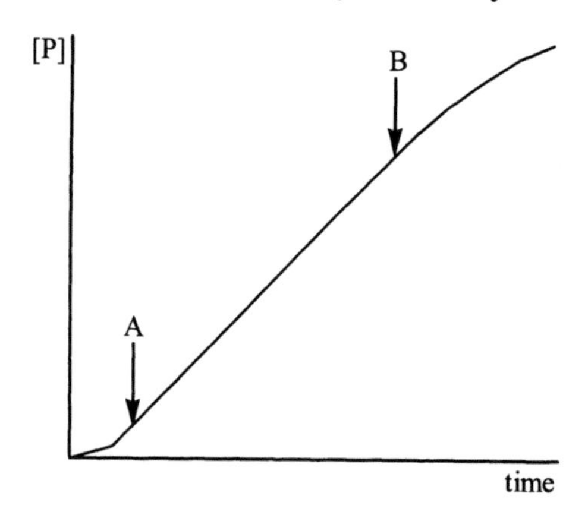

Fig. 18.3 – Typical plot of product concentration against time for an enzyme-catalysed reaction. Conditions for a two-point procedure should be chosen by preliminary experiments so that the readings fall between A and B.

18.2.2 Fixed-time methods

In fixed-time methods of analysis, the change in experimental reading (e.g. absorbance) over a fixed time interval is recorded. If the reaction has linear characteristics over this time interval, then the observed change is directly proportional to the initial velocity, and hence can be related to the concentration of the substance being analysed (see section 7.2.2).

Such analysers may be termed **discrete**, when each sample is mixed with reagents by means of automatic pipettes and passed into a detector at an appropriate time, according to the programmed instructions. Present-day discrete analysers (e.g. the Hitachi 7180, the Bayer Advia 1650 and the Olympus AU5400 systems) can work efficiently on large batches for single tests (e.g. analysing glucose in several hundred samples per hour) or can be programmed to carry out 12 or more different tests on each sample with a throughput of more than a thousand tests per hour. The stress on automatic pipetting systems is considerable, and early systems (e.g. the Vickers M300 of the 1970s) were susceptible to frequent mechanical breakdown. However, such problems have now been overcome and discrete analysers are now employed in most high-throughput laboratories. Detection is usually by spectrophotometry, but ion-selective electrode modules may be fitted to some types. Most modern discrete analysers are **selective**, i.e. more tests can be done on some samples than on others, as considered appropriate. Another development is to have a **random access** facility, allowing the operator, through the linking computer, to be in control throughout. For instance, samples added while the instrument is already in operation can be given priority, if required. The computer may also be able to arrange the quickest way of carrying out all the required tests on the samples.

An alternative approach, widely used between the 1960s and the 1990s (in analysers such as the Technicon Autoanalyser SMA and CHEM 1^+ systems), involves the use of **continuous-flow** methods. If a solution is pumped through a tube at a fixed flow-rate, each molecule will take the same time to pass between two given points on the tube.

In continuous-flow analysis, a mixture of reagents is forced continuously through a tube, by means of a peristaltic or other type of pump, and, at regular time intervals, one of a series of samples is automatically introduced into the reagent line. Suitable mixing coils and constant-temperature heating-baths ensure that each sample is mixed with the reagent and that, as they are pumped along the tube, a reaction takes place under controlled conditions. Eventually, the reaction mixture reaches a detector and the progress of the reaction is determined. The set-up ensures that a fixed time elapses between the mixing of each sample with reagent and the arrival of the resulting mixture at the detector. Often air (or other gas) is pumped into the reaction tube along with the sample and reagents, producing large bubbles which segment the reaction mixture into lengths of no more than a few centimetres each. These pass along. the tube separately, with gas in between, and so diffusion between reaction mixtures from different samples is prevented. Alternatively, tubing of a very narrow bore may be used to minimize diffusion.

In some applications, the concentration of a product of the primary, enzyme-catalysed reaction is determined by a separate, indicator reaction. For example, in an assay of **alkaline phosphatise (ALP)** involving the use of disodium phenyl phosphate as substrate:

$$\text{disodium phenyl phosphate} \xrightarrow[\text{pH 10}]{\text{ALP}} \text{phenol} + \text{sodium phosphate} \qquad (18.3)$$

the phenol liberated must be analysed after the primary reaction has been stopped because of the nature of the indicator reagents. This can be done satisfactorily if a one-point procedure is being employed, whereas such a coupled system could not be used for monitoring *continuously* the course of the primary reaction. Employing the King and Armstrong (1934) method, the phenol is made to react with Folin-Ciocalteu reagent and the absorbance determined at 680 nm (see section 3.2.1). However, this reagent also reacts with the phenolic amino acid, tyrosine, which is present in a bound form in alkaline phosphatise and other plasma proteins, so these must be removed prior to treatment with the reagent. In continuous-flow analysis procedures, this is usually achieved by placing a **dialyser** in the system, allowing the phenol produced during the primary reaction, but not any of the protein present, to mix with the indicator reagent.

The procedure may be calibrated by placing phenol standards of known concentration amongst the batch of samples. The incubation time of the enzyme-catalysed reaction is effectively the flow-time from the mixing of sample and primary reagents to the introduction of the indicator reagent. This may easily be determined by introducing a coloured dye into the sample line at an appropriate point and observing .its movement. However, it is usually considered unnecessary to know the precise incubation time, particularly since it is often in the order of 10 minutes, meaning that the reaction is likely to have proceeded beyond the linear phase. Hence it is best to calibrate the procedure for each analysis by the use of standards of known alkaline phosphatase activity, this being a relatively stable enzyme. Such standards (e.g. standard sera) may be obtained commercially, but of course have to be stored carefully to avoid loss of enzyme activity (section 17.3).

Many continuous-flow procedures have an advantage over the above in that there is no need to perform a separate indicator reaction. The reaction catalysed by **glucose oxidase**, for example, can be monitored by coupling with a peroxidise-mediated dye formation, both reactions taking place within the same system, rather than requiring consecutive processes (see section 15.2.3). This is also true for enzyme-catalysed reactions which do not need coupling to others. For example, the assay of **alkaline phosphatase** by the use of colourless *p*-nitrophenyl phosphate as substrate involves the formation of a product (*p*-nitrophenol) which is yellow in alkaline conditions, the absorbance being read at 410 nm. The system may be calibrated with *p*-nitrophenol standards and the incubation time determined as the time taken for a coloured dye to travel from the point of mixing to the colorimeter. However, as before, the system may be calibrated by the use of alkaline phosphatase standards, eliminating errors due to non-linearity.

In procedures where enzymes are being used as **reagents** (e.g. the determination of blood glucose by the use of the glucose oxidase/peroxidase coupled method mentioned above), a dialyser is often included at a point in the sample line before it merges with the enzyme line. This ensures that the only protein in the reaction mixture is the enzyme being used as a reagent, since no protein from the sample would be able to pass through the dialysis membrane. Calibration of such procedures is easier than for enzyme assay, since there is not usually any problem in obtaining standards of reliable, precisely known composition. The enzyme reagent may also be used in an immobilized form in an **enzyme analytical reactor**, which would replace the mixing coil in the system (see section 20.2.3). For example, immobilized glucose dehydrogenase has been used to determine blood glucose, NAD^+ being reduced in the reaction.

Continuous-flow procedures of this type can analyse in the region of 60 samples per hour, which is low compared to the performance of a discrete analyser. However, the sample stream may be split and identical fractions pumped simultaneously into a variety of reagent streams. Thus each sample may be analysed for several different factors at the same time, with little extra labour involved. However, such a system is wasteful of reagents and each test will be carried out on each sample, whether necessary or not. Hence there was a marked reduction in the popularity of continuous-flow analysers towards the end of the twentieth century as the performance of discrete systems improved in significant fashion.

18.2.3 Fixed-concentration methods

In fixed-concentration methods of analysis, the time for an experimental reading (relating to the concentration of some substance participating in the reaction) to change by a specified amount is measured automatically. As with fixed-time methods, the readings should ideally be restricted to the linear-phase of the reaction, as indicated by preliminary experiments. Under these conditions, the time taken is inversely proportional to initial velocity. Procedures based on this principle must involve discrete rather than continuous-flow analysis. However, they have only been used to a very limited extent.

18.2.4 Methods involving continuous monitoring

Several automated rate-monitoring systems, specifically designed for enzymatic analysis, became available during the 1970s. These included the LKB Kinetic Analysis System and the Beckman Enzyme Activity Analyser. Although they differed in design, all were discrete analysers which could handle in excess of 40 samples per hour, working sequentially. The main difference from other types of discrete analyser (see section 18.2.2) is that when a sample was mixed with reagents, the ensuing reaction was monitored continuously (or almost continuously, readings being taken at no more than 3-second intervals) by means of a spectrophotometer, the results being fed directly into a computer. Constant temperature was maintained throughout.

Any lag-phase reaction (see section 18.2.1) could be eliminated from the calculations either by imposing a delay before the first reading was taken or by programming the computer to recognize the various phases of the reaction. When the linear-phase had been observed for 15-20 seconds, enough readings would have been taken for the computer to calculate its slope and give a statistical assessment of the results, after which the reaction mixture could be ejected from the detector and a new one introduced. Warnings were often printed if there was any significant departure from linearity over the period of interest, or if the optical limits of the instrument were being exceeded. However, the use of such specific systems has been overtaken by more general developments in automation.

With the development of microprocessor technology, several multi-purpose **spectrophotometers** were provided with optional kinetic attachments. The kinetics analysis facility of the Beckman DU-70 spectrophotometer (introduced during the late 1980s) could make up to 1200 readings per minute on a single sample or could monitor simultaneously the reaction in a series of samples, making readings every few seconds. The absorbance against time curve curves could be visualized on a screen, and linear regression analysis carried out on selected portions of the curves. The more recent Beckman DU Series 500 spectrophotometers included kinetics in the standard applications software package, and the present-day Beckman Coulter 800 spectrophotometer allows for the simultaneous measurement and analysis of up to 12 rate-reactions.

Centrifugal analysers can also be used for enzymatic analysis, and some, e.g. the Roche Cobas Bio and the Baker Centrifichem systems, were widely employed during the 1980s, although their use has since declined. In such systems, 15-30 samples could be analysed simultaneously in cuvettes on a spinning, horizontal centrifuge plate. The cuvettes were on the outer portions of the plate, and each was aligned with specific depressions on the inner portion which originally contained separated sample and reagents. The inner plate could be removed for ease of loading, and to allow one batch of sample to be loaded while another was being analysed

When centrifugation commenced, each sample was mixed with its reagent as they were carried together by centrifugal force into a cuvette, thus starting a reaction. As the plate revolved, each cuvette passed through a detecting system (usually a spectrophotometer) linked to a computer, so readings could be taken and stored for each reaction mixture on each circuit (i.e. at about 100 ms intervals).

Also on each circuit, a dark current reading was taken, when the light beam was interrupted by a solid part of the rotor. An oscilloscope could be linked to the detector, to enable the progress of the reaction in each cuvette to be visualized. Regardless of this, a suitable computer could process the readings to give the best estimate of the initial velocity of each reaction, and print out the results. A trap could then be opened at the bottom of each cuvette, and the contents emptied by centrifugal force, preparing the instrument for the next batch of samples. It was possible to analyse in the order of 200 samples per hour by such a procedure.

The use of such systems declined as the reliability, throughput and sophistication of discrete analysers continued to advance. For example, the Beckman Cx7 Synchron discrete analyser of the 1990s, with a throughput of up to 825 results per hour and a random-access facility, incorporated a kinetics capability. Such a feature is also present in a successor, the Beckman Coulter Unicell DxC Synchron 800 system which, in its routine mode, can perform up to 1440 tests per hour.

18.3 HIGH-THROUGHPUT ASSAYS (HTA)

The advent of genomics and proteomics (see chapter 21) has identified many new enzyme (protein) targets for therapeutic intervention. Pharmaceutical, agrochemical and academic institutions possess vast depositories of synthesised chemicals which may or may not modulate the activity of these target enzymes (e.g. kinases). To screen the large and increasing number of available chemicals, that may form the basis of drug discovery, robotics are required to simultaneously test the compounds in functional or binding assays, a process called **high throughput screening (HTS)**. Multi-well plastic plates (typically 96, 384 or 1536 wells) provide the means to interact small volumes (typically $\leq 10 \mu l$) of sample (e.g. enzyme or cells) with stock solutions of the compound library. Chromogenic, fluorescent, luminescent and radiochemical techniques can be used as a means of detecting the occurrence of enzyme inhibition.

The vast numbers of chemicals involved (in some cases $>2 \times 10^6$) means that a successful HTS should involve the minimum of reagent handling, have little or no processing (e.g. heating or cooling to stop a reaction) and provide clear-cut evidence of enzyme inhibition. A good example of this type of HTS is the PDELight™ HTS cAMP phosphodiesterase (PDE) assay (Cambrex). PDEs catalyse the hydrolysis of cAMP to AMP; this results in the termination of cAMP stimulation of various signalling cascades. Inhibition of PDE can have a beneficial cellular effect by maintaining cAMP levels in tissues involved in vasodilation, thus facilitating improved passage of air or blood, which is important to alleviate asthma and heart disease symptoms. The PDE HTA mixes small volumes of a target PDE with cAMP and an inhibitor from a chemical library. Following a period of incubation, a reagent mixture is added which enzymically converts the AMP produced by the PDE to ATP. In the presence of luciferin and luciferase, the ATP is then reconverted to AMP with the emission of light (of energy = $h\nu$, h and ν being as defined in section 2.6), a process which can be quantified in a luminometer. The luciferase-catalysed reaction is:

$$\text{luciferin} + \text{ATP} + O_2 \xrightarrow{\text{luciferase/Mg}^{2+}} \text{oxyluciferin} + CO_2 + \text{AMP} + (\text{PP})_i + h\nu \qquad (18.4)$$

HTA provides an ideal means of screening large numbers of potential inhibitors, but it can only be effective if it is managed by modern laboratory robotics. These high capacity robotic machines can offer automated plate preparation, analysis and data handling, resulting in the screening of up to 100 000 compounds in a day. The compounds that are identified as being targets in the first screen can be then be investigated further using a more detailed secondary screen.

SUMMARY OF CHAPTER 18

The course of enzyme-catalysed reactions may be monitored by a variety of techniques, of which the most common at the present time are spectrophotometric, spectrofluorimetric and electrochemical procedures. Radiochemical procedures may also be used, but although they are extremely sensitive, they do not permit the rate of the reaction to be monitored continuously.

Fully-automatic procedures of enzymatic analysis may involve fixed-time methods (either discrete or continuous-flow analysis), fixed-concentration methods or continuous-monitoring methods. The last-mentioned type includes the use of discrete, rate-monitoring analysers and centrifugal analysers, both of which may be coupled to computers to allow automatic processing of results. However, discrete analysers have become increasingly dominant because of their flexibility and performance.

High-throughput assays are becoming increasingly important.

FURTHER READING

Bisswanger, H. (2004), *Practical Enzymology*, Wiley–VCH (Chapter 3).

Burtis, C. A., Ashwood, E. R. and Bruns, D. C. (2005), *Tietz Textbook of Clinical Chemistry*, 4th edn., Saunders (Chapters 3-12).

Fersht, A. (1999), *Structure and Mechanism in Protein Science,* Freeman (Chapter 6). Freitag, R. (ed.) (1996), *Biosensors in Analytical Biotechnology,* Academic Press. Killard, A. J. and Smyth, M. R. (2000), Creatinine biosensors - principles and design, *Trends in Biotechnology,* **18**, 433-437.

Luong, J. H. T., Bouvrette, P. and Male, K. B. (1997), Development and application of biosensors in food analysis, *Trends in Biotechnology,* **15,** 369-377.

Methods in Enzymology, **246** (1995): Biochemical spectroscopy; **278** (1997), Fluorescence spectroscopy; Academic Press.

Newman, J. D. and Turner, A. P. F. (1992), Biosensors - principles and practice, *Essays in Biochemistry,* **27**, 147-159, Portland Press.

Reed, R., Holmes, D. *et al* (1998), *Practical Skills in Biomolecular Sciences,* Longman. Schoeff, L. E. and Williams, R. H. (eds.) (1993), *Principles of Laboratory Instruments,* Mosby.

Sheehan, D. (2000), *Physical Biochemistry - Principles and Applications,* Wiley.

Wilson, K. and Walker, J. (eds.) (2000), *Principles and Techniques of Practical Biochemistry,* 5th edn., Cambridge University Press.

19

Applications of Enzymatic Analysis in Medicine, Forensic Science and Industry

19.1 APPLICATIONS IN MEDICINE

19.1.1 Assay of plasma enzymes

Assays of enzymes present in blood plasma or serum have been carried out routinely in clinical chemistry laboratories for more than half a century and, although in recent years the use of some enzyme assays has declined as new techniques have been developed, others continue to play an important role in diagnosis. (For those not familiar with medical terminology, plasma is the fluid from unclotted blood, serum the fluid from clotted blood: in most respects, the two are identical.) Although a few plasma enzymes have a clearly defined role in that particular location (e.g. in coagulation), the majority do not: they are simply enzymes which have leaked out of blood or tissue cells. For each such enzyme present in plasma, a balance is normally set up between its rate of arrival by leakage from cells and its rate of removal by catabolism or excretion. Therefore, for each plasma enzyme, there is a normal concentration range and hence a **normal range of activity**, which can be determined.

If the cells of a particular tissue are affected by disease in such a way that many of them no longer have intact membranes, then their contents will leak out into the bloodstream at an increased rate and the enzymes associated with those cells will be found in the plasma in elevated amounts. Since many enzymes or isoenzymes are characteristically associated with the cells of certain tissues, plasma enzyme assay can help to identify the location of damaged cells. The results should, of course, correlate with the symptoms and case history of the patient, and with other biochemical findings. Some important examples of the application of plasma enzyme assay to diagnosis will now be discussed.

Lactate dehydrogenase (LDH) is easily assayed, since it catalyses a reaction involving NAD^+/NADH as coenzyme.

It is found in most cells of the human body, and so an increased activity of this enzyme in plasma does not, in itself, give much clue as to what is wrong with the patient. However, the total activity is in fact the sum of the activities of five isoenzymes (see section 5.2.2), and these are associated with different tissues. In particular, liver and skeletal muscle cells contain mainly the M_4 form, while those of heart muscle contain mainly H_4 and MH_3. All of the isoenzymes are found in kidney cells and erythrocytes (red blood cells), the H_4 form being the most prominent. If plasma is subjected to electrophoresis (e.g. on cellulose acetate strips) at pH 8.6, the LDH isoenzymes may then be located by means of a specific stain, e.g. a mixture of lactate, NAD^+ and a chromogen, which will only form a coloured product where LDH is present to catalyse the first step in the reaction:

$$\text{lactate} + NAD^+ \xrightarrow{\text{LDH}} \text{pyruvate} + \text{NADH}; \quad \text{NADH} + \text{chromogen} \rightarrow NAD^+ + \text{coloured product} \qquad (19.1)$$

In this way, plasma is found to contain all five isoenzymes (Fig. 19.1), and an abnormal pattern can be of help in establishing a diagnosis.

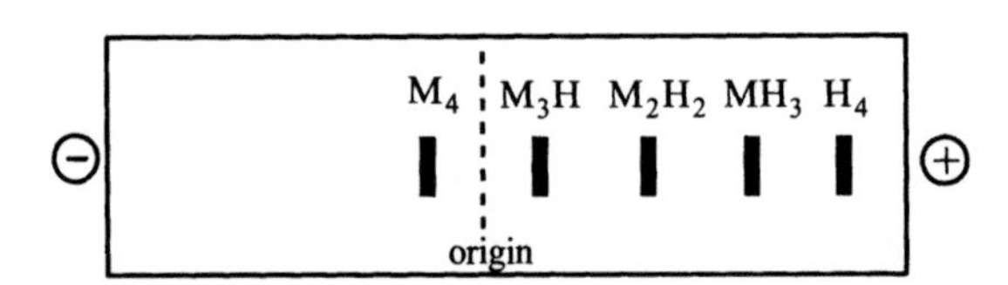

Fig. 19.1 – Separation of the isoenzymes of LDH achieved by electrophoresis at pH 8.6.

Separation may also be achieved by ion exchange (e.g. QAE-Sephadex) chromatography. Various different properties of the LDH isoenzyme can enable their relative proportions in plasma to be assessed without separation having to be performed. H_4 is heat-stable and its activity is unaffected if it is kept at 60°C for 30 minutes, whereas that of M_4 is largely destroyed. H_4 can act as a hydroxybutyrate dehydrogenase (HBD), i.e. can use hydroxybutyrate as substrate, whereas M_4 has little activity in that respect. Also, H_4, but not M_4, is strongly inhibited by oxalate (0.2 mmol l^{-1}), whereas M_4, but not H_4, is markedly inactivated by urea (2.0 mol l^{-1}). Immunological reagents may, for instance, be used to precipitate all isoenzymes containing M sub-units, leaving only H_4 in solution. Hence, if the total LDH activity is found to be higher than normal, there are several ways of establishing whether this is predominantly due to an excess of H_4 (as could be seen in the case of heart disease, haematological disorders or renal disease) or to an excess of M_4 (which would indicate skeletal muscle or liver disease).

Aspartate transaminase (AST), formerly known as glutamate oxaloacetate transaminase (GOT), an enzyme widely distributed throughout the body, catalyses the reaction: aspartate + 2-oxoglutarate $\rightleftharpoons$ oxaloacetate + glutamate. It may be assayed by coupling, via the product, oxaloacetate, to an indicator reaction catalysed by malate dehydrogenase which involves NAD^+/NADH as coenzyme.

Markedly raised plasma activities (10-100 times normal) of AST usually indicate severe damage to the cells of heart (as in myocardial infarction) or liver (as in viral hepatitis or toxic liver necrosis). Moderate increases of activity are found in many diseases.

Alanine transaminase (ALT), formerly known as glutamate pyruvate transaminase (GPT), catalyses the reaction: alanine + 2-oxogluturate $\rightleftharpoons$ pyruvate + glutamate.

This enzyme may be assayed by coupling, via the product, pyruvate, to an LDH-catalysed indicator reaction. ALT is found in high concentrations in liver cells, and in much smaller concentrations elsewhere. Hence, a markedly raised plasma activity indicates a severe liver disease, usually viral hepatitis or toxic liver necrosis.

Alkaline phosphatise (ALP), the name given to a group of isoenzymes which catalyse the hydrolysis of organic phosphates at alkaline pH (equation 1.9), is found principally in bone, liver, kidney, intestinal wall, lactating mammary gland and placenta. Different isoenzymes are associated with each of these, and they may be partially separated by electrophoresis at pH 8.6 and visualized with a specific stain such as calcium α-naphthyl phosphate mixed with a diazonium salt (Fig. 19.2).

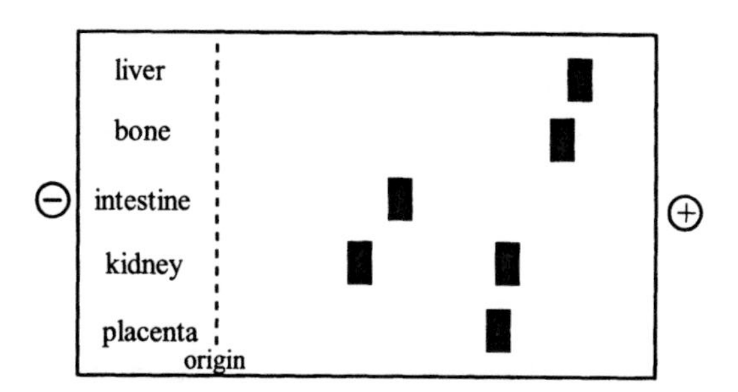

Fig. 19.2 – Separation of the isoenzymes of alkaline phosphatise achieved by electrophoresis at pH 8.6.

The kidney isoenzymes are not normally present in plasma to any appreciable extent, even in patients with kidney disease, and the placental isoenzyme is only found in the plasma of pregnant women. The total activity of ALP is easy to determine, because the relatively low specificity of each of the isoenzymes means that an artificial substrate whose hydrolysis is easy to monitor can be used in the assay procedure (see section 18.2.2). As with LDH, if the total activity is found to be in excess of normal, the isoenzyme or isoenzymes responsible may be identified from the electrophoresis pattern, considered together with the results of other tests. Unlike LDH, no hybrid isoenzymes are found, and at least some of the ALP isoenzymes may be products of the same gene, the differences being due to post-genetic modification, particularly glycosylation. Hence, some isoenzymes, especially those associated with liver and bone, have very similar characteristics and are difficult to distinguish from each other. Fortuitously, mild digestion of the carbohydrate residues by neuraminidase increases the differences between the liver and bone isoenzymes, which improves their electrophoretic separation.

Placental ALP is very heat-stable, as is the similar Regan isoenzyme found in some patients with cancer, whereas the liver and bone isoenzymes are much less stable at elevated temperatures

Hence the effect of heat (e.g. 65°C for 30 minutes) on the total enzyme activity can sometimes help to indicate which are the predominant isoenzymes present. Inhibition studies may also be used, for the intestinal and placental forms are strongly inhibited by L-phenylalanine. Also, urea (2 mol l^{-1}) inactivates the bone isoenzyme more strongly than the liver form, and this in turn more strongly than the ones from intestine and placenta.

ALP in the liver is found associated with the cell membrane which adjoins the biliary canaliculus, and so high plasma concentrations of the liver isoenzyme indicate cholestasis (i.e. blockage of the bile duct, with a resulting back flow through the liver into the bloodstream) rather than simply damage to the liver cells. In view of the generally similar characteristics of the bone and liver isoenzymes, confirmation of high plasma concentrations of the liver form may be achieved by the investigation of other enzymes whose plasma activities are increased in cholestasis. These include **5′-nucleotidase** and **γ-glutamyl transferase (GGT)**, also called γ-glutamyl transpeptidase. High plasma activities of the latter enzyme are also found in other forms of liver disease.

Creatine kinase (CK), also known, with a degree of tautology, as creatine phosphokinase (CPK), catalyses the reaction: creatine + ATP $\rightleftharpoons$ phosphocreatine + ADP. It may be assayed by a variety of procedures, e.g. by coupling via the product, ADP, to a reaction catalysed by pyruvate kinase, and thence via the further product, pyruvate, to an LDH-catalysed indicator reaction.

The enzyme is found mainly in the heart and skeletal muscle, and in brain. It is a dimer and, like the tetrameric LDH, may be made up from two types of polypeptide chain (B or M) in any combination. Thus, three isoenzymes may be found: BB, the main form of brain, but also found to some extent in nerve tissue, thyroid, kidney and intestine; MB, found in heart muscle and the diaphragm, but not normally present in significant amounts in skeletal muscle; and MM, found in both heart and skeletal muscle. On electrophoresis, the BB isoenzyme travels nearest to the anode, but this form is not normally present in plasma. In addition to electrophoresis, immunological techniques and triazine dye affinity chromatography are proving useful in the analysis of the individual CK isoenzymes.

An increased plasma activity of the MB isoenzyme (accompanying MM) is a good indicator of a possible myocardial infarction, whereas the MM form alone is found in high concentrations in plasma when skeletal muscle cells are damaged (e.g. in the early stages of muscular dystrophy). Skeletal muscle disorders of this type are also characterized by increased plasma activities of the appropriate isoenzyme (isoenzyme A) of **fructose-l,6-bisphosphate aldolase**, often called simply **aldolase** (see section 10.1.1).

Hence, to summarize our discussion so far, liver cell damage is indicated by increased plasma activities of AST, ALT, LDH (M_4) and GGT, the magnitude of the increases reflecting the degree of damage. ALP and 5′-nucleotidase are useful mainly as indicators of cholestasis, when the GGT activity will also be raised. Skeletal muscle damage may result in the presence in the plasma of excess CK (MM), aldolase (A), and, to a lesser extent, LDH (M_4), ALT and AST. Levels are usually higher in the early stages of muscular dystrophy than they are when the muscle has become wasted.

Severe damage to heart cells, as in myocardial infarction, is characterized by increased plasma activities of CK (MM + MB), AST and LDH (H_4). They start to rise in that order, and subsequently fall in the same order. Within a week of an infarction, total plasma activities of each of these three enzymes may be back to normal, but the effects may still be seen, with H_4 abnormally prominent amongst the LDH isoenzymes, resulting in the HBD activity still being above normal limits.

A carboxypeptidase normally present in plasma can remove the terminal lysine residues from CK-M polypeptide chains to produce isoforms of the enzyme, differing in electrophoretic mobility but not in enzyme activity. Thus, CK-MM (isoform CKMM3) released from cells is progressively converted to CK-MM2 and CK-MMl in plasma, while CK-MB2 is converted to CK-MBl. Alan Wu (1989) argued that an increased plasma CK-MM3:CK-MMl ratio might be the first detectable biochemical indication of myocardial infarction, 2-5 hours after the onset of chest pains.

Cardiac failure may cause secondary liver damage, so there could be some confusion as to the interpretation of the results of plasma enzyme assay. However, it would be apparent from the HBD and CK-MB activities (together with non-enzymic protein markers such as **Troponins I and T**) whether or not a myocardial infarction had occurred, whereas the extent of the liver damage would be indicated by ALT.

Diseases in parts of the body not previously mentioned may also be diagnosed with the help of plasma enzyme assay.

The digestive enzymes, **trypsin, triacylglycerol lipase** and **α-amylase** are produced in the pancreas and these are found in increased amounts in plasma in certain diseases of the pancreas, particularly acute pancreatitis. Of the three, α-amylase is the simplest to assay, since it utilizes starch as substrate and produces reducing sugars as products. An unusual feature of pancreatic α-amylase is that it is one of the few proteins small enough to pass through the glomerulus of the kidney and be excreted in the urine. In a rare condition called macro-amylasaemia, the plasma α-amylase forms complexes with other proteins (possibly immunoglobulins) and can no longer be excreted in this way, which causes the plasma α-amylase activity to rise. However, macro-amylasaemia could not be confused with acute pancreatitis, because of the very different clinical symptoms.

An **acid phosphatase** isoenzyme is found in large amounts in the prostate gland, and its assay in plasma has been used in the diagnosis of prostatic carcinoma, although the glycoprotein, **prostate-specific antigen (PSA)**, is now regarded as a better marker. Acid phosphatase isoenzymes are also found in liver, red cells, platelets and bone. They catalyse the same reaction as alkaline phosphatase, but at lower pH values, and may be distinguished from each other by immunoassay techniques, or by the action of various inhibitors. The prostate and red cell forms of the enzyme are inactivated by ethanol; the red cell (but not the prostate) form by formaldehyde, and the prostate (but not the red cell) form by L-tartrate. The total plasma acid phosphatase activity is raised in a variety of conditions, but this is of little use in diagnosis, as there are better markers for each of them.

Cholinesterase is another enzyme which is assayed in plasma only in specific circumstances. Although plasma concentrations are low in liver disease, this is also the case with the routinely-measured protein, albumin, so assay of cholinesterase would provide no additional information.

The enzyme catalyses the hydrolysis of choline esters, e.g. acetylcholine + H_2O $\rightleftharpoons$ acetic acid + choline. It may conveniently be assayed by pH-stat or by electrochemical procedures. The 'true' **acetylcholinesterase** is found in nerve tissues and red blood cells. However, it is not this isoenzyme but the **pseudocholinesterase** of liver, heart and intestine which is normally present in blood plasma. Acetylcholinesterase will act on acetyl-β-methyl choline as substrate, but not on benzoylcholine; pseudocholinesterase has the opposite specificity.

The main clinical significance of plasma cholinesterase is that its presence is assumed when the muscle relaxant **scoline** (suxamethonium; succinyl dicholine) is administered, for otherwise the effect would be permanent. Patients with low plasma activities of cholinesterase (e.g. as a result of liver disease or of poisoning by organophosphorus insecticides) experience severe breathing difficulties for several hours after treatment with scoline, so a preliminary assay may be carried out to see if it is safe to administer the drug. One further problem is that a rare condition exists where the subject has a modified pseudocholinesterase as a result of an inherited genetic mutation. The plasma activity in such individuals is usually low, and they react adversely to scoline administration. The condition may be investigated by determining the **dibucaine number** (i.e. the percentage inhibition by dibucaine) of the plasma cholinesterase: subjects with the normal enzyme have a high dibucaine number (75-85), while those with one type of abnormal enzyme have a low one (15-30). Alternatively, some patients may show reduced inhibition by fluoride. Such inhibition studies are useful in detecting homozygotes and heterozygotes (see section 19.1.2) in the family of a patient known to have a cholinesterase deficiency, with a view to identifying those individuals at risk from scoline administration.

In general, most assays of plasma enzymes are carried out by kinetic methods, since these are the most convenient. However, radioimmunoassay procedures (section 15.2.4) have become increasingly important, because they are specific for a particular isoenzyme.

19.1.2 Enzymes and inborn errors of metabolism

An inborn error of metabolism is characterized by the loss of activity of a specific enzyme as a result of a genetic mutation. Inborn errors are rare but possibly severe conditions, often causing mental retardation or even death in infancy.

The **preliminary diagnosis** is usually made by observing a build-up in plasma or urine of the metabolic intermediate which is the substrate for the defective enzyme. It cannot be metabolized in the usual way, and so leaks out of the cell in relatively large amounts. The diagnosis may then be **confirmed** by assay of the enzyme in a sample of the tissue where it is metabolically most important (often the liver). In some instances, it may be possible to assay the same enzyme in more accessible cells (e.g. in red or white blood cells, or in skin), but, because of the known existence of isoenzymes (see section 19.1.1), it can never be assumed that the defect will be expressed wherever activity of that enzyme is usually found.

If, because of previous family history, it is considered possible that a particular pregnancy might result in the birth of a baby with an inborn error of metabolism, it may be possible to obtain a **pre-natal diagnosis** by assaying the enzyme in amniotic fluid cells, which are mainly derived from skin.

Among the many disorders which have been diagnosed in this way before birth is **Tay-Sachs disease**, a deficiency of ***β-N*-acetylhexosaminidase** (also known as **hexosaminidase A**) activity.

The application of **recombinant DNA techniques** to the investigation of gene structure now offers a diagnostic procedure even more reliable than enzyme assay (see section 21.1.2).

With most inborn errors, the disease is transmitted from generation to generation by an **autosomal recessive** mode of inheritance. The main features of autosomal inheritance are that every individual has a gene for the synthesis of the relevant enzyme on each of a pair of chromosomes, one coming from the father and one from the mother, and that the inheritance pattern does not depend on the sex of the individual. If the mode is recessive, the affected patient must have inherited the same abnormal gene from each parent, for otherwise the disease would not express itself as a clinical abnormality. Such a **homozygous** patient would usually have less than 10% of the normal activity of the enzyme, whereas both **heterozygous** parents, each having one normal and one abnormal gene, would be able to synthesize active enzyme from their normal gene and would thus have about 50% of the normal activity, an amount usually sufficient for a normal life. There is a 1 in 4 chance that a child of heterozygous parents will be homozygous for that particular defect, and a 1 in 2 chance that the child would be heterozygous like the parents. Probably the best-known example of an autosomal recessive inborn error is **phenylketonuria**, where the patient has a severely limited ability to convert phenylalanine to tyrosine, because of a reduced activity of **phenylalanine 4-monooxygenase**, better known as **phenylalanine hydroxylase**.

Autosomal dominant diseases, in contrast to autosomal dominant characteristics such as eye colour, are *not* found, presumably because they are self-limiting. With such a mode of inheritance, the heterozygous individual would be clinically affected, so would be unlikely to grow to sufficiently normal maturity to have children and pass on the abnormal gene.

However, **X-linked dominant** inborn errors *are* found. In these, the mutated gene forms part of the chromosome pair which determines the sex of the individual. Females possess two X chromosomes, whereas males possess one X chromosome and one, much shorter, Y chromosome. Therefore, where a gene on the X chromosome has no counterpart on the Y chromosome, it would seem likely that females would normally produce twice as much as males of the protein coded for on that gene. This does not happen, because, as argued by Mary Lyon, a random cut-off procedure operates in females. Shortly after conception, when the female embryo consists of relatively few cells, one of the X chromosomes is inactivated in each cell. If it happens that the girl-to-be is heterozygous for an X-linked disease, that means that some normal genes will be inactivated, as well as some abnormal ones. The process is quite random. So, if most of the inactivated genes are abnormal ones, the girl who develops from those few cells will produce her enzyme largely from normal genes, and will therefore be clinically normal (termed **favourable Lyonization**). However, it is just as likely that most of the inactivated genes are normal ones, a situation which could ultimately lead to the birth of a severely affected patient (termed **unfavourable Lyonization**)

So, X-linked dominant diseases are inherited through the mother alone, who is heterozygous but not necessarily clinically abnormal. However, a heterozygous daughter may not be so lucky. Furthermore, a boy who inherits the abnormal gene from his mother is hemizygous because the gene has no counterpart on his Y chromosome, so he must be severely affected. Each daughter of a heterozygous mother has a 1 in 2 chance of being heterozygous too, and each son has a 1 in 2 chance of being hemizygous.

An example of an inborn error which appears to be transmitted by an X-linked dominant mode is **ornithine carbamoyltransferase (OCT)** deficiency, also known as **hyperammonaemia**, a defect of urea cycle metabolism.

Another X-linked, but recessive, disease is the **Lesch-Nyhan syndrome**, a disorder of purine metabolism resulting from a deficiency of **hypoxanthine phosphoribosyl transferase** activity. With such an X-linked recessive disease, only the hemizygous male exhibits clinical symptoms.

In most inborn errors, it is likely that the enzyme is present in an altered form as a result of the mutation, rather than being absent altogether. In the case of the Lesch-Nyhan syndrome, James Wyngaarden, in 1974, used radioimmunoassay to demonstrate that the enzyme-protein was present, even though no catalytic activity could be detected.

In contrast to the situation with the inherited disorders of haemoglobin (section 2.4.5), it was not possible to determine the structures of any of these abnormal enzymes by direct means. That was because the human body contains much smaller amounts of each enzyme than it does of haemoglobin, and in much less accessible places. However, indirect determinations showed that the same general features are found, i.e. that the mutation often results in the substitution of only a single amino acid residue by another, and that, since not all mutations will be identical, different substitutions in a particular protein may be found in different families, causing symptoms of different severity (see section 21.1.2). The latter phenomenon is termed **genetic heterogeneity**.

Therefore, although a group of patients may be classified as having the same disease because they have a reduced activity of the same enzyme, the nature of the alteration may not always be the same. For example, it is known that the characteristics (e.g. K_m) of the residual enzyme activity in OCT deficiency varies from patient to patient. Although it might be argued that different proportions of normal and abnormal enzyme in different heterozygous females might account for this, at least to some extent, only genetic heterogeneity could explain the fact that some affected males have relatively mild symptoms whereas most have died shortly after birth. In the case of autosomal recessive diseases, a 'homozygous' patient could have *two* forms of the abnormal enzyme, since the relevant gene from each parent might have different mutations. Direct evidence of genetic heterogeneity is now being obtained by recombinant DNA technology.

Treatment of inborn errors of metabolism usually involves reducing the dietary intake of substances which cannot be metabolized in the normal way because of the defect. The aim is to prevent their concentrations building up to high levels within cells, this being one of the main factors responsible (directly or indirectly) for the clinical symptoms.

Such methods of treatment can be extremely effective in certain instances (e.g. with **phenylketonuria**), but of course the special diets have to be maintained throughout life, to a greater or lesser extent. In many other inborn errors, dietary treatment has no effect. This may be due to endogenous synthesis of the metabolites in question, or it could be because the end-product of the blocked metabolic pathways is vitally important to normal body function and cannot be provided in any other way.

Ideally, complete treatment would be achieved if some normal enzyme could be introduced at the appropriate site. In the case of the enzymes of the gastrointestinal tract, this may easily be achieved by oral administration. For example, where there is a **lactase** (β-galactosidase) deficiency in the cells of the intestine, which may occur as an inborn error or as a secondary feature of other diseases, ingested lactose (in milk) cannot be broken down to galactose and glucose, and severe discomfort results. This condition may be treated by removing milk products from the diet, or, more conveniently if more expensively, by adding to it β-galactosidase extracted from micro-organisms to perform the task normally done by the body's lactase.

However, where the main site of action of the enzyme in question is not in such direct contact with the external environment (if, for example, it is in the liver or the central nervous system), the problems involved in its introduction are immense. Protein is not absorbed intact into the body from the gastrointestinal tract, and although it could be infused directly into the bloodstream, it would be fairly quickly destroyed by antibodies if it was of non-human origin. In any case, it would not usually be taken up by the cells of the appropriate tissue. In some cases, the **attachment of sugar units** to enzymes (particularly lysosomal ones) prior to infusion increases their chances of reaching the appropriate destination: the sugars presumably trigger some specific transport system in membranes.

Another possible approach would be to trap the enzyme in artificial membrane-enclosed structures called **liposomes** (see section 20.2.1). Cells of a particular tissue might then be able to take these up from the bloodstream, provided their membranes were of suitable composition. Yet another possible long-term treatment for inborn errors of metabolism involves **organ** or **tissue transplantation**: for example, fibroblasts from a normal subject of identical tissue-type to the patient may be transplanted by subcutaneous injection.

The various approaches to enzyme-replacement therapy are still in the developmental stage, although some successes have been reported, e.g. in the treatment of Gaucher's Disease, a disorder characterised by an accumulation of glycolipid in lyosomes, resulting from a deficiency of acid β-glucocerebrosidase. Also in the development stage is the even more fundamental **gene therapy**, where the normal gene is introduced by recombinant DNA techniques. However, some promising results are being obtained (see section 21.1.2).

Note that enzymes may also be used in the treatment of diseases other than inborn errors of metabolism. For example, enzymes such as **u-plasminogen activator** (formerly called **urokinase**), extracted from human urine, can be infused into the bloodstream of patients at risk from a **pulmonary embolism** (a fragment of blood-clot lodging in the pulmonary artery). These enzymes stimulate a **cascade system** (see section 14.2.5) responsible for the production of active **plasmin**, a proteolytic enzyme which digests **fibrin**, the main structural component of blood-clots.

Some enzymes may also be used to restrict the growth of **cancer** cells by depriving them of essential nutrients. For example, **L-asparaginase** may be used in the treatment of several types of **leukaemia**, since the tumour cells, in contrast to normal cells, have a requirement for exogenous L-asparagine.

19.1.3 Enzymes as reagents in clinical chemistry

D-glucose in blood and other physiological fluids is commonly analysed by means of procedures involving glucose oxidase (see sections 15.2.3, 18.1.5 and 18.2.2). These methods are specific for the β-form of D-glucose, but, except when solutions are freshly prepared, β-D-glucose and α-D-glucose will be present as an equilibrium mixture in samples and standards alike, so this apparent complication can be ignored. The reaction catalysed by glucose oxidase is utilized in a test-strip (Clinistix) for the screening of urine specimens: glucose oxidase and the reagents for a colour-producing indicator reaction are impregnated into a plastic strip, so that the intensity of the blue colour obtained when it is dipped into a specimen gives an indication of the D-glucose content (see section 18.1.8). This can be of use in the diagnosis of **diabetes mellitus**. D-glucose may also be analysed by a hexokinase-catalysed reaction, although this enzyme can utilize any D-hexose as substrate, or by use of glucose dehydrogenase (see sections 18.1.6 and 18.2.2).

Blood **lactate** and **pyruvate** are usually determined by means of LDH-catalysed methods (see section 19.1.1), and blood **urea** is sometimes analysed by procedures involving urease (see sections 18.1.5 and 18.1.8). Blood **cholesterol** may be determined by a reaction catalysed by cholesterol oxidase, the product 4-cholesten-3-one having an absorption maximum at 240 nm; and **triglycerides** and **urate** may also be analysed by enzymatic methods.

Luciferase, a Mg^{2+}-activated enzyme, utilizes luciferin (firefly extract) as substrate (see equation 18.4). The product, oxyluciferin exhibits a green **chemiluminescence**, so the reaction may be followed on a spectrofluorimeter with the light source blocked off, or in a luminometer. Luciferase-catalysed procedures are used chiefly for the analysis of **ATP** (e.g. from blood platelets), but may also be used to determine the Mg^{2+} concentration or the pO_2.

19.2 APPLICATIONS IN FORENSIC SCIENCE

Enzymes have long been used in forensic science as a means of detecting body fluids and as markers of genetic individuality. After the onset of puberty, **human semen** contains high levels of seminal acid phosphatases. These levels remain high until a male reaches 40, after which they show a gradual decline. The seminal acid phosphatase level appears to be relatively constant at a particular age, irrespective of sperm count or viability. For this reason a test for seminal acid phosphatase activity is still used as a presumptive identification of semen. In forensic laboratories, α-napthyl phosphate is generally used as the substrate. Upon hydrolysis, the product reacts with a diazonium salt (Brentamine Fast Blue) to produce an intense purple colour. Confirmatory tests for the presence of sperm may include microscopic visualization (Christmas-tree stain), detection of prostate-specific antigen and DNA fingerprinting (see below).

Traditionally, detection of the occurrence of **saliva** in forensic samples has relied upon the presence of α-amylase, an enzyme which catalyses the hydrolysis of starch and glycogen. The simple starch/iodine test has been used but lacks sensitivity and the presence of α-amylase is more commonly detected using an insoluble starch/dye complex (e.g. amylopectin/Procion Red). The hydrolysis of the insoluble complex liberates a soluble coloured compound which can be measured in a spectrophotometer. The levels of α-amylase are fifty times greater in saliva than in other body fluids. However, repeated deposition of other fluids (vaginal, semen or perspiration) in an area may elevate levels of the enzyme to that found in saliva, which limits the value of the test.

Alcohol dehydrogenase has been used to monitor **ethanol** levels in forensic samples, by the measurement of changes in absorbance at 340nm as the coenzyme, NAD^+, is reduced (see section 6.5.1). RIA, EMIT and ELISA procedures (see section 17.2.3) have been used to monitor the presence of **agrochemicals and pharmaceuticals** in biological samples. To confirm death by a suspected **insect attack**, the assay of serum tryptase (a proteolytic enzyme which is released from mast-cells as a consequence of the allergic response to insect venoms) may be employed, together with an ELISA for the detection of a specific IgE. ELISA-based assays are also used to monitor **serum proteins** and for the post-mortem confirmation of **HIV infections**.

For the past 20 years, **DNA fingerprinting**, which utilizes a number of enzymes as reagents (see section 21.1.2), has played an extremely important role in forensic science. However, even before the arrival of this technique, there was considerable interest in **polymorphic enzyme-systems** as **markers of genetic individuality**. The enzymes adenosine deaminase, adenylate kinase, carbonic anhydrase, esterase D, glucose-6-phosphate dehydrogenase, peptidase A, glyoxalase and phosphoglucomutase were important forensic polymorphic markers. The use of IEF (see chapter 16.2.1) and *in situ* activity-staining produce characteristic isoenzyme patterns. Several polymorphic enzymes can then be used to establish an individual's identity within a population. However, enzyme polymorphism compares unfavourably to the information that can now be gained from extremely small amounts of DNA.

19.3 APPLICATIONS IN INDUSTRY

We will now mention just a few of the many applications of enzymatic analysis in industry.

In the foodstuff industry, the activity of certain enzymes may be determined before and after **pasteurization** and **sterilization** procedures, to check whether these have been properly carried out. For example, **alkaline phosphatase** and **invertase** present in milk are inactivated within the same temperature range as is required for pasteurization, so the activities of these enzymes at the end of the process give an indication of its effectiveness.

Similarly, the degree of **bacterial contamination** of foodstuffs can be estimated by the assay of microbial enzymes not normally present in food.

For example, milk should contain only small amounts of **reductases**, but bacteria produce large amounts. Reductases may be easily assayed because they catalyse the reduction of methylene blue to colourless leuco-methylene blue under anaerobic conditions. A test-strip (Bactostrip) incorporating 2,3,4-triphenyltetrazolium salts provides a convenient way of testing for the presence of bacteria, a red formazan dye being produced as a result of the action of reductase.

Enzyme assay may also be used to determine whether **stored plant products** are suitable for use as foodstuffs. For example, **α-amylase** should be present in relatively low amounts in stored wheat seeds. However, if sprouting (i.e. germination of the stored crop) takes place, as may happen if the crop is left to stand under conditions which are too moist, or if there is prolonged rain at harvest time, then there is a greatly increased production of α-amylase and some proteolytic enzymes. This causes breakdown of the reserve starch and proteins, so flour produced from the sprouted wheat is not suitable for baking purposes. Hence the α-amylase activity of the seeds may be determined (see section 19.1.1) to give an indication of the degree of sprouting. Similarly, the amylase activity of malt (sprouted barley) is often determined. Once a flour has been produced, its amylase content may again be assayed to give an indication of the amount of starch breakdown which can be expected to take place when a dough is prepared (see section 20.3.2).

The freshness of meat may be determined by the use of **monoamine oxidase** to detect amines formed during degradation. Also, EIA (see section 17.2.3) is used to detect food adulteration, e.g. the addition of horse to beef products.

Enzyme assay may be applied to the investigation of **diseases in plants**. For example, it has been found that an injury (either mechanical or pathogenic) results in a marked, localized increase in the activity of **glucose-6-phosphate dehydrogenase**, but not of **glucose phosphate isomerase**, indicating diversion of glucose breakdown from glycolysis to the pentose phosphate pathway.

Enzyme assay is also used for research into such processes as the **browning of plant products**, which presents problems during the conversion of fruits and vegetables into drinks and preserves. The browning process involves the cyanide-resistant uptake of oxygen, which oxidizes phenols to quinones, thus initiating the sequence of reactions resulting in the formation of dark melanins.

The determination of D-glucose by enzymatic analysis (see section 19.1.3) is widely performed in the foodstuff and other industries. The method involving **glucose oxidase** is specific for D-glucose, whereas most chemical methods would not be able to distinguish D-glucose from any other monosaccharides present. The concentrations of various other sugars, and individual amino acids, may also be determined by specific methods of enzymatic analysis (e.g. the disaccharide sucrose by an **invertase**-catalysed procedure). In winemaking, the concentration of malic acid is sometimes determined by a method involving **malate dehydrogenase**.

Many industrial processes utilize micro-organisms, and enzymatic analysis may be employed to ensure that the concentrations of vital ingredients in the growth media are within required limits (see section 20.1.1). For example, **glycerol kinase** is obtained commercially from *Candida mycoderma,* and it is essential that the glycerol content of the growth medium is high enough throughout to induce glycerol kinase synthesis by the organism.

Hence the glycerol concentration may be monitored, and more glycerol added as required. A suitable procedure involves the reaction catalysed by **glycerol dehydrogenase**, in which glycerol is oxidized to dihydroxyacetone by NAD^+.

Finally, an important example of the use of enzymatic analysis for the determination of inhibitors is the **cholinesterase**-catalysed procedure for the analysis of organophosphorus insecticides (see section 19.1.1).

SUMMARY OF CHAPTER 19

In medicine, the assay of blood plasma enzymes can be extremely useful in helping to establish a diagnosis in a sick patient. The concentrations, and hence activities, of many enzymes in plasma increase when the cells of a tissue are damaged, because there is an increased rate of leakage of enzymes out of these cells. Since many enzymes or isoenzymes are associated with particular tissues, the pattern of plasma enzyme activities can indicate which tissues are damaged. The diagnosis of an inborn error of metabolism can be established by demonstrating the reduced activity of a particular enzyme in a tissue, or by revealing a gene abnormality by recombinant DNA technology. In some instances, it may be possible to obtain a pre-natal diagnosis. Enzymes are becoming increasingly important in the treatment of inborn errors and other diseases. Enzyme assay has played a significant role in forensic science in detecting body fluids and indicating genetic individuality, and enzyme-catalysed reactions are involved in the key technique of DNA fingerprinting. Enzyme assay is also important in the foodstuff industry in monitoring the processing and storage of food, and as a means of detecting contamination by micro-organisms. Enzymes are widely used as reagents in clinical chemistry, forensic science and industry.

FURTHER READING

Brady, R. O. (2006), Enzyme replacement for lysosomal diseases, *Annual Review of Medicine*, **57**, 283-291.

Burtis, C. A., Ashwood, E. R. and Bruns, D. C. (2005), *Tietz Textbook of Clinical Chemistry,* 4th edn., Saunders (Chapters 8, 21).

Gerhartz, W. (2003), *Enzymes in Industry: Production and Applications*, Wiley–VCH.

Goldberg, D. M. (2005), Clinical enzymology: an autobiographical history, *Clinica Chimica Acta*, **357**, 93-112.

James, S. J. and Nordby, J. J. (eds.) (2003), *Forensic Science: An Introduction to Scientific and Investigative Techniques*, CRC Press.

Payne-James, J., Byard, R. *et al* (eds.) (2005), *Forensic and Legal Medicine, Volumes 1-4*, Elsevier.

Price; N. C. and Stevens, L. (1999), *Fundamentals of Enzymology,* 3rd edn, Oxford University Press (Chapter 10).

Scriver, C. R. and Sly, W. S. (eds.) (2001), *The Metabolic and Molecular Bases of Inherited Disease,* 8th edn., McGraw-Hill.

Sudbery, P. (1998), *Human Molecular Genetics,* Longman.

Wu, A. H. B., Smith, A. *et al* (2004), Evaluation of a point-of-care assay for cardiac markers for patients suspected of acute myocardial infarction, *Clinica Chimica Acta*, **346**, 211-219.

20

Biotechnological Applications of Enzymes

20.1 LARGE-SCALE PRODUCTION AND PURIFICATION OF ENZYMES

20.1.1 Production of enzymes on an industrial scale

The general principles of enzyme extraction and purification were discussed in Chapter 16. Here we will be concerned chiefly with factors affecting the choice of source material and with those involved in the design of large-scale procedures, where the enzyme is extracted from at least 1 kilogram of cells.

Although enzymes have been used in certain industrial processes for centuries, their precise role, or even their identity, was not known over most of this period. They were often utilized as components of intact cells (e.g. yeasts in the baking and brewing industries) or as extracts containing a mixture of many enzymes (e.g. malt, containing **amylase** and **proteolytic enzymes**, also used in baking and brewing). The first enzyme to be made commercially available in a partially-purified form (by Christian Hansen, 1874) was the acid protease, **rennin (chymosin)**: this was as rennet, a crude preparation obtained from the fourth stomach of young calves, used to curdle milk in cheese production. Despite the importance of enzymes from animal and plant sources, their large-scale extraction poses technical, economical and possibly ethical problems. Hence the trend has been towards increased utilization of micro-organisms as sources of enzymes.

Because of the large number, and wide range, of micro-organisms available, a species can usually be found which contains an enzyme bearing a reasonable resemblance to one from animal or plant sources, or even one more advantageous for a particular application (see section 20.3.2). If there is not, it may be possible to use genetic manipulation to make a micro-organism produce the required enzyme (see section 21.1).

Such procedures, involving **recombinant DNA technology**, can be used to increase the yield of enzymes already produced by the micro-organism (e.g. **β-galactosidase** and **aspartate carbamoyltransferase** by *E. coli*).

They may also be used to produce completely different enzymes, including ones normally synthesized by eukaryotic cells. These techniques may even be used to produce enzymes of modified structure from suitably modified or synthesized genes: this is termed **protein engineering.**

However, it is not usually sufficient simply to find a micro-organism capable, naturally or otherwise, of synthesizing the required enzyme: a mutant strain is often sought which will have a minimum of controls restricting the synthesis of the enzyme (see below), and will not synthesize certain other substances.

Once a suitable strain of a suitable micro-organism has been found (or produced), a culture may be grown to produce a large supply of the enzyme (loosely termed **fermentation**). In the same way, animal or plant cells could be cultivated (see section 15.1) to enable large-scale extractions of their enzymes to be performed, but this has yet to be done on a widespread, commercial basis.

Some enzymes, e.g. *Aspergillus* **proteases**, may be obtained from a **semi-solid culture**, where the low water content and high degree of aeration (at the surface) apparently favours production of these enzymes. However, the vast majority are produced in aseptic **submerged culture**, where the conditions of pH, temperature, degree of aeration etc. can be finely controlled. The synthesis of most of the enzymes of commercial significance is related to cell growth, but can lag behind this, particularly in the case of hydrolytic enzymes and/or enzymes whose production is controlled by plasmids. In general, in a single-batch fermentation procedure, the number of cells present may increase for about a day (termed the **growth phase**) until a limiting value is reached, but enzyme production is likely to continue at significant levels for several days after this.

The culture medium must contain the essential nutrients needed by the micro-organism, together with any **inducer** necessary for the synthesis of the required enzyme (see section 3.1.5). Alternatively, it may be possible, by growing cells under conditions where the substrate of the enzyme is present in only limited amounts throughout, to isolate and culture a mutant strain which does not require induction. However, regardless of induction, the synthesis of many catabolic enzymes is **repressed** if the cells are grown rapidly on readily utilizable carbohydrate or protein. This **catabolic repression** can be overcome if the rate of utilization of the carbon source is controlled, either by feeding small amounts into the system at regular intervals, or by adding it in a form (e.g. an ester) which is slowly hydrolysed to yield the nutrient. Similarly, **feedback repression** of biosynthetic enzymes (and some others) may occur in the presence of metabolic end-products, so the build-up of these should be avoided if the synthesis of such an enzyme is required. The problems of repression may also be overcome by the use of mutants.

Oxygen is often a limiting nutrient in fermentation procedures, and the capacity for absorbing oxygen can be assessed by filling a fermenter with a solution of reducing agent (e.g. sodium sulphite) and operating it under the normal conditions. In general, it is important that the culture medium is homogeneous with respect to pO_2, pH, temperature and nutrient concentrations, so the mixing procedure (usually by propeller or flat-bladed turbine) must be adequately designed. This presents one of the main problems in converting a small-scale procedure to a large-scale one.

Most enzymes obtained commercially from microbial fermentation procedures are hydrolases. These are usually **extracellular enzymes**, i.e. they pass into the culture medium and do not have to be extracted from the cell. Increased yields of these enzymes may be obtained by introducing surfactants (e.g. Tween 80) into the culture medium, although the precise reason for this is not known.

Extracellular enzymes may be separated from the cells by filtration or centrifugation, once fermentation is complete. Continuous fermentation processes could be employed, but these are more difficult to design and control than are single-batch ones. **Intracellular enzymes** (e.g. **glucose isomerase**) must be released by cellular disruption. Many of the techniques discussed in section 16.1.2 are unsuitable for large-scale procedures, and those most commonly employed in this context are **liquid shear**, obtained by passing a cell suspension through a needle valve at very high pressure (termed **high pressure homogenization**), and grinding with abrasives in a **bead mill**. The cell debris can then be removed, e.g. by centrifugation.

The types of centrifuges commonly seen in laboratories, involving tubes in fixed-angle or swing-out rotors (see section 15.3.2) are not suitable for scaling up for use in industrial processes. This is because the mechanical stress on the centrifuge rotor is proportional to the square of the radius, so problems (including ones involving safety) arise if bigger centrifuges and rotors are made to the same basic design. In addition, starting-up and slowing-down procedures are time-consuming, so single-batch operations are inefficient, particularly if the capacity is low. Hence, systems which allow continuous or semi-continuous operation are preferred for industrial applications. Common designs include tubular bowl, disc bowl (disc stack) and decanter (scroll) centrifuges.

Tubular bowl centrifuges (of which Sharples centrifuges are well-known examples) have a rotor which is essentially a long tube, usually placed vertically with constrictions at the top and the bottom. The feed material comes in continuously at the bottom and the clarified liquid passes out at the top, while the solid material collects at the side of the tube because of centrifugal effects. With such a design, centrifugation has to be stopped at intervals to enable the solid material to be removed. **Disc bowl centrifuges** have a rotor containing a stack of discs, to increase the effective length of passage through the centrifuge, and may be designed to allow the sludge to be forced out continuously through a valve. **Decanter centrifuges** are orientated horizontally and incorporate an Archimedes screw to scrape the sludge from the walls of the rotor and force it in the direction opposite to that of the flow of clarified liquid, enabling both fractions to be removed continuously. Selection of the type of centrifuge to be used depends on the size of particles to be separated and the volume of particles in the feed.

In general, centrifugation is most useful when the solid fraction is the one containing the enzyme of importance, while filtration (using stationary or rotary systems) provides a convenient way of removing unwanted solid when the important enzyme is in solution. Another procedure for removing unwanted solid is **aqueous biphasic partition**, which involves a two-phase aqueous system (e.g. dextran/polyethylene glycol). Conditions are generally chosen so that the enzyme of interest remains largely in the upper phase, making it easy to separate from the cell debris in the lower phase. The two phases may be separated from each other in continuous fashion, for example, by use of a disc bowl centrifuge.

20.1.2 Large-scale purification of enzymes

Once an enzyme has been separated from cells or cellular debris, it is usually concentrated, e.g. by evaporation or ultrafiltration. The enzyme could also be purified by the procedures discussed in section 16.2.2, but it is often unnecessary and uneconomical to do this, provided no materials are present in the solution which affect the enzyme's usefulness. Stabilizers are often added, as are biocides, to deal with contaminating micro-organisms, and sometimes the solution is spray-dried to yield an impure solid. Obviously, the precise treatment given to the enzyme solution will be determined largely by the nature of the application for which it was prepared. For example, enzymes which will be used for processing food must be free from toxic impurities.

If further purification (termed **downstream processing**) is required, a procedure is usually worked out on a pilot scale before a large-scale process is put into operation. First, the enzyme preparation should be analysed to determine parameters such as enzyme activity (and, ideally, concentration), total protein concentration, conductivity, pH, and an assessment of nucleic acid and other non-protein contaminants, which can suggest how next to proceed. For example, if the enzyme concentration is low (less than 1 g l^{-1}), it may need to be increased by ultrafiltration or by means of a chromatographic step. However, the processing of large volumes of viscous crude extract presents problems that are not always compatible with column chromatography. The crude extract can be stirred with a resin (e.g. IEX) for a period of time, after which the resin is separated from the particulate liquid by filtration or sedimentation and then the target enzyme can be eluted. This process is not efficient and larger amounts of resin are required to achieve the same binding efficiency as a column. The crude extract can be pumped through resin suspended as a fluidized bed (see section 20.2.2) but turbulence and channelling may result in poor adsorption of protein. Alternatively, crude extracts can be pumped with an upward flow of buffer which expands the resin particle bed, comprised of beads (e.g. from GE Healthcare or Pall) of different sizes and densities, which become suspended in the liquid flow. The largest dense particles remain at the bottom of the column and the lighter, smaller particles are held in suspension by the flow of the liquid, resulting in a stable (expanded) bed of resin. Cellular debris passes through the expanded bed unimpaired and the target enzyme binds to the functional groups (IEX, HIC, IMAC or affinity) attached to the beads. The beads are washed before being allowed to settle and the elution conditions applied in the reverse direction. Adsorptive membranes (2-3mm thin discs, hollow-fibres or spiral-wound membranes) can also be used instead of conventional bead-based resins. Membranes can have a variety of functional groups attached (IEX, HIC, IMAC and affinity) and may have better flow-properties than the equivalent columns. This speeds up the processes of washing, binding and elution, in doing so reducing the possible denaturation of the target enzyme.

There are no clear-cut rules for devising a purification procedure, but generally a sequence of chromatography steps is sought where the products of one step are suitable to be the feed for the next. So, for example, a sample applied at low salt concentration to an ion exchange column can be eluted in high salt, and then introduced without further treatment to a hydrophobic interaction system.

Similarly, it is sensible to try to alternate steps leading to a dilution of a product with ones causing its concentration. Thus, SEC, leading to product dilution, might be followed by a concentration step involving affinity chromatography or IEX.

Once a suitable sequence has been identified, each step should be optimized in terms of factors such as volume load, mass load, flow-rate and cost. The system is then ready to be scaled up. In general, for a 100-fold scale-up, the procedure is to increase sample load, volumetric flow-rate and column volume by a factor of 100, while column bed height, linear flow-rate and sample concentration remain unchanged. If the results do not correspond to those obtained in the pilot system, then of course some adjustments must be made. The components of a typical scaled-up process are tank modules, together with delivery, separation, monitoring, fractionation and control modules. The **first tank module** consists of containers storing sample and buffers. The **delivery module** has tubes, control valves and pumps, linking the first tank module to the **separation module**, where chromatography is carried out. The **monitoring module** contains devices to monitor various properties of the column eluate, such as UV absorption, conductivity, pH and flow-rate. After this comes the **fractionation module**, consisting of control valves and tubes leading to the containers of the **second tank module**, where product is collected. The **control module** is programmed to open and close valves and make other adjustments; so, for example, only fractions containing protein (as indicated by the UV monitor) are collected, the rest running to waste.

20.1.3 Synthesis of artificial enzymes

A relatively new and completely different approach to the production of catalysts for industry involves synthesizing molecules which **mimic** the action of natural enzymes because of the incorporation of some features of their active sites. Such artificial enzymes have sometimes been given the name **synzymes**.

The starting material for a synzyme need not necessarily be a protein. Thus, for example, synzymes with chymotrypsin-like characteristics have been obtained based on cyclodextrins, which consist of 6 to 10 D-glucose units linked head-to-tail in a ring. Glycosidic oxygen atoms and -CH linkages point inwards, creating a hydrophobic environment which can act as a binding-pocket. Catalytic activity is provided by the attachment of imidazole, hydroxyl and carboxyl groups, as found in the active sites of serine proteases. The resulting synzyme resembles chymotrypsin in its esterase activity and, furthermore, it is more stable. Cyclodextrins may also be linked to pyridoxal coenzymes, producing synzymes with transaminase activity.

Another example of an enzyme mimetic is a structure with methylmalonyl-CoA mutase and methylaspartate mutase activity synthesized by Yukito Murakami and colleagues (1991). They combined a hydrophobic vitamin B_{12} with a bilayer membrane composed of an aggregate of synthetic peptide-lipid molecules, and found it to have catalytic activity if used in combination with a substrate-activation process involving vanadium trichloride and atmospheric oxygen.

One more approach involves making antibodies with the characteristics of enzymes. If, for example, an analogue of the supposed transition-state of a particular reaction (see section 4.6) is injected into a mouse, the immune system will make antibodies against it, and some of these may be able to catalyse the reaction in question.

These catalytic antibodies (or **abzymes**), like any other antibodies, may be proliferated by the techniques of monoclonal antibody production (see section 15.2.4). Recombinant DNA technology may also be used in the production of abzymes, which is likely to become increasingly significant over the next few years.

20.2 IMMOBILIZED ENZYMES

20.2.1 Preparation of immobilized enzymes

An immobilized enzyme is one which has been attached to or enclosed by an insoluble support medium (termed a **carrier**) or one where the enzyme molecules have been cross-linked to each other, without loss of catalytic activity (Fig. 20.1).

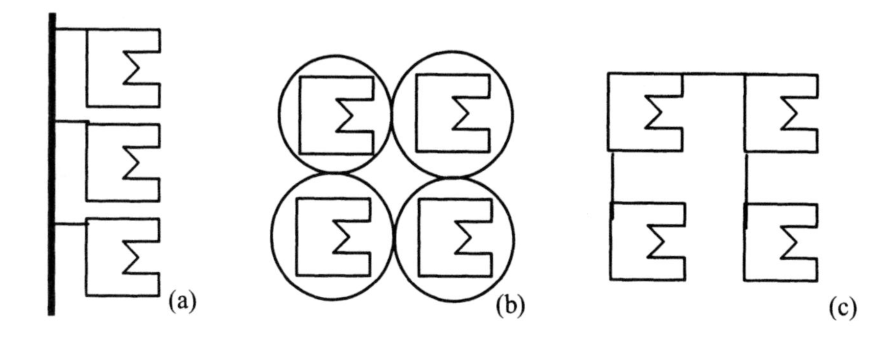

Fig. 20.1 – Diagrammatic representation of enzyme immobilization by: (a) attachment to an insoluble support medium; (b) entrapping by an insoluble support medium; and (c) cross-linking of enzyme molecules. Note that, in some instances, more than one of these factors may be involved.

Physical adsorption on to an inert carrier is a very simple procedure for immobilizing an enzyme, for it requires just the mixing of the enzyme solution with the carrier. In 1916, J. M. Nelson and Edward Griffin showed that **invertase** could be adsorbed onto activated charcoal without any change in enzymic activity, thus producing the first immobilized enzyme, although they made no subsequent use of it. Other inorganic materials which may be used as carriers include clay, alumina and silica. The weak linkages established between enzyme and carrier (mainly van der Waals and hydrogen bonds) have little effect on catalytic activity. However, because the bonds are so weak, the enzyme can easily be desorbed from the carrier. This can be brought about by changes in pH, ionic strength or substrate concentration. Also, the adsorption process is non-specific, so many other substances will become attached to the carrier as the immobilized enzyme is used.

Ionic binding provides a slightly more specific way of attaching an enzyme to a carrier: therefore, many ion exchange resins, e.g. DEAE-Sephadex and CM-cellulose, have been used as support media. The enzyme will remain bound to the carrier provided pH and ionic strength are maintained at suitable values.

Covalent binding can provide even more permanent linkages between enzyme and carrier.

However, it is essential that the conditions used for the formation of the covalent bonds are sufficiently mild that little, if any, catalytic activity is lost. Also, the active site of the enzyme must remain free from covalent attachments, so it is sometimes protected by a substrate or substrate-analogue during the immobilization procedure.

The enzyme functional groups most commonly linked by covalent bonds to a carrier are free α- or ε-amino groups, but sulphydryl, hydroxyl, imidazole or free carboxyl groups may also be involved.

Many procedures depend on the coupling of phenolic, imidazole or free amino groups on an enzyme to a **diazonium** derivative of a carrier. For example, the method developed by Dan Campbell and colleagues in 1951 for linking albumin to diazotized *p*-aminobenzyl cellulose to form an immobilized antigen has been widely used to prepare immobilized enzymes:

$$\text{cellulose}-OCH_2-C_6H_4-NH_2 \xrightarrow{NaNO_2/HCl} \text{cellulose}-OCH_2-C_6H_4-\overset{+}{N}\equiv N$$

p-aminobenzyl cellulose (diazonium derivative)

$$\xrightarrow{\text{enzyme}} \text{cellulose}-OCH_2-C_6H_4-N{=}N-\text{enzyme} \qquad (20.1)$$

(immobilized enzyme)

In 1953, N. Grubhofer and L. Schleith, using the same principle, but starting with diazotized polyaminopolystyrene, were the first to immobilize enzymes with a view to their future application: the enzymes in question were **pepsin, ribonuclease, carboxypeptidase** and others.

A **peptide** bond may be formed between a free amino group in an enzyme and a carboxyl group in an insoluble carrier, if the latter group is present as an isocyanate, acid azide or other reactive derivative. For example, H. Brandenberger, in 1957, prepared an isocyanate derivative of polyaminopolystyrene by treating it with phosgene ($COCl_2$), and then used it to immobilize **catalase**:

$$[-CHCH_2-(C_6H_4-NH_2)]_n \xrightarrow{COCl_2} [-CHCH_2-(C_6H_4-NCO)]_n \xrightarrow[\text{(enzyme)}]{H_2N.E} [-CHCH_2-(C_6H_4-NHCONH-E)]_n \qquad (20.2)$$

polyamino-polystyrene (isocyanate derivative) (immobilized enzyme)

Many enzymes have been immobilized by reacting with the azide derivative of a carrier, whose formation may involve treatment with hydrazine (NH_2NH_2). For example:

$$\underset{\text{CM-cellulose}}{\text{cellulose}-OCH_2CO_2H} \xrightarrow{CH_3OH/HCl} \underset{\text{(methyl ester)}}{\text{cellulose}-OCH_2CO_2CH_3} \tag{20.3}$$

$$\xrightarrow{NH_2NH_2} \underset{\text{(hydrazide derivative)}}{\text{cellulose}-OCH_2CONHNH_2} \xrightarrow{NaNO_2/HCl} \underset{\text{(azide derivative)}}{\text{cellulose}-OCH_2CON_3} \xrightarrow{H_2N.E} \underset{\text{(immobilized enzyme)}}{\text{cellulose}-OCH_2CONH-E}$$

Peptide bonds between enzyme and carrier may also be formed by the use of condensing reagents such as carbodiimides. For example:

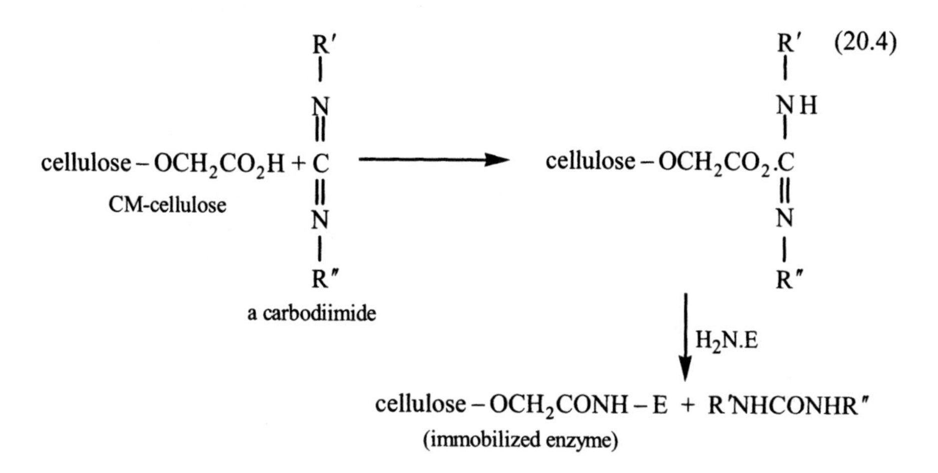

Another category of covalent bond formation is the **alkylation** of phenolic, sulphydryl or free amino groups by reactive groups in the carrier. For example, bromoacetyl cellulose has been used to immobilize a variety of enzymes:

$$\underset{\text{bromoacetyl cellulose}}{\text{cellulose}-OCOCH_2Br} \xrightarrow{\text{enzyme}} \underset{\text{(immobilized enzyme)}}{\text{cellulose}-OCOCH_2-\text{enzyme}} \tag{20.5}$$

Yet another widely used method for enzyme immobilization was developed by Rolf Axén and colleagues in 1967, and involves the reagent, cyanogen bromide (CNBr). This reagent, which is also used for the preparation of immobilized ligands for affinity chromatography (see section 16.2.2), activates polysaccharide hydroxyl groups. The most significant of several reaction sequences which may take place is as follows:

$$\begin{array}{l}|-OH\\|-OH\end{array} \xrightarrow{CNBr} \begin{array}{l}|-O\\|-O\end{array}\!\!>C{=}NH \xrightarrow{H_2N.E} \begin{array}{l}|-O\overset{\overset{NH}{\|}}{C}.NHE\\|-OH\end{array} \qquad (20.6)$$

polysaccharide (imidocarbonate derivative) (immobilized enzyme)

An enzyme which possesses an accessible sulphydryl group not essential for activity can easily be linked to a similar group on a carrier by the formation of a **disulphide bridge**. The commercial significance of this procedure is that it is reversible, so inactivated or unwanted enzyme may be removed from the carrier under reducing conditions, and then be replaced by fresh enzyme. The only other immobilization technique where such easy replacement is possible is ionic binding.

In general, the conditions required for the covalent attachment of an enzyme to an insoluble carrier are such that some loss of activity is inevitable. However, very little activity change often takes place if a covalent attachment is brought about by means of **chelation** rather than by a chemical reaction. A wide range of carriers (e.g. cellulose, glass and nylon), if treated with salts of transition metals (e.g. titanium, vanadium or iron chlorides), then washed and dried, can chelate enzymes. Strong metal bridges (see section 11.4.1) are formed between hydroxyl oxygen atoms of the carrier and amino nitrogen atoms on the enzyme.

The **entrapping** (or **occlusion**) of an enzyme is commonly achieved within the lattice of a **polymerized** gel. The most widely used is polyacrylamide, which may be synthesized from an aqueous solution of acrylamide and N,N-methylene-bisacrylamide (BIS) in the presence of initiators and accelerators, as outlined (without stoichiometry) below:

$$\underset{\text{acrylamide}}{\begin{array}{c}CH_2=CH\\|\\CO\\|\\NH_2\end{array}} + \underset{\text{BIS}}{\begin{array}{c}CH_2=CH\\|\\CO\\|\\NH\\|\\CH_2\\|\\NH\\|\\CO\\|\\CH_2=CH\end{array}} \xrightarrow[\text{accelerators}]{\text{initiators}} \underset{\text{polyacrylamide}}{\begin{array}{l}-CH_2.CH.CH_2CH.CH_2.CH.CH_2-\\ \quad\;\; |\qquad\;\; |\qquad\quad |\\ \quad\; CO\quad\; CO\qquad CO\\ \quad\;\; |\qquad\;\; |\qquad\quad |\\ \quad\; NH\quad NH_2\quad\; NH\\ \quad\;\; |\qquad\qquad\qquad |\\ \quad\; CH_2\qquad\qquad\; CH_2\\ \quad\;\; |\qquad\qquad\qquad |\\ \quad\; NH\qquad\qquad\;\; NH\\ \quad\;\; |\qquad\qquad\qquad |\\ \quad\; CO\qquad\qquad\;\; CO\\ \quad\;\; |\qquad\qquad\qquad |\\-CH_2.CH.CH_2CH.CH_2.CH.CH_2-\\ \qquad\qquad\quad |\\ \qquad\qquad\; CO\\ \qquad\qquad\quad |\\ \qquad\qquad\; NH\\ \qquad\qquad\quad |\end{array}} \qquad (20.7)$$

The degree of cross-linking, and hence the average pore size, is dependent on the proportions of acrylamide and BIS present initially, for only the latter can cross-link. If an enzyme is present in the solution, it will be entrapped as the gel is formed.

Alternatively, an enzyme may be entrapped within the semi-permeable membrane of a **microcapsule.** This is most commonly done by the **interfacial polymerization** method developed by Thomas (T. M. S.) Chang and colleagues around 1964. An aqueous solution containing the enzyme and a hydrophilic monomer is added to a water-immiscible organic solvent and an emulsion is formed. More of the same organic solvent, but containing a suitable hydrophobic monomer, is then added and the solutions are well mixed. Polymerization of the monomers will take place at each interface between organic and aqueous solvents, and, in this way, enzyme molecules in aqueous solution are enclosed in polymer membranes. For example, nylon microcapsules may be formed if the hydrophilic monomer is 1,6-hexamethylene diamine, the hydrophobic monomer sebacoyl chloride, and the organic solvent a cyclohexane-chloroform mixture. Such microcapsules are typically 10-100 μm in diameter.

The entrapping of an enzyme, either by gel formation or microencapsulation, is of general application and should not affect the activity of the enzyme. However, in some instances, free radicals generated during the polymerization procedure may cause some loss of enzymic activity. Also, since the entrapped enzymes cannot escape because of their size, it follows that a very large substrate will not be able to diffuse in to reach the enzymes. Hence, this immobilization procedure is not suitable for proteolytic enzymes or others whose substrates are macromolecules. Each polymerized gel will inevitably contain a range of pore sizes, for the cross-linking cannot be exactly regular: therefore, in contrast to the situation with macro capsules, there is likely to be some degree of leakage of enzyme. A further advantage of microencapsulation is that each enzyme is in much closer contact with the surrounding solution than are those entrapped in the interior of gels.

Another encapsulation technique which can be used for enzyme immobilization is the formation of **liposomes**: amphipathic lipids such as phosphatidyl choline and cholesterol (see section 16.1.3) are dissolved in chloroform and spread as a film over the walls of a rotating flask; an aqueous solution of enzyme is added and rapidly dispersed, and liposomes are formed as lipid membranes enclose the water droplets. Some of the enzyme will remain in the aqueous solution within the liposomes, while the rest will be incorporated into the membrane.

Immobilization by **cross-linking** molecules of enzyme is commonly brought about by the action of glutaraldehyde, whose two aldehyde groups form Schiff base linkages with free amino groups:

$$\begin{matrix} \mathrm{E.NH_2} \\ \\ \\ \\ \mathrm{E.NH_2} \\ \text{enzyme} \end{matrix} \quad + \quad \begin{matrix} \mathrm{CHO} \\ | \\ \mathrm{(CH_2)_3} \\ | \\ \mathrm{CHO} \\ \text{glutaraldehyde} \end{matrix} \quad \longrightarrow \quad \begin{matrix} \mathrm{E.N} \\ \| \\ \mathrm{CH} \\ | \\ \mathrm{(CH_2)_3} \\ | \\ \mathrm{CH} \\ \| \\ \mathrm{E.N} \end{matrix} \qquad (20.8)$$

Since several free amino groups are likely to be present on each enzyme molecule, a cross-linked network will be formed. This procedure was first used by Florante Quiocho and Frederic Richards, in 1964, to immobilize **carboxypeptidase A**. Other reagents with two functional groups (i.e. **bifunctional reagents**) of relevance to enzyme immobilization include derivatives of bis-diazobenzidine,

$$N\equiv\overset{+}{N}-C_6H_4-C_6H_4-\overset{+}{N}\equiv N$$

which act by means of diazo coupling.

Although cross-linking between identical enzyme molecules can result in immobilization at high enzyme concentrations, it is not an ideal method since many molecules simply act as supports for others. Hence, cross-linking is often performed in conjunction with other methods of immobilization. It can, for example, be used to prevent the leakage of enzymes from a polymerized gel (see above), or to trap the enzyme around pre-formed polymer molecules (e.g. starch). Also, bifunctional reagents may be used, not only to link molecules of enzyme to each other, but also to link them to an inert carrier: thus, glutaraldehyde has often been used to attach enzymes to amino groups of carriers such as aminoethyl (AE)-cellulose or aminoalkylated porous glass.

As with any other method which involves the formation of covalent bonds by chemical reaction, cross-linking is usually performed under conditions which cause some loss of enzymic activity.

Immobilized enzymes are usually prepared in particle form, but enzymes may be attached to, or entrapped within, carriers in the form of membranes, tubes or fibres, according to the requirements of a given application. Also, in the case of intracellular enzymes, it may be more economical to **immobilize** the **intact cell** rather than to perform an extraction step. This is satisfactory provided none of the other enzymes present affect the proposed application, and provided the appropriate substrates and products can freely pass through the cell membranes. The most common method for immobilizing microbial cells is to entrap them in polyacrylamide gel, as done by Klaus Mosbach and colleagues in 1966. The use of immobilized cells is particularly advantageous where it is known that the enzyme of importance would be unstable during or after extraction, a not uncommon situation.

20.2.2 Properties of immobilized enzymes

The properties of an immobilized enzyme may be different from those of the same enzyme in free solution, and depend both on the method of immobilization and on the nature of the insoluble carrier. A reduction in specific activity may occur as an enzyme is immobilized, particularly if a chemical process is involved, since the conditions might cause some denaturation to take place (see section 20.2.1). Furthermore, the carrier creates a new micro-environment for the enzyme, and could thus influence its activity in a variety of ways. For example, the characteristics of the enzyme might be altered if the active site undergoes some conformational change as a result of chemical or physical interactions between enzyme and carrier.

Also, the carrier could affect the characteristics of the enzyme-catalysed reaction by imposing some steric hindrance, by preventing free diffusion of substrate to all molecules of the enzyme, or by forming electrostatic interactions with molecules of substrate or product (see below).

The **stability** of an enzyme on heating or storage can increase, decrease or stay the same when it is immobilized, depending on how the new micro-environment affects its tendency to denature: few clear-cut trends have been observed. Another relevant factor is the ease of attack by substances which might degrade the enzyme. For example, autodigestion of proteolytic enzymes is reduced if the molecules are protected from each other by immobilization. Similarly, steric hindrance protects many other immobilized enzymes from attack by proteolytic enzymes.

The **pH optimum** can change by as much as 2 pH units when the enzyme is immobilized, largely owing to the effect of the new micro-environment. Leon Goldstein and colleagues (1964-70), and others, have demonstrated that the pH optimum of many enzymes moves to a more alkaline value if the carrier is anionic, and to a more acid value if it is cationic, due to changes in the degree of ionization of amino acid residues at the active site. It has been proposed that the change in pH optimum, ΔpH, is given by: $\Delta pH = 0.43F\psi/RT$, where F is the Faraday constant, ψ the electrostatic potential, R the gas constant and T the absolute temperature. The effect is not observed if the ionic strength is high, presumably because the salt ions interfere with the **electrostatic field** produced by the carrier.

The **apparent K_m** may be similarly affected by the **electrostatic field** of the carrier, being significantly decreased when a carrier is used which is of opposite charge to that of a substrate. For example, when positively charged benzoyl-L-arginine ethyl ester is used as substrate, the apparent K_m of **ficain** immobilized by covalent attachment to CM-cellulose, a polyanionic carrier because of the presence of unreacted carboxyl groups, is ten times less than that of the free enzyme. Goldstein and colleagues explained this on the basis of the electrostatic interaction causing the substrate to be present in the region of the carrier, and hence of the enzyme, at higher concentrations than elsewhere in the solution. The converse is true for the situation where the carrier has the same charge as the substrate: the apparent K_m for **creatine kinase** immobilized on CM-cellulose is ten times greater than that for the free enzyme, the substrates ATP and creatine being negatively charged. Theoretical considerations suggest that the change in K_m is given by: $K'_m = K_m.e^{-zF\psi/RT}$, where z is the electrovalency of the substrate. Again, these effects are not seen at high ionic strength.

The **apparent K_m** may also be affected by **diffusion factors**. Malcolm Lilly and colleagues, in 1968, postulated that around each particle or membrane is an unstirred layer (or layers) of solution, 10-100 μm thick. The substrate contained in such a layer will be quickly utilized by an enzyme immobilized within the particle or membrane, so the subsequent reaction velocity will depend on the rate at which substrate from the bulk solution can diffuse through the unstirred layer and reach the enzyme. In general, the steady-state substrate concentration in the vicinity of the enzyme will be less than that in the bulk solution, and so the reaction velocity will be less than that expected from the external substrate concentration.

The apparent values of K_m and V_{max} will both change, with the new K_m value possibly being given by: $K'_m = K_m + V'_{max}.\delta l / D$, where δl is the thickness of the unstirred layer and D the diffusion constant. The thickness of the diffusion layer depends on the rate at which the bulk solution is stirred, so that it may be possible to reduce the thickness of the layer and hence increase the reaction velocity by more vigorous stirring. However, restricted diffusion of the substrate through a membrane or gel lattice can also reduce the rate of reaction. Furthermore, the ease of diffusion of the product away from the enzyme can be another significant factor, particularly if the reaction being catalysed is subject to product inhibition.

For one or more of these reasons, the apparent K_m of an immobilized enzyme is often significantly higher than for the same enzyme in free solution. For example, Leon Goldstein, Rachel Goldman, Ephraim Katchalski and colleagues (1964-71) showed that this was true for **papain** and **alkaline phosphatase** immobilized in collodion membranes, and, furthermore, the apparent K_m increased with increasing membrane thickness.

The rate of diffusion of substrate to the enzyme will rise to a limiting value as the substrate concentration in the bulk solution is increased. If this limiting value is reached before an immobilized enzyme is completely saturated with substrate, then its **apparent V_{max}** value will be less than that in free solution.

20.2.3 Applications of immobilized enzymes: general principles

The immobilized enzyme system chosen for a given application should fit the requirements in terms of stability, activity, pH optimum and other characteristics: the factors discussed in section 20.2.2 should all be considered.

Sometimes the changes in properties which take place when an enzyme is immobilized may be used to advantage. For example, if an enzyme-catalysed reaction cannot be linked directly to another because of incompatible pH-activity ranges, it may be possible to immobilize the enzymes in such a way that their pH-activity ranges now overlap, thus allowing them to be used in a single rather than a two-step process.

Regardless of this, component enzymes of a coupled system may be immobilized together (e.g. entrapped in a gel) or, if immobilized separately, may later be placed in close proximity. This increases the efficiency of the coupled process by ensuring that the concentration of the substance which is the product of one of the enzyme-catalysed reactions and the substrate of the other is kept at a higher concentration in the neighbourhood of the enzyme than in the bulk solution. This may be used in an industrial application, or as a model system to investigate the properties of enzymes located closely together in the living cell (e.g. the **malate dehydrogenase/citrate synthase** system of mitochondria). Another example of the value of immobilized enzymes as model systems is the use of liposomes to investigate *in vitro* the effect of a lipid environment on the activity of enzymes which are associated with membranes *in vivo*.

In general, the property of immobilized enzymes which is of greatest industrial importance is the ease with which they can be separated from reaction mixtures. Hence, in contrast to systems involving soluble enzymes, the reaction can be stopped by physical removal of the immobilized enzyme, without requiring such procedures as heat inactivation which might affect the products of the reaction.

Furthermore, the enzyme will still be active and largely uncontaminated, so can be used again. For these reasons, immobilized enzymes are ideal for use in continuously operated processes. At the present time, continuous industrial processes involving immobilized enzymes are usually carried out in simple **stirred-tank reactors** or in **packed-bed reactors** (Fig. 20.2).

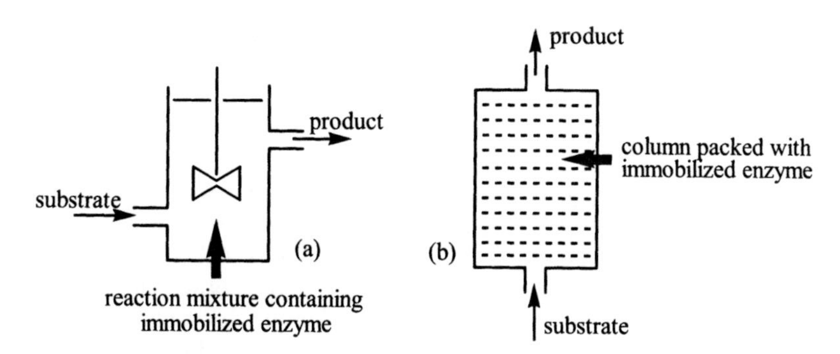

Fig. 20.2 – Diagrammatic representation of: (a) a continuously-operated stirred-tank reactor (CSTR); and (b) a continuously-operated packed-bed, or plugged-flow, reactor with an upwards direction of flow.

The direction of flow in packed-bed reactors may be in an upwards or a downwards direction, but the former is preferable in applications where a downward flow would cause compression or clogging of the packed-bed. A more recent development is the **fluidized-bed reactor**, which incorporates some of the features of both stirred-tank and packed-bed reactors: this is an upward flow system which is ideal for processes where the substrate solution is highly viscous and the product is gaseous. Finally, as mentioned in section 20.2.1, some processes now make use of enzymes immobilized in membranes, hollow-fibres or tubes.

For obvious commercial reasons, all these processes involve substantial conversion of substrate to product, so any mathematical description requires an integrated form of the Michaelis-Menten equation (see section 20.3.1).

Immobilized enzymes may also be used as components of analytical systems, either in dry-reagent or automated techniques (see Chapter 18). For example, tubular and packed-bed reactors have been incorporated into continuous-flow analysers. Here, normal Michaelis-Menten kinetics will be more applicable, as when free-enzymes are used as reagents.

20.3 ENZYME UTILIZATION IN INDUSTRY

20.3.1 Introduction

Enzymes may be used in industry as components of living cells or after isolation, in free or immobilized forms. All of these may be referred to as **biocatalysts**.

The use of whole cells (mainly micro-organisms) rather than isolated enzymes to generate a required product has attractions in that the problems (and cost) of enzyme isolation do not have to be faced

However, nutrients have to be added to enable the cells to grow in culture, and a large number of reactions will be taking place which have nothing to do with process of interest, or which might interfere with it. So, specific enzyme inhibitors might be added to minimize the chances of metabolities participating in wasteful side-reactions, or to prevent the further metabolism of the desired product. For example, the yield of glycerol produced by fermentation in *Saccharomyces cerevisiae* is greatly increased if sodium bisulphite is added to inhibit the conversion of acetaldehyde to ethanol: this step utilizes NADH (see section 20.3.2) and, if operating, would deplete the cell of the NADH required for the synthesis of glycerol phosphate, from which glycerol is obtained by dephosphorylation.

Regulatory mechanisms also need to be taken into account. So, for example, in the production of lysine by fermentation in *Corynebacterium glutamicum,* the first enzyme in the pathway, aspartate kinase (also known as aspartokinase), is subject to concerted feedback inhibition by L-Iysine and L-threonine (i.e. both must be present in significant amounts to cause inhibition). Since, under normal circumstances, the production of lysine would be accompanied by the production of threonine, concerted inhibition of aspartate kinase would take place and prevent accumulation of each of these amino acids. The solution is to use a mutant of *C. glutamicum* which lacks an enzyme essential for the synthesis of threonine, but not for lysine. Sufficient threonine is added to allow a limited amount of growth but not enough for significant inhibition of aspartate kinase to occur, so lysine production can continue unchecked. Processes involving isolated enzymes should be more straightforward, as only a limited number of reactions will be taking place, and it will be known exactly what these are. If the process involves just a single enzyme-catalysed reaction, which is often the case, then the kinetic features may be indicated by the Michaelis-Menten reaction. Of course, Michaelis-Menten kinetics are investigated under steady-state conditions, when product concentration remains insignificant (see section 7.1.2), whereas the whole point of using enzyme-catalysed reactions in industrial processes is to give rise to considerable amounts of product. Hence, the Michaelis-Menten equation is applied to batch industrial processes in an integrated form: $V_{max}.t = [S_0] - [S] + 2.303K_m \log_{10}([S_0]/[S])$, where [S] is the substrate concentration at time *t.* By means of this equation, the concentration of product formed in a specified period of time, which should be the same as the decrease in substrate concentration ([So] – [S]), can be calculated if the relevant values of K_m and V_{max} are known. It should be remembered, however, that, like the Michaelis-Menten equation itself, the integrated form is only valid provided [S] » [E_0] and the reaction is effectively irreversible. Also, the enzyme must be stable over the time required, there must be no substrate or product inhibition taking place, and the solution must remain well mixed.

Up to the present time, industrial processes have generally utilized enzymes in aqueous solution. However, much attention is now being given to possible applications involving enzymes which function in non-aqueous media.

20.3.2 Applications in food and drink industries

Some applications of enzymatic analysis in industry were discussed in section 19.3. Here we will consider some industrial processes where enzymes have a more direct involvement.

The traditional use of yeasts (e.g. *Saccharomyces carlsbergensis)* in the baking and brewing industries arose because they contain the enzymes for alcoholic fermentation. In common with other organisms, they metabolize hexose sugars to produce pyruvate, but, whereas animals convert this to lactate under anaerobic conditions, the anaerobic end-product in yeasts is ethanol, with carbon dioxide being evolved:

$$\underset{\text{pyruvic acid}}{CH_3.CO.CO_2H} \xrightarrow[Mg^{2+},\ TPP]{\text{pyruvate decarboxylase}} \underset{\text{acetaldehyde}}{CH_3.CHO} \xrightarrow[\text{NADH} \rightarrow \text{NAD}^+]{\text{alcohol dehydrogenase}} \underset{\text{ethanol}}{CH_3.CH_2OH} + CO_2 \quad (20.9)$$

In the **baking of bread**, the preliminary process involves the mixing of wheat flour (mainly starch and protein) with yeast and water. Starch consists of D-glucose units linked by α-1,4 glycosidic bonds, with α-1,6 bonds at branching points. The enzymes **α-amylase** and **β-amylase** present in the flour cleave some of the α-1,4 bonds, the eventual products being glucose, maltose (a disaccharide) and some oligosaccharides which cannot be broken down further because of the presence of α-1,6 bonds. Glucose and maltose can then be metabolized by the yeast, and carbon dioxide is formed, which distends the protein framework of the dough, ready for baking. However, wheat flour often has a low α-amylase content, so it may be supplemented with malt flour (see below) or, a more recent development, by fungal α-amylase (from *Aspergillus oryzae).* Wheat α-amylase is very heat stable, so it continues to act for a time during the baking process. This may lead to too much starch breakdown taking place, and cause the bread to be somewhat soggy. For this reason, wheat α-amylase is sometimes inactivated by a brief treatment with superheated steam prior to supplementation of the flour with the more heat-labile fungal α-amylase. Proteases from *Aspergillus oryzae* may be introduced to cause strictly limited breakdown of the wheat protein (gluten), thus shortening mixing time and enabling a smoother, more uniform dough to be obtained. Malt flour also contains proteases, but in variable amounts, so its use might lead to excessive gluten breakdown.

In **brewing**, the main starting material is malt, produced by allowing barley seed to germinate in moist conditions. The reserve starch is broken down by the amylase present (as in wheat, discussed above and in section 19.2) to give, among other products, glucose and maltose. The grains may then be roasted to prevent further growth and to add flavour, after which the soluble material present is extracted by water to produce the wort. This is further flavoured with hops, and then yeast is added to produce ethanol by alcoholic fermentation of the glucose and maltose.

Bacterial α-amylase (from *Bacillus subtilis),* which is even more heat stable than wheat α-amylase, is of increasing importance in the brewing industry. It may be used to produce a wort from unmalted barley (and maize), although some malt (a minimum of 20%) is usually added. Fungal ***exo*-1,4-α-glucosidase**, better known as **amyloglucosidase** or **glucamylase** (e.g. from *Aspergillus niger)* may also be added, since this can cleave α-1,6 as well as α-1,4 glycosidic bonds and so increase the yield of glucose and maltose from starch. Malt contains some **protease** activity, so bacterial proteases are usually added when a wort is being prepared from unmalted barley.

In the industrial production of glucose from starch, the latter is first solubilized and partly degraded by bacterial **α-amylase** (or, but less commonly than hitherto, by HCl), and then treated with fungal amyloglucosidase (free or immobilized). As mentioned above, this can cleave both types of glycosidic bond found in starch, so gives a good yield of glucose. Glucose may also be obtained from cellulose-containing waste products by treatment with **cellulase** (e.g. from *Aspergillus niger*). As a further possibility, it may be produced, together with galactose, by the action of **β-galactosidase (lactase)** on lactose, which is present in whey and so is a major by-product of cheese manufacturing.

Invert sugar, a mixture of glucose and fructose, is produced from sucrose by the action of yeast **β-fructofuranosidase**, better known as **invertase** (an enzyme which can only be extracted by disruption of the yeast cell wall). Invert sugar is less likely to crystallize from solution at high concentrations than is sucrose, giving it an advantage in jam-making. In confectionery, crystalline sucrose is coated with chocolate and invertase, causing the subsequent conversion of sucrose to invert sugar, and hence liquification to form a 'soft centre'.

Invert sugar may also be produced from glucose by the action of **glucose isomerase** (bacterial or fungal), an enzyme now thought to be identical to **xylose isomerase**. The first commercial production of such a high-fructose syrup from glucose, using soluble glucose isomerase, was reported by Yoshiyuki Takasaki in 1966; one using glucose isomerase immobilized on DEAE-cellulose was described by K. N. Thompson and colleagues in 1974. At the present time, conversion of glucose to fructose is the biggest industrial application of immobilized enzymes. Fructose is much sweeter than glucose, so high-fructose syrup can be used to replace sucrose as a sweetener, where it is economical to do so. It may also find increased use in this respect because of fears about the adverse effects of sucrose on health, regardless of whether fructose can be proved to be any safer.

Certain **amino acids** essential for animal growth (e.g. methionine) are found in low amounts in the proteins of some vegetable-based foodstuffs, so may have to be added to the diet. However, animals can only utilize L-amino acids, whereas D- and L-amino acids, in equal amounts, are produced by chemical processes. Dietary supplements usually consist of such mixtures, so half is wasted, even if no other harm is done. Hence the procedure introduced in 1969 by Ichiro Chibata and colleagues for the production of L-amino acids from racemic mixtures is of great significance: acetyl D- and L-amino acids are passed through a system containing **aminoacylase** immobilized on a DEAE-Sephadex (or other) carrier, causing the deacylation of the L-form alone, which can then be separated; the unchanged D-form is then subjected to racemization, and the process repeated.

The **clarification of cider, wines and fruit juices** (e.g. apple) is usually achieved by treatment with fungal **pectinases**. The pectins of fruit and vegetables play an important role in jam-making and other processes by bringing about gel formation. However, they cause fruit drinks to be cloudy by preventing the flocculation of suspended particles. Pectinases are a group of enzymes including **polygalacturonases**, which break the main chains of pectins, and **pectinesterases**, which hydrolyse methyl esters. Their action releases the trapped particles and allows them to flocculate.

Cheese production involves the conversion of the milk protein, κ-casein, to para-casein by a defined, limited hydrolysis catalysed by **chymosin (rennin)**. In the presence of Ca^{2+}, para-casein clots and may be separated from the whey, after which the clot is allowed to mature under controlled conditions, still in the presence of chymosin, to form cheese. Since chymosin can only be extracted from calves killed before they are weaned (pepsin is produced instead of chymosin after weaning), the enzyme is in short supply, so because of this, and also ethical issues, there has been a large-scale search for an acceptable substitute. Proteases from animals (**pepsin**), plants (**ficain** and **papain**) and over a thousand micro-organisms have been tried, either on their own or mixed with calf chymosin. Several have been largely, if not completely, successful (e.g. a protease from *Mucor rouxii)*.

Papain is sometimes used as a **meat tenderizer**; some South American natives have traditionally wrapped their meat in leaves of papaya, the fruit from which papain is extracted. Papain (and other proteases) may also be used in the brewing industry to prevent **chill hazes**, caused by precipitation of complexes of protein and tannin at low temperatures.

20.3.3 Applications in other industries

The use of enzymes in medical diagnosis and therapy was discussed in section 19.1. To give another example, immobilized enzymes may be components of artificial kidney machines, which are used to remove urea and other waste products from the body, where kidney disease prevents this being done by natural processes. Urea enters the machine from the blood, by dialysis (termed **haemodialysis**), and is converted to CO_2 and NH_4^+ by immobilized urease. Toxic NH_4^+ is then either trapped on ion exchange resins or incorporated into glutamate by the action of immobilized **glutamate dehydrogenase** linked to **alcohol dehydrogenase** to ensure coenzyme recycling, before the fluid is returned to the blood stream.

Some applications of enzymes in the chemical industry were discussed in section 20.3.2, e.g. the production of L-amino acids as possible supplements to foodstuffs.

Enzymes may also be of great value in the pharmaceutical industry, e.g. for the conversion of naturally occurring penicillin G (benzyl penicillin) to 6-amino penicillanic acid (6-APA). This reaction, which is catalysed by bacterial **penicillin amidase**, cleaves the side chain of the substrate at mildly alkaline pH values:

$$\text{C}_6\text{H}_5\text{-CH}_2\text{CONH-(penam: S, CH}_3\text{, CH}_3\text{, N, CO}_2\text{H, O)} \rightleftharpoons \text{C}_6\text{H}_5\text{-CH}_2\text{CO}_2\text{H} + \text{H}_2\text{N-(penam: S, CH}_3\text{, CH}_3\text{, N, CO}_2\text{H, O)} \qquad (20.10)$$

benzyl penicillin — phenylacetic acid — 6-APA

Many commercial processes have been developed for the synthesis of 6-APA, including the use of *E. coli* immobilized in polyacrylamide gel, and the enzyme entrapped in fibres of cellulose acetate. The importance of 6-APA is that new side chains can be attached to give a variety of semisynthetic penicillins, e.g. ampicillin (α-methyl benzyl penicillin), a broad-spectrum antibiotic. This may be done by chemical processes or by the use of the enzyme under slightly acidic conditions, to favour the back reaction.

Enzymes from different bacteria may be applied to catalyse the forward and back reactions, to make use of their slightly different characteristics and specificities. For example, a mutant of *Kluyvera citrophila* has been used for 6-APA production, and one of *Pseudomonas melanogenum* for ampicillin synthesis.

Washing powders incorporating bacterial proteases have been available for many years. Their commercial importance waned because of fears about the effect of enzyme dust on the respiratory system, but this problem was overcome by containment in granules which rupture only on contact with water. The enzymes in question, **subtilisins** from *Bacillus subtilis* mutants, are stable to alkali, high temperature (e.g. 65°C), detergents and bleaches. They will attack blood and other protein stains.

Bacterial proteases are also used in the **leather and textile industries** to loosen hair (or wool) and enable it to be separated from hide.

SUMMARY OF CHAPTER 20

Enzymes are extracted on a large scale mainly from microbial sources, because a wide range of enzymes are available in micro-organisms, and because cells producing the required enzymes can be easily cultivated.

Most fermentation processes are carried out in submerged culture, where the conditions can be carefully controlled. The cells must be grown in the presence of suitable inducers, and in the absence of the relevant repressors, if the required enzyme is to be produced. Most enzymes of commercial importance are extracellular ones: unlike intracellular enzymes, they leak out into the culture medium and so do not require extraction from the cells when the fermentation is finished. The degree to which the enzyme is then purified depends both on economic factors and on the intended application.

Enzymes may be immobilized by being bound to an insoluble carrier (by physical adsorption, ionic binding or covalent binding), by being entrapped in a gel or membrane, or by being cross-linked (often, but not always, in addition to being bound or entrapped). Immobilization can affect the stability, pH optimum, apparent K_m and apparent V_{max} of an enzyme. This depends on the method of immobilization and the nature of the carrier, so these factors must be considered when an immobilized enzyme system is being chosen for a particular application.

Immobilized enzymes are easily removed from a reaction mixture, thereby stopping the reaction and making the enzyme available to be used again. They are ideal for incorporation into continuous processes.

Enzymes have been used for centuries in the baking and brewing industries, as components of yeast cells and malt. More recently, many applications have been found in these and other industries for purified enzymes.

FURTHER READING

As for Chapter 16, plus:

Akkara, J. A., Ayyagari, S. R. and Bruno, F. F. (1999), Enzymatic synthesis and modification of polymers in non-aqueous solvents, *Trends in Biotechnology*, **17**, 67-73.

Arden, N. and Betenbaugh, M. J. (2004), Life and death in mammalian cell culture: strategies for apoptosis inhibition, *Trends in Biotechnology*, **22**, 174-180.

Barredo, J. L. (2005), *Microbial Enzymes and Biotransformations*, Humana Press.

Bell, G., Halling, P. I. *et al* (1995), Biocatalyst behaviour in low-water systems, *Trends in Biotechnology,* **13**, 468-473.
Bickerstaff, G. F. (ed.) (1996), *Immobilization of Enzymes and Cells,* Humana Press
Bisswanger, H. (2004), *Practical Enzymology*, Wiley–VCH (Chapter 5).
Breslow, R. (2005), *Artificial Enzymes*, Wiley–VCH.
Crabb, W. D. and Mitchinson, C. (1997), Enzymes involved in the processing of starch to sugars, *Trends in Biotechnology,* **15**, 349-352.
Dalbøge, H. and Lange, L. (1998), Using molecular techniques to identify new microbial catalysts, *Trends in Biotechnology,* **16**, 265-272.
DeGrado, W. E, Summa, C. M. *et al* (1999), *De novo* design and structural characterization of proteins and metalloproteins, *Annual Review of Biochemistry,* **68**, 779-819.
Deran, P. M. (1995), *Bioprocess Engineering Principles,* Academic Press.
Doyle, A. and Griffiths, J. B. (1998), *Cell and Tissue Culture - Laboratory Procedures in Biotechnology,* Wiley.
Faber, K. (2000), *Biotransformations*, Springer-Verlag.
Fessner, W. D. (2000), *Biocatalysis: From Discovery to Application*, Springer-Verlag.
Gerhartz, W. (2003), *Enzymes in Industry: Production and Applications*, Wiley–VCH.
Godfrey, T. and West, S. (eds.) (1996), *Industrial Enzymology,* 2nd edn., Macmillan.
Guisan, J. M. (2006), *Immobilization of Enzymes and Cells*, Humana Press.
Harrison, R. G. (ed.) (1994), *Protein Purification Process Engineering,* Dekker.
Hou, C. T. (2005), *Handbook of Industrial Biocatalysis*, Dekker.
Hummel, W. (1999), Large-scale applications of NAD(P)-dependent oxidoreductases - recent developments, *Trends in Biotechnology,* **17**, 487-492.
Kirchning, A. (2004), *Immobilized Catalysts*, Springer-Verlag.
Klibanov, A. M. (1997), Why are enzymes less active in organic solvents than in water?, *Trends in Biotechnology,* **15**, 97-101.
Lye, G. 1. and Woodley, J. M. (1999), Applications of *in situ* product removal techniques to biocatalytic processes, *Trends in Biotechnology,* **17**, 395-402.
Martin, I., Wendt, D. and Heberer, M. (2004), The role of bioreactors in tissue engineering, *Trends in Biotechnology*, **22**, 80-86.
Methods in Enzymology, **178** (1989): Antibodies, antigens and molecular mimicry; **135, 136** (1987), **137** (1988): Immobilised enzymes; **152-155** (1987), **216** (1992), **217, 218** (1993); Academic Press.
Nixon, A. E., Ostermeier, M. and Benkovic, S. J. (1998), Hybrid enzymes - manipulating enzyme design, *Trends in Biotechnology,* **16**, 258-264.
Orive, G., Hernández, R. M. *et al* (2004), History, challenges and perspectives of cell microencapsulation, *Trends in Biotechnology*, **22**, 87-92.
Phillips, R. S. (1996), Temperature modulation of the stereochemistry of enzymatic catalysis - prospects of exploitation, *Trends in Biotechnology,* **14**, 13-16.
Pierratos, A. (2004), New approaches to haemodialysis, *Annual Review of Medicine*, **55**, 179-189.
Price, N. C. and Stevens, L. (1999), *Fundamentals of Enzymology,* 3rd edn., Oxford University Press (Chapters 3, 10, 11).

Scouten, W. H., Luong, J. H. T. and Brown, R. S. (1995), Enzyme and protein immobilization techniques for applications in biosensor design, *Trends in Biotechnology,* **13**, 178-185.
Sofer, G. and Hagel, L. (1997), *Process Chromatography,* Academic Press.
Spirin, A. S. (2004), High-throughput cell-free systems for synthesis of functionally active proteins, *Trends in Biotechnology*, **22**, 538-545.
Stanbury, P. F., Whittaker, A. and Hall, S. J. (1995), *Principles of Fermentation Technology,* 2nd edn., Butterworth-Heinemann.
Straathof, A. J. J. and Adlercreutz, P. (eds.) (2000), *Applied Biocatalysis,* 2nd edn., CRC Press.
Street, G. (ed.) (1994), *Highly Selective Separations in Biotechnology,* Blackie
Tischer, W. and Kasche, V. (1999), Immobilized enzymes: crystals or carriers?, *Trends in Biotechnology,* **17**, 326-335.
Uhlig, H. (1998), *Industrial Enzymes and their Applications,* Wiley.
Whitehurst, R. (2002), *Enzymes in Food Technology*, CRC Press.
Wilson, K. and Walker, J. (eds.) (2000), *Principles and Techniques of Practical Biochemistry,* 5th edn., Cambridge University Press.
Wiseman, A. (ed.) (1995),.*Handbook of Enzyme Biotechnology,* 3rd edn., Ellis Horwood.
Zwanenburg, B. (2001), *Enzymes in Action: Green Solutions for Chemical Problems*, Kluwer.

PROBLEMS

20.1 An enzyme with a pH optimum of 7.5 was bound in a positively charged support. Estimate the pH optimum of the immobilized enzyme at 310 K, given that the electronic potential was found to be 0.10 V. Assume that R = 8.314 J K^{-1} mol^{-1} and F = 96487 J V^{-1}.

20.2 For an enzyme-catalysed reaction under specified conditions, K_m = 4.0 mmol l^{-1}, V_{max} = 0.50 mmol l^{-1} min^{-1} and the initial substrate concentration is 20.0 mmol l^{-1}. How long will it take for the product concentration to rise from zero to 5.0 mmol l^{-1}, assuming that the Michaelis-Menten equation is valid throughout?

21

Genomics, Proteomics and Bioinformatics

21.1 ENZYMES AND RECOMBINANT DNA TECHNOLOGY

21.1.1 Introduction

Recombinant DNA technology (sometimes called **genetic engineering)** makes use of a variety of enzymes, particularly restriction endonucleases (from bacteria) and DNA ligases (from bacteria or bacteriophages), to insert extra genes into cells with the help of vehicles termed **vectors**.

One important group of vectors are **plasmids** which are small, circular, cytoplasmic molecules of DNA, acting as extrachromosomal genes in bacteria. Sometimes, but not always, they may be able to attach themselves to chromosomes, when they are known as **episomes**. Plasmids are not essential for the functioning of a particular organism but, when they are present, they confer additional properties upon it (e.g. resistance to certain antibiotics). It is possible to extract and purify plasmids, and to insert extra genes into the circle. These altered plasmids can then be taken up again from the medium by the micro-organism.

For example, the plasmid pBR322 can be cleaved at a single specific site by the *Eco*RI restriction enzyme. This makes a staggered cut in the double helix, leaving complementary single-stranded ends. If the DNA fragment to be inserted was removed from its original location on a large DNA molecule by means of the same enzyme, then it would have single-stranded ends complementary to those of the cut plasmid, and the two can be spliced by means of DNA ligase (Fig. 21.1).

The new circular plasmid so formed can be taken up to a small degree by *E. coli* bacteria, and since the plasmid contains genes which convey resistance to the antibiotics ampicillin and tetracycline, the bacteria containing the plasmid can easily be selected. In this and similar ways, synthetic, prokaryotic and eukaryotic genes can be incorporated into *E. coli* and other bacteria. Then, as the bacterium is grown in culture, the inserted gene will be replicated together with the vector. The production of identical copies is termed **cloning**.

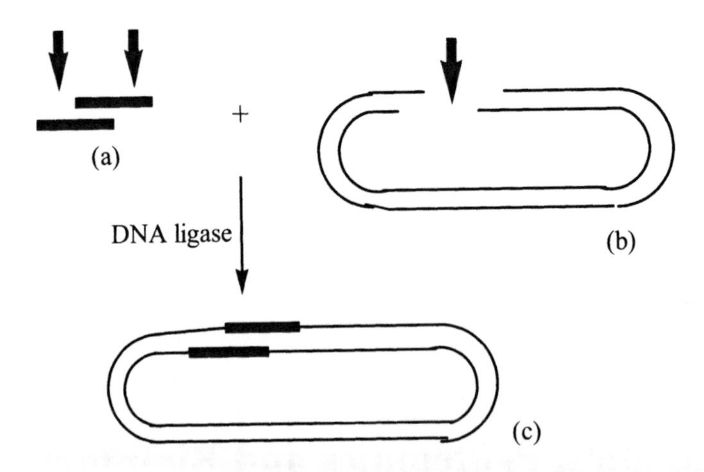

Fig. 21.1 – Simplified procedure for the production of a recombinant DNA molecule (c) by insertion of a gene fragment (a) into a plasmid (b). The gene fragment would have been removed from the chromosome and a cut made in the circular plasmid, by means of the same restriction endonuclease (e.g. *Eco*RI). Sites of cleavage are indicated by thick arrows.

Another group of vectors are variants of a bacterial virus known as **λ phage**. The phage DNA is about 45 kb (i.e. 45 000 base pairs) long, of which the middle third has no role in the infection process and can be replaced by another piece of DNA of about the same length (again by means of restriction endonuclease and ligase enzymes) without affecting the ability of the phage to infect bacteria and be reproduced in abundance. Genes or gene fragments of about 15 kb length can be inserted and cloned in this way, in contrast to the limit of about 10 kb when plasmids are used as vectors. Even so, many eukaryotic genes are longer than this.

Longer genes may be inserted by the combined use of plasmids and phages in the technique known as **cosmid cloning.** When λ DNA is synthesized, multiple copies are linked through the **cos sites** located at both ends of each to form a single long chain. Cleavage subsequently takes place at cos sites 35 to 45 kb apart by the action of λ packaging enzymes, after which the λ DNA is packaged into phage heads. Cosmid cloning involves the insertion of cos sites into a plasmid such as pBR322, after which eukaryotic DNA may also be inserted by means of restriction and ligase enzymes. Ideally a 35-45 kb segment of DNA will be inserted between two cos sites, with an intact tetracycline-resistance gene in close proximity. The DNA segment may then be removed and put into a phage head by λ packaging enzymes, prior to being introduced into *E. coli* by phage attack. Furthermore, if the tetracycline-resistance gene has remained attached, treatment with tetracyline will enable bacteria containing the inserted DNA to be selected and cultured. There is of course insufficient of the original phage DNA present for phage reproduction to take place, but the inserted DNA will behave as a plasmid and be replicated as such.

21.1.2 Applications

Whatever vector is used to insert new DNA into bacteria, this DNA can specify the synthesis of a protein by transcription/translation.

As we saw in section 20.1.1, this can be of value in the **large-scale production of enzymes**, if the gene for a useful but relatively inaccessible enzyme can be incorporated into a bacterium such as *E. coli*. However, a major problem in using this procedure to synthesize eukaryotic enzymes is that eukaryotic genes have regions called introns whose structural information is not expressed during protein synthesis, being lost at the mRNA stage (see section 3.1.4). *E. coli* does not possess systems which can excise non-coding regions, so would synthesize proteins of incorrect structure from eukaryotic genes.

The way round this problem is to start with the appropriate mRNA, in which the structural information from the introns will already have been discarded, and use this to synthesize a **complementary DNA (cDNA)** which can be inserted into the bacterium to produce a protein of the correct structure. For example, a single-strand cDNA possessing a hairpin loop at one end can be synthesized complementary to the mRNA by the use of a viral reverse transcriptase, the process requiring deoxynucleoside triphosphates (dATP etc.) to provide the building-blocks and a short oligodeoxynucleotide primer which can bind to the 3′-end of the mRNA. After it has served its purpose as template, the mRNA may be removed by NaOH, after which the hairpin loop can act as a primer for DNA polymerase I to synthesize the second strand of cDNA complementary to the first. The two strands are still linked by the hairpin loop at one end, but this is easily removed by cleavage with S1 nuclease. However, S1 nuclease may also remove some of the important bases from the cDNA, so an alternative approach is to avoid the need for this enzyme by employing a separate short oligodeoxynucleotide primer to start the synthesis of the second strand. Strangely, in view of the fact that the template is now DNA, reverse transcriptase may be used to catalyse the synthesis of this second strand. In yet another approach to producing cDNA, the mRNA template is not removed by NaOH but treated with ribonuclease to give rise to random nicks in the chain. DNA polymerase I can then remove the remaining ribonucleotides, because of its $5' \rightarrow 3'$ exonuclease activity, and replace them with the corresponding deoxy-ribonucleotides. Finally the breaks in the chain are sealed by DNA ligase.

Obviously, the source material would be one known to contain the appropriate mRNA but, even so, many other mRNA molecules would be present. Usually, therefore, a **cDNA library** is cloned from the mixture of mRNA molecules, and the required cDNA then selected by use of a suitable **probe** possessing a complementary base sequence. For example, after bacteria have been allowed to take up plasmids containing different components of the cDNA library, they are spread on an agar plate and cultures grown. A nitrocellulose sheet is then placed over the plate, so that part of each colony is transferred and part left behind. The nitrocellulose sheet is treated with NaOH and heated to lyse the bacteria and denature and bind the DNA. A ^{32}P-labelled probe (e.g. a previously purified single-strand cDNA or mRNA) which will bind only to complementary sequences in the required cDNA, is then applied, and the location of the bound radioactive probe determined by **autoradiography**, using an X-ray film. The culture in the corresponding position on the agar plate is thus identified as the one to use for further cloning of the required cDNA.

Providing the primary structure of the desired protein is known, a **synthetic oligodeoxynucleotide** can be used as a probe, although difficulties could arise because several base sequences code for the same amino acid (see Fig. 3.5). Oligodeoxynucleotides of specific structure are easily and quickly synthesized by the solid-phase **phosphoramidite method**. Each cycle of this involves: removal (by acid) of the dimethoxytrityl protecting group at the 5′-end of the growing chain; coupling (with the aid of tetrazole) this liberated 5′-end to the 3′-phosphoramidite derivative of the next deoxynucleoside to be added; and oxidizing the phosphate triester group so formed to the required phosphotriester. Commercial systems are available for this purpose, e.g. the Applied Biosystems 3400 DNA Synthesizer.

So, cDNA corresponding to the required protein may be obtained in a variety of ways. However, even if this cDNA rather than the complete gene is inserted into *E. coli* via a suitable vector, the final structure of the protein produced may still not be quite correct, because of the inability of the bacterium to bring about required post-translational modifications such as addition of sugar units. A partial solution is to clone DNA into **shuttle vectors**, plasmids that can replicate both in *E. coli* and in yeast cells. Since yeast cells are eukaryotic, they are more likely to be able to achieve a reasonable if not total degree of success with the post-translational modifications. Thus, calf chymosin (see section 20.3.2) has been produced by inserting the appropriate DNA into the yeast *Kluyveromyces* sp. as well as into *E. coli.*

Recombinant DNA techniques may also be used in **genomics**, i.e. the **scientific investigation of the genome** (the total genetic material in a cell), which dominated biochemistry during the latter part of the twentieth century. However, if a particular investigation of genetic structure or function involved a requirement to insert and clone an entire eukaryotic gene, it would be unlikely that any of the vectors considered in section 21.1.1 could cope with the size of the insert. Eukaryotic genes may be several hundred kb in length, this mostly consisting of intron regions.

Yeast artificial chromosomes (YACs), linear DNA segments derived from plasmids, into which all the basic functional units of yeast chromosomes (centromeres, telomeres and autonomous-replication sequences) have been incorporated, are one answer to this problem. DNA segments ten times bigger than could be cloned using other vectors can be inserted into YACs and successfully replicated in yeast. YACs made an important contribution to the **Human Genome Project**, which was started in 1988 with the aim of determining the structure and chromosomal location of every human gene, and successfully completed in 2000.

The genome may be investigated in a variety of ways. Treatment with a restriction enzyme will cause cleavage at specific sites and the production of DNA fragments which can be separated according to size by electrophoresis in agarose gels. DNA molecules from two individual humans differ in sequence on average by one base pair every 200-500, and a difference in sequence at any point may add or delete a cleavage site for a particular restriction enzyme, resulting in the production of a DNA fragment of different size. When this occurs it is called a **restriction-fragment length polymorphism (RFLP)**. It may be investigated in easily accessible cells such as white blood cells.

Treatment of the genome with a restriction endonuclease would produce something in the order of a million DNA fragments, so analysis by electrophoresis alone is clearly not practicable. However, even without knowing their location in the genome, genes of interest may be investigated by the use of probes, in ways similar to those discussed above for the selection of cDNA molecules. Edwin Southern (1975) showed that one or more fragments corresponding to a particular gene could be identified by transferring the electrophoresis products to a nitrocellulose sheet (a process subsequently termed **Southern blotting**) and treating with an appropriate ^{32}P-labelled probe (e.g. cDNA). Autoradiography may then be used to reveal which fragments have sequences complementary to parts of the probe, and the size of these fragments can be estimated by reference to their positions on the electrophoresis strip (Fig. 21.2).

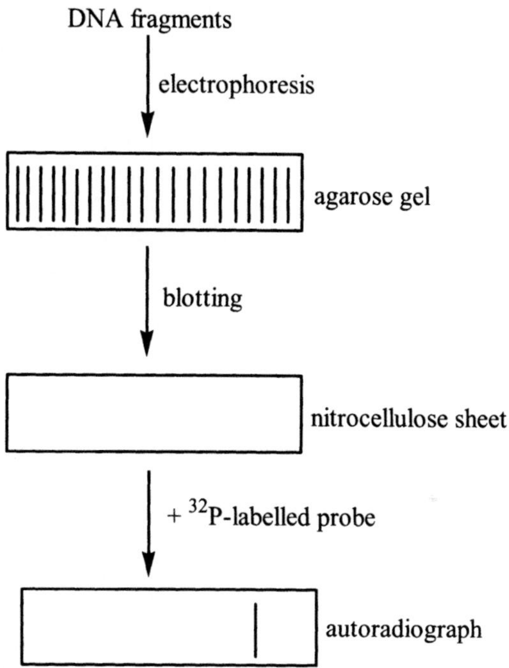

Fig. 21.2 – The Southern blotting procedure. Restriction fragments of DNA are subjected to agarose gel electrophoresis. (The bands produced may be visualized by staining, for example, with ethidium bromide.) The separated fragments are transferred to a nitrocellulose sheet, and treated with a ^{32}P-labelled probe (e.g. cDNA). Autoradiography reveals which of the fragments have sequences complementary to, and hence able to bind, the radioactive probe.

The introduction of this technique provided an important new approach to the **diagnosis of inherited diseases** (see section 19.1.2). For example, Judy Chang and Yuet-Wai Kan (1982) showed that treatment with restriction enzyme *Mst*II produces 0.2 and 1.15 kb fragments from the normal β-globin gene, but a single 1.35 kb fragment from the same gene in **sickle cell anaemia** (see section 2.4.5). Presumably an *Mst*II cleavage site is lost as a direct consequence of the mutation.

An alternative but related approach is to investigate patients' mRNA instead of restriction-fragments of DNA: this is termed **northern blotting** to distinguish it from Southern blotting. For example, Tsung-Sheng Su, Arthur Beaudet and William O'Brien (1982) showed that a cDNA probe corresponding to the gene for argininosuccinate synthase identified a 1550 base mRNA in extracts from normal skin fibroblasts, whereas in those from some patients with **citrullinaemia**, a deficiency of argininosuccinate synthase inherited by an autosomal recessive mode, two bands, of length 240 and 1240 bases, were detected. All three bands were found when heterozygotes for the disease were investigated. However, in contrast to sickle cell anaemia, which is a specific disease, genetic heterogeneity in most inherited disorders is likely to mean that not all affected families will show exactly the same features, and this is indeed the case with citrullinaemia. Hence, when specific diagnostic features have been established in a family by means of techniques involving Southern or northern blotting, these can be applied to any member of the same family with a high degree of certainty, but cannot automatically be assumed to be applicable to a member of an unrelated family.

It is convenient when a mutation causing an inherited disorder also creates or destroys a cleavage site for a restriction enzyme, but this will not necessarily be the case. However, that does not prevent restriction enzymes from being used in the diagnosis of such disorders. Most of the eukaryotic genome consists of non-coding sequences, and it appears that these regions can tolerate mutations more easily than exon regions, allowing such mutations to become well established in populations. Hence RFLPs resulting from mutations in non-coding regions are relatively common. This, allied to Southern blotting techniques, enables a section of the chromosome which includes a particular gene sandwiched somewhere between two cleavage sites to be tracked through a family, from one generation to another. If the gene contains a mutation responsible for an inherited disorder, there should be an association between the inheritance of the disorder and of a particular distance between cleavage sites. The use of DNA probes to carry out such linkage studies is termed **gene tracking**; the situation discussed earlier, where probes are used to obtain direct evidence of mutations, is termed **gene detection**.

Gene tracking is most easily carried out for X-linked diseases, where males only have one version of the relevant chromosome, thus reducing complexities. For example, Rima Rozen and colleagues (1985) investigated **ornithine carbamoyl transferase deficiency** (see section 19.1.2) by Southern blot techniques after treatment of DNA with the restriction enzyme *Msp*I and found two sets of polymorphic bands in affected males, giving four combinations in all, which they assigned as haplotypes. Haplotype A showed 6.6 and 5.1 kb bands; haplotype B 6.6 and 4.4 kb bands; haplotype C 6.2 and 5.1 kb bands; and haplotype D 6.2 and 4.4 kb bands. Hemizygous males could carry only one haplotype, whereas females would either have two of the same (e.g. AA) or two different (e.g. AB) haplotypes. Gene tracking studies showed that the disease was associated with haplotype B in one family, but with haplotype D in a different family.

Investigation of autosomal recessive conditions is inevitably more complicated. Many different haplotypes are possible at the phenylalanine hydroxylase locus, based on combinations of fragments obtained by digestion of the DNA with eight different restriction enzymes, but only a relatively small number are seen in patients with **phenylketonuria**, an inborn error of this enzyme. Anthony DiLella and colleagues (1987) found that over half of northern Europeans suffering from this disorder carried just two of these haplotypes (2 and 3).

RFLP analysis has also contributed to important advances in **forensic science** (see section 19.2). At various sites on human chromosomes **(minisatellite regions)** are non-coding sequences that are repeated many times. The number of such **tandem repeats** varies greatly from person to person, and so the length of restriction-fragments from such a region of the chromosome will also vary from person to person, except in the special case of identical twins. Therefore, probes to detect **variations in the number of tandem repeats (VNTRs)** can be used to identify individuals. Alec Jeffreys and colleagues (1985) obtained a probe for the minisatellite region of the human myoglobin gene (which forms a separate, or satellite, band to the main DNA peak on buoyant density ultracentrifugation) and used it in a Southern blotting procedure with human DNA treated with the restriction enzyme *Hinf*I. They found that the probe detected many other such regions, all of different length, to produce a series of bands unique to an individual (a **DNA (or genetic) fingerprint**). Later, they developed probes specific for particular loci which would produce just two bands, one derived from a chromosome inherited from the father and the other from the equivalent chromosome inherited from the mother. In this way, a hair root, or a trace of blood or semen, found at the scene of a crime can be identified with a high degree of certainty as coming from a particular individual.

Very little source material may be available, but here as in an increasing number of situations the DNA present may be rapidly amplified by means of a powerful procedure devised by Kary Mullis (1983), the **polymerase chain reaction (PCR)**. In its original form, two oligodeoxynucleotide primers are synthesized to bind to the ends of the sequence to be copied (up to 6 kb apart), one primer to each DNA strand. This binding takes place after the DNA strands have been separated by heat treatment (to something like 95°C) and allowed to cool (to about 50°C). A heat-stable DNA polymerase from *Thermus aquaticus* (Taq) is added, together with the four deoxynucleoside triphosphates, and new chains complementary to the DNA strands are synthesized in a 5′ to 3′ direction at about 70°C, adding to the 3′-end of each primer (see Fig. 21.3). When it is thought that synthesis of the required sequence is complete, the DNA strands (now four of them) are separated by heat, and then the cycle starts all over again when the system cools. Because of its heat-stable nature, no new enzyme has to be added, and each new cycle can be initiated simply by changing the temperature, e.g. using an Applied Biosystems DNA Thermal Cycler. The process is exponential, so a million copies will be made in about 20 cycles, each cycle lasting just a few minutes. In the multiplex PCR procedure, more than one pair of primers are introduced, allowing a number of gene sequences to be amplified at the same time.

One exciting possibility for the PCR procedure is to amplify traces of DNA obtained from bodies of extinct animals. Already such a process has been carried out successfully with DNA extracted from mammoths and from human bones tens of thousands of years old. There has also been much speculation about the application of the technique to DNA from insects preserved in amber for millions of years, including blood cells from animals the insects may have fed upon. The idea that dinosaur DNA may be obtained and amplified in this way was central to the *Jurassic Park* film trilogy.

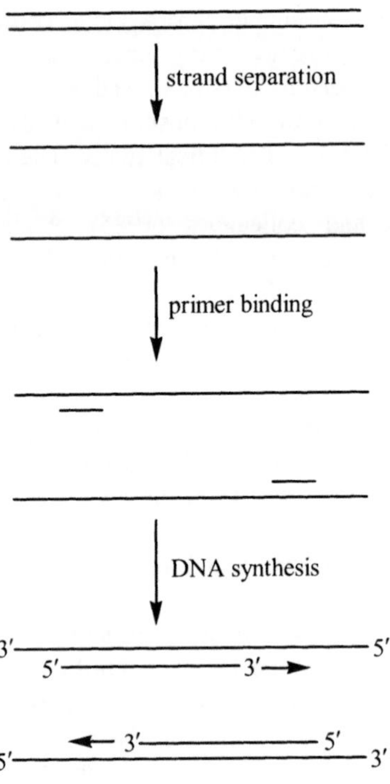

Fig. 21.3 – Diagrammatic representation of the three stages in each cycle of the polymerase chain reaction (PCR) for DNA amplification. Each stage takes place at a different temperature, so the process may be controlled by temperature programming. A million copies of a section of double-stranded DNA may be produced in just 20 cycles.. Each separated strand serves as template for the synthesis in the 5′ to 3′ direction of a complementary DNA strand, catalysed by DNA polymerase.

Information about **evolutionary relationships** may be obtained by applying restriction enzyme techniques to DNA from a variety of races and species, living or dead. For example, James Wainscoat and colleagues (1986) investigated DNA polymorphisms in the β-globin gene cluster of individuals from eight diverse human populations: North European, Cypriot, Italian, Indian, Polynesian, Melanesian, Thai and African. They found that non-Africans showed a number of common haplotypes, while the Africans had predominantly a different haplotype. This is consistent with the view of some, but not all, palaeoanthropologists that modern humans arose in Africa, and a group migrated from that continent to give rise to all the other populations.

Similarly, a study by Rebecca Cann and colleagues (1987) of mitochondrial DNA, which is inherited solely through the maternal line, was interpreted as showing that all living humans are descended from a woman who lived in Africa around 200 000 years ago. An investigation by Svante Pääbo and colleagues (1997) of DNA extracted from fossil bones and amplified by PCR apparently confirmed that modern humans are not descended from Neanderthals, who were well-established in Europe before the supposed African exodus could have taken place.

If the **complete base sequence** of a gene is to be determined, the first requirement is to clone the gene. YACs (see above) offer the possibility of being able to clone a eukaryotic gene as a complete unit but, with other vectors, this has to be done in a series of fragments. The usual procedure is to cleave the genome with one or more restriction enzymes in an incomplete fashion so that fragments with overlapping sequences will be produced. These are all cloned (the process has been termed shotgun cloning) to form a **genomic DNA library**. Part of the gene of interest is then identified by means of a specific probe, as discussed previously for selection from a cDNA library.

When a section of the gene has been identified, cloned further and purified, some of the sample may in turn be used as a probe to detect overlapping sequences on the next section, and so on. This is termed **chromosome walking**.

The base sequence of a purified section of gene may be determined, for example, by the chain-terminator method of Fred Sanger. Single-strand DNA is required, and this is readily available by cloning in **filamentous bacteriophage M13**, a vector whose DNA is circular but single-stranded. The DNA is isolated and allowed to link with an oligodeoxynucleotide primer which has a sequence complementary to the phage DNA just to the 3′-side of the inserted fragment. The sequencing procedure may then be carried out as outlined in section 3.1.6.

The sequence of bases in a section of gene amplified by the PCR technique provides a **sequence tagged site (STS)**, to act as a marker on the larger DNA molecule. An STS derived from a cDNA molecule can be used as an **expressed sequence tag (EST)**, to identify a section of a gene whose encoded information is expressed.

As an example of sequencing following gene cloning, Anthony DiLella and colleagues (1987) investigated DNA structure in the two common haplotypes associated with phenylketonuria in northern Europe. They showed that the mutation in patients homozygous for haplotype 2 is invariably a C to T transition in an expressed region, resulting in the substitution of tryptophan for arginine at residue 408 of phenylalanine hydroxylase. With haplotype 3, the mutation is always a single base substitution at a specific site in an intron region. The surprising finding that two single mutations are so well established in the population could not have been made without the use of restriction enzymes, and its significance is still uncertain. Subsequently, PCR techniques were utilized in the investigation of mutant genes at the molecular level. Thus, Yoshiyuki Okano and colleagues (1991) demonstrated a correlation between various specific mutations of the phenylalanine hydroxylase gene and the severity of phenylketonuria, in patients from Denmark and Germany. Similar findings were later obtained by others, e.g. Sang-Wun Kim, Jongson Jung and colleagues, who in 2006 reported the results of a study of Korean patients.

With the development of the gene cloning and PCR techniques, it has now become much more feasible to investigate mutant genes directly, rather than have to make deductions from gene tracking studies. Thus, a screening procedure now exists which, from a single drop of blood, can detect the mutation ΔF508, the most common cause in European populations of **cystic fibrosis**, an inherited disorder affecting ion transport across membranes.

The techniques of DNA sequencing are now widely used as an **indirect means of determining the primary structure of a protein** (see section 3.1.6). For example, a cDNA molecule may be synthesized from mRNA, as described above, and the base sequence of this is directly related, by means of the genetic code, to the amino acid sequence of a protein. If there are any doubts about the identity of the protein, the cDNA may be used as a probe to bind the corresponding mRNA which, after purification, is used to specify the synthesis of the protein, and allow it to be characterized.

Recombinant DNA techniques are also used in **site-directed mutagenesis** (see section 10.1.5). For example, the relevant gene may be cloned in phage Ml3, and the single-strand circular DNA isolated. An oligodeoxynucleotide is synthesized with a sequence complementary to that of a section of the gene, except for one or two bases, which have been deliberately changed. This oligodeoxynucleotide is then allowed to bind to the gene, and serves as a primer for the synthesis of a complementary strand of DNA, catalysed by the Klenow fragment of *E. coli* DNA polymerase I, as discussed above.

When synthesis is complete, the circle is closed by means of DNA ligase and used to infect *E. coli.* Both normal and mutated DNA will be cloned in this way, but the latter can be selected by its greater ability to bind to the original oligodeoxynucleotide, because of its perfect match, under conditions of high stringency (temperature and salt concentration). It can then be used for further cloning, and for the production of a protein whose structure will be different from the natural form because of the mutation.

Site-directed mutagenesis is an essential feature of **protein engineering**, to enhance the usefulness of enzymes in science, medicine and industry by modifying specificity, regulation, stability or other characteristics. Trypsin, subtilisin, carboxypeptidase, lactate dehydrogenase and alcohol dehydrogenase are amongst the enzymes that have been ‘engineered’ in this way. So, for example, lactate dehydrogenase from *Bacillus stearothermophilus* may effectively be converted into a malate dehydrogenase by replacing threonine-246 with glycine and glutamate-l02 with arginine; while replacement of arginine-173 by glutamine in the same enzyme results in the removal of a requirement for activation by fructose-1,6-bisphosphate. Similarly, subtilisin BPN′ from *Bacillus amyloliquefaciens* is made more stable if a cysteine, serine or alanine residue is made to substitute for methionine-222.

Again, as demonstrated in 2000 by Myriam Altamirano and colleagues, modification of the α/β-barrel scaffold (see section 11.3.5) of indole-3-glycerol-phosphate synthase can give rise to a new phosphoribosylanthranilate isomerase whose catalytic properties are similar to those of the natural enzyme, but with even greater specificity.

Site-directed mutagenesis may also be extremely useful in investigating the active sites of enzymes (see sections 10.1.5 and 10.2), and can be employed to study the consequences of inherited disorders. Thus, DiLella and colleagues (see above) used the technique to confirm that an amino acid substitution at position 408 of phenylalanine hydroxylase produces enzyme characteristics consistent with those seen in patients with phenylketonuria.

Recombinant DNA technology also offers possibilities for the treatment of inherited disorders by **gene therapy**, although this is still at an experimental stage. For example, Richard Mulligan and colleagues (1989) inserted the human adenosine deaminase (ADA) gene into mouse haemopoietic stem cells, with the aid of a retrovirus, and then incorporated these manipulated cells into mice by means of bone marrow transplants. Six months later, they found that various mouse blood cells were still synthesizing ADA. It seemed possible that a similar procedure could be used to treat ADA deficiency in humans, which is a cause of **severe combined immunodeficiency (SCID)**. Trials in humans started in 1990, using an inactivated retrovirus as vector to introduce the normal ADA gene into a patient's T lymphocytes. However, most of the patients failed to show any significant clinical improvement.

Better results were reported in 2000 by Marina Cavazzana-Calvo, Salima Hacein-Bey and colleagues for a gene therapy trial involving patients with a different form of SCID, an X-linked condition caused by a γ-chain deficiency. The retroviral transfer of a normal γc gene into bone marrow cells of patients with this disorder had apparently cured 9 of the 10 patients in the trial. Unfortunately, by 2003, two of the treated children had developed T-cell leukaemia, and one subsequently died. As the leukaemia was most likely a consequence of the application of retroviruses in the trials, much work is now being carried out to find suitable non-viral vectors or other procedures which can be used safely as well as effectively in human gene therapy. Much remains to be done, but hopes of eventual success remain high.

In **agriculture** too, genetic engineering of this kind, although controversial, could bring about significant advances, such as making plants resistant to viruses and insect pests. This, for example, would reduce the necessity of having to spray crops with toxic insecticides.

Notwithstanding the progress over the past 25 years, which has resulted in Nobel Prizes for Werner Arber, Paul Berg, Kary Mullis, Michael Smith and others, the use of restriction endonucleases and other enzymes in recombinant DNA technology is clearly just beginning.

21.2 PROTEOMICS

21.2.1 The application of mass spectrometry to the investigation of the proteome

As we noted in the previous section, the total genetic material in a cell is termed the **genome**, and the study of the genome is called **genomics**. Similarly, the aggregate of all the expressed proteins in a cell is known as the **proteome**, and the study of the proteome is termed **proteomics**.

A key technique for the identification and sequence analysis of proteins and peptides in proteomics is **mass spectrometry (MS)** (see section 2.4.3), used in conjunction with 2-dimensional polyacrylamide gel electrophoresis (2D-PAGE), chromatography or capillary electrophoresis (see section 21.2.2).

A mass spectrometer produces gas ions from a sample, separates them according to their mass-to-charge ratio (*m*/*z*) and records the relative abundance of each of the ions to obtain a mass spectrum. There are many different MS methods for the generation and detection of ions but, in general, most are restricted to analysing ions with a *m*/*z* of less than 2000, which limits their application to biological macromolecules. **Electrospray ionization (ESI)** (see section 11.3.4) and **matrix-assisted laser desorption/ionization (MALDI)** are the most widely-used techniques for the production of gas phase ions from biological molecules in the liquid and solid phase respectively.

ESI has proved to be a very useful technique for the analysis of biological molecules in that they can be extracted from solution, ionized, transferred to the gas phase and then subjected to mass analysis. In ESI, a solution of sample is delivered to the tip of a needle, which is maintained at a high potential (1-5 kV). The electrical field at the needle tip charges the surface of the emerging solvent and sample flow, producing a fine aerosol of charged droplets. The droplets begin to reduce in size through evaporation until they reach the Rayleigh limit (which occurs when droplet surface charge density equals the liquid surface tension); at this point, the droplet breaks up as a result of electrostatic forces and smaller droplets are produced. The process is repeated until a further decrease in the size of the droplets results in the formation of highly-charged positive or negative molecular ion species. The ions formed are sampled into the mass spectrometer vacuum system and directed towards the mass detector. Under ESI, macromolecules such as proteins and peptides receive protons at their basic sites, yielding ions with multiple charges. The resultant *m*/*z* ratios fall within the mass range of simple mass analysers such as quadrupoles (generally *m*/*z* <2000), although time-of-flight (TOF) and hybrid (quadrupole and TOF) detectors are also used. The requirement that a peptide or protein is in an ionized state in solution prior to transfer to the gas phase makes the ESI technique compatible with reversed-phase HPLC and capillary electrophoresis (CE). ESI is a concentration-dependent technique and improvements in the levels of sensitivity can be achieved using nanospray ionization sources coupled to capillary HPLC/CE columns operating a low flow-rates (e.g. 200 nl min^{-1}).

In MALDI, samples are mixed and crystallized with an excess of matrix (e.g. 2,5-dihydrobenzoic acid or α-cyano-4-hydroxycinnamic acid) which acts as a chromophore for laser irradiation (usually a nitrogen laser which pulses every 3 ns at 337.1 nm). The major effect of the laser is to supply heat to the matrix, some of which transfers to the gas phase containing neutrally and predominantly singly-charged sample ions. Sensitivity of detection in MALDI lies in the fact that the ions are formed in a vacuum and then transferred through to a TOF detector. The pulsed mode of operation of TOF analyzers is also compatible with the pulsed production of ions in MALDI and does not have the same mass range limit of quadrupole detectors.

Both MALDI and ESI are routinely used in the analysis of the proteome (see section 21.2.2) but, because MALDI is less susceptible to interference by salts and detergents, it is therefore routinely used for tryptic peptide identification following 2D electrophoresis. The peptide masses obtained can be fed into a tryptic fragment database for possible identification of the polypeptide. The exact sequence of the peptide fragment can be ascertained in a **mass spectrometer with tandem mass capability (MS-MS)**. Ions of a known mass number selected by the primary mass analyser can be subjected to additional collisions resulting in decomposition of the ion, generating a ladder of sequence ions which can be detected by a second mass analyser (Fig. 21.4).

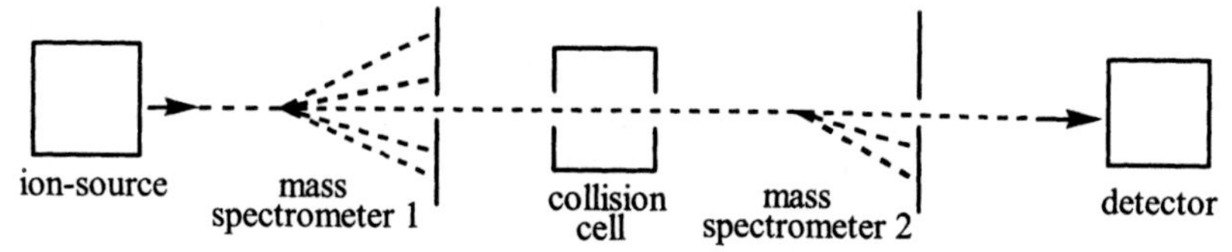

Fig. 21.4 – The main features of tandem mass spectrometry (MS-MS).

In such a process, the most frequently-observed ions are the N-terminal type b_1 acylium ions (NH_2. CHR′CO…NHCHR″CO^+) or the C-terminal ions, y ions ($^+NH_3$.CHR″CO…NHCHR′CO_2H) formed by H rearrangement on the carboxyl terminus (see Fig. 21.5). The masses of these fragments can be used to construct the amino acid sequence of the fragment peptide and a database can be used to search for the parent protein's identity (see section 21.2.2).

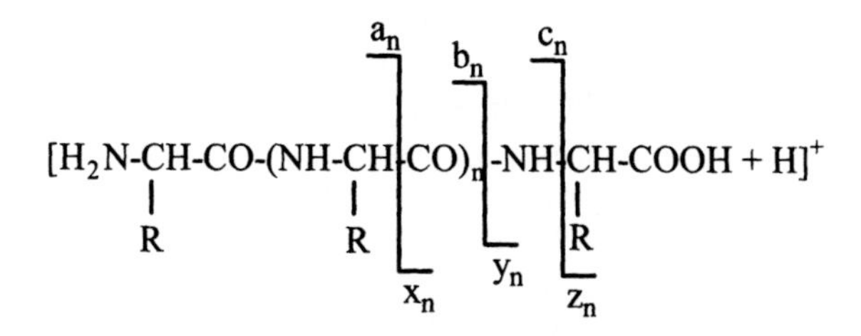

Fig. 21.5 – The common fragment ions that are produced in mass spectrometry when peptide ions are subjected to collisionally-activated dissociation (CAD).

A problem associated with ESI and MALDI is quantification of the data. MALDI produces a heterogeneous mixture of matrix and sample which may separate during crystallization and, although the spectra from ESI are more reproducible, the signal is not necessarily proportional to the concentration of sample in solution. The use of **isotopically-coded affinity tags (ICAT)** such as ^{13}C to differentially-label proteins can overcome these problems.

21.2.2 Proteomics research

During the genomic era (see section 21.1.2), research focused upon collecting sequence data from the genome, providing the analysis that the human genome contains only approximately 30 000 different genes.

However, it has been estimated that there are approximately 400 000 different proteins, due to alternative splicing of genes and many different post-translational modifications e.g. glycosylation and phosphorylation. The genome provides the information to manufacturer a protein, but a study of the genome (genomics) cannot provide information on the activity or function of that protein once it has been synthesized. Proteomics is the study of all the proteins present within a cell at a given time, including their post-translational modifications, protein-protein interactions and differential display (e.g. what proteins are unregulated in various disease-states).

To isolate and characterise the vast number of proteins present in proteomic research, a number of different approaches have been used. **Two-dimensional polyacrylamide gel electrophoresis (2D-PAGE)** has been a popular choice for proteomic analysis. Complex protein mixtures are separated in the first dimension on the basis of their surface charge (pI) using isoelectric focusing (see section 16.2.2) and in the second dimension by their relative molecular mass using SDS-PAGE. The technique has sufficient resolving power to separate thousands of different proteins in a single experiment and resolved polypeptides can be visualised by stains (silver or fluorescent) or immunoblotting (see section 15.3).

A 2D gel image of the proteins present in a disease-state sample can be compared to the 2D gel image of the proteins present in a control sample using computer-based image analysis software (e.g. Nonlinear Dynamics Phoretix 2D-Evolution; Bio-Rad PDQuest; PerkinElmer ProFINDER; or GE Healthcare DeCyder). To identify proteins which have been differentially expressed, protein spots from the 2D gel can be isolated, digested with trypsin and the masses of the peptide fragments measured using mass spectrometry (ESI-MS or MALDI). To obtain the tentative identification of the protein in a 2D gel spot, the peptides' masses can be fed into a tryptic fragment database (containing the theoretical tryptic fragments of all the proteins in that database, e.g. MS-Fit, MOWSE, ProFound, PepSea or PepIdent). Problems can arise from the complexity of the sample, especially when two peptide fragments have the same mass but different sequence (e.g. ABCD or BACD) or due to post-translational additions to amino acids in the peptide sequence. The use of MS-MS can reveal each tryptic peptide's amino acid sequence (including post-translational modifications, if required) and searches using amino acid sequence databases (e.g. Mascot, SEQUEST or Sonar ms/ms) can identify the target protein. However, certain groups of proteins (high or low pI, M_r <10,000 and >100,000) can be difficult to identify using 2D-PAGE without prior enrichment.

A major problem in proteomics is the analysis time required to process the data produced. The complexity of the data can be reduced by **surface enhanced laser desorption/ionization (SELDI)**, which binds proteins to a surface-activated chip (e.g. IEX, affinity, IMAC, immuno- or reversed-phase). Unbound proteins are washed from the surface and, after the application of a matrix solution, the presence of bound proteins can then be analysed. Patterns of fragments generated have been used as a means of diagnosing disease-states and is useful for the analysis of proteins of low M_r. Different combinations of chromatography (e.g. reversed-phase; IMAC + reversed-phase; and IEX + reversed-phase) and capillary electrophoresis prior to mass spectrometry have also been used to analyse complex protein mixtures.

21.3 ENZYMES AND BIOINFORMATICS

21.3.1 Introduction

In addition to the major advances which have taken place in enzymology in recent years as a result of developments in DNA technology and proteomics, significant progress has also been made, and will continue to be made, because of the rapid increase in the power of computer technology. Indeed, so much information has become available from laboratories investigating the structure and function of genes and proteins, that this could not have been adequately stored, communicated and made use of without the parallel developments in computing.

Apart from the enormous amount of DNA sequence data collected during the course of the Human Genome Project (see section 21.1.2), the complete sequences of the genomes of the bacteria, *Haemophilus influenzae* and *Mycoplasma genitalium,* were announced in 1995, and that of the yeast, *Saccharomyces cerevisiae,* in the following year. In total, sequence databases now contain more than a billion base pairs.

Bioinformatics is the name generally given to the collection, storage, communication, analysis and interpretation of biological data. Some prefer to regard the processing aspects as belonging to a separate field, termed **computational biology**, but, regardless of that, they are heavily dependent on the information in the databases, communicated by means of the **Internet**.

One of the largest databases of DNA sequence information is maintained at the **European Molecular Biology Laboratory (EMBL)** Outstation, Cambridge. This can be accessed through the **World Wide Web** pages of the **European Bioinformatics Institute** http://www.ebi/ac/uk). Also linked to the European Bioinformatics Institute is the **SWISS-PROT** annotated protein sequence database, which is available on the ExPASy (Expert Protein Analysis System) proteomics server of the **Swiss Institute of Bioinformatics** (http://expasy.cbr.nrc.ca/sprot). For three-dimensional structures of proteins, determined by X-ray crystallography and NMR, the major international database is the **Protein Data Bank (PDB)**, operated by the **Research Collaboratory for Structural Bioinformatics (RCSB)**, supported by the U.S. National Science Foundation. The PDB can be accessed at http://www.rcsb.org/pdb).

Similarities ("homologies") between a newly-obtained **open reading frame** (an expressed sequence) and structures in **expressed sequence tag (EST)** libraries (see section 21.1.2) held on databases can be searched for and assessed by means of programs such as **BLAST** and **FASTA**, both obtainable through the European Bioinformatics Institute. In this way, for example, an unexpected connection was made between the enzyme, **peptidyl-prolyl isomerase**, which is involved in protein folding, and the protein, **cyclophilin**, which has a role in immunosuppression. Predictions about three-dimensional protein structure can be made on the basis of sequence data, and structures can be visualised using programs such as **Rasmol**.

Turning to applications, comparisons of closely-related structures could help to predict the characteristics of a variant form of an enzyme, possibly one produced by site-directed mutagenesis (see section 21.1.2) for use in biotechnology. Similarly, such comparisons could indicate which amino acid residues were essential to the functioning of a particular enzyme, and which could be varied.

This could be used, for example, in the production of drugs to combat the **HIV-l virus**, generally though to be the cause of **acquired immune deficiency syndrome (AIDS)**. HIV-l is well-known for its ability to become resistant to drugs through mutation, so sites on viral proteins which are essential to function, and hence likely to be conserved while others change, could be effective targets for therapeutic drugs. With that in mind, comparative studies have been carried out on the structures of variant forms of a protease from the HIV-1 virus.

Another example of a practical application of bioinformatics concerns **cathepsin K**, a serine protease involved in bone resorption.

Searches of EST databases by Fred Drake and colleagues (1996) showed that, in contrast to other libraries, a significant number of clones from a human osteoclast library expressed the cathepsin K gene. This suggested that inhibitors to cathepsin K might be effective in the treatment of diseases such as osteoporosis, which are characterized by excessive bone loss. Clearly, such studies are still in their early stages. Nevertheless, there is every reason to think that bioinformatics will be of increasing importance in the study of enzymes and their applications.

21.3.2 Systems biology and microarrays

The large amount of information gained from genomic, proteomic and bioinformatic studies has allowed scientists to take a wider picture of the interactions within cellular biology. **Systems biology** has been established to investigate the integration of the gene and protein networks involved in cell signalling, metabolism and communication between cells. These biological processes within the cell are influenced by many external and internal parameters which control the levels of mRNA expression and the interactions between different proteins. The discovery of protein biomarkers has been the province of 2D-PAGE and/or mass spectrometry (see section 21.2). However, neither technique is particularly adept at processing in a high-throughput format the small amounts of sample that may be available from clinical samples. **Microarray technology** allows the simultaneous analysis of thousands of parameters within a single experiment. Thousands of different probes can be immobilized as surface spots (150-200 μm) on a template, enabling their simultaneous interaction with cellular targets in a format which requires little sample and allows rapid analysis.

DNA microarrays (DNA microchips) have been used to great effect in the analysis of gene expression in a living cell, providing information on gene and drug discovery as well as disease diagnosis. The arrays are constructed robotically by placing cDNA (oligonucleotide) probes from different genes in a predetermined pattern onto a glass or nylon plate. The mRNA isolated from a sample can be converted into fluorescently-labelled cDNA and incubated on the microarray plate. Unbound DNA can be removed by washing and the presence of complementary bound DNA visualised using an appropriate laser. The cDNA produced from the mRNA of two different treatments or disease states can be differentially labelled with red or green fluorescent dyes before being mixed and incubated on the microarray. Following washing, the plate can be sequentially scanned with a red and green laser. The information obtained can then be combined to reveal both common and unique gene expression.

However there is no absolute correlation between mRNA levels and the corresponding protein levels. **Protein microarrays** have been developed to provide some insight into global changes in protein interactions. Protein microarrays consisting of recombinant proteins or peptides fragments have been used to screen patient blood sera for antibodies to cancer or autoimmune diseases. The recombinant proteins or peptide fragments can be spotted onto a plate and the interaction with the antibodies present in the sera of patients can then be visualised by a fluorescently-labelled secondary antibody. Potential enzyme targets (e.g. kinases) can also be screened by a similar approach using radiochemical, fluorescent, chemiluminescent and mass spectrometric techniques as detection methods. Related techniques include antibody microarrays and reverse protein microarrays, which have been used to determine a variety of different cellular protein interactions.

SUMMARY OF CHAPTER 21

Restriction endonucleases and other enzymes are essential components of recombinant DNA technology. This, together with advances in mass spectrometry and computing, has led to highly-significant developments in genomics, proteomics and bioinformatics, which has greatly expanded our knowledge of protein interactions within the cell. In the future, the challenge for scientists will be in the integration of that knowledge to provide major advances in science, medicine and industry.

FURTHER READING

As for Chapter 3, plus:

Akay, M. (2006), *Genomic and Proteomic Engineering in Medicine and Biology*, Wiley–IEEE.

Alcamo, I. E. (2001), *DNA Technology,* 2nd edn., Academic Press.

Altamirano, M. M., Blackburn, J. M. *et al* (2000), Directed evolution of new catalytic activity using the α/β-barrel scaffold, *Nature,* **403**, 617-622.

Aris, A. and Villaverde, A. (2004), Modular protein engineering for non-viral gene therapy, *Trends in Biotechnology*, **22**, 371-377.

Arnold, F. H. (2003), *Directed Enzyme Evolution: Screening and Selection Methods*, Humana Press.

Baxevanis, A. D. and Ouellette, B. F. F. (1998), *Bioinformatics - A Practical Guide to the Analysis of Genes and Proteins,* Wiley.

Benkovic, S. J. and Ballesteros, A. (1997), Biocatalysts - the next generation, *Trends in Biotechnology,* **15**, 385-386.

Benton, D. (1996), Bioinformatics - principles and potential of a new multidisciplinary tool, *Trends in Biotechnology,* **14,** 261-272 (see also 273-321).

Berry, A. (2001), *From Protein Folding* to *New Enzymes*, Portland Press.

Bishop, M. J. and Rawlings, C. 1. (eds.) (1997), *DNA and Protein Sequence Analysis - A Practical Approach,* IRL.

Brakmann, S. and Johnsson, K. (2002), *Directed Molecular Evolution of proteins: Or How to Improve Enzymes for Biocatalysis*, Wiley–VCH.

Brown, T. A. (1995), *Gene Cloning,* 3rd edn., Chapman and Hall.

Cavalli-Sforza, L. L. and Feldman, M. W. (2003), The application of molecular genetic approaches to the study of human evolution, *Nature Genetics*, **33** (supplement), 266-275.

Cavazzano-Calvo, M., Lagresle, C. *et al* (2005), Gene therapy for severe combined immunodeficiency, *Annual Review of Medicine*, **56**, 585-602.

Chatterjee, R. and Yuan, L. (2006), Directed evolution of metabolic pathways, *Trends in Biotechnology*, **24**, 28-38.

Cleland, J. L. and Craik, C. S. (eds.) (1996), *Protein Engineering - Principles and Practice,* Wiley-Liss.

Dieffenbach, C. W. and Dveksler, G. S. (1995), *PCR Primer - A Laboratory Manual,* CSHLPress.

Drake, F. H., Dodds, R. A. *et al* (1996), Cathepsin K, but not cathepsin B, Land S, is abundantly expressed in human osteoclasts, *Journal of Biological Chemistry,* **271**, 12511-12516 (see also 12517-12524).

Drlica, K. (1996), *Understanding DNA and Gene Cloning,* 3rd edn, Wiley.

Figeys, D. (ed.) (2005), *Industrial Proteomics: Applications for Biotechnology and Pharmaceuticals*, Wiley.

Gelehrter, T. D., Collins, F. S. and Ginsburg, D. (1998), *Principles of Medical Genetics,* 2nd edn., Williams and Wilkins.

Glover, D. H. and Hames, B. D. (eds.) (1996), *DNA Cloning - Expression Systems, a Practical Approach,* Oxford University Press.

Gray, P. M. D., Kemp, G. J. L. *et al* (the EU Bridge DataBase Project Consortium) (1996), Macromolecular structure information and databases, *Trends in Biochemical Sciences,* **21**, 251-256.

Hacein-Bey-Abina, S., von Kalle, C. *et al* (2003), LM02-associated clonal T-cell proliferation in two patients after gene therapy for SCID-X1, *Science*, **302**, 415-419.

Hamdan, M. H. and Righetti, P. G. (2005), *Proteomics Today*, Wiley.

Hultschig, C., Kreutzberger, J. *et al* (2006), Recent advances in protein microarrays, *Current Opinion in Chemical Biology*, **10**, 4-10.

Irvine, D. V., Shaw, M. L. *et al* (2005), Engineering chromosomes for delivery of therapeutic genes, *Trends in Biotechnology*, **23**, 575-583.

Kim, S.-W., Jung, J. *et al* (2006), Structural and functional analyses of mutations of the human phenylalanine hydroxylase gene, *Clinica Chimica Acta*, **365**, 279-287.

Kolkman, A., Slijper, M. and Heck, A. J. R. (2005), Development and application of proteomics technologies in *Saccharomyces cerevisiae*, *Trends in Biotechnology*, **23**, 598-604.

Kuchner, O. and Arnold, F. H. (1997), Directed evolution of enzyme catalysts, *Trends in Biotechnology,* **15**, 523-530.

Laskin, A. I., Li. G.-X. and Yu, Y.-T. (2006), *Enzyme Engineering XIV*, New York Academy of Sciences.

Mann, A. M., Hendrickson, R. C. and Pandey, A. (2001), Analysis of proteins and proteomes by mass spectrometry, *Annual Review of Biochemistry*, **70**, 437-473.

Methods in Enzymology, **152-155** (1987), **216** (1992), **217, 218** (1993): Recombinant DNA; **202, 203** (1991), Molecular design and modelling; **183** (1990), **224** (1993), **395** (2005): Molecular evolution; **266** (1996): Computer methods for macromolecular sequence analysis; **346** (2002): Gene therapy methods); **393, 384** (2004): Numerical computer methods; **388** (2004): Protein engineering: **402** (2005): Biological mass spectrometry; Academic Press.
Marshall, E. (2002), Gene therapy: what to do when clear success comes with an unclear risk, *Science*, **298**, 510-511.
Mingarro, I., von Heijne, G. and Whitley, P. (1997), Membrane-protein engineering, *Trends in Biotechnology,* **15**, 432-437.
Moss, D. K. (ed.) (2005), *Essays in Bioinformatics*, IOS Press.
Mountain, A. (2000), Gene therapy - the first decade, *Trends in Biotechnology,* **18**, 119-126.
Nishimura, N., Nakayama, T. *et al* (2000), Direct polymerase chain reaction from whole blood without DNA isolation, *Annals of Clinical Biochemistry,* **37**, 674-680.
Noordewier, M. O. and Warren, P. V. (2001), Gene expression microarrays and the integration of biological knowledge, *Trends in Biotechnology*, **19**, 412-415.
Okano, Y., Eisensmith, R. C. *et al* (1991), Molecular basis of phenotype heterogeneity in phenylketonuria, *New England Journal of Medicine,* **324**, 1232-1238.
Old, R. W. and Primrose, S. B. (1994), *Principles of Gene Manipulation,* 5th edn., Blackwell.
Ovchinnikov, I. V., Gőtherstrőm, A. *et al* (2000), Molecular analysis of Neanderthal DNA from the northern Caucasus, *Nature*, **404**, 490-493.
Pakendorf, B. and Stoneking, M. (2005), Mitochondrial DNA and human evolution, *Annual review of Genomics and Human Genetics*, **6**, 165-183.
Panno, P. (2004), *Gene Therapy*, Facts on File.
Peracchi, A. (2001), Enzyme catalysis: removing chemically 'essential' residues by site-directed mutagenesis, *Trends in Biochemical Sciences*, **26**, 457-503.
Pingoud, A. (2004), *Restriction Endonucleases*, Springer-Verlag.
Plebani, M. (2005), Proteomics: the next revolution in laboratory medicine, *Clinica Chimica Acta*, **377**, 113-122.
Sandvik, A. K., Alsberg, B. K. *et al* (2006), Gene expression analysis and clinical diagnosis, *Clinica Chimica Acta*, **363**, 157-164.
Schwartz, M. J. (1998), DNA diagnosis of cystic fibrosis, *Annals of Clinical Biochemistry,* **35,** 584-610.
Scriver, C. R. and Sly, W. S. (2001), *The Metabolic and Molecular Bases of Inherited Disease,* 8th edn., McGraw-Hill.
Stoughton, R. B. (2005), Applications of DNA microarrays in biology, *Annual Review of Biochemistry*, **74**, 53-82.
Sudbery, P. (1998), *Human Molecular Genetics,* Longman.
Turnpenny, P. and Ellard, S. (2004), *Emery's Elements of Medical Genetics*, 12th edn., Churchill-Livingstone.
Urnov, F. D., Miller, J. C. *et al* (2005), Highly efficient endogenous human gene correction using designed zinc-finger nucleases, *Nature*, **435**, 646-651.
Veenstra, T. D. and Yates, J. R. (2006), *Proteomics for Biological Discovery*, Wiley.

Venkatasubbarao, S. (2004), Microarrays: status and prospects, *Trends in Biotechnology*, **22**, 630-637.
Verma. I. M. and Weitzman, M. D. (2005), Gene therapy: twenty-first century medicine, *Annual Review of Biochemistry*, **74**, 711-738.
White, T. J. (1996), The future of PCR technology - diversification of technologies and applications, *Trends in Biotechnology,* **14,** 478-483.
Zhu, H., Bilgin, M. and Snyder, M. (2003), Proteomics, *Annual Review of Biochemistry*, **72,** 783-812.

PROBLEMS

21.1 DNA from a pregnant woman known to be heterozygous for ornithine carbamoyltransferase deficiency was digested with restriction enzyme *Msp*I and probed by Southern blotting, haplotypes B and D both being detected (see section 21.1.2). Her sister, who was free of the X-linked disorder, was shown to carry haplotypes B and C. Their mother was dead, but was known to have been heterozygous for ornithine carbamoyltransferase deficiency. Pre-natal investigation of the foetus showed it to be male, and carry haplotype B. Deduce the status of the foetus with regard to ornithine carbamoyltransferase deficiency, and the haplotypes carried by the woman's parents. The chain-terminator sequencing procedure was applied to a single-strand portion of recombinant DNA isolated from bacteriophage M13, and a diagram of the resulting electrophoresis pattern is shown below. Lane A represents the contents of the flask to which dideoxy-ATP had been added, lane C that containing dideoxy-CTP, and so on. The smallest fragment, representing the 5′-end of the growing DNA chain (the primer plus the first nucleotide to be attached to it, is on the left, and the largest fragment on the right.

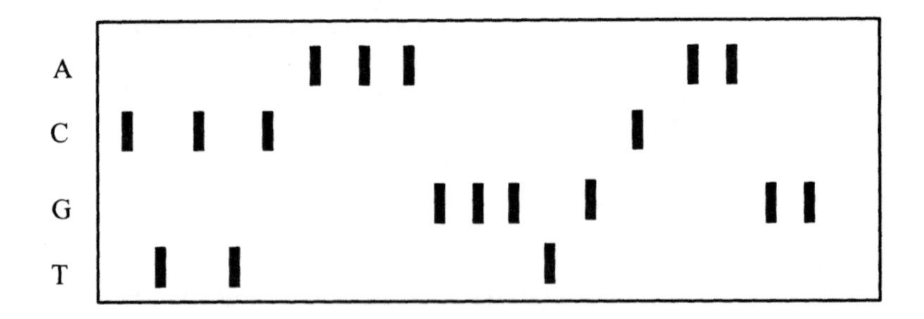

The primer was complementary in structure to the DNA of the vector immediately adjacent to the 3′-end of the inserted segment.

(a) Deduce the sequence of the isolated section of the DNA which had been inserted into the vector (i.e. the sequence of the single-strand of DNA used as template in the chain-terminator procedure, excluding the portion which bound to the primer).

(b) Assuming that the base at the 3′-end of this DNA segment is the first in a coding sequence, deduce the amino acid sequence specified.

Answers to Problems

1.1 (a) acylcholine acyl-hydrolase (3.1.1).
(b) carbamoyl phosphate: L-ornithine carbamoyltransferase (2.1.3).
(c) D-alanine: D-alanine ligase (ADP forming) (6.3.2)
(d) glycerol: NAD^+ 2-oxidoreductase (1.1.1).
(Note: not 1.6.99. Oxidoreductases involving NAD^+ are classified in the direction where NAD^+ is the electron acceptor rather than NADH the electron donor, except where another redox catalyst is involved. Here, therefore, the reaction is classified in the opposite direction to that given.)
(e) D-fructose-l,6-bisphosphate D-glyceraldehyde-3-phosphate-lyase (4.1.2).
(f) NADH: ferricytochrome b_5 oxidoreductase (1.6.2).
(g) UDP glucose-4-epimerase (5.1.3).

1.2 (a) 6.4.1.1
(b) 1.8.1.2
(c) 2.7.4.3
(d) 5.3.1.1
(e) 3.4.13.11
(f) 3.2.1.1
(g) 4.3.2.1

2.1 Gly-Ala-Thr-Arg-Ser-Phe-Val-Leu-Lys-Ala-Phe-Ser-Lys-Ile-Trp-*Gln*-Thr-Ser

2.2 Probable structure:

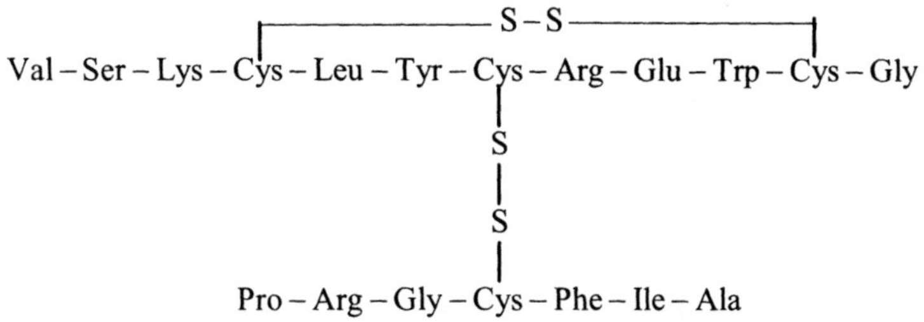

```
                      ┌────────────S–S────────────┐
Val – Ser – Lys – Cys – Leu – Tyr – Cys – Arg – Glu – Trp – Cys – Gly
                                     |
                                     S
                                     |
                                     S
                                     |
              Pro – Arg – Gly – Cys – Phe – Ile – Ala
```

3.1 Reading the electrophoresis strip from left to right (small to large fragments), the sequence of the complementary strand can be deduced to be:

5′-AGCCGTGCACAGTTCAGT-3′

Hence, the sequence of the original strand is:

3′-TCGGCACGTGTCAAGTCA-5′.

3.2 (a) Reading the electrophoresis strip from left to right (small to large fragments, the sequence of the DNA can be deduced to be:

3′-GAATCTAGTGTACCCCGA-5′.

(b) The sequence of the mRNA corresponding to this DNA sequence is:

5′-CUUAGAUCACAUGGGGCU-3′.

By reference to the genetic code, this can be seen to correspond to a peptide of amino acid sequence:

Leu-Arg-Ser-His-Gly-Ala.

3.3 (a) (i) 4.74 (ii) 5.04.
(b) Concentration of acetic acid in tube 4 = (0.04 × 5/6) mol dm^{-3}.
Concentration of acetate in tube 4 = (0.1 × 1/6) mol dm^{-3}.
pH of tube 4 (and hence approximate isoelectric pH of casein) = 4.44.

3.4 (a) 99.9%; (b) 90.9%; (c) 9.09%; and (d) 0.10%.

6.1 (a)

Reaction	$\Delta G^{\ominus\prime}$(kJ mol^{-1})
argininosuccinate + $H_2O \rightleftharpoons$ aspartate + citrulline	– 34.4
arginine + fumarate $\rightleftharpoons$ argininosuccinate	– 11.7
NH_4^+ + citrulline $\rightleftharpoons$ arginine + H_2O	+ 30.5
aspartate + $H_2O \rightleftharpoons$ malate + NH_4^+	+ 12.6
fumarate + $H_2O \rightleftharpoons$ malate	– 3.0

$\Delta S^{\ominus\prime} = -43.9$ J $K^{-1}mol^{-1}$.

(b) $\Delta G = +2.7$ kJ mol^{-1} (note H_2O concentration is set at 1.0 mol l^{-1}).
K'_{eq} 3.2; [$fumarate_{eq}$] = 2.4×10^{-4} mol l^{-1}; and [$malate_{eq}$] = 7.6×10^{-4} mol l^{-1}.

6.2 (a) $E'_{\ominus}(NAD^+/NADH + H^+) - E'_{\ominus}$(malate/oxaloacetate) = – 0.145 V.
(b) $\Delta G = +31.4$ kJ mol^{-1}

([H^+] = 10^{-7} mol l^{-1} at pH 7, but, since $\Delta G^{\ominus\prime}$ and not $\Delta G^{\ominus}$ is being used here, this is not required).

(c) $K'_{eq} = 1.3 \times 10^{-5}$; [$oxaloacetate_{eq}$] = [$NADH_{eq}$] = 0.0002 mol l^{-1}.

6.3 Activation energy for catalysed reaction = 43 kJ mol^{-1} and for uncatalysed reaction = 69 kJ mol^{-1}.

6.4 $[A_0]$ (mmol l^{-1}): 10.0 8.0 6.0 4.0 2.0
v_0 (absorbance units min^{-1}): 0.107 0.085 0.064 0.042 0.021

The reaction is first order with respect to A. (Note that graphs of absorbance against time do not necessarily pass through the origin, even when no product is present at zero time. This could be due to the instrument cells not matching, the reagent blank not being included in the reference or to the inaccurate assessment of mixing time. Since it is the slope which is being measured, this will not affect results provided the same procedure is used throughout.)

6.5 (a)

Reaction	$\Delta G^{\ominus\prime}$ (kJ mol^{-1})
glucose-6-P ⇌ glucose-1-P	+ 7.2
UDP-glucose + $2P_i$ ⇌ fructose-6-P + UTP	+ 21.7
glucose-1-P + UTP ⇌ UDP-glucose + $(PP)_i$	0
$(PP\}_i$ ⇌ $2P_i$	– 26.8
glucose-6-P ⇌ fructose-6-P	+ 2.1

$\Delta G = -1.8$ kJ mol^{-1}.

(b) Activation energy for catalysed reaction = 43 kJ mol^{-1} and for uncatalysed reaction = 68 kJ mol^{-1}.

7.1 $K_m = 13$ mmol l^{-1}; $V_{max} = 540$ μmol l^{-1} min^{-1}.

7.2 $[NAD^+]$ (mmol l^{-1}): 1.5 2.0 2.5 3.33 5.0 10.0
v_0 (absorbance units min^{-1}): 0.046 0.053 0.059 0.065 0.074 0.085

$K_m = 1.7$ mmol l^{-1}; $V_{max} = 0.10$ absorbance units min^{-1}.

7.3 $k_1 = 500$ mol^{-1} l s^{-1}; $k_2 = 1.5$ s^{-1}; $k_{-1} = 6$ s^{-1}.

7.4
$$v_0 = \frac{k_1 k_2 k_3 [S_0][E_0]}{k_2 k_3 + k_{-1} k_3 + k_{-1} k_2 + (k_1 k_3 + k_1 k_{-2} + k_1 k_2)[S_0]}$$

8.1 With A: competitive inhibition.
With B: non-competitive inhibition.
Note that the 'uninhibited' reaction exhibits substrate inhibition.

8.2 Non-competitive inhibition; $[I_0] = 1.3$ mmol l^{-1}.

8.3 Uncompetitive inhibition ($K_i = 2.1$ mmol l^{-1}); $V_{max} = 0.04$ absorbance units min^{-1} in presence of 3.0 mmol l^{-1} oxaloacetate.

8.4 Non-competitive inhibition (K_i = 2.0 mmol l^{-1}); K'_m = 22 mmol l^{-1}, V'_{max} = 360 µmol l^{-1} min^{-1}, in presence of 3.0 mmol l^{-1} inhibitor.

8.5 *Linear* competitive inhibition (K_i = 4.5 mmol l^{-1}).

8.6 Mixed inhibition (competitive-non-competitive); secondary plots are not linear, so inhibition is possibly partial.

9.1 A primary plot of $1/v_0$ against $1/[AX_0]$ is drawn. From a secondary plot of intercept against $1/[B_0]$, V_{max} = 1000 µmol $l^{-1}min^{-1}$ and K_m^B = 4.0 mmol l^{-1}; and from a secondary plot of slope against $1/[B_0]$, K_s^{AX} = 1.5 mmol l^{-1} and K_m^{AX} = 2.0 mmol l^{-1}.

9.2 Primary plot excludes ping-pong bi-bi mechanism; inhibition studies with I show a competitive pattern, which excludes the compulsory-order mechanism where AX binds first; product inhibition studies with A similarly excludes the compulsory-order mechanism where B binds first. Conclusion: probably a random-order ternary-complex mechanism.

9.3 Primary plot excludes ping-pong bi-bi mechanism; product inhibition studies with oxaloacetate show a mixed pattern, which excludes a random-order mechanism and the compulsory-order mechanism where malate binds first. Equilibrium isotope exchange studies indicate the compulsory-order ternary-complex mechanism where NAD^+ binds first, so this is probably the mechanism which operates.

9.4 Use of the dead-end inhibitor shows an uncompetitive pattern with respect to varying [AX] (K_i = 2.0 mmol l^{-1}); and product inhibition by A gives a competitive pattern with respect to varying [AX], which excludes the compulsory order mechanism where B binds first. In the experiments with varying [B], the primary plot excludes a ping-pong bi-bi mechanism, and product inhibition by BX gives a mixed pattern with respect to varying [B], which excludes a random-order mechanism, as well as the compulsory-order mechanism where B binds first. Equilibrium isotope exchange studies also exclude the compulsory-order mechanism where B binds first.

Conclusion: probably the compulsory-order ternary-complex mechanism where AX binds first.

9.5

$$\phi_0 = \frac{k_{-3}k_5 + k_4k_5 + k_3k_5 + k_3k_4}{k_3k_4k_5} \qquad \phi_{AX} = \frac{1}{k_1}$$

$$\phi_B = \frac{k_{-2}k_{-3} + k_{-2}k_4 + k_3k_4}{k_2k_3k_4} \qquad \phi_{AXB} = \frac{k_{-1}(k_{-2}k_{-3} + k_{-2}k_4 + k_3k_4)}{k_1k_2k_3k_4}$$

9.6 There is a mixed inhibition pattern with respect to varying [lysine], a competitive inhibition pattern with respect to varying [2-oxoglutarate] and an uncompetitive pattern with respect to varying [NADPH]. Cleland's rules may be applied as for a two-substrate reaction, so saccharopine and 2-oxoghitarate compete for a site on the same enzyme form, and there is no reversible link between saccharopine and NADPH. The simplest mechanism consistent with these results is a compulsory-order one with 2-oxoglutarate binding first, then lysine and then NADPH, to form a quaternary complex. $NADP^+$ is the first product to leave (as an irreversible step) and then saccharopine.

10.1 The variation of V_{max}/K_m with pH at fixed [E_0] indicates the involvement of two residues, of approximate pK_a values 4.2 and 8.2; the variation of V_{max} with pH at fixed [E_0] indicates the involvement of a residue whose pK_a is approximately 4.1. Hence, two essential ionizing residues are indicated: the degree of ionization of the one of lower pK_a values is important in both the free enzyme and the enzyme-substrate complex, whereas the ionization of the other is important only in the free enzyme (n.b. other types of investigation identify the two residues as histidine-159 and cysteine-25; the substrate forms a covalent link with cysteine in much the same way as with the essential serine residue in serine proteases).

12.1 Negative cooperativity.

12.2 Positive cooperativity; $h = 1.8$.

12.3 Negative cooperativity; four binding sites.

13.1 Positive homotropic cooperativity is indicated. From the Hill plot, $h = 4$. From the sedimentation data, $n = 4$ (4 sub-units). Hence the Monod-Wyman-Changeux model may be operating.

13.2 (a) Binding data show that there is no cooperativity; (b) the primary plot excludes a ping-pong bi-bi mechanism, and also shows there are some departures from linearity; product inhibition by NAD^+ gives a competitive pattern with respect to varying NADH, which excludes the compulsory-order mechanism where B binds first; and (c) equilibrium isotope exchange results exclude the compulsory-order mechanism where NADH binds first.

Conclusions: probably a random-order mechanism. (If this is the case, our hypothetical NADH-utilizing enzyme is atypical, since the coenzyme would normally be expected to bind first.)

The departures from linearity must be due to kinetic factors rather than to cooperative binding.

13.3 Positive homotropic cooperativity is indicated, with A an allosteric inhibitor. From the Hill plot, $h = 3$ for the inhibited reaction and $h = 2$ for the uninhibited reaction. The two molecular weights determined were presumably those for oligomer and monomer, so $n = 3$. Therefore $h = n$ when L is large (in presence of an allosteric inhibitor), so the Monod-Wyman-Changeux mechanism is indicated.

15.1 Specific activity ≈ 11.5 μmol min^{-1}mg $protein^{-1}$ in each case, the good agreement between the results suggesting that the assay procedure is valid. (Note that many students make the mistake of calculating specific activity as μmol $l^{-1}min^{-1}$mg $protein^{-1}$; the incubation volume (5 cm^3 in this problem) must be taken into account, since this is what is acted upon by the enzyme.)

16.1 Specific activity of precipitate = 1900 μmol min^{-1}mg $protein^{-1}$.
Specific activity of supernatant = 2000 μmol min^{-1}mg $protein^{-1}$.
Hence the purification step has *not* been successful.

17.1 The time required is 2.1 minutes.

17.2 (a) Fig. 7.3a; (b) linear as long as substrate concentration is non-limiting; (c) Fig. 7.1 at fixed concentrations of other substrates (NADH being a co-substrate); (d) Fig. 6.3 (identical for an enzyme-catalysed reaction; see also section 15.2.2); (e) Fig. 8.1a; (f) Fig. 8.12b; (g) Fig. 17.1; and (h) Fig. 17.2a.

20.1 The pH optimum = 7.5 – 1.6 = 5.9.

20.2 The product concentration will be 5.0 mmol l^{-1} when the substrate concentration is 15 mmol l^{-1}. From the integrated form of the Michaelis-Menten equation, this will be the situation 12.3 minutes after the start of the reaction.

21.1 Comparison of the woman with her sister suggests that the disorder is associated with haplotype D, so the foetus is normal. The father of the sisters must have passed his only haplotype on to each, so this must be B. His heterozygous wife must have carried haplotypes C and D.

21.2 (a) 3′-GAGAGTTTCCCACGTTCC-5′
(b) Leu-Ser-Lys-Gly-Ala-Arg
(The mRNA would be 5′-CUCUCAAAGGGUGCAAGG-3′.)

Abbreviations

ACTase	aspartate carbamoyltransferase
ADH	alcohol dehydrogenase
ADP	adenosine-5′-diphosphate
ALP	alkaline phosphatase
ALT	alanine transaminase
AMP	adenosine-5′-monophosphate
AST	aspartate transaminase
ATP	adenosine-5′-triphosphate
CK	creatine kinase
CM-	carboxymethyl-
CTP	cytidine-5′-triphosphate
DEAE-	diethylaminoethyl-
DFP	diisopropylphosphofluoridate
DNA	deoxyribonucleic acid
$[E_0]$	initial (and usually total) enzyme concentration
E64	L-*trans*-epoxysuccinyl-leucylamide-(4-guanidino)-butane
EDTA	ethylene diamine tetra-acetate
EIA	enzyme immunoassay
ELISA	enzyme-linked immunosorbent assay
EMIT	enzyme multiplied immunoassay technique
FAD	flavin adenine dinucleotide (oxidized form)
$FADH_2$	flavin adenine dinucleotide (reduced form)
FBPase	fructose-1,6-bisphosphatase
GDP	guanosine-5′-diphosphate
GTP	guanosine-5′-triphosphate
Hb	haemoglobin
HCIC	hydrophobic charge induction chromatography
HIC	hydrophobic interaction chromatography
HPLC	high-performance liquid chromatography
HTS	high-throughput screening
IEF	isoelectric focusing
IEX	ion exchange chromatography
IMAC	immobilized metal ion affinity chromatography
LDH	lactate dehydrogenase
K_m	Michaelis constant
Mb	myoglobin
2-ME	2-mercaptoethanol
M_r	molecular weight (relative molecular mass)
MMC	mixed mode chromatography

MS	mass spectrometry
NAD^+	nicotinamide adenine dinucleotide (oxidized form)
NADH	nicotinamide adenine dinucleotide (reduced form)
$NADP^+$	nicotinamide adenine dinucleotide phosphate (oxidized form)
NADPH	nicotinamide adenine dinucleotide phosphate (reduced form)
OCT	ornithine carbamoyltransferase
PEG	polyethylene glycol
PFK	phosphofructokinase
P_i	inorganic orthophosphate $\left(HO-PO_3^{2-}\right)$
$(PP)_i$	inorganic pyrophosphate $\left(HO-PO_2^{-}-O-PO_3^{2-}\right)$
PAGE	polyacrylamide gel electrophoresis
PDE	phosphodiesterase
PMSF	phenylmethylsulphonyl fluoride
QAE-	diethyl-(2-hydroxypropyl)aminoethyl-
RIA	radioimmunoassay
RNA	ribonucleic acid
$[S_0]$	initial substrate concentration
SEC	size-exclusion chromatography
SDS	sodium dodecyl sulphate
THF	tetrahydrofolate
TPP	thiamine pyrophosphate
UDP	uridine-5′-diphosphate
UTP	uridine-5′-triphosphate
v_0	initial (steady-state) reaction velocity
V_{max}	maximum possible v_0 at fixed $[E_0]$

Index

Lightning Source UK Ltd.
Milton Keynes UK
UKOW031824030513

210138UK00002B/16/P